Biodiversity and Conservation

Biodiversity and Conservation

Edited by **Anne Offit**

SYRAWOOD
PUBLISHING HOUSE
New York

Published by Syrawood Publishing House,
750 Third Avenue, 9th Floor,
New York, NY 10017, USA
www.syrawoodpublishinghouse.com

Biodiversity and Conservation
Edited by Anne Offit

International Standard Book Number: 978-1-68286-277-3 (Hardback)

Printed in China.

Contents

Preface

The purpose of the book is to provide a glimpse into the dynamics and to present opinions and studies of some of the scientists engaged in the development of new ideas in the field from very different standpoints. This book will prove useful to students and researchers owing to its high content quality.

Theories, research, practical applications, and case studies related to all contemporary topics of relevance to biodiversity and conservation, have been included in this book. The chapters included in this book on environmental management and planning, green technology, environmental health, sustainable development, etc. provide an exhaustive insight into this field. A number of latest researches have been included to keep the readers up-to-date with the global concepts in this area of study.

At the end, I would like to appreciate all the efforts made by the authors in completing their chapters professionally. I express my deepest gratitude to all of them for contributing to this book by sharing their valuable works. A special thanks to my family and friends for their constant support in this journey.

Editor

Potential of parasitoid *Trichogramma chilonis* (Ishii) (Hymenoptera: Trichogrammatidae) against the sugarcane stem borer, *Chilo infuscatellus* (Lepidoptera; Pyralidae) under field conditions

Shahbaz Ahmad*, Muhammad Ashfaq, Mansoor-ul-Hassan and Shahbaz Talib Sahi

Department of Agriculture-Entomology, University of Agriculture Faisalabad, Pakistan.

The present experiment reports the role of egg parasitoid, *Trichogramma chilonis* (Ishii) (Hymenoptera; Trichogrammtiade) as a pest management tool for the control of sugarcane stem borer (*Chilo infuscatellus*) under natural field conditions. The prime focus of study was to find out the effectiveness of egg parasitoid as a primary tool after its field releases at different time intervals. The experiment was conducted at Rahim Yar Khan, Punjab during 2008 to 2009. The trails were laid out in randomized complete block designed with five treatments of field releases an interval of 10, 15 and 20 days. The intensity of damage showed that minimum infestation of 9.31% was recorded in treatment T_1 (72,000) *Trichogramma* cards followed by T_2, T_3 and T_4 with infestation percentage of 13.60, 17.07 and 19.44, respectively. The highest infestation was recorded at 25.95% in control treatment. The lowest infestation (16.57%) was observed after 10 days intervals of field releases followed by recording after 15 and 20 days interval with infestation of 17.14 and 17.52%.

Key words: *Trichogramma chilonis*, egg parasitoid, sugarcane, *Chilo infuscatellus*.

INTRODUCTION

Trichogramma chilonis (Ishii) (Hymenoptera; Trichogrammitidae) is an important egg parasitoid of lepidopterous pests with effective control as against of the sugarcane stem borers. It has been in use in sugarcane, corn, cotton, vegetables and fruits crops for the control of pests since 20 years. The *Trichogramma* has been considered as one of the most important parasitoid for more than 100 years and it was also reported that *T. chilonis* can be more effective when it is parasitizing potential and searching ability is well adopted in the field (Smith, 1996). *Trichogramma* specie is most effective against the lepidopterous pest particularly for the sugarcane stem borers (Metcalfe and Breniere, 1969; Nagarkatti and Nagaraja, 1977). The inundate releases of bio-agent for the control of lepidopterous pests are being practiced in more than 32 million ha each year around the world (Hassan, 1993). The release of *T. chilonis* in China, Switzerland, Canada and former USSR reduced the damage up to 70 to 92% on sugarcane (Li 1994), while in Asia, it reduced the incidence of pests in sugarcane by early shoot borer (*Chilo infuscatellus)* from 43 to 82% (Bharati et al., 2002). Rafique et al. (2007) reported that *T. chilonis* give very good results against the sugarcane stalk borer. The application of *T. chilonis* at 60,000 eggs reduces the pest infestation at 83% against the *C. infuscatellus* (Table 1). Zia et al. (2007) reported that the application of *T. chilonis* against *C. infuscatellus* showed a negative correlation with an increase in number of eggs which indicated that it can successfully be used to control stem borer and it reduces the borer infestation at 2.74%.

Mustafa et al. (2006) studies that infestation of *C. infuscatellus* was reduced at 52.04% during 2000 and

*Corresponding author. E-mail: sacsharif@yahoo.com.

Table 1. Analysis of variance of pest infestation.

Source of variation	Degrees of freedom	Mean squares	F-value
Replication	2	1.502	4.14
Day	2	3.430	9.46**
Variety	4	352.925	973.36**
Day × variety	8	0.292	0.81 ^{NS}
Error	28	0.363	

** = Highly significant (*P*<0.01).

Table 2. Borer's infestation after different time interval.

Treatment	Interval (days)			Mean
	10	15	20	
T_1	8.90±0.18	9.61±0.67	9.41±0.26	9.31±0.24 A
T_2	12.88±0.13	13.89±0.33	14.04±0.05	13.60±0.21 B
T_3	16.79±0.10	17.13±0.54	17.27±0.40	17.07±0.21 C
T_4	18.93±0.37	19.38±0.29	20.02±0.34	19.44±0.23 D
$T_{5\ (Control)}$	25.35±0.55	25.66±0.32	26.86±0.55	25.95±0.34 E
Mean	16.57±1.49 A	17.14±1.45 B	17.52±1.57 C	

Means ± SE, sharing similar letter in a row or in a column are statistically non-significant (*P*>0.05).

60.03% during 2001. The damage ranged from 3.9 to 10.5% in treated and 9.6 to 18.6% in untreated plots. The internodes damaged reduction due to *Trichogramma* treated ranged from 28.6 to 59.7%. Shenhmar et al. (2003) reported that *T. chilonis* was released at 50000 per acre for 10 days interval from July to October. In terms of borer infestation, the damaged was 52.04% during 2000 and 60.03% during 2001; so the aim of the present study is to find out the efficiency of *Trichogramma* with most suitable dose against the sugarcane stem borer.

MATERIALS AND METHODS

Present field trial was conducted in sugar cane crop under natural conditions at Akramabad Distt, Rahim Yar Khan. Homogeneous sets of a standard commercial sugarcane variety US-718 was planted by maintaining rows at a distance of 2.5 feet R × R and 1 foot P × P distance between set to set. The experiment was laid out in a randomized complete block design RCBD and replicated thrice with a plot size of 5 × 10 m^2 with five treatments in each block. Standard agronomic practices as irrigations, fertilizers, hoeing were adopted through out the crop period as per crop requirements. Releases of Trichogramma were made according to treatments having number of parasitoids population as by into the 72,000, 60,000, 48,000, 40,000 and untreated control as indicated by T_1, T_2, T_3, T_4, T_5, respectively (Table 2). Parasitoids of Trichogramma were taken from the insectary at Department of Agri. Entomology, University of Agriculture Faisalabad; reared and releases were made accordingly into field site. Prepared parasitized cards of *T. chilonis* were released into each treatment by installing them on ventral surface of the leaves in order to avoid the direct exposure of the sunlight. Infestation of the sugarcane stem borer was monitored on the basis of randomly selected sample from each treatment.

Each treatment consisted of five tillers of each experiment and the number of damaged plants by *C. infuscatellus* as well *as* the percentage of its infestation was calculated as follows:

$$\text{Infestation (\%)} = \frac{\text{Number of damaged tillers}}{\text{Total tillers}} \times 100$$

The percent reduction in the stem borer after the releases of *T. chilonis* was determined by the following formula:

$$\text{Reduction (\%)} = \frac{\text{Damaged tillers in control plot} - \text{damaged tillers in treated plot}}{\text{Total tillers in control plot}} \times 100$$

The effectiveness of the biological control agent of *T. chilonis* was evaluated by means of the sugarcane stem borer infestation and compared with that of the control treatment plot. The data obtained was statistically analyzed by using Tukey HSD test (Mstate Package).

RESULTS AND DISCUSSION

The results revealed a significant difference among all the treatments (P<0.01). The lowest infestation (9.31%) was recorded in the treatment where releases of *T. chilonis* were made with 72000, 60000, 48000 and 40,000 parasitoids egg of *Trichogramma* with infestation percentage of 13.60, 17.07 and 19.44, respectively. The lowest infestation was recorded in control treatment (25.95%). Among the time interval, the lowest infestation (16.57%) was observed after 10 days followed by 15 and 20 days interval with 17.14 and 17.52%, respectively. The present findings are partially compared with the results reported by Rafique et al. (2007) and Zia et al. (2007) who stated that when 60000 eggs of *Trichogramma* per acre were released causing 83% reduction of *C. infuscatellus* and showed significant results and also reported that population of *C. infuscatellus* showed a negative correlation with an increase in number of eggs which indicated that it can be successfully used to control stem borer. Our present findings are agreed to those of Muhammad and Muhyuddin (1987) who reported that by the use of *Trichogramma*, favorably affect natural enemies of sugarcane stem borer *C. infuscatellus*. Similarly, Ashraf et al. (l993) and Mustafa et al. (2006) reported that by using *Trichogramma,* infestation was reduced from 52.04 to 60.03% during 2000 to 2001 and damage ranged from 3.9 to 10.5% in treated area and 9.6 to 18.6% in untreated plots.

The internodes damaged reduced due to releases of *Trichogramma,* infestation ranged from 28.6 to 59.7%. The present findings are also in line to those of Rajendran and Hanifa (1996), Shenhmar et al. (2003) and Nadeem and Hamed (2011) who reported that *Trichogramma* is very effective against the sugarcane borers. The present results are in line with the results reported by Nadeem et al. (2008) and Nadeem et al. (2009) who stated that the temperature for *T. chilonis* at 20, 25, 28, 31, 35 and 40°C in the incubators. The biological parameters such as parasitism (95.6%), emergence (98.0%), developmental period (7.3 days) and adult longevity (9.0) days were very favorable at 28°C. It was also reported that 40°C and low 20°C temperature did not support the development period while temperature at 28°C was most favorable for the rearing of *T. chilonis.*

ACKNOWLEDGEMENTS

The authors are thankful to Mr. Abdul Ghaffar, Ex. Director General, Agriculture extension for his co-operation, suggestion and provision of facilities to conduct the present research trial.

REFERENCES

Ashraf M, B Fatima B, and N Ahmad N (1993). Control of sugarcane by inundative releases of *Trichogramma chilonis* (Ashii). Pak. J. Zool., 25(1): 23-25.

Bharati NB, S Ramprasad S, N Mathivanan N, K Srinivisan K, and S Chelliah S (2002). Impact of bio-agents in the management of soil born diseases and insect pests. In: Pl. Prot. During 21 Century, Hyderabad. India, (2): 19-25.

Hassan SA (1993). The mass rearing and utilization of *Trichogramma* to control lepidopterous pests, achievements and outlook. Pestic. Sci., 37: 387-391.

Li LY (1994). Worldwide use of *Trichogramma* for biological control on different crops. Survey, pp. 37-51.

Metcalfe JR and, J Breniere J (1969). Egg parasites (*Trichogramma* spp)for the control of sugarcane moth borers. Pests of Sug. Eldevier Publishing Company, the Netherland. Pp. 81-115.

Muhammad A, and Al Mohyuddin Al (1987). Invasion of Gurdaspur borer *Acigona Steniellus* (Hamp.) in NWFP (Pakistan) and its biological control. Proc. 21st. Ann. Conv. Pak. Soc. Tech. Rawalpindi,. pp.82-85.

Mustafa G,, A Ghani A, and Z Sikandar Z (2006). Performance of *Trichogramma* for the control of sugarcane borers in farmers fields at various districts of Punjab (Pakistan). J. Pak. Sug., 211(5): 5-7.

Nadeem, S, and M Hamed M (2008). Comparative development and parasitization of *Trichogramma chilonis* Ishii and *Trichogrammatoidea bactrae* Nagaraja under different temperature conditions. Pakistan J. Zool., 40(6): 431-434.

Nadeem S, and M. Hamed M (2011). Biological control of sugarcane borers with inundative releases of *Trichogramma chilonis* (Ishii) (Hymenoptera: Trichogrammatidae) in farmer fields. Pakistan J. Agric. Sci., 48(1): 71-74.

Nadeem S, M Ashfaq M, M Hamid M, S Ahmed S, and M K Nadeem MK (2009). Comparative rearing of *Trichogramma chilonis* at different temperatures conditions. Pak. Entomol., 31(1): 33-36.

Nagarkatti S, and H Nagaraja H (1977). Biosystematics of *Trichogramma and* Trichogrammatoidae species. Ann. Rev. Entomol., 22: 157-176.

Rafique MS, A Suhail A, MJ Arif MJ, MD Gogi MD, and A Munir A (2007). Effectiveness of stem borer (*Chilo infuscatellus*) (Lepidoptera: Pyralidae). *Trichogramma chilonis* (Ishii) (Hymenoptera: Tricogrammitidae) against sugarcane J. Pak. Entomol., 27(2):141-144.

Rajendran B, and AM. Hanifa AM (1996). Efficacy of different releases techniques for *Trichograrnrna chilonis* (Ishii). Parasitoid on sugarcane internode borer. Ind. J. Pl. Protect., 2(24): 98-101.

Shenhmar M, J Singh J, SP Singh SP, KS Barar KS, D Singh D, PL Tadon PL, CR Balal CR, SK Jalali SK, and RJ Rabindra RJ (2003). Effectiveness of *Trichogramma chilonis* Ishii. For the management of *Chilo auricilius* on sugarcane in different sugar mills areas. Proc. Symp. Biol. Cont. Lepidopt. Pests. Bang. Ind., pp. 333-335.

Smith SM (1996). Biological control with *Trichogramma adances* success and potential of their use. Ann. Rev. Entomol. 41: 375-406.

Zia H, A Naheed A, and S Rizwana S (2007). Biocontrol of insect pests of sugarcane (Saccharum sp). J. Pak. Sug., 22(5): 13-22.

2

Traditional uses and cultivation of indigenous watermelons (*Citrullus lanatus*) in Mali

Aminata Dolo Nantoumé[1], Sidiki Traoré[1], Jorgen Lindskrog Christiansen[2], Sven Bode Andersen[2] and Brita Dahl Jensen[2]*

[1]Institut d'Economie Rurale/CRRA-Sotuba, Programme Fruits andLégumes, BP 262 Bamako, Mali.
[2]Department of Agriculture and Ecology, Faculty of Life Sciences, University of Copenhagen, Thorvaldsensvej 40, 1871 Frederiksberg C, Denmark.

Traditional knowledge associated with uses and cultivation of indigenous watermelon landraces was surveyed in five regions of Mali. Production of local types of watermelon plays an important role for many households. Farmers in charge of the production were mostly men, while women were in charge of seed extraction and processing. In total, ten types of landraces were differentiated, of which seven were seed types. The frequency of the types differed in the five regions, reflecting different agro-ecological conditions and traditions. Red, sweet flesh types used for dessert dominated in the more humid areas of West and Central Mali and were used as a cash crop. White flesh types, mainly grown for their seeds, were more important towards the desert region in the North. Here, a wild type was also used for human consumption. The seed was processed into sauces, snacks, and porridge, contributing fat and protein to the diet. Seed for sowing was sourced on-farm. In West and Central Mali, watermelons were intercropped with cereals. In the North, watermelon was cultivated as a sole crop in the sandy soils. Some of the landraces possess high heat and drought tolerance, a trait of interest to enhance food security in arid and desert areas.

Key words: *Citrullus*, genetic resources, Mali, survey study, traditional knowledge.

INTRODUCTION

Watermelon (*Citrullus lanatus* (Thunb.) Matsum. and Nakai) is an important and successful crop worldwide, popular for its juicy and sweet dessert type fruits. However, other types of watermelon, typically with white flesh, can be used as a multipurpose vegetable and the seed may serve as a source of protein, fat and carbohydrates as well as minerals (El-Adawy and Taha, 2001; Loukou et al., 2007). In addition, some types of watermelons can be stored for a long time and may in this way serve as a food and water source during drought prone periods (van der Vossen et al., 2004; Knight, 1995).

The origin, wild distribution and domestication of *C. lanatus* are still under debate. One theory is that the cultivated and wild watermelons have diverged independently from a common ancestor, possibly *C. ecirrhosus*, found in Namibia (Dane and Liu, 2007). It is believed that the largest genetic diversity is found among farmers to the north of the Kalahari Desert (van der Vossen et al., 2004) and that the domestication of watermelon began with the selection of non-bitter types by the Bushmen. Archaeobotanical findings of 5000 year old seeds of wild *C. lanatus* in the Sahara of Libya (Wasylikowa and van der Veen, 2004) has raised the possibility that the distribution in the past was much more extensive, and that the domestication also could have occurred somewhere in North Africa. This would tie up with the earliest evidence of agriculture in Africa and early records of domesticated watermelons in Egypt (Wasylikowa and van der, 2004). The watermelon subsequently spread over Africa and to other continents

*Corresponding author. E-mail: dahl@life.ku.dk.

(Wehner, 2008). The domestication has led to the development of indigenous landraces, defined as 'varieties with a high capacity to tolerate biotic and abiotic stress, resulting in a high yield stability and an intermediate yield level under a low input agricultural system' (Zeven, 1998).

Planned cultivar breeding programmes in sweet dessert types of watermelon began in the US around 1880 to 1900 (Wehner, 2008) and has since been adopted in different parts of the world. Improved seed types have been developed in Asia, for instance in China, where black and red seed types represent a significant market value (Yang and Walters, 1992). As for other vegetable crops in Africa, little breeding has been carried out to produce locally adapted cultivars (van der Vossen et al., 2004), and only a fraction of the genetic potential has supposedly been exploited. Modern varieties of overseas origin are frequently available in the formal seed market in bigger cities in Africa, including some new dessert type cultivars of watermelon. However, it is unknown whether or how much these imported genotypes have influenced the local germplasm in use by African farmers.

The use of local, indigenous landraces of cooking and seed type watermelons in Nigeria (Badifu, 1993), the Northern part of Namibia (Maggs-Kölling et al., 2000), Cote d'Ivoire (Loukou et al., 2007) and Benin (Achigan-Dako et al., 2008), indicates that watermelon plays a more important role for the diet than anticipated from the use of modern sweet dessert types alone. Given the drought tolerant nature of watermelon, it can be anticipated that its importance as a food source is related to agro-ecological conditions.

In Mali, very sporadic information exists on the importance, cultivation and use of indigenous watermelons, although the cultivation and use of seed type watermelon has been reported from Tombouctou (Jensen et al., 2011). With an increasing commercial production of sweet watermelon, dessert types with red flesh during the last decades (FAOSTAT, 2012), there may be a potential risk that cultivation of indigenous, well adapted landraces may be abandoned as farmers wish to try modern cultivars, unless their own landraces are valued. This ethnobotanical survey study was designed to gain insight into the traditional knowledge related to the importance, uses and cultivation of indigenous genotypes of watermelon. The study also aimed to document and describe the distribution and frequency of local watermelons in Mali, with focus on categorical identification of landraces.

METHODS

Survey areas

The survey was carried out in June to July 2008 in five important watermelon producing regions of Mali: Kayes in the West, Segou and Mopti in the Central part, and Gao and Tombouctou in the populated part of Northern Mali. In each of the five regions, surveys were carried out in two villages, which were chosen upon recommendations from local village chiefs. The 10 villages are mapped in Figure 1. The altitude of the villages were recorded using a hand held GPS device (eTrex Venture GPS navigator, Garmin Corporation, Taiwan) and was between 192 to 314 m above sea level (Table 1).

Climate and agro-ecological conditions in the survey areas

The five regions represent different climate types: sub-humid (Kayes), semi-arid (Segou), arid (Mopti) and semi-desert (Gao) and desert (Tombouctou) with a decreasing precipitation from 1000 mm/year to less that 200 mm/year in the respective regions (UN-WWAP, 2006). Average annual temperatures range from 26° to 29°C from South to North of Mali with a minimal temperature < 10°C and a maximal temperature > 45°C registered for Mali (Anonymous, 2001). The growing season of 5 to 7 months is more unpredictable in the Northern regions, at the border of the Sahara desert than in the regions further South (Table 1). Subsistence farming is dominating in all regions. Soils in Kayes, Segou and Mopti are ferrous tropical soils with low to medium fertility. Towards the North, soils are sandy with a low content of organic matter (UN-WWAP, 2006). In the Kayes region, cultivation of field crops is relatively intensive with use of fertilisers and pesticides. In the Northern regions, production is only possible in connection with rainy seasons or near the Niger River after the rainy period, or where irrigation is possible. Harvested products are commonly stocked during dry seasons for consumption through low production periods.

Survey methodology

Ten individual farmers (men and women) were interviewed in each of the 10 villages to collect as much variation possible with available resources. Survey questions were clustered to obtain information about the following themes:

1. Socio-demographic information concerning the respondents' ethnicity, age and involvement in watermelon production,
2. The indigenous landraces and their characteristics,
3. Watermelon uses and socio-economic importance, and
4. Seed sources and cropping systems.

Besides the information obtained from respondents, visits were made to 5 farmers' fields in Gao and 10 farmers' fields in Tombouctou to back up the interview data related to cultivation practices. Fields were not visited in Kayes, Segou and Mopti as the interviews took place after harvest.

The interviews were carried out in French, Bambara or Sonraï language with interpretation from local languages by research scientists. The interviews were mostly carried out around central village meeting points or in farmers' homesteads, all with individual respondents isolated from each other. In the village Makana (region Kayes) the respondents were all men, and other women provided information about the uses.

Seeds of local landraces of watermelon were collected during the survey from farmers and village markets and used as the focus of informal group discussions. Such discussions aimed to understand farmers' perception of types of landraces with the view to make a categorical description, and to study the frequency of the different

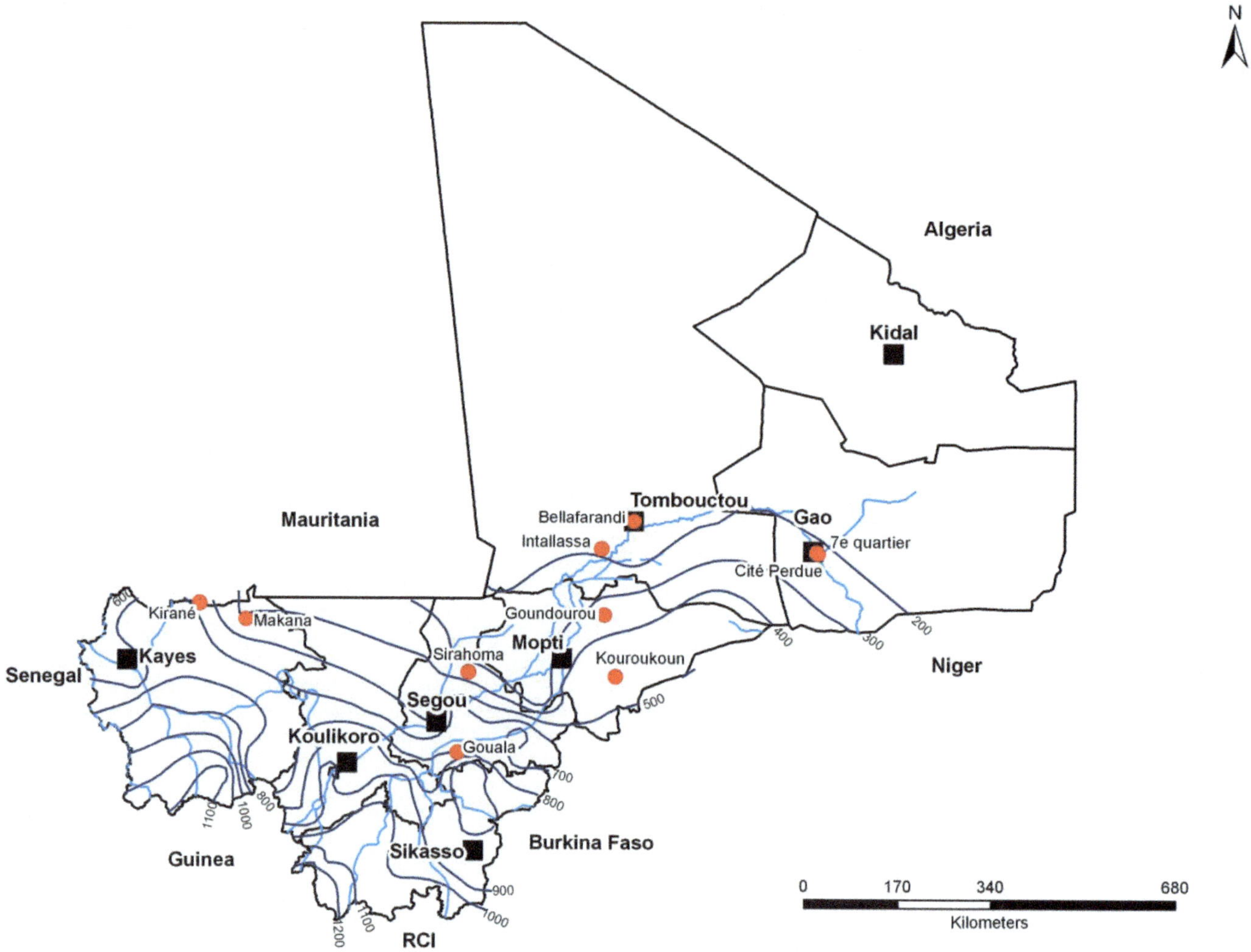

Figure 1. Map of Mali with regions and the capital Bamako, the Niger River and isohyets (precipitation in mm). The 10 villages in the five regions where the watermelon survey was carried out are situated on the map.

Table 1. Agro-ecological characterization of survey areas in Mali with cultivation of local watermelons.

Description/Region	Kayes	Segou	Mopti	Gao	Tombouctou
Area[1]	West	Central	North (Southern part)	North	North
Climate zones[2]	Soudanian-sahelian Soudanian south	Soudanian north	Sahelo-saharian	Sahelian	Saharian
Agro ecologic zones	Sub-humid	Semi-arid	Arid	Semi- desert	Desert
Altitude (m.a.s.l)	192 to 278	278 to 314	258 to 303	244 to 254	260 to 275
Precipitation (months/year)	5 to 7	3 to 5	3 to 4	1 to 3	Only dry months
Precipitation (mm/year)	700 to 800	600 to 800	400 to 600	200 to 400	< 200

[1]The description is related to the populated part of Mali, South of the Sahara desert. [2]According to UN-WWAP (2006).

Table 2. Socio-demographic information about the respondents in five regions in Mali with cultivation of local watermelons.

Parameter	Kayes [1]N=20	Segou N=20	Mopti N=20	Gao N=20	Tombouctou N=20	Total N=100	[2]P
Gender							0.079
Male	18	13	10	14	12	67	
Female	2	7	10	6	8	33	
Average age	48.0	53.9	48.5	52.0	56.7	51.8	
Average experience, years							
Experience as farmer	30.5	34.3	33.6	29.0	43.1	34.1	
Experience with watermelons	22.3	21.7	27.4	21.6	42.4	27.1	
Status of land							0.010
Owner or spouse owner	17	12	18	20	15	82	
Rent	3	8	2	0	5	18	
Main staple crops[3]	Peanut, sorghum, maize, millet, bean	Millet, rice, bean, sorghum, peanut	Millet, bean, peanut, sorghum, maize, rice, vegetables	Millet, bean, sorghum, vegetables, rice	Millet, rice, bean, vegetables, sorghum		

[1]N, The number of respondents per region; [2]P, Fisher's exact test, [3]Only watermelon growers were interviewed. The most important crop is mentioned first.

types of landraces in the five regions.

Data analysis

Data collected during the survey were analysed using Fisher's exact test, allowing frequency observations below 5, in the statistical programme R, version 2.8.1 (R Development Core Team, 2008) to test for significant differences among regions.

RESULTS

Respondents and their involvement in watermelon production

Data related to the respondents and their involvement in watermelon production is presented in Table 2. Among respondents, 67% were men and 33% women, with no significant differences among regions, and they were on average 52 years of age (20 to 78 years) and had 34 years of experience as a farmer. The experience with watermelon cultivation was on average 27 years. The respondents belonged to several of the many ethnic groups in Mali: 'Peul' and 'Sarakole' in Kayes, 'Bambara' in Segou, 'Dogon' in Mopti, and 'Tamacheck' in Gao, and 'Sonraï' and 'Bella' in Tombouctou regions (data not shown). The status of land for watermelon production (own or rented) varied among regions (P=0.01, Table 2). Most respondents produced watermelons on land owned by the household, inherited through generations (82%), while 18% rented land to cultivate their crops, and this was most prominent in Segou. The main staple crops were millet, rice, bean, sorghum, peanut, maize and vegetables, with a different order of importance in the five regions reflected by the precipitation pattern and dietary habits.

The indigenous watermelon landraces and their characteristics

A total of 132 landrace accessions of watermelon were collected in the form of fruits and seeds from farmers and local markets. The accessions represented sweet and non-sweet types with white, light red to red flesh. They were categorised and named on the basis of seed characteristics, which for the farmers were important to identify landraces (Figure 2). Fruit shape and rind pattern to some extent differentiated the types, but it was not until harvest, when the fruits were cut, that the exact type was revealed. Ten cultivated types of landraces were differentiated. Red flesh types were few (2) and white flesh types were more abundant (8), and most of them (7) were produced for their seeds. In addition to the cultivated types, a wild growing type called 'Birkinda' was found spontaneously in the fields in the region of Tombouctou. The named types were typically region specific. In the region of Kayes, only the 'Podé' type was produced. In Segou, two types, 'Kissè Filani' and 'Zèrè' were present. 'Zèrè', 'Gamou Karignan' and 'Kandé' were found in Mopti, and two types were present in Gao,

Figure 2. Major watermelon landrace types in Mali region with some fruit and seed characteristics. a. Type 'Fombou', b. Type 'Gamou Karignan', c. Type 'Kandé', d. Type 'Kaneye', e. Type 'Kankani blanc', f. Type 'Kankani rouge', g. Type 'Kissè Filani', h. Type 'Musa Musa', i. Type 'Podé', j. Type 'Zèrè', **Column 1,** Fruit; **Column 2,** Cut fruit; **Column 3,** seeds.

Table 3. Watermelon accessions (132) collected during the survey in the different regions assigned to types and major uses.

Region	Kayes	Segou	Mopti	Gao	Tombouctou	Total	[1]P
Watermelon type							<0.001
Sweet flesh	1	10	2	0	10	23	
Non sweet flesh	13	7	22	23	44	109	
Watermelon use							<0.001
Cooking	0	0	9	0	0	9	
Seed	13	7	13	23	44	100	
Dessert	1	10	2	0	10	23	
Total	14	17	24	23	54	132	

[1]P, Fisher's exact test.

namely 'Fombou' and 'Musa Musa'. These two types were also present in Tombouctou, where additionally three more types were found, namely 'Kankani blanc', 'Kankani rouge', and 'Kaneye', the latter also known as 'Ekanaye' or 'Bara Bara', depending on the language of the ethnic group growing it. Tombouctou was the region with most types, and included the types identified in Gao. Respondents in Gao told that the types grown in Gao have been introduced from Tombouctou by boat on the Niger River.

Four of the types, identified in different regions, were similar in seed characteristics, all with a cream colour: 'Podé', 'Kissè Filani', 'Kandé' and 'Kaneye'. These types and also 'Musa Musa' seeds were covered by a fleshy pericarp. When white flesh watermelons are harvested before maturity, then the flesh is soft and can be cooked. These fruits are then called 'Moli' in Gao and Tombouctou. Two of the identified type names ('Zèrè', Bambara language, and 'Kankani', Sonraï language) appeared to be used more widely locally as a collective term for watermelons. The type 'Zèrè' was used to describe a subgroup of several watermelon genotypes. In the village of Gouala, 17 km from the main road to Segou, 'Zèrè' genotypes included: 'Zèrè Mémé', 'Libanais', and 'Americain', and at a market in Segou other names were 'Zèrè Dotemeka' (village name in Segou) and 'Zèrè Camerounais'. 'Zèrè Meme' is believed to originate from Mali, but it is uncertain whether production is now based on improved seed. The names of some of the other 'Zèrè' genotypes suggest that part of the production is based on introduced seed (Zèré Libanais, Americain and Camerounais).

The frequency of the sweet flesh and non sweet flesh types collected and their uses were significantly different among regions (P<0.001, Table 3). A majority of non-sweet flesh types were collected in all regions, except for the region of Segou, where sweet flesh types used for dessert were more common (Table 3). Seed of types used for cooking were only collected in Mopti. In total, 76% of the 132 accessions collected were used for their seed, including the wild growing type 'Birkinda', only found in the Tombouctou Region.

Watermelon uses and socio-economic importance

Respondents revealed that the particular use, socio-economic importance and perception of diet contribution of the watermelons in general differed among regions (P=0.044 to P<0.001, Table 4), except that all respondents used the watermelons for food and all used the fruits. A majority of the respondents (85%) used the plant residues for fodder. It was mainly leaves, stems and what remains from the fruits which were fed to the animals. Watermelon was also used to produce soap (15%), mainly in Kayes and Mopti, and for cosmetics (3%) in the Kayes region. All households were using the fruits for food. Seeds were used by 77% of the households in all regions, but in particular, in the Northern regions where the white flesh types were dominating. In Segou, a white flesh seed type, 'Kissè Filani' was dominating in the village of Sirahoma, whereas the red flesh type 'Zèrè' was grown as one of the principal crops for cash in the village of Gouala, and the seeds were not used. In Goundourou in Mopti they grew the white flesh type 'Gamou Karignan' only to consume the flesh. Watermelon was mainly consumed cooked (85%), but also fresh (56%), although with little fresh consumption in Mopti. The seeds were processed into flour (44%), oil (32%) or high energy snacks (18%), a practice mainly developed in the Tombouctou region. In Gao, snacks were also consumed, although, this was not mentioned by respondents. For most households (70%) consumption was more important than sale, and 26% of the households did not sell at all. Sales exceeded consumption for 30% of the respondents, particularly in regions with red flesh, sweet types (Segou and Kayes). However, in Gao and Tombouctou, where the white flesh

Table 4. Watermelon uses and socio-economic importance in five regions in Mali with cultivation of local watermelons.

Region	Kayes	Segou	Mopti	Gao	Tombouctou	Total	[2]P
	[1]N=20	N=20	N=20	N=20	N=20	N=100	
Use							
Food	20	20	20	20	20	100	ns
Fodder	17	12	18	19	19	85	<0.001
Soap	7	1	7	0	0	15	<0.001
Cosmetics	3	0	0	0	0	3	0.035
Plant part used for food							
Fruit	20	20	20	20	20	100	ns[1]
Seed	18	9	10	20	20	77	<0.001
Preparation for food							
Fresh	16	12	1	12	15	56	<0.001
Cooked	16	9	20	20	20	85	<0.001
Flour	1	9	10	12	12	44	0.044
Oil	0	5	10	11	6	32	<0.001
Snack	3	0	0	0	15	18	<0.001
Consumption/sale							<0.001
Home consumption only	5	6	11	2	2	26	
Home consumption > Sale	6	3	9	15	11	44	
Sale > Home consumption	9	11	0	3	7	30	
Diet contribution							<0.001
50 to 80%	0	7	8	10	12	37	
5 to 49%	0	13	11	5	5	34	
No idea	20	0	1	5	3	29	

[1]N, Number of respondents; [2]P, Fisher's exact test; ns, Not significant at 5% level.

types were dominating, the seeds had particular importance for cooking and consumption, and only surplus seed was sold. The importance of the watermelons for home consumption was also reflected in the perception of its importance for the diet. Thirty-four percent of the respondents were of the opinion that watermelons contributed with 5 to 49% to their diet, and 37% thought it contributed with 50 to 80% to their diet, and the remaining respondents had no idea about the diet contribution. The contribution to the diet as a source of protein and fat was in particular reported to be high in the areas where the seed types were grown (Gao, Tombouctou).

The recipes used to prepare watermelon based products differed from region to region, and depended on the ethnic groups in the villages. Red flesh, sweet tasting types were consumed raw as a dessert and a thirst quencher everywhere. The flesh of the white flesh types may be eaten raw, but were mainly used for cooking of various dishes, either as the principal ingredient or as a condiment. Flesh is frequently cooked to produce various thick sauces for cereal based meals. An example of a unique dish was the preparation of the white flesh from 'Gamou Karignan', which was cooked and consumed as a dessert or snack, sometimes flavoured with leaves of hibiscus. Seeds were used raw or roasted to various snack products with salt or sugar, or pounded to flour as a basis for drinks, sauces, porridges and oil extraction. Seeds were also boiled with ash to produce soap. The wild growing type 'Birkinda', found spontaneously in the fields in the region of Tombouctou, was mainly used to feed animals, but it was also used for human consumption.

Seed sources and cropping systems

Most farmers interviewed used their own seeds (85%) or seeds received from family, friends or neighbours (10%) with no significant differences among regions (Table 5). Seed storage, cropping systems, and intercrop species varied among regions (P<0.001, Table 5). Seeds were

Table 5. Watermelon seed sources, storage and cropping systems in five major regions of Mali.

Region	Kayes	Segou	Mopti	Gao	Tombouctou	Total	[2]P
	[1]N = 20	N = 20	N = 20	N = 20	N = 20	N = 100	
Source of seeds							ns
Owner	15	15	17	19	19	85	
Family/Neighbour	3	2	3	1	1	10	
Markets (informal and formal)	2	3	0	0	0	5	
Seed storage	N = 18	N = 17					<0.001
House (in bag, bottle, tin, gourd)	17	17	0	20	17	71	
Traditional seed storage house	1	0	10	0	3	14	
Other (organic manure)	0	0	10	0	0	10	
Cropping system							<0.001
Sole crop	2	11	0	20	20	53	
Intercropped	18	9	20	0	0	47	
Intercrop							<0.001
Millet	5	0	20	0	0	25	
Maize	1	0	0	0	0	1	
Sorghum	12	9	0	0	0	21	
Fertilizer and plant protection							
Organic manure	7	15	20	3	7	52	<0.001
Chemical fertiliser	0	10	0	0	2	12	0.014
Herbicide	0	0	0	0	0	0	ns
Pesticide	4	10	0	0	0	14	0.096
None	9	0	0	17	11	37	0.026

[1]N, Number of respondents; [2]P, Fisher's exact test; ns, Not significant at 5% level.

mainly stored in the main house in a container (71%), but in the village Goundourou (Mopti) seeds were kept in traditional, stilted seed storage houses, which also accommodated other seeds. In the village Kouroukoun (Mopti), farmers stored the seeds by mixing with household waste and animal manure in a pile inside the yard. The seeds were then spread with this manure during the rainy season and then germinated.

Five percent of respondents (Segou and Kayes) had brought seed from the market, and one respondent in the village of Gouala, Segou, mentioned that he bought improved seed of the red flesh type from the market. Visits to commercial farmers' fields and to a market in Segou, one of the main centers for commercial sweet watermelon production, suggested that the use of improved seed in this region perhaps was more common than respondents from the surveyed villages indicated.

The watermelon crops were grown either in sand or sandy loam. The variation recorded in cropping systems and management were related to the agro-ecological conditions and local traditions in the different regions. Most crops were sown at the start of the rains, typically in July, but in Segou sowing was carried out throughout a longer period, also after the rains had started. In Tombouctou, a second crop was established late after the rains using a different cropping practice. The crop was sown at the shore of the Niger river, when the river was withdrawing, in some places using additional bucket irrigation with water from the river. Dry sowing, before the rain, was rare and only practiced by one respondent.

In the Western and Central part of Mali (Kayes, Segou and Mopti) intercropping with cereals prevailed. In Kayes, with a relatively high precipitation, watermelon was intercropped with sorghum, millet and maize. In the slightly drier region of Segou, watermelon was intercropped with sorghum, and with millet in the drier Mopti (Table 5). In the semi desert and desert part of Mali (Gao and Tombouctou), watermelons were grown in pure stands, and in some sandy dry areas, there were no crop alternative. When grown as a sole crop, it was frequently observed to be grown in strips of land with different watermelon types next to each other.

When intercropping was practised, the watermelon seeds were mixed with cereal seeds, or with manure (Mopti), and the seeds were then sown in the same holes as the cereals, leading to a random establishment of watermelon plants in the field, but with a lower density than the cereals. After germination, replanting from areas with many plants to open areas with failed germination was sometimes practised to secure an even stand of watermelon. Farmers' demonstrations in the field showed that the typical planting distance was approximately 0.75 to 1.0 x 1.0 to 1.5 m. Two to five-six seeds were sown per hole. Drought tolerance was reported as the main attribute of watermelon crops in all regions (68% of respondents).

The use of fertilisers was in general related to the region and types of watermelon grown ($P<0.001$ for organic manure, $P=0.014$ for chemical fertiliser, Table 5). Organic manure was used in all the regions (52% of all farmers) but particularly in Mopti (100%) and in Segou (75%) where it was related to intercropping with cereals. Chemical fertilisers were mostly used in Segou for the dessert watermelon crop. Overall, only 12 and 14% of the respondents used chemical fertiliser and pesticides, respectively. No use of herbicides was recorded and 37% used neither fertiliser nor plant protection. The watermelon crops were harvested 3 to 4 months after sowing, beginning in September and ending in December, with the major harvest during the months of September and October (76%). In Tombouctou, a late harvest follows during December and January from the late crop grown near the river. Fruits were harvested before maturity for cooking purpose, whereas the majority were left to ripen before seeds were extracted.

DISCUSSION

From a socio-economic perspective, the present study reveals that production of local genotypes of watermelon plays an important role for many households in Mali for food security and also as a cash crop, in particular in the region of Segou.

In our study, the farmers on average had 27 years of experience with watermelons. This indicates that farmers involved in watermelon cultivation are from the older generation, and this could perhaps become a bottleneck in relation to the transfer of traditional knowledge to future generations, both related to the cultivation, mainly carried out by men, but also related to the post harvest seed extraction management and food processing, mainly performed by women. A visit in the Tombouctou region during the harvest season revealed that women possess a considerable knowledge about processing of seed into snacks and other food products, which may pave the way for a more sophisticated form of income generation on the basis of the local seed types. In Benin, women maintain a similar knowledge about transformation of seeds into value added products (Achigan-Dako et al., 2008), also supporting that seed types have an undeveloped potential for income generation. Special action and attention to the watermelon crop and its potential will most likely be needed at the community level to ensure that the knowledge related to the crop and its uses will sustain.

Both white and red flesh types of local watermelon landraces are grown in Mali. The present study shows that fruits and seeds of these watermelons are used in various ways and contribute significantly to the diet, and that watermelons generate income when fruits or seeds are sold or after processing. The types, their importance and uses were region specific, reflecting different agro-ecological conditions and traditions. When seed was collected, types used for cooking were only reported in Mopti, but when the survey was carried out, cooking was practiced in all regions. This suggest that true cooking types are found in Mopti, but that other types, such as seed types, also are used for cooking.

In the less arid areas of Kayes, Segou and Mopti, the watermelons are often intercropped with the main cereals, and may secure food supply, even if the cereal crop fails, and add to food and diet diversity. In the desert area towards the North, watermelon is in some places the only crop, which can grow on rain-fed, sandy soils. In this way it has a more central role for food security for some households, for which reason the seed types may have become more prominent.

The recipes developed for the use of the seed differed in the regions, and may reflect different ethnic traditions or that different recipes have evolved because of isolation between regions. Traditionally, ethnicity related to occupation. For instance, the Bambara in the region of Segou were traditionally farmers, and this may explain that a more intensive cash crop production near Segou has evolved. A relatively short distance to the regional market in Segou and easy access by tarmac road to the capital Bamako may be another factor influencing the domination of the 'Zere' cash crop dessert type in this region.

Similar to the present study, results from a survey in Benin highlighted that local genotypes, knowledge and practices were region specific (Achigan-Dako et al., 2008), reflecting traditions of different ethnic groups and the climatic conditions prevailing. The importance of *C. lanatus* among the Cucurbit seed crop species is believed to relate to the relatively short growing season of 3 months (Achigan-Dako et al., 2008). This combined with the ability to grow under low rainfall conditions makes watermelon an interesting plant for desert communities.

In general, the perception of sufficient supply with watermelons was associated with less health problems. The use of immature fruits of some types, 'Musa Musa' or

'Bara Bara', was even reported as a remedy to cure fever. In the North, it was a common knowledge that drinks based on crushed watermelon seeds were rich in energy, and could keep people working for the whole day in the field. Studies of the nutritive value of *Citrullus* spp. and other curcurbits have revealed that seeds of these species are important for supply of oil and protein (Badifu, 1993; Loukou et al., 2007), and the latter report of a higher content of oil and protein in *C. lanatus* seed than in peanut, *Arachis hypogaea*. This supports the value of watermelon seeds as a cheap source of protein in comparison to meat. In rural Mali, a study has shown that dietary diversity is positively associated with nutrient adequacy (Torheim et al., 2004). This supports that it is crucial to maintain the diversity of locally adapted, indigenous, often neglected and underutilised crops, such as the watermelon, both from a biodiversity and a nutritional point of view.

The commercial sweet watermelon acreage in Mali has more than quadrupled during the last 30 years, from 5.500 ha in 1980 to 23.847 ha in 2010 with an estimated production of 150.000 T in 1980 to 430.995 T in 2010 (FAOSTAT, 2012), and cause a potential risk that cultivation of local landraces, in particular the red flesh, sweet types, slowly may be abandoned, leading to loss of interesting germplasm. This is supported by observations in the region of Segou, where a significant commercial production is practiced. Here, some respondents from Gouala village producing watermelon for cash indicated that they moved from cultivation of local to introduced types, because they prefer big, sweet fruits. Generally, white flesh, seed watermelons are produced in more remote areas, and because of their special seed traits, which cannot be matched by modern, commercial cultivars, there is perhaps less of a threat for loss of these genotypes. However, watermelon grown as a cash crop receives attention from the extension service, whereas local seed watermelons in the region of Tombouctou for instance, so far have received less attention from the extension service than other irrigated cash crops cultivated along the Niger River. This may be because they are mainly used for home consumption and therefore not valued in the same way. In this way, there is a risk that cultivation of the local types may be substituted with other types or crop species where these can grow, unless local research stations and extension services start to play a positive role by including the seed watermelons in their palette of serviced crops. An active role of the extension service could be a strategy to highlight the value of the crop in the local community that may diminish the risk of loss of genotypes. Some respondents indicated that this was the only crop they could grow, as they had no access to fields near the river. This implies on the other hand that the watermelon seed types may be sustained with no additional sensitisation, because they are regarded as an important food source.

The knowledge, in combination with an interest in selling value added, processed seed products in local communities may pave the way and be the driver for more organised income generation based on seed watermelons. Indications from young people were that the processing of seeds is cumbersome and time consuming, and may be lost, unless more modern processing techniques are developed. Support from extension services and others should therefore, in particular encourage to modernise the transformation processes, perhaps by forming small cooperatives for more efficient seed processing. This would raise awareness of the potential of the crop and thereby indirectly lead to in situ on farm conservation. In the Tombouctou region, cultivating different types in strips next to each other within the same fields may be a practice, which together with the practice of seed exchange at local markets, could help to maintain the diversity of the material. In Peru, farmers rely in some areas still on indigenous maize landraces, and the survival of these seems to rely on a non-restricted movement of seeds (Stromberg et al., 2010). A collective, organised action of farmers to conserve seed types may be a way forward to secure diversity of important genotypes for the future.

Climate change has been projected to have impact on the future food security in Mali, and it is foreseen that the percentage of the population in risk of hunger will rise (Butt et al., 2005), that Malian agriculture does not have a significant potential for reducing undernourishment by 2015 (Butt and Mccarl, 2005), and that crop yield changes are in the range of minus 17% to plus 6% at the national level (Butt et al., 2005). Since 1970, the registered precipitation has declined, and the isohyets have moved approximately 200 km south in Mali (Anonymous, 2001). Cultivation of locally adapted, heat and drought tolerant crop species and cultivars may be a way to adapt to the effects of climate change. Watermelon landraces, which in the present study was yielding fruits under desert conditions in the Tombouctou and Gao regions, with no input of chemical fertilisers or pesticides, may be a plant genetic resource, which could be a brick in a puzzle to improve food security and diet diversity even in other similar heat and drought prone areas.

Conclusion

This study confirms the persistence of traditional knowledge related to use and production of watermelon and the persistence of indigenous landraces in Mali. Ten different types of landraces were categorised, and the presence, frequency and use of them differed in the five regions, reflecting different agro-ecological conditions and traditions. The study has revealed that local types of watermelon, in particular seed types, play an important role for the nutritional status in households, in particular in the desert regions in the North, where watermelon in some places is the only crop, which can grow in the dessert sand. It would be desirable to study the diversity of the crop to enable an *in-situ* conservation strategy, and to study the potential of the types further to value and identify their potential under arid conditions, especially seen in the light of a need for drought tolerant plant species in a world with climate change.

ACKNOWLEDGEMENTS

This study was carried out on the basis of a research grant from Danish International Development Agency (Danida, Project no. 104.Dan.8-919). We thank all colleagues at regional centers of Institute d'Economie Rurale (IER) for valuable assistance and information about watermelon cultivation. We are grateful to Amadou Sidibé and Kadiatou Touré Gamby, IER Sotuba, for coordinating the study in Mali. Lassana Touré and Souleymane Traoré, IER, are thanked for designing the map in Figure 1. We thank the staff at Mali Folkecenter for collaboration and the budget that made the second trip to Tombouctou possible. We are grateful to our terrific drivers, Sinaly Quattara, IER Gao, and Ibrahim Traoré for taking us safely through the sandy tracks of the desert. A special thank to farmers and their families for introducing us to the world of their watermelons, bringing life to the desert.

REFERENCES

Achigan-Dako EG, Fagbemissi R, Avohou HT, Vodouhe RS, Coulibaly O, Ahanchede A (2008). Importance and practices of Egusi crops (*Citrullus lanatus* (Thunb.) Matsum. & Nakai, *Cucumeropsis mannii* Naudin and *Lagenaria siceraria* (Molina) Standl. cv. 'Aklamkpa') in sociolinguistic areas in Benin. Biotechnol. Agron. Soc. Environ. 12:393-403.

Anonymous (2001). Situation générale de la diversité biologique au Mali. Tome I. Ministère de L'Equipement de l'Aménagement du Territoire de l'Environnement et de l'Urbanisme. Republique du Mali p. 139.

Badifu GIO (1993). Food potentials of some unconventional oilseeds grown in Nigeria - A brief review. Plant Foods Hum. Nutr. 43:211-224.

Butt TA, Mccarl BA (2005). An analytical framework for making long-term projections of undernourishment: A case study for agriculture in Mali. Food Policy 30:434-451.

Butt TA, Mccarl BA, Angerer J, Dyke PT, Stuth JW (2005). The economic and food security implications of climate change in Mali. Clim. Change 68:355-378.

Dane F, Liu J (2007). Diversity and origin of cultivated and citron type watermelon (*Citrullus lanatus*). Genet. Resour. Crop Evol. 54:1255-1265.

El-Adawy TA,Taha KM (2001). Characteristics and composition of watermelon, pumpkin, and paprika seed oils and flours. J. Agric. Food Chem. 49:1253-1259.

FAOSTAT (2012). Production. Crops. FAOSTAT. Food and Agriculture Organisation of the United Nations, Rome. http://faostat.fao.org/

Jensen BD, Touré FM, Hamattal MA, Touré FA, Nantoumé AD (2011). Watermelons in the sand of Sahara : Cultivation and use of indigenous landraces in the Tombouctou Region of Mali. Ethnobot. Res. Appl. 9:151-162.

Knight MH (1995). Tsama melons, *Citrullus lanatus*, a supplementary water supply for wildlife in the southern Kalahari. Afr. J. Ecol. 33:71-80.

Loukou AL, Gnakri D, Djè Y, Kippré AV, Malice M, Baudoin J-P, Zoro Bi IA (2007). Macronutrient composition of three cucurbit species cultivated for seed consumption in Côte d'Ivoire. Afr. J. Biotechnol. 6:529-533.

Maggs-Kölling GL, Madsen S, Christiansen JL (2000). A phenetic analysis of morphological variation in *Citrullus lanatus* in Namibia. Genet. Resour. Crop Evol. 47:385-393.

R Development Core Team (2008). R: A language and environment for statistical computing. R Foundation for Statistical Computing, Vienna, Austria. http://www.R-project.org.

Stromberg PM, Pascual U, Bellon MR (2010). Seed systems and farmers' seed choices: The case of maize in the Peruvian Amazon. Hum. Ecol. 38:539-553.

Torheim LE, Ouattara F, Diarra MM, Thiam FD, Barikmo I, Hatløy A, Oshaug A (2004). Nutrient adequacy and dietary diversity in rural Mali: association and determinants. Eur. J. Clin. Nutr. 58:594-604.

UN-WWAP (2006). Rapport National sur la mise en valeur des ressources en eau : Mali Une étude de cas du WWAP préparée pour le 2ème rapport mondial des Nations Unies sur la mise en valeur des ressources en eau: L'eau, une responsabilité partagée. Direction nationale de l'hydraulique, Ministère des mines, de l'énergie et de l'eau. Mali, p. 212. http://www.unesco.org/ulis.

van der Vossen HAM, Denton OA, El Tahir IM (2004). *Citrullus lanatus* (Thunb.) Matsum. and Nakai. In Grubben GJH and Denton OA (eds), Plant Resources of Tropical Africa 2.Vegetables. PROTA Foundation, Wageningen, Netherlands / Backhuys Publishers, Leiden, Netherlands / CTA, Wageningen, Netherlands. pp. 185-191.

Wasylikowa K, van der Veen M (2004). An archaeobotanical contribution to the history of watermelon, *Citrullus lanatus* (Thunb.) Matsum. & Nakai (syn. *C. vulgaris* Schrad.). Veg. Hist. Archaeobot. 13:213-217.

Wehner TC (2008). Watermelon. In Prohens J and Nuez F (eds), Vegetables I. Asteraceae, Brassicaceae, Chenopodicaceae, and Cucurbitaceae. Springer Science+Business Media, New York. pp. 381-418.

Yang SL, Walters TW (1992). Ethnobotany and the economic role of the Cucurbitaceae of China. Econ. Bot. 46:349-367.

Zeven AC (1998). Landraces: A review of definitions and classifications. Euphytica 104:127-139.

Influence of insecticide treatment on ant diversity in tropical agroforestry system: Some aspect of the recolonization process

Zéphirin Tadu[1], Champlain Djiéto-Lordon [1], Regis Babin [4], Yede [1], Edith Blandine Messop-Youbi[1, 3] and Abraham Fomena [2]

[1]Laboratory of Zoology, Faculty of Science, University of Yaounde I, Po. Box 812, Yaounde, Cameroon.
[2]Laboratory of Parasitology and Ecology, Faculty of Science, University of Yaounde I, P . Box 812, Yaounde, Cameroon.
[3]Department of Zoology and Animal Physiology, Faculty of Science, University of Buea, PO Box 63 Buea, Cameroon.
[4]CIRAD, UPR Bioagresseurs Analyse et Maîtrise du Risque, F-34398 Montpellier, France.

In order to assess the influence of insecticides treatment on ant diversity and recolonization processes in cocoa based agroforestry systems, sampling was performed using chemical knock-down technique during two consecutive cocoa campaigns (2006 and 2007) in six smallholding cocoa farms located in tree localities of the Centre Region of Cameroon. These localities are located along the South-North agroecological benchmark ranging from Ngomedzap in the humid tropical rainforest in the south through Obala in mixed degraded forest / savannah in the transitional zone to Bokito in savannah landscape. From sampled ants, community diversity was measured using Shannon index and similarities between communities calculated using the Bray Curtis distance. After one-year recolonization following insecticide treatment, species richness increased in all the study sites, from 23 to 48 species at Ngomedzap, from 39 to 46 at Obala, and from 37 to 40 at Bokito. The diversity index increased from 2.15 to 2.29 at Obala while decreasing respectively from 2.12 to 2.04 at Ngomedzap and from 2.0 to 0.89 at Bokito. After one year recolonization following treatment, a total of 35 newly establish ant species were observed. Among them 11 were arboreal and two ground-nesting species. Comparison of the community composition and population size between the two sample periods suggested that certain arboreal nesting species such as *Oecophylla longinoda, Tetramorium aculeatum* and some species of the genus *Crematogaster* may survive insecticide treatment, thanks to their nesting behaviour. Also, recolonization process may be strongly influenced by border effects.

Key words: Ant, Bokito, cocoa farm, Cameroon, diversity, insecticides treatment, Ngomedzap, Obala, recolonization process.

INTRODUCTION

Ants and termites are two major arthropod groups in tropical humid rainforest. Both are usually cited as the numerically dominant and the most rich-species groups of artropods (Hölldobler and Wilson 1990 p.1; Watt et al., 2002; Fisher, 2005); also, they are known to play key role in various ecological processes. Despite these considerations, anthropogenic disturbances that are known to negatively affect arthropod community in these regions (Fitzherbert et al. 2008) are progressively intensifying worldwide. In the most tropical region of the world, forest loss caused by timber and fuel wood exploitations or charcoal production and continuing expansion of agriculture is alarming (Widodo et al., 2004; Fayle et al., 2010). These disturbances affect species richness and ant abundance in native ecosystems (Floren et al., 2001; Watt et al., 2002), as well as the dynamic of the whole ecosystem (Floren et al., 2001). Meanwhile, low diversity is not always related to anthropogenic disturbance, but may also be related either to lower species density (Schonberg et al., 2004) or to the presence of aggressive dominant ant species (Majer, 1990) or invasive species (Lebreton et al., 2003).

Even though it is well established that anthropogenic disturbances affect arthropod diversity, their impact on species richness is still under discussion. For instance, it have been demonstrated that species richness do not differ along disturbance gradient in Cordillera de Tilara'n in Costa Rica (Schonberg et al., 2004) and in the Lahn-Dill-Bergland in Germany (Dahms et al., 2005). Also, some studies have demonstrated that certain agricultural systems like agroforest give a glimmer of hope for biodiversity preservation in disturbed area. In this context, rustic cocoa farms appeared as the best systems that provide favorable condition for biodiversity preservation (Rice and Greenberg, 2000; Merijn et al., 2007) as they, support high levels of biodiversity, even approaching that of undisturbed tropical forests for some groups (Merijn et al., 2007).

In the center region of Cameroon, cocoa are grown along an agroecological benchmark that cover three vegetation types: forest in the south and savannah in the north. Between the two areas there is a forest - savannah transition zone. Along this ecological gradient, environmental conditions in cocoa agroforest are highly diversified and they usually associated cocoa trees with forest and planted fruit trees that presents and socio-economic and ecological importance (Sonwa et al., 2007). This variability induced variations in pest and disease pressures and consequently variation in pesticides treatments regime. In Cameroonian cocoa farms, the control of the main insect pest, *Shalbergella singularis* (Hemiptera, Miridae) and disease, black pod disease (mildew) caused by *Phytophtora megakarya* (Fungi) by farmers involved various insecticide and fungicides. About 60% of cocoa farmer in southern Cameroon yearly apply pesticide in their farms. These included, mainly copper based fungicides (such as Nordox, Kocide, Cacaobre and Ridomil), and various insecticides (including Azinphos methyl or Cypercal, Dursban and Aldrin). Among insecticides, some belonged to "Class 1b" of the World Health Organization classification, and are known to be "highly dangerous" (Sonwa et al., 2008) as they represent serious threat for the hole ecosystem. Consequently, they are no longer recommended in agriculture. In the center region of Cameroon, cocoa are grown along an agroecological benchmark that cover three vegetation types: forest in the south and savannah in the north. Between the two areas there is a forest - savannah transition zone. The aim of the present study was to evaluate in these areas, how arthropod communities react to the insecticides treatments. Base on ant model, the study evaluate variations in the community composition and structure after one year recolonization following an insecticide treatment in small-holding cocoa farms. Data are discussed with reference to environmental conditions and anthropogenic disturbances.

MATERIALS AND METHODS

Study sites

The study was conducted within an area situated between the latitudes 2°35' - 4°15'N and the longitudes 11°48' - 11°15'E (Figure 1). The Altitude varied between 450 and 715 m. The climate is subequatorial with a bimodal rainfall regime, and an annual rain fall of about 1900 mm. The annual mean of temperature is around 25°C with weak thermal variations (Bisseleua and Vidal, 2008). The latitudinal gradient induced progressive variation in rainfall, temperature and vegetation characteristics.

Considering this, three localities were chosen for sampling the first in the evergreen forest at Ngomedzap, the second in the semi-deciduous forest/savannah transitional zone at Obala and the third in the bushy savannah at Bokito (Letouzey, 1985).

Ngomedzap, in the southern part of the study area is covered by evergreen forest, weakly logged thanks to low human population density and low intensity of agricultural activities. Annual rainfall is around 1700 to 1800 mm (Santoir and Bopda, 1995). Cocoa agrosystems are located near the forest margin or in the dense forest environment. They presented the most diversified and dense shade trees, dominated by forest trees maintained at the garden creation mixed with some planted fruit trees.

Obala, in the centre of the area is covered by semi-deciduous forest strongly degraded because of high human population density and intense agricultural activities. Cocoa agrosystems are located in these strongly degraded forests around houses (Bisseleua and Vidal, 2008), and showed relatively less dense and less diversified shade trees communities compared to Ngomedzap, in which native forest trees have been cute and replaced by domestic trees. Annual rainfall vary between 1400 - 1700 mm/year (Santoir and Bopda, 1995).

Bokito, located in the northeast part of the area, is covered by bushy-savannah vegetation with patches of gallery-forest and man-made agroforest (Letouzey 1985). The annual rainfall is the lowest (1300 to 1500 mm/year) (Santoir and Bopda, 1995). Cocoa plantations showed less dense and less diversified shade tree communities, composed of man-made mixture of planted forest and domestic trees (Table 1).

The study farms at Ngomendzap were the older (60 years) compared to Obala (20 years) and Bokito (20 to 60 years) (Table 1). From different points of view (geographical, environmental and agronomical), Obala site appeared to be intermediate between Bokito and Ngomedzap. The choice of plantations was guide on one hand by its geographical location with a view to covering contrasting agroecological conditions and on the other by the absence of any insecticide treatment for at least three years. Table 1 gives the main geographical and agroecological characteristics of the chosen plantations.

Sampling method

During 2006 and 2007 respectively, ants were sampled in selected plots by chemical knock down. For each plantation, ants were sampled on 100 adjacent trees in an area with heterogeneous

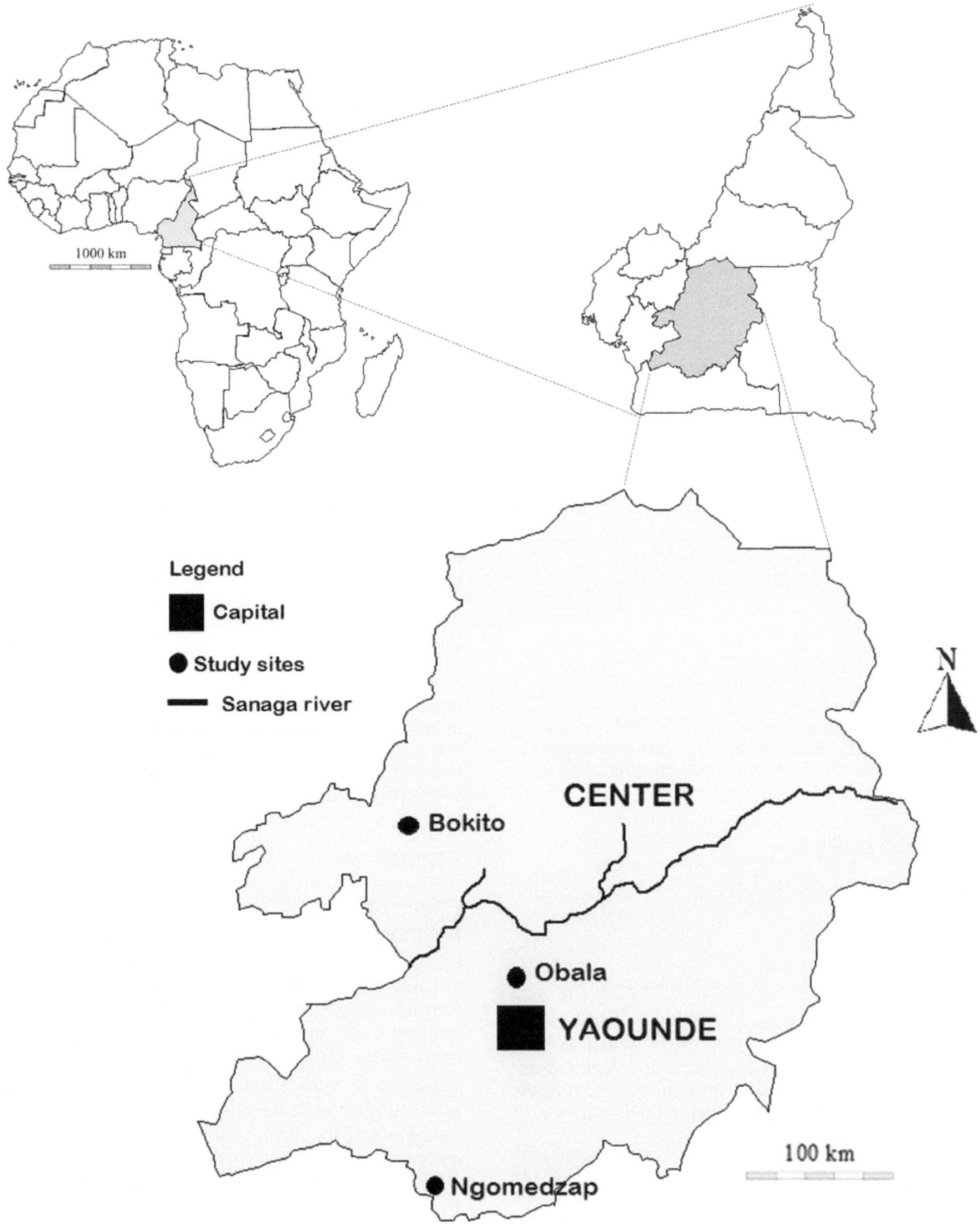

Figure 1. Map indicating the location of the study sites.

shade level. Plots measured about 2025 m² (45 × 45 m). To achieve this, Plastic sheets measuring 4 × 4 m were spread at the foot of each cocoa tree. Plants were subsequently sprayed with an endosulfan-based insecticide using a motorized mistblower (Solo type 40123; Solo Kleinmotoren, Germany) at 100 mL/ha. Endosulfan was chosen because it has a broad action-spectrum and a sufficient 'shock' effect, to kill most of the insects within few hours. Cocoa trees canopies were sprayed around 6 A.M, when there was no wind. Ten hours after the spraying, all the arthropods felt on the plastic sheet were collected and preserved in hemolysis

Table 1. Geographic point and maintenance activity of the six plots sprayed in southern Cameroon cocoa farms during 2006 - 2007 campaign.

Parameter	Site					
	Bokito		Obala			Ngomedzap
Farms	BO.PI	BO.PII	OB.PI	OB.PII	OB.PIII	NG.PI
Latitude	04°29'24"	04°34'29"	04°06'19"	04°06'14"	04°08'58"	03°16'10"
Longitude	11°04'42"	11°10'45"	11°28'40"	11°28'52"	11°23'02"	11°13'21"
Altitude (m)	450	450	600	650	550	700
Ages/year	20	60	20	60	60	60
Density/ha	1000	1300	1400	1300	1300	1500
Prunning/year	0	0	0	0	0	1
Weeding/year	2	1	0	0	1	2

tubes containing 70% ethanol. Ants were later sorted in the laboratory and kept apart for identifications. These were conducted in the Laboratory of Zoology at the University of Yaounde 1, based on the dichotomic keys (Hölldobler and Wilson, 1990; Bolton, 1994; Taylor, 2002). Samples were later compared to the collection of the Belgium "Musée Royal d'Afrique Centrale" at Tervuren. Voucher specimens are preserved in the Laboratory of Zoology at the University of Yaounde 1 (Cameroon).

Data analysis

Ant diversity

Prior to analyzes data from each locality were pooled. In order to characterize ant community, Shannon diversity index was compute using Estimate 8.2 software (Colwell, 2006) and the Equitability index (E) calculated using the following formulae:

$$H' = -\sum_{i=1}^{N} P_i \mathrm{Log}_2\, P_i$$

$$E = \frac{H'}{\log_2 S}$$

Where P_i (= n_i/N) is proportion of i species in the population.

The effect of insecticide on species richness and ant diversity was evaluated by comparing the community of 2006 to that of 2007 using Friedman chi squared test (Q) and associated Wilcoxon test corrected with sequential Bonferroni procedure for pairwise comparison. The analyses were conducted with R software (R 2.13.1, 2011).

To evaluate the impact of insecticide treatment on ant community composition, Bray Curtis index also known as Sorensen quantitative index was used to determine similarities between the community of the first year to that of the second and the cluster was performed using Ward aggregation method. Bray Curtis index is given by the following formula:

$$C_n = \frac{2jN}{N_a + N_b}$$

Where N_a: is the total number of individual in site A; N_b: the total number of individual in site B and 2jN: is the sum of the lower of the abundance for species found in both sites.

Variation of the ant community composition

Between the 2006 and the 2007 campaigns, ant species collected were separated into pre-establish (present in the 2006 sample) and newly establish ant species (new species that appeared in 2007). The effect of treatment on each pre-establishes ant population was assessed by comparing the abundance between the two samples using Friedman chi squared test (Q) and associated Wilcoxon test corrected with sequential Bonferroni procedure for pairwise comparisons. The analysis was done using R software (R 2.13.1, 2011).

Also to assess geographic variation of the re-colonizing community in relation to locality and nesting behavior, newly established ant species were organized in presence/absence matrices data according on one hand, to locality and on the other hand to nesting habits and clusters were performed using Euclidian distance base on Ward aggregation method using XLSAT software (version 2010.4.02; Addinsoft, 2010).

RESULT

Effect of insecticide treatment on ant richness and diversity

A total of 60 ant species belonged to 7 subfamilies, 24 genera were identified from a sample of 52911 workers (Appendix 1) during the two collection campaigns on the three sites. After one year recolonization, data obtained indicated a significant increase of species richness and ant abundance in each study site. Contrarily, the diversity decreased in both Bokito and Ngomedzap, while increasing at Obala (Table 2).

In 2006, the community parameter varied between the sites, with Ngomedzap showing the highest diversity, but the lowest abundance and species richness. It was followed by Obala that also showed the highest species richness but not the highest abundance, while the lowest diversity and the highest abundance were observed at Bokito (Table 2).

One year after insecticide treatment, were observed significant variations of the three community parameters on each site (Table 2). According to the site, there are significant variation of mean number of workers per tree

Table 2 Variation of Shannon diversity index under treatment effect in center region of Cameroon (2006-2007).

Parameter	Site						Q test
	Bokito		Obala		Ngomedzap		
	2006	2007	2006	2007	2006	2007	
Sample size (N)	13239	17788	7253	8924	402	5305	
Mean number of workers/tree	66.20±8.71[a]	88.94±24.41[b]	23.63±10.32[a]	29.07±10.84[a]	4.02±2.82[a]	53.05±10.44[b]	Q= 198.99; df= 5; P<0.0001***
Specific richness (S)	37[a]	40[a]	39[a]	46[b]	23[a]	48[b]	Q= 199.17; df= 5; P<0.0001***
Shannon index (H')	2.0[a]	0.89[b]	2.15[a]	2.29[b]	2.12[a]	2.04[a]	Q= 199.17; df= 5; P<0.0001***
Equitability (E)	0.55[a]	0.24[b]	0.59[a]	0.60[a]	0.68[a]	0.53[b]	Q= 199,17 ; df= 5; P<0.0001***

Q test: Friedman Chis-square test; ***significant difference with P< 0.001; the different letter in each site translate a significant difference at 5% interval confidence level between 2006 and 2007 samples base on Wilcoxon pairwise test corrected with sequential Bonferonni procedure.

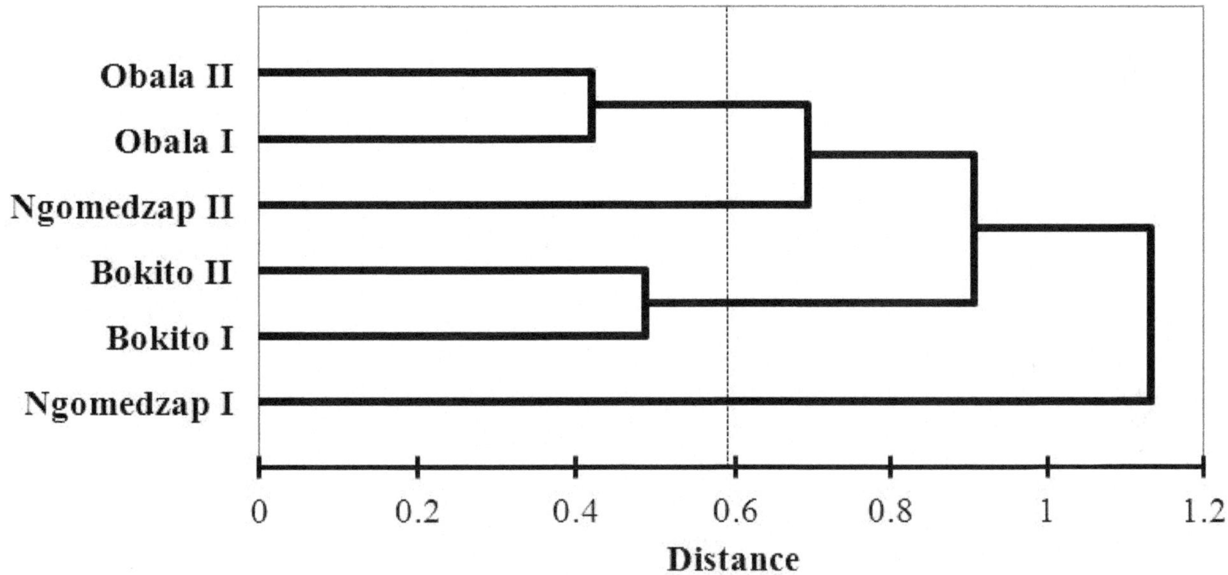

Figure 2. Similarity between site after treatment and recolonization (Cluster analysis was performing with XLSTAT 2010.4.02 base on Bray Curtis distance use Ward aggregation method). I and II represent respectively 2006 and 2007 sampling.

(P < 0.001), species richness (P < 0.001) and diversity index (P < 0.001) between the two sampling period (Table 2). For instance, at Ngomedzap, despite a highly significant increase of species richness and mean number of workers per tree, there was non-significant decrease of the diversity index. The same trend was observed at Bokito where the increased of the mean number of workers per tree and species richness were followed by a significant decreased of the diversity index. Contrarily, at Obala, the increase of mean number of workers par tree and diversity were subsequent to a significant increase of the diversity index (Table 2).

Cluster analysis based on species composition perform on the base of Bray Curtis distance revealed no significant variation between the ant communities of the two samples both at Bokito and Obala. Contrarily, ant community at Ngomedzap was highly disturbed and new establishes community tends to be more similar to that of Obala (Figure 2).

Influence of treatment on community structure

Bokito ant community

Before the first treatment the community was composed of 37 ant species, numerically dominated by, *Oecophylla longinoda* (42.87%), *Cataulacus guineensis* (13.14%), *Polyrachis decemdentata* (11.74%), *Crematogaster gabonensis* (7.83%) and *Tetramorium aculeatum* (7.37%). After one year of re-colonization process, there were an increase of the number of species to 40, with 11 newly established species (*Axinidris bidens, Tapinoma* sp.2, *Camponotus* sp.3, *Phasmomyrmex aberrans, Polyrachis monista*, *Polyrachis* sp.1, *Polyrachis* sp.3, *Crematogaster pulchella, Myrmicaria opaciventris, Pheidole* sp.1, and *Pachycondyla* sp.1) and seven *(Cataulacus* sp.2, *Tetramorium* sp.3, *Anochetus* sp.1, *Plathytyrea modesta*, *Tapinoma melanocephalum*, *Technomyrmex* sp.1, and *Dorylus nigricans)* that disappeared were disappear.

In terms of abundance, there was significant increase of the abundance of *O. longinoda* that remained the dominant ant species in the community. The same trend was encountered with *Camponotus acvapimensi* whose relative abundance increased from (0.86 to 5.95%). Contrarily, there was a drop in the abundance of some dominant species such as *T. aculeatum* (from 7.37 to 1.78 %) and *P. decemdentata* (from 11.74 to 1.25 %), *Cataulacus guineensis* from 13.14 to 0.07), while those of *Crematogaster gabonensis* remained relatively stable (from 7.37 to 5.379%) (Appendix 1).

Obala ant community

At Obala, the first community was made of 39 ant species, numerically dominated by *Oecophylla longinoda* (38.80%) and *Crematogaster gabonensis* (13.33%). After one year of recolonisation process, six ant species desapeared (*Crematogaster* sp1., *Pachycondyla sorror*, *Pa. tarsata*, *Ta. melanocephalum*, *Technomyrmex* sp1., and *Technomyrmex* sp2), while 13 new species (*Anoplolepis tenella*, *Camponotus* sp.2, *Polyrachis monista*, *Polyrachis* sp.1, *Polyrachis* sp.3, *Polyrachis* sp.5, *Atopomyrmex mocquerysi*, *Cataulacus* sp.1, *Pheidole* sp.1, *Tetramorium* sp.2, *Anochetus* sp.2, *Platythyrea* sp. and *Tetraponera ophthalmica* were observed.

Concerning the community structure, there was a decrease in the abundance of *O. longinoda* (from 38.80 to 18.13%), *Camponotus vividus* (from 13.10 to 4.90%) and polyrrhachis decemdentata (from 6.31 to 2.38%) while there was increase in abundance of *Crematoagster gabonensis* (from 13.33 to 29.98%), Crematogaster striatula (from 0.04 to 13.58%), *Cr. gabonensis* (from 5.79 to 29.98. and *P. laboriosa* (from 0.15 to 2.29%) (Appendix 1).

Ngomedzap ant community

For the first treatment a community composed of 23 ant species, numerically dominated respectively *Crematogaster striatula (*23.88 %) and *Cr. clariventris (*33.83 % workers) was sampled one year after the treatment, three of these ant species were no longer found (*Pheidole* sp.2, *Anochetus* sp.1 and *Pa. tarsata*), while 28 newly established species were obtained.

Concerning the relative abundance, with exception of few species that showed significant increase of the abundance: *Cr. striatula* from 23.88 to 26.22% *Tetramorium aculeatum* from 2.99 to 7.65 and *Camponotus vividus* from 4.48 to 8.37%) many species showed reduction of their relative abundance (Appendix 1).

Globally, one year after insecticide treatment the population of *C. acvapimensis* ($P < 0.0001$), *C. vividus* ($P < 0.0001$), *Cr. clariventris* ($P < 0.0001$), *Cr. gabonensis* ($P < 0.0001$), *Cr. striatula* ($P < 0.0001$), *O. longinoda* ($P < 0.0001$), *P. decemdentata* ($P < 0.0001$) and *T. aculeatum* ($P < 0.0001$) were significantly affected (Table 3).

For instance *C. acvapimensis* population increased at Bokito and Obala and felt at Ngomedzap. *Oecophylla longinoda* population increased at Bokito and decreased at Obala. *Crematogaster* population decreased at Bokito same as *T. aculeatum* (Table 3). Differences obtain with pairwise comparison were consigned in Table 3.

Ant recolonization and distribution

After treatment and recolonization 35 ant species were new established in prospected farms. Among them, 11, 13 and 29 ant species were collected respectively at Bokito, Obala and Ngomedzap. Among the 35 species identified, 21 are site-specific (15 at Ngomedzap, four at Bokito and two at Obala), and only three species are common to the three localities (Figure 3).

In relation to their nesting behavior, new establish species are grouping inside specialize arboreal ant species, ground dwelling ant species with arboreal habit, ground nesting species and finally undetermined nesting behavior (grouping ants species that may indifferently be encountered in various opportunistic structure on the ground or in trees) (Figure 4).

DISCUSSION

Influence of insecticide treatment on ant richness and diversity

Influence of treatment on ant species richness with a total of 60 ant species were collected between 2006 and 2007 in small-holding cocoa farms located along the South-North agro ecological gradient in southern Cameroon, the diversity appeared relatively diversifies. Highest species richness was observed respectively the forest (Ngomedzap) and the transitional (Obala) landscapes compared to savannah (Bokito). This may related to either the natural heterogeneity in relation with availability of various habitat (Schonberg et al., 2004; Ramesh et al., 2010), phytophysiognomies (Cardoso et al., 2010) or to the variations in the level of anthropogenic disturbance (Folgarait, 1998; Floren et al., 2001; Savitha et al., 2008; Ramesh et al., 2010).

The complexity of the structure in natural forest at Ngomedzap and the intermediate character of vegetation structure of Obala offered a great diversity of habitat,

Table 3. Variation of ant abundance between numerical dominant species under treatment effect in center region of Cameroon (2006-2007).

Species	Site						Q test
	Bokito		Obala		Ngomedzap		
	2006	2007	2006	2007	2006	2007	
Camponotus acvapimensis	114 (0.86)[a]	1058 (5.95)[b]	204 (2.81)[a]	290 (3.25)[a]	18 (4.48)[a]	94 (1.77)[b]	Q= 63.61; df= 5 ; P< 0.001***
Camponotus vividus	694 (5.24)[a]	223 (1.25)[b]	950 (13.10)[a]	437 (4.90)[a]	18 (4.48)[a]	444 (8.37)[b]	Q= 211.88; df= 5 ; P< 0.001***
Oecophylla longinoda	5676 (42.87)[a]	14336 (80.59)[b]	2814 (38.80)[a]	1618 (18.13)[a]	0 (0.00)	12 (0.23)	Q= 184.54; df= 3 ; P< 0.001***
Polyrachis decemdentata	1554 (11.74)[a]	222 (1.25)[b]	458 (6.31)[a]	212 (2.38)[b]	1 (0.25)[a]	2 (0.04)[b]	Q= 248.41; df=5 ; P< 0.001***
Crematogaster clariventris	39 (0.29)[a]	18 (0.10)[a]	199 (2.74)[a]	31 (0.35)[a]	136 (33.83)[a]	273 (5.15)[a]	Q= 53.96; df=5 ; P< 0.0001***
Crematogaster gabonensis	977 (7.38)[a]	1030 (5.79)[a]	967 (13.33)[a]	2675 (29.98)[b]	1 (0.25)[a]	111 (2.09)[b]	Q= 305.81; df=5 ; P< 0.0001***
Crematogaster striatula	250 (1.89)[a]	7 (0.04)[b]	696 (9.60)[a]	1212 (13.58)[b]	96 (23.88)[a]	1391 (26.22)[b]	Q= 81.18; df=5 ; P< 0.0001***
Tetramorium aculeatum	976 (7.37)[a]	316 (1.78)[b]	14 (0.19)[a]	86 (0.96)[b]	12 (2.99)[a]	406 (7.65)[b]	Q= 93.94; df= 5 ; P< 0.0001***

Q test: Friedman Chi-square test ; ***significant difference with P< 0.001; the different letter in each site translate a significant difference at 5% interval confidence level between 2006 and 2007 samples base on Wilcoxon pairwise test corrected with sequential Bonferonni procedure.

favorable to the establishment of various species. This may explain the high species richness recorded in both sites. The vegetation structure also relies on the variation of anthropogenic disturbances level based on farm management strategies. Thanks to climatic variations, there were site variation in pest and diseases pressure, induction variation in pesticide application model, weeding frequencies and mechanical destruction of arboreal ant-nests during pruning activities.

Specific richness increase under treatment effect increased gradually from savannah to forest. In fact, at climax stage, ant mosaic is known as very compact structure with a few gaps or no ant's lands between colonies (Hölldobler and Lumsden, 1980; Majer et al., 1994 b). Under treatment effect, ant mosaic was broken; population size on dominant ant colonies reduced and thus, marginal space was created, along with declining of intensity of interspecific competition between numerical dominant ant species (Sensus Davidson, 1997). This contributed to the establishment of submissive species that arise from neighboring vegetation.

Similarity between sites

Forest (Ngomedzap) and transitional (Bokito) areas are separated from savannah (Bokito) by Sanaga River. Both sites were more homogenous on a point of view of their vegetation structure and other factor like proximity. Therefore, ant communities remain more similar between Ngomedzap and Obala farms compared to Bokito cocoa farms.

The proximity between sites has been also suggested by Cardoso et al. (2010) to explain variation of similarity in ant communities in a Brazilian Restinga. Despite anthropogenic disturbance mainly insecticide use, both communities (Obala and Ngomedzap) can evolve differently but may not change drastically.

Influence of treatment on ant diversity

Insecticides treatment significantly affected diversity at Ngomedzap and Bokito but not at Obala. According to his intermediate characters, Obala ant community is liable to combined effect of natural environmental pressure and anthropogenic disturbance encounter in forest and savannah areas. In response of permanently high disturbance link to insecticide and fungicide used more than another sites, various ant species are developing mechanism to resist stressing and restoring rapidly their population. Evidence of negative influence of anthropogenic disturbance on ant diversity was also demonstrated by Floren et al. (2001) along disturbance gradient at Kinabalu National Park in Malaysia.

Insecticide treatment and pre-establish ant communities

Evidence of harmful formerly established ant communities are well demonstrating according to achieve result. If some species' colonies are able to survive to insecticide treatment, others could not and their population significantly decreased. It was the case of *O. longinoda*, *T. aculeatum* and some species of the genus *Crematogaster*. The pesticide treatment may negatively or positively affect each species population depending of various intrinsic as well as extrinsic factors.

In the last cases, the ability of a species to rapidly colonized liberated spaces from adjacent areas or to escape insecticide treatment thanks to nest-structure. *Oecophylla longinoda* colonies live in multiple nest nests are made of host plant leaves pull together and stitch with silk produced by a larva (Leston, 1973; Dejean et al., 1997; Djieto-Lordon and Dejean, 1999). *Tetramorium aculeatum* colonies live in carton nest build under host plant leaves (Dejean, 2000 a) this structure may provide

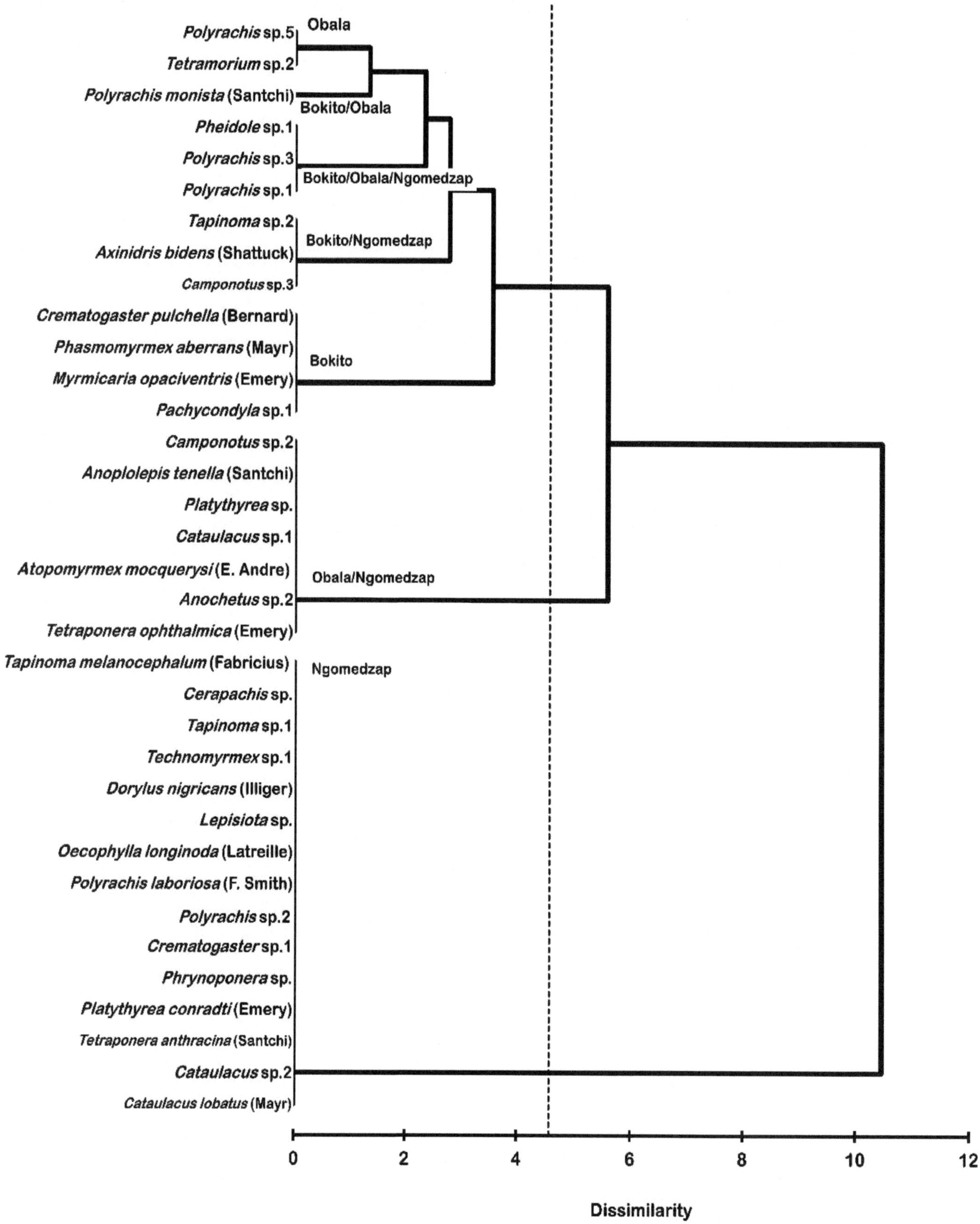

Figure 3. Common and proper distribution of new ant species according to site (Cluster analysis was performing using Euclidian distance with Ward aggregation method, XLSTAT 2010.4.02).

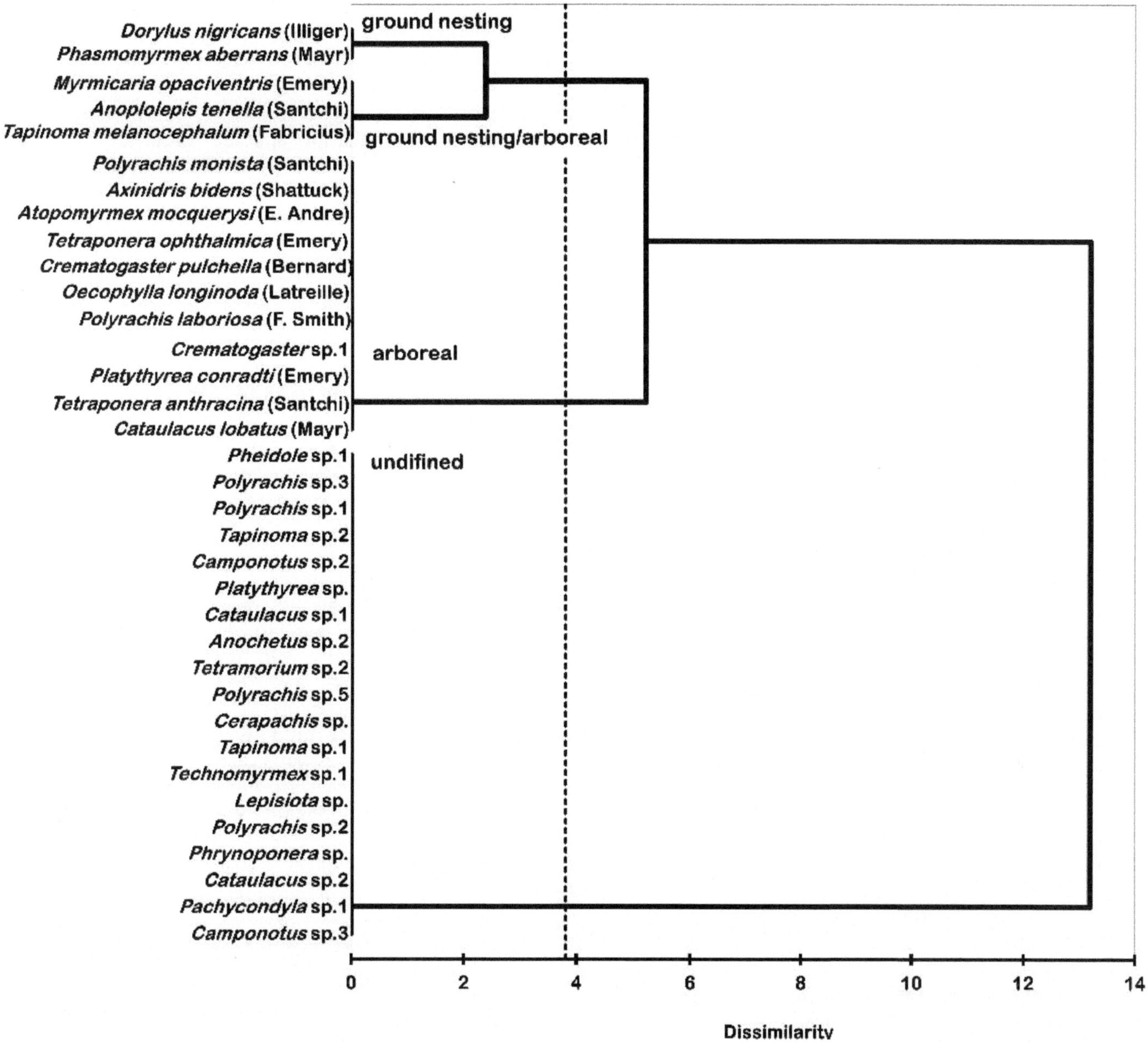

Figure 4. Distribution of new establish species according to their nesting behavior (Cluster analysis was performing using Euclidian distance with Ward aggregation method, XLSTAT 2010. 4.02)

waterproof to a nest. For some *Crematogaster* spp. nesting behavior habitually include carton nest located on shade tree, hollow twigs or branches, crevices and old pods (Leston, 1973). Species such as *Cr. striatula* nest opportunistically in the twigs, crevices and old pods, this nesting behavior may provide a best protection to the colony and just the workers that forage outside the nest may be affected. Similar was the situation of *Cr. clariventris* (carton nest) who built their nest habitually on shade tree (Leston, 1973). These structures may keep the center of colony safe during insecticide treatment (Leston, 1973; Majer, 1978).

The success of ecological *Crematogaster* species in agrosystems was also explained by pruning activities during which nests of *Tetramorium aculeatum* and *Oecophylla longinoda* were destroyed, favoring the expansion of the formers whose nest were located out of the sprayed areas. Impact of pruning is more evident in Ngomedzap farm where one year after our first treatment, both an increase of *T. aculeatum* abundance and arrival of *O. longinoda* occurred. According to the fact that during the one-year recolonization period, there were no management activity conducted by farmers on the pots, the environment were favorable to the establishment an expansion of *T. aculeatum* and *O. longinoda* colonies.

Recolonization process

With respect to nesting behavior, three categories of ant were distinguished among the species that were appeared in the community only during the second sampling: (1) arboreal ant species, (2) Ground-dwelling ant species with arboreal foraging, who exploit homopterans for honeydew and (3) ground-nesting species. Despite the widespread on insecticide applications in agrosystems in Cameroon, very few studies have been conducted on their impact on the arthropods communities. Kenne et al. (2003) in an experimental orchard at Minkoa Meyos demonstrated that regular and persistent application of insecticide treatments favored proliferation of ground-nesting arboreal-foraging ant species that may monopolised the tree crowns. Dominant arboreal nesting species such as *Oecophylla longinoda* that newly colonized studied plots at Ngomedzap exhibit a slow recolonization rate small size colony indicating the establishment phase. This is in conformity with Kenne et al. (2003) who cited this species among the pioner arboreal ant species that colonied tree crown afetr insecticide treatment.

In the studied communities, potential beneficial ant species represented by *O. longinoda* and *T. aculeatum* may be affected by the treatment; the same trend was observed by Leston (1973) and Majer (1978) in Ghana cocoa farms. These authors hypothesized that even a single spray may cause considerable damages to the population of *O. longinoda* and *T. aculeatum,* two potential biological agents. Their abundance decrease in favor of increase of abundance of species belonged to *Crematogaster* genus (Majer, 1978) and other dominant ground-dwelling arboreal foraging and ground-nesting species. These species usually are positively associated with Hemipterans and scale insect that pose problem in agricultural areas (Blüthgen and Fiedler, 2002; Kenne et al., 2003). Nevertheless, another aspect to look at is the phytopathological implication or recolonization process of tree by ground nesting species after treatment. In fact, several ground species to protect their trophobionts transport soil particles that contains fungus spore. Passively, these species may contribute to vertical propagation of fungus disease agents in cocoa farm.

Conclusion

In favor of the disturbance due to insecticide effect, ant species richness increased in all the study sites, while diversity decrease both at Bokito and Ngomedzap, contrarily to Obala where it increased. High turnover observed at Obala may be explained by the high level of anthropogenic disturbance observed in the locality. At Ngomedzap and Bokito, area with less anthropogenic disturbance, turnover may probably take a longer times-period. On the three sites, treatment favored the proliferation of newly established arboreal ant species, ground-dwelling species and ground species. Nesting behavior of various dominant ant species procures protection to colony against chemical treatment and other mechanical factors of disturbance. Border effect due to the proximity with adjacent farms, fallows or native vegetation is an important factor that may affect recolonization process of farm. Though, anthropogenic disturbance may alter the distribution and abundance of certain ant species, they contributed to the ecological success of other ant species. Agronomical importance of tree crowns recolonization by ground-dwelling ant species was investigated, in order to understand their implication in the propagation of fungus diseases in cocoa farms. In fact, protections of trophobionts by these species are required for transportation of soil particles that contain fungus spores on the tree to build a habit.

ACKNOWLEDGEMENTS

We thank Bokito, Obala and Ngomedzap cocoa farmers for permitting us to work in their cocoa farms. This work was funded by the REPARAC project. We are also grateful to "La Direction Générale de la Coopération Belge au Développement" who gave the opportunity to Mr. Tadu through a scientific visit to assess the collection of the "Muséum Royal de l'Afrique Centrale" of Belgium and confirm ant identifications.

REFERENCES

Bisseleua DHB, Vidal S (2008). Plant biodiversity and vegetation structure in traditional cocoa forest gardens in southern Cameroon under different management. Biodiv. Consev. 17:1821-1835.

Blüthgen N, Fiedler K (2002). Interaction between weaver ants Oecophylla smaragdina, homopterans, trees and lianas in an australian rain forest canopy. J. Anim. Ecol. 71:793-801.

Bolton B (1994). Identification guide to the ant genera of the world. Havard University Press, Cambridge, Massachusetts London, England. p. 222.

Cardoso DC, Sobrinho TG, Shoereder JH (2010). Ant community composition and its relationship with phytophysiognomies in a Brazilian Restinga. Insect. Soc. 57:293-301.

Colwell RK (2006). EstimateS: Statistical estimation of species richness and shared species from samples. In: <purl.oclc.org/estimates> V.P.U.

Dahms H, Wellstein C, Wolter V, Dauber J (2005). Effect of management practices on ant species richness and community composition in grasslands (Hymenoptera: Formicidae). Myrmecolo. Nach. 7:9-16.

Davidson DW (1997). The role of resource imbalances in the evolutionary ecology of tropical arboreal ants. Biol. J. Linn. Soc. 61:153-181.

Davidson DW (1998). Resource discovery versus resource domination in ants: a functional mechanism for breaking the trade-off. Ecol. Entomol. 23:484-490.

Dejean A, Djieto-Lordon C, Lenoir A (1994). Mosaic ant territories in an african secondary rain forest (Hymenoptera: Formicidae).

Sociobiology 23(3):1-17.
Dejean A (2000 a). Ant protection (Hymenoptera: Formicidae) of two pioneer plant species against the variagate locust. Sociobiology 36:217-226.
Dejean A, Djieto-Lordon C, Durand JL (1997). Ant mosaic in oil palm plantation of the southwest province of Cameroon: impact on leaf miner beetle (Coleoptera: Chrysomelidae). Apic. Soc. Insect. 90(5):1092-1096.
Dejean A, McKey D, Gibernau M, Belin M (2000 b). The arboreal ant mosaic in a Camerounian rain forest (Hymenoptera: Formicidae). Sociobiology 23(3):403-423.
Djieto-Lordon C, Dejean A (1999). Tropical arboreal ant mosaics: innate attraction and imprinting determine nest site selection in dominant ants. Behav. Ecol. Sociobiol. 45:219-225.
Fayle TM, Turner EC, Snaddon JL, Chey VK, Chung AYC, Eggleton P, Foster WA (2010). Oil palm expansion into rain forest greatly reduces ant biodiversity in canopy, epiphytes and leaf-litter. Basic Appl. Ecol. 11:337-345.
Fisher BL (2005). A model for a global inventory of ants: a case study in Madagascar. Proceedings of the California Academia of Sciences 56:86-97.
Fitzherbert EB, Struebig MJ, Morel A, Danielsen F, Bruhl CA, Donald PF, Phalan B (2008). How will oil palm expansion affect biodiversity? Trends Ecol. Evol. 23(10):538-545.
Floren A, Freking A, Bielil M, Linsenmair EK (2001). Anthropogenic disturbance changes the structure of arboreal tropical ant communities. Ecography 24:547-554.
Folgarait PJ (1998). Ant biodiversity and its relationship to ecosystem functioning: a review. Biodiv. Conserv. 7:1221-1244.
Hölldobler B, Lumsden CJ (1980). Territorial strategy in ants. Science 210:732-739.
Hölldobler B, Wilson ED (1990). The Ants. The Belknap of Havard University Press Cambridge, Massachusetts. p. 732.
Kenne M, Djieto-Lordon C, Orivel J, Mony R, Fabre A, Dejean A (2003). influence of insecticide treatments on ant-hemiptera associations in Tropical Plantations. J. Econ. Entomol. 96(2):251-258.
Lebreton J, Chazeau J, Jourdan H (2003). Immadiate impacts of invasion by *Wasmannia auropunctata* (Hymenoptera: Formicidae) on native litter ant fauna in a New Caledonian rainforest. Aus. Ecol. 28:204-209.
Leston D (1973). The ant mosaic-tropical tree crops and the limiting of pest and diseases. Pans 19(3):311-339.
Majer JD (1972). The ant mosaic in Ghana cocoa farm. Bull. Entomol. Res. 62:151-160.
Majer JD (1976). The influence of ants and ant manipulation on the cocoa farm fauna. J. App. Ecol. 13:157-175.
Majer JD (1978). The influence of blanket and selective spraying on ant distribution in a west african cocoa farm. Theobroma 8:87-93.
Majer JD (1990). The abundance and diversity of arboreal ants in Northern Australia. Biotropica 22(2):191-199.
Majer JD, Delabie JHC (1994a). Comparison of the ant communities of annualy inundated and terre firme forest at Trombetas in the Braziliann amazon. Insect. Soc. 41:343-359.
Majer JD, Delabie JHC, Smith MRB (1994b). Arboreal ant community patterns in Brazilian cocoa farms. Biotropica 26:73-83.
Majer JD, Kitching RL, Heterick BE, Hurley K, Brennan KEC (2001). North-South patterns within arboreal ant assemblages from rain forest in eastern Australia. Biotropica 33(4):643-661.
Merijn MB, Ingolf SD, Teja T (2007). The contribution of cocoa agroforests to the conservation of lower canopy ant and beetle diversity in Indonesia. Biodiv. Conserv. 16(8):2429-2444.
Ramesh T, Hussain J, Satpathy KK, Selvanayagam M, Prasard MVR (2010). Diversity, distribution and species composition of ants fauna at Department of Atomic Energy (DAE) campus Kalpakkam, south India. W. J. Zool. 5:56-65.
Rice RA, Greenberg R (2000). Cocoa cultivation asnd the conservation of biological diversity. AMBIO 29:167-173.
Savitha S, Barve N, Davidar P (2008). Response of ants to disturbance gradients in and around Bangalore, India. Trop. Ecol. 49(2):235-243.
Schonberg LA, Longino JT, Nadkarni NM, Yanoviak SP (2004). Arboreal ant species richness in primary forest, secondary sorest, and pasture habitats of a tropical montane landscape. Biotropica 36:402-409.
Santoir C, Bopda A (1995). Atlas regional Sud-Cameroun. Orstom, Paris.
Sonwa DJ, Nkongmeneck BA, Weisse SF, Tchatat M, Adesina AA, Janssens MJJ (2007). Diversity of plants in cocoa agroforests in the humid forest zone of Southern Cameroon. Biodiv. Conserv. 16:2385-2400.
Sonwa DJ, Coulibaly O, Weise SF, Akinwumi Adesina A and Janssens MJJ (2008). Management of cocoa: constraints during acquisition and application of pesticides in the humid forest zones of southern Cameroon. Crop Protect. 27:1159-1164.
Taylor B (2002). The ants of West African and the Congo basin (Hymenoptera: Formicidae). In: Grazingfield W., Nottingham, 11, NG11 7FN, U.K. Visiting Academic in the Department of Life Science, University of Nottingham 1903 screens.
Watt AD, Stork NE, Bolton B (2002). The diversity of ant in relation to forest disturbance and plantation establishment in southern Cameroon. J. App. Ecol. 39:18-30.
Widodo ES, Naito T, Mohamed M, Hashimoto Y (2004). Effect of selective logging on arboreal ants of a Bornean rainforest. Entomol. Sci. 7:341-349.

Appendix 1. Diversity, distribution and variation of ant abundance in camerounian cocoa farms before (2006) and after treatment (2007) and plot recolonization.

Site	Bokito		Obala		Ngomedzap	
Sample	2006	2007	2006	2007	2006	2007
Cerapachiynae						
Cerapachis sp. *	0 (0.00)	0 (0.00)	0 (0.00)	0 (0.00)	0 (0.00)	4 (0.08)
Dolichoderinae						
Axinidris bidens (Shattuck)*	0 (0.00)	2 (0.001)	0 (0.00)	0 (0.00)	0 (0.00)	2 (0.04)
Tapinoma melanocephalum (Fabricius)*	2 (0.02)	0 (0.00)	21 (0.29)	0 (0.00)	0 (0.00)	52 (0.98)
Tapinoma sp1. *	10 (0.08)	15 (0.08)	20 (0.28)	90 (1.01)	0 (0.00)	55 (1.04)
Tapinoma sp2. *	0 (0.00)	1 (0.01)	0 (0.00)	0 (0.00)	0 (0.00)	1 (0.02)
Technomyrmex sp1.	1 (0.01)	0 (0.00)	19 (0.26)	0 (0.00)	0 (0.00)	1 (0.02)
Technomyrmex sp2.	0 (0.00)	0 (0.00)	1 (0.01)	0 (0.00)	0 (0.00)	0 (0.00)
Dorylinae						
Dorylus nigricans (Illiger)*	30 (0.23)	0 (0.00)	1 (0.01)	1048 (11.74)	0 (0.00)	1919 (36.17)
Formicinae						
Anoplolepis tenella (Santchi)*	0 (0.00)	0 (0.00)	0 (0.00)	4 (0.04)	0 (0.00)	1 (0.02)
Camponotus acvapimensis (Mayr)	114 (0.86)	1058 (5.95)	204 (2.81)	290 (3.25)	18 (4.48)	94 (1.77)
Camponotus brutus (Forel)	126 (0.95)	88 (0.49)	61 (0.84)	105 (1.18)	25 (6.22)	28 (0.53)
Camponotus maculatus (Forel)	10 (0.08)	2 (0.01)	3 (0.04)	27 (0.30)	1 (0.25)	6 (0.11)
Camponotus pompeius (Forel)	1 (0.01)	1 (0.01)	11 (0.15)	3 (0.03)	1 (0.25)	1 (0.02)
Camponotus sp1.*	17 (0.13)	3 (0.02)	34 (0.47)	18 (0.20)	24 (5.97)	21 (0.40)
Camponotus sp2. *	0 (0.00)	0 (0.00)	0 (0.00)	8 (0.09)	0 (0.00)	3 (0.06)
Camponotus sp3.*	0 (0.00)	9 (0.05)	5 (0.07)	23 (0.26)	0 (0.00)	33 (0.62)
Camponotus vividus (F. Smith)	694 (5.24)	223 (1.25)	950 (13.10)	437 (4.90)	18 (4.48)	444 (8.37)
Lepisiota sp. *	1 (0.01)	2 (0.01)	0 (0.00)	0 (0.00)	0 (0.00)	35 (0.66)
Oecophylla longinoda (Latreille)*	5676 (42.87)	14336 (80.59)	2814 (38.80)	1618 (18.13)	0 (0.00)	12 (0.23)
Phasmomyrmex aberrans (Mayr)*	0 (0.00)	23 (0.13)	4 (0.06)	6 (0.07)	0 (0.00)	0 (0.00)
Polyrachis decemdentata (Andre)	1554 (11.74)	222 (1.25)	458 (6.31)	212 (2.38)	1 (0.25)	2 (0.04)
Polyrachis laboriosa (F. Smith)*	19 (0.14)	9 (0.05)	11 (0.15)	204 (2.29)	0 (0.00)	2 (0.04)
Polyrachis militaris (Fabricius)	12 (0.09)	1 (0.01)	46 (0.63)	31 (0.35)	1 (0.25)	11 (0.21)
Polyrachis monista (Santchi)*	0 (0.00)	1 (0.01)	0 (0.00)	9 (0.10)	0 (0.00)	0 (0.00)
Polyrachis sp1. *	0 (0.00)	4 (0.02)	0 (0.00)	48 (0.54)	0 (0.00)	20 (0.38)
Polyrachis sp2. *	62 (0.47)	52 (0.29)	53 (0.73)	38 (0.43)	0 (0.00)	8 (0.15)
Polyrachis sp3. *	0 (0.00)	5 (0.03)	0 (0.00)	1 (0.01)	0 (0.00)	1 (0.02)
Polyrachis sp5. *	0 (0.00)	0 (0.00)	0 (0.00)	2 (0.02)	0 (0.00)	0 (0.00)
Polyrachis weissi (Santchi)	53 (0.40)	40 (0.22)	138 (1.90)	221 (2.48)	8 (1.99)	8 (0.15)
Myrmicinae						
Atopomyrmex mocquerysi (E. Andre)*	349 (2.64)	81 (0.46)	0 (0.00)	215 (2.41)	0 (0.00)	1 (0.02)
Cataulacus erinaceus (Stitz)	43 (0.32)	1 (0.01)	9 (0.12)	18 (0.20)	0 (0.00)	0 (0.00)
Cataulacus guineensis (F. Smith)	1739 (13.14)	12 (0.07)	65 (0.90)	42 (0.47)	3 (0.75)	21 (0.40)
Cataulacus lobatus (Mayr)*	174 (1.31)	1 (0.01)	69 (0.95)	2 (0.02)	0 (0.00)	2 (0.04)
Cataulacus sp1. *	1 (0.01)	56 (0.31)	0 (0.00)	23 (0.26)	0 (0.00)	5 (0.09)
Cataulacus sp2. *	3 (0.02)	0 (0.00)	21 (0.29)	1 (0.01)	0 (0.00)	9 (0.17)
Crematogaster clariventris (Mayr)	39 (0.29)	18 (0.10)	199 (2.74)	31 (0.35)	136 (33.83)	273 (5.15)
Crematogaster gabonensis (Emery)	977 (7.38)	1030 (5.79)	967 (13.33)	2675 (29.98)	1 (0.25)	111 (2.09)
Crematogaster pulchella (Bernard)*	0 (0.00)	7 (0.04)	0 (0.00)	0 (0.00)	0 (0.00)	0 (0.00)
Crematogaster sp1.	89 (0.67)	10 (0.06)	143 (1.97)	0 (0.00)	0 (0.00)	2 (0.04)
Crematogaster striatula (Emery)	250 (1.89)	7 (0.04)	696 (9.60)	1212 (13.58)	96 (23.88)	1391 (26.22)
Myrmicaria opaciventris (Emery)*	0 (0.00)	2 (0.01)	66 (0.91)	11 (0.12)	3 (0.75)	12 (0.23)
Pheidole megacephala (Fabricius)	59 (0.45)	108 (0.61)	9 (0.12)	16 (0.18)	29 (7.21)	189 (3.56)

Appendix 1. Contd.

Pheidole sp1.*	0 (0.00)	2 (0.01)	0 (0.00)	2 (0.02)	0 (0.00)	23 (0.43)
Pheidole sp2.	4 (0.03)	9 (0.05)	0 (0.00)	0 (0.00)	1 (0.25)	0 (0.00)
Tetramorium aculeatum (Mayr)	976 (7.37)	316 (1.78)	14 (0.19)	86 (0.96)	12 (2.99)	406 (7.65)
Tetramorium sp1.	1 (0.01)	3 (0.02)	7 (0.10)	1 (0.01)	5 (1.24)	11 (0.21)
Tetramorium sp2.	0 (0.00)	0 (0.00)	0 (0.00)	10 (0.11)	1 (0.25)	3 (0.06)
Tetramorium sp3.	2 (0.02)	0 (0.00)	8 (0.11)	10 (0.11)	2 (0.50)	24 (0.45)
Ponerinae						
Anochetus sp1.	1 (0.01)	0 (0.00)	80 (1.10)	72 (0.81)	10 (2.49)	0 (0.00)
Anochetus sp2. *	0 (0.00)	0 (0.00)	0 (0.00)	14 (0.16)	0 (0.00)	43 (0.81)
Odontomachus troglodytes (Santchi)	47 (0.37)	19 (0.11)	36 (0.50)	19 (0.21)	0 (0.00)	0 (0.00)
Pachycondyla soror (Emery)	0 (0.00)	0 (0.00)	1 (0.01)	0 (0.00)	0 (0.00)	0 (0.00)
Pachycondyla sp1.	0 (0.00)	1 (0.01)	0 (0.00)	0 (0.00)	0 (0.00)	0 (0.00)
Pachycondyla tarsata (Fabricius)	5 (0.04)	0 (0.00)	1 (0.01)	0 (0.00)	2 (0.50)	0 (0.00)
Phrynoponera sp.*	0 (0.00)	0 (0.00)	0 (0.00)	0 (0.00)	0 (0.00)	1 (0.02)
Platythyrea conradti (Emery)*	0 (0.00)	0 (0.00)	4 (0.06)	1 (0.01)	0 (0.00)	3 (0.06)
Platythyrea modesta (Emery)	11 (0.08)	0 (0.00)	3 (0.04)	12 (0.13)	4 (1.00)	4 (0.08)
Platythyrea sp. *	0 (0.00)	0 (0.00)	0 (0.00)	2 (0.02)	0 (0.00)	2 (0.04)
Pseudomyrmecinae						
Tetraponera anthracina (Santchi)*	87 (0.66)	8 (0.04)	1 (0.01)	5 (0.06)	0 (0.00)	4 (0.08)
Tetraponera ophthalmica (Emery)*	0 (0.00)	0 (0.00)	0 (0.00)	1 (0.01)	0 (0.00)	1 (0.02)
Total	13239	17788	7253	8924	402	5305

*New establish ant species.

4

Impact of *Ageratum conyzoides* L. on the diversity and composition of vegetation in the Shivalik hills of Himachal Pradesh (Northwestern Himalaya), India

Kuldip S. Dogra[1], Ravinder K. Kohli[2], Sarvesh K. Sood[1] and Praveen K. Dobhal[2]

[1]Department of Biosciences, Himachal Pradesh University Summer Hill Shimla (India) – 171005.
[2]Department of Botany, Panjab University Chandigarh (India) – 160014.

The flora of the Shivalik hills of Himachal Pradesh, India is under threat due to the rapid invasion of invasive species. Invasive species means an alien species, which becomes established in natural or semi-natural ecosystems and threatens native biological diversity. In their new regimes they show diverse life forms, habit, morphology, reproductive biology, grow fast, have the ability to grow under different habitats, produce enormous number of very small, light weight seeds that and can - survive in soil for years helping long distance dispersal and spread. They can out-compete native species, reduce wildlife habitat potential, alter natural ecosystem processes and limit overall biodiversity. *Ageratum conyzoides* is one such widely adaptive weed from sub-tropical America that has entered in the Shivalik hills of Hamrpur district of Himachal Pradesh. It has grown as monocultures, in grasslands, forests, agricultural, plantations and horticultural fields in Himachal Pradesh. Hence, it was decided to evaluate the impact *of A. conyzoides* on the diversity and floristic compositions of native species. It was found that as compared to control, in the Ageratum invaded area; the average number of plant species has reduced by 32.10%; the α diversity has reduced by 41.21% and the dry biomass of plants has also reduced significantly. It was concluded that invasion of *A. conyzoides* is drastically affecting the productivity and diversity of the invaded areas in Shivalik hills of Hamirpur district.

Key words: Invasive species, biological diversity, shivalik hills.

INTRODUCTION

The biota of the world is being homogenized as a result of the decline of native species and their replacement by a relatively small number of alien species that either deliberately or accidentally moves beyond their natural ranges (McKinney and Lockwood, 1999). These invasive alien species have encroached into many ecosystems and communities throughout the world, disrupting ecosystem structure, function and reducing native biodiversity (Borgmann and Rodewald, 2005). These out-compete native species or occupy the available niches in alien environment (Cowie, 1998a; 1998b) and cause major economic loss in countries around the world, by decreasing growth and productivity of useful species (Pimentel et al., 2000).

Increase in the rate of invasion and deliberate introduction of aliens into an area by man is the by-product of the globalization of regional economics. Large parts of the world are currently dominated by human modified ecosystems that often comprise a greater biomass of introduced than native organisms (Vitousek et al., 1997). Besides human actions, several other factors contribute

*Corresponding author. Email: dograks6@yahoo.co.in;

Figure 1. Photo of *Ageratum conyzoides* at flowering stage.

Figure 2. Monoculture of *Ageratum conyzoides* in Shivalik hills of Himachal Pradesh.

to successful invasion by alien plants. The climatic and edaphic similarities between the original and new habitats are very important factors for the establishment of alien species (Holdgate, 1986). Biological invasion are clearly a potent force of change, operating on a global scale and affecting many dimensions of society (Wilcove et al., 1998; Ohlemuller et al., 2006).

Ageratum conyzoides (Figure 1 & 2) is a native of tropical America. It has spreads world wide in the tropical and subtropical areas (Wagner et al., 1999). In India, it was introduced in 1860 as an ornamental plant (National Focal Point for APFISN, India, 2005). Later it escaped as a weed in various habitats throughout India. In Himachal Pradesh, the weed is established dominantly up to 1800

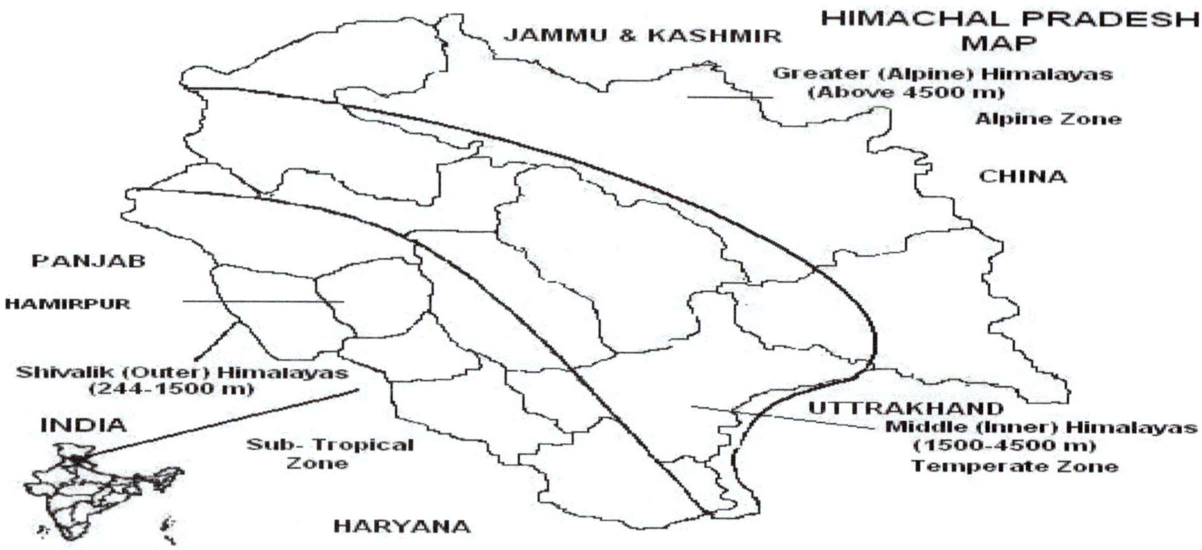

Figure 3. Map of Himachal Pradesh showing Hamirpur district.

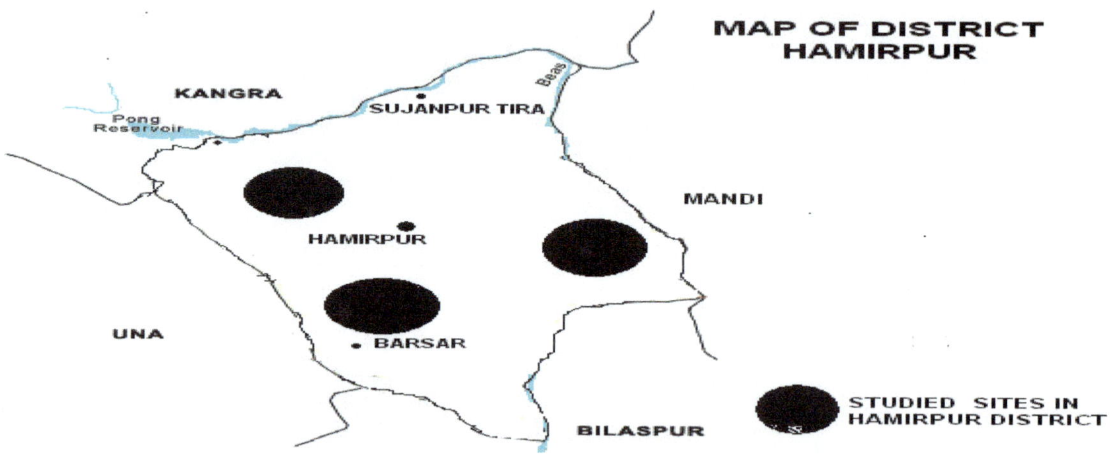

Figure 4. Map showing studied sites in Hamirpur district of Himachal Pradesh.

m. It is an annual aromatic herbaceous plant and gaining height up to 2 m in the Shivalik hills (Dogra, 2008; Dogra et al. 2009). It has a pubescent stem and ovate to ovate-rhomboid leaves. The plant has a purplish blue coloured inflorescence which produces an enormous number of seeds (*Cypsela*; more then 40,000 from a single plant). These seeds easily dispersed into wide areas in the hilly tracts of Shivalik hills by wind and water, which will help in their establishment in a wide range of climatic conditions (Kohli et al., 2006).

In Shivalik hills *A. conyzoides* L. has invaded predominantly in the grasslands, agricultural fields, forests, wastelands and in pastures. The invasion of bill goat weed created a pressure on the existence of useful medicinal plant species and as a result drastically affects the diversity and composition of vegetation in the Shivalik hills.

METHODOLOGY

Study Area

The lower Himalaya of Northwestern Indian state of Himachal Pradesh is also known as Shivalik hills (Figure 3). The Hamirpur district (area 1118 Km^2) is situated between 31°52' and 31°30' North latitude and 76°18' and 76°44' East longitude in the Shivalik hills of Himachal Pradesh (Figure 4). The altitude in Hamirpur district varies from 400 to 1100 m above mean sea level. The climate in the area is subtropical to mild warm temperate. The average rainfall in this area is about 1500 to 1800 mm. The minimum temperature in the Shivalik hills is 5°C in January (winter) and maximum in June up-to 40°C (summer). The potential vegetation of the area is commonly includes tropical thorn forests, northern tropical dry forests, subtropical broad leaved forests and tropical pine forests (Balokhra, 1999).

Vegetational Analysis

Vegetation analysis was done by random-systematic design and gradsect methods (Barbour et al., 1999; Singh and Singh, 1992). Three sites invaded with *A. conyzoides* were selected at random in the Shivalik hills of Hamirpur district. A parallel control (non-invaded with *Ageratum*) was also selected to compare the species richness, diversity and composition of vegetation in the invaded and non-invaded areas. In each invaded and non-invaded site an area of 200 m^2 was selected and 20 quadrats of 2 m^2 were laid randomly in both areas. All the plant species appeared in the invaded and non-invaded areas sampled, identified and their importance value index was calculated (Mishra, 1968). Further, species richness, diversity, index of dominance, similarity, dissimilarity index and evenness of invaded and non invaded areas was calculated and compared to find out the loss due to invasion of *A. conyzoides*. A variety of indices are available but to avoid conceptual and technical problems and to get precision, only a few such as Margalef's richness, Hill's evenness, Shannon's diversity, Simpson's index of dominance are applicable (Ludwig and Reynolds (1988). The vegetation other than invasive species under study per square meter of the area from any 3 quadrats from each site was uprooted and their fresh and dry biomass (after oven drying) measured.

To study the impact of *A. conyzoides* on other plant species a vegetational analysis was done during October 2006 to March 2007.

The plants were identified with the help of herbaria of the Department of Botany, Punjab University Chandigarh and YSP University of Horticulture and Forestry, Nauni, Solan, Himachal Pradesh. Confirmation of names was done by comparing with herbaria specimens maintained in Punjab University, Department of Botany; Forest Research Institute, Dehradun; Botanical Survey of India, Dehradun. The plants were categorized according to their habits like tree, shrub, herb, sedges, climber and vine.

Soil Analysis

The changes in phenolics, soil pH, electric conductivity, amount of organic carbon, organic matter, available nutrients e.g. nitrogen (N), phosphorus (P), potassium (K), sodium (Na), calcium (Ca), magnesium (Mg) and chloride (Cl) were compared between the *A. conyzoides* invaded habitats with their respective controls (weed free areas). The soils were collected from the *A. conyzoides* invaded habitats and from their respective controls. They were collected after removing litter from the top surface layer (4 - 6 cm) in each habitat. The sample of the soils were sieved, shade dried and stored for further use.

The pH and electric conductivity of the soil was measured after preparing a soil extract in the ratio of 1:2, soil: water (w/v) by using EcoScan digital pH meter and EcoScan Con 5 digital conductivity meter respectively. The phenolic content of the soils was estimated by the method of Swain and Hillis (1959) using Folin- ciocalteu reagent. Organic carbon and organic matter were estimated by the rapid titration method of Walkey and Black (1934). Available N was estimated using alkaline potassium permanganate as per the method of AOAC (1960). Available P was determined by the method of Olsen et al. (1954). Available Na and K were estimated using ammonium acetate as per Bower and Gschwend (1952). The available Ca, Mg and Cl were determined by the methods given by Black (1973).

Statistics Analysis

For each experiment, statistical analysis was done using software programmes like SPSS ver. 10.0, Origin 6 and Micorstat. For determining the significance of a single treatment with control (paired treatment), student's 2 sample *t*-test was applied. Significance of difference between soil characteristics of control and weed invaded sites was determined using 2 sample *t*-test at $P < 0.05$ or $P < 0.01$.

RESULTS

Impact of *A. conyzoides* on the plant diversity

Invasion of *Ageratum conyzoides* drastically affected the diversity and composition of vegetation in the Shivalik hills of Himachal Pradesh. As per the survey conducted during present study there were 81 species found in the control or un-invaded area as compared to 55 in *Ageratum* invaded area (Table 1). The number of species decreased by 32.10% in the *Ageratum* invaded area as compared to control. Further, Margalef's index of species richness, alpha species diversity and evenness index were reduced by 37.01, 41.21 and 15.48%, respectively in the *Ageratum* invaded areas. Likewise, the number of abundant species (N_1) and very abundant species (N_2) were also in the *Ageratum* invaded areas. The higher value of index of dominance in the invaded areas predicts that communities were homogenous in nature and dominated by a single species. On the other hand communities were showing more heterogeneity in the un-invaded areas. The similarity index of the communities in the invaded and un-invaded areas was only 52.12%, which shows clear indication of loss of species due to invasion of *Ageratum*. The comparison between fresh and dry biomass of invaded and un-invaded areas also show significant change as these decreased by 52.12 and 47.88%, respectively in the invaded areas.

Impact of *Ageratum* invasion on the floristic composition

The floristic composition describes the pattern of distribution of species in any particular habitat. *Ageratum* invasion heavily disturbs the composition and structure of species in the invaded habitats. In the present study floristic composition between *Ageratum* invaded and un-invaded areas was compared. A total of 94 plant species belonging to 37 families of flowering plants were collected during the present study. The number of plant families was 35 in the control area in contrast to only 25 in the *Ageratum* invaded areas. Among 94 plant species, 42 species were found common in control and invaded areas (Table, 2,). Thirty nine species [*A. scandens*, *A. viridis*, *A. arvensis*, *A. scoparia*, *A. vulgaris*, *B. paniculata*, *C. asiatica*, *C. ambrosioides*, *C. indica*, *C. arvense*, *D.*

Table 1. Impact of *Ageratum conyzoides* invasion on the plant diversity and biomass.

S.N.	Parameters	Control	Invaded	% decrease over control
1	Total Species	81	55	(–) 32.10
2	Average Fresh Biomass (g/m^2)	612.83 ± 36.01	372.68 ± 40.41	(–) 39.19
3	Average Dry Biomass (g/m^2)	449.26 ± 19.97	231.54 ± 17.08	(–) 48.46
4	Margalef Index of Richness (R_1)	6.29 ± 0.46	3.96 ± 0.50	(–) 37.04
5	Simpson's Index of Dominance (λ)	0.06 ± 0.03	0.15 ± 0.004	(+) 60.00
6	Shannon's Index of Diversity (H')	3.81 ± 0.31	2.24 ± 0.24	(–) 41.21
7	Diversity Number (N_1)	20.97 ± 0.97	10.30 ± 0.85	(–) 50.88
8	Diversity Number (N_2)	14.29 ± 1.30	6.09 ± 0.15	(–) 57.38
9	Index of Evenness (Es)	0.84 ± 0.01	0.71 ± 0.002	(–) 15.48
10	Similarity Index	52.12 ± 7.64		
11	Dissimilarity Index	47.88 ± 7.64		

All values significant at 5% significance level after applying two population *t* test; (–) show less value and (+) show high value in invaded site.

sissoo, D. indica, D. viscosa, E. alba, E. microphylla, F. indica, G. indicum, G. optiva, I. carnea, I. pentaphylla, I. quamoclit, L. esculentum, M. lupulina, N. linearis, P. minima, P. annua, P. arillata, P. plebium, R. arvensis, R. capitata, R. hastatus, S. officinalis, S. candicans, S. orientale, S. media, T. bartramia, U. labota, V. hirsute and *V. tetrasperma*] were found absent in the *Ageratum* invaded areas. Thirteen plant species (*B. wightiana, B. diffusa, C. bursa-pastoris, C. arvense, D. repens, I. nil, L. stivus, L. cephalotus, M. alba, R. dentatus, S. nigrum, T. emodi*) were growing only in the *Ageratum* invaded areas.

A. conyzoides was found to be a major plant species in the invaded area besides *C. dactylon, D. annulatum, T. repens, B. pilosa, S. glauca* and *C. carandas*; while *M. koenigii* was well established plant species in the control areas along with *C. dactylon, D. annulatum, B. pilosa, C. carandas, A. vasica, A. aspera,* and *T. repens*.

The number of herbal species was quite high as compared to other life forms in both types of areas. But the number of herbs decreased sharply as compared to other life forms in the invaded areas. It significantly shows that herbal vegetation was affected maximally as compared to other type of vegetative (Table 2). The species sequence of commonly found species in both type of areas plotted in the decreasing order of their IVI values shows that the species become less stable in the *Ageratum* invaded areas (Figure 5).

Major medicinal plants in the control areas were *M. koenigii, D. annulatum, C. dactylon, T. repens, A. aspera, A. vasica* and *C. carandas* (Figure 6). These were the most affected medicinal plants due to the invasion of *A. conyzoides*. The IVI values of these species were found very less in the invaded areas as compared to un-invaded areas. The IVI of *M. koenigii* and *A. vasica* in the control area was 9.89 and 4.47%, respectively; and both the species were not recorded in the top 10 plants in the invaded areas. The IVI of *C. dactylon* was reduced by 42.34% in the invaded area. *Dichanthium annulatum* and *C. carandas* had 59.67 and 63.88% less IVI in the invaded area. The IVI of *A. aspera* was reduced by 53.08%. Further, species sequence of commonly found species in both type of areas plotted in the decreasing order of their IVI values shows that the species become less stable in the *Ageratum* invaded areas (Figure 7).

Impact on soil nutrients

The amount of phenolic in the soil of *A. conyzoides* invaded area was 40.90% more as compared to the control area (Table 3). The pH in the control and invaded soil was near neutral; in the control area it was slightly less whereas in *A. conyzoides* invaded area it was slightly more than 7. The conductivity of the ions was 31.20% more in the *A. conyzoides* invaded area as compared to control. Percent organic carbon and organic matter also increased in the invaded area by nearly 49%. The increase in the available nitrogen content was highest among all other nutrients. It increased by 56.85% in the invaded area. The amount of available phosphorus, potassium and sodium were more by 47.51, 37.91 and 25.29% respectively in the *A. conyzoides* invaded soil as compared to the control soil. Similarly, the amount of available calcium, magnesium and chloride also increased in the *A. conyzoides* invaded soils. The increase was 37.48, 31.54 and 32.67% respectively in the invaded soil as compared to the control soil.

DISCUSSION AND CONCLUSION

Being primary producers, the plants are the major components of the ecosystem. So, it is very important to

Table 2. Floristic composition of the vegetation in the control and *A. conyzoides* invaded areas (alphabetical order).
Total species = 94; Total species in control site = 81.
Total species in invaded site = 55; Common species = 42.

Name Of Plants	Control	Invaded
Achyranthes aspera L.	+	+
Adhatoda vasica Nees	+	+
Aerva scandens L.	+	-
Ageratum conyziodes L.	-	+
Ajuga bracteosa Wall. ex Benth.	+	+
Amaranthus viridis L.	+	-
Anagallis arvensis L.	+	-
Anisomeles indica (L.) Kunt.	+	+
Argemone mexicana L.	+	+
Artemisia scoparia Waldst and Kit.	+	-
Artemisia vulgaris L.	+	-
Asparagus facemasks Willd.	+	+
Bidens pilosa L.	+	+
Blumea membranacea DC.	+	+
Blumea wightiana DC.	-	+
Boerhavia diffusa L.	-	+
Buddleia paniculata L.	+	-
Cannabis sativa L.	+	+
Capsella bursa-pastoris (L.) Medic.	-	+
Carissa carandas L.	+	+
Cassia angustifolia L.	+	+
Cassia occidentalis L.	+	+
Cassia tora L.	+	+
Centella asiatica L. Urb.	+	-
Chenopodium ambrosioides L.	+	-
Cirsium arvensis (L.) Scop.	-	+
Cocculus indica L.	+	-
Convolvulus arvensis L.	+	-
Cynodon dactylon (L.) Pers.	+	+
Dalbergia sissoo Roxb.	+	-
Datura stramonium L.	+	+
Dichanthium annulatum (Forssk.) Stapf.	+	+
Dillenia indica L.	+	-
Dodonaea viscosa (L.) Jacq.	+	-
Duchesnea indica (Andr.) Focke	+	+
Duranta repens L.	-	+
Eclipta alba L.	+	-
Euphorbia granulates L.	+	+
Euphorbia hirta L.	+	+
Euphorbia microphylla B. Heyne ex Roth	+	-
Flacourtia ramontchi L. Herit	+	+
Fumaria indica (Hausskn.) Pugsley	+	-
Gnaphalium indicum L.	+	-
Grewia optiva Drumm. ex Burret	+	-
Indigofera trifoliata L.	+	+
Ipomoea carnea Jacq.	+	-
Ipomoea nil (L.) Roth.	-	+

Table 2 contd.

Ipomoea pentaphylla Cav.	+	-
Ipomoea quamoclit L.	+	-
Lantana camara L.	+	+
Lathyrus sativus L.	-	+
Launaea fallax (Jaub. and Spach) Kunt.	+	+
Leucas cephalotes Spreng.	-	+
Lycopersicon esculentum Mill.	+	-
Mallotus philippensis (Lam.) Muell.-Arg.	+	+
Malvastrum coromandelianum (L.) Garcke	+	+
Medicago lupulina L.	+	-
Mimosa pudica L.	+	+
Morus alba L.	-	+
Murraya koenigii (L.) Spreng.	+	+
Nepeta linearis Royle ex Benth.	+	-
Oxalis corniculata L.	+	+
Parthenium hysterophorus L.	+	+
Phyllanthus amarus Schumach and Thonn	+	+
Physalis minima L.	+	-
Poa annua L.	+	-
Polygala arillata Buch.-Ham ex D. Don	+	-
Polygonum plebium R. Br.	+	-
Ranunculus arvensis L.	+	-
Rhynchosia capitata (Roth.) DC.	+	-
Rubus ellipticus Sm.	+	+
Rumex dentatus L.	-	+
Rumex hastatus D. Don	+	-
Salvia officinalis L.	+	-
Salvia plebeia R. Br.	+	+
Saussurea candicans C.B. Cl.	+	-
Sesamum orientale L.	+	-
Setaria glauca (L.) Beauv.	+	+
Sida cordifolia L.	+	+
Solanum indicum L.	+	+
Solanum nigrum L.	-	+
Sonchus oleraceus L.	+	+
Stellaria media (L.) Cirillo	+	-
Trichodesma indicum L. Sm.	+	+
Tridax procumbens L.	+	+
Trifolium repens L.	+	+
Trigonella emodi Benth.	-	+
Triumfetta bartramia L.	+	-
Urena labota L.	+	-
Vernonia cinerea L. Less	+	+
Vicia hirsuta (L.) S.f. Gray.	+	-
Vicia tetrasperma L. Schreb.	+	-
Xanthium strumarium (Mill.) Torrey and A. Gray	+	+
Ziziphus jujuba Mill.	+	+

save the plant kingdom from various threats to sustain all other living beings. The higher loss in case of the plants growing in invaded areas shows that they become less productive in comparison with the plants in the control

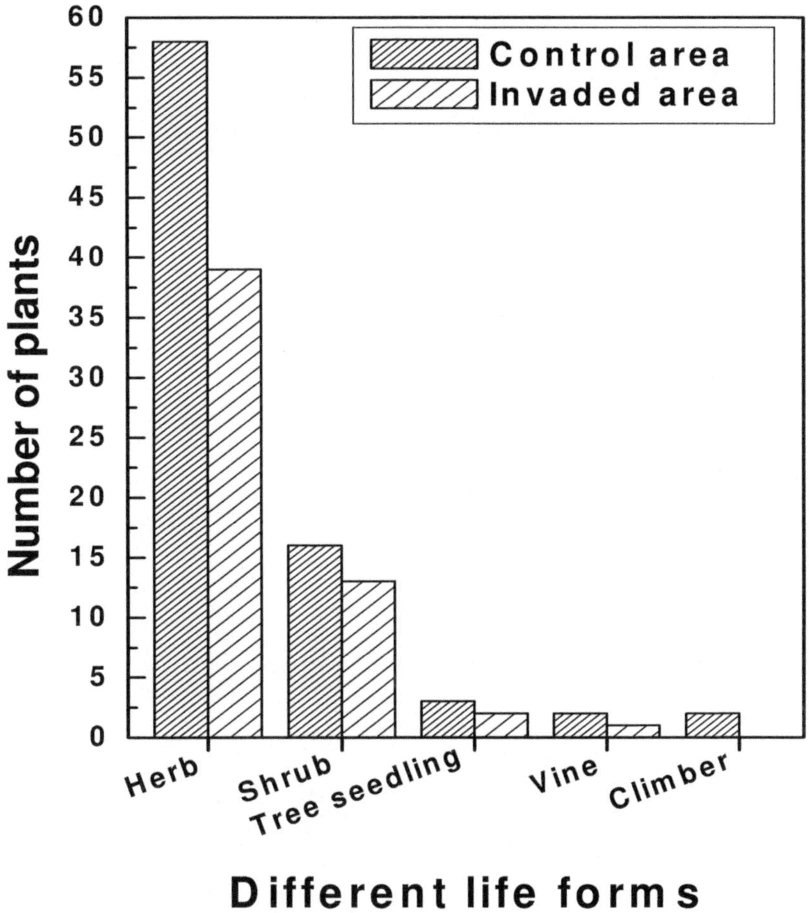

Figure 5. Comparison of different forms of plants in control and *Ageratum* invaded areas.

area. Biomass is directly related to the productivity of the plant species as they lose their normal physiological functions because of the interference by the invading species. The pressure created by the invasive species in the invaded habitats disturbs the functions of biological communities and reduce the diversity of species and dependent fauna (Kinzig et al., 2001). To respond effectively to the invasive species problems, quantitative measurements of the impact of invasion on diversity are required (Schooler et al. (2006).

Ageratum conyzoides has become a very strong invader in the Shivalik hills in Himachal Pradesh and it has increased its density and abundance in the invaded habitats as a result posing a threat to the extinction of native species. An increasing abundance of the invaders can decrease the diversity of species (Kercher and Zedler, 2004). Much effort has been put into identifying determinants constraining broad-scale variability in species richness (Francis and Currie, 2003; Rahbek, 2005). It is apparent that the factors influencing patterns of species richness vary with the geographical extent and sample resolution (Willis and Whittaker, 2002). Therefore, only by multiple analyses scales for different locations and at various spatial scales general explanations of broadscale species richness, diversity and distribution patterns can be derived (Zhao and Fang, 2006).

It's concluded from the studies that *A. conyzoides* is a strong invader in these areas and its increased abundance, cover and density poses a threat to the native species which also include medicinally important species. The decrease rate of biomass and various ecological indices (∝-diversity, abundant species, Margalef's index etc.) in the *Ageratum* invaded habitats clearly signifies that these become less productive and stable as compared to non-invaded habitats. There were some species which totally lost in the invaded areas while some other species preferred to grow there. The absence of seedlings of tree species like *Dalbergia sissoo* and *Acacia catechu* in the invaded areas showed that it inhibited their seedling growth. Thus, *Ageratum cony-*

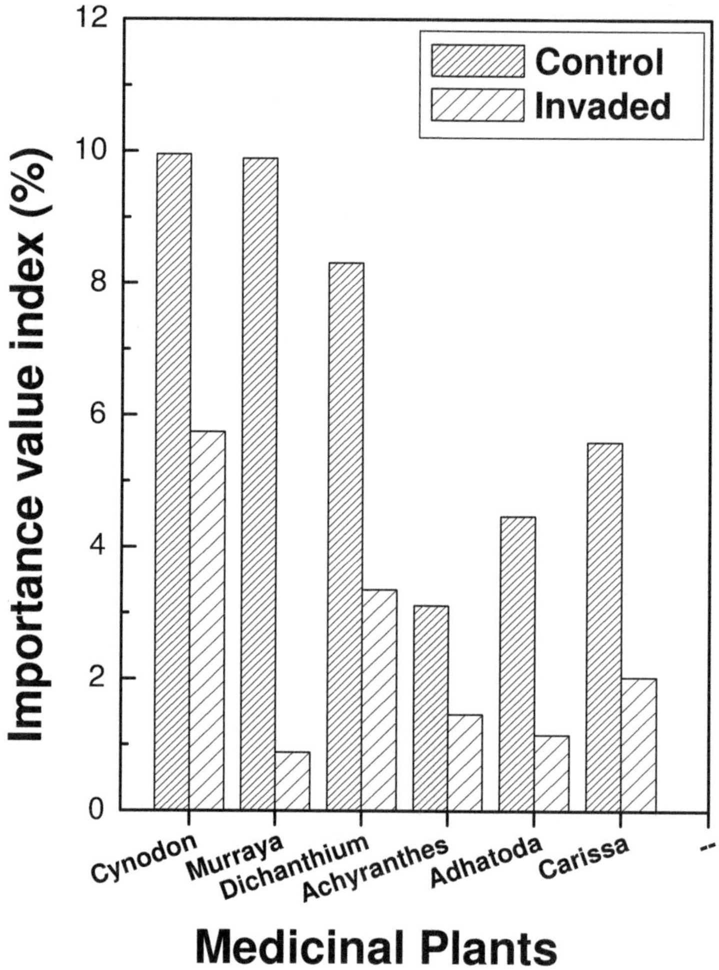

Figure 6. Impact of *A. conyzoides* on the IVI (%) of some medicinal plant species. All values are significant at 5% significance level after applying two population *t*-tests.

zoides directly alters the growth of other plant species by forming its own monocultures.

The invasion of *Ageratum conyzoides* also alters the physico-chemical properties of soils in the invaded areas. The soils in the invaded areas become nutrient rich which generally help in the growth of invasive species. It was clear from the results that the values of all soil nutrients found to be higher in the *Ageratum* invaded areas as compared to control. Minimum change was observed in case of pH compared to other parameters. Likewise, the phenolics- a well known group of allelochemical- were found to be more in weed-invaded soil as compared to the control soil. These phenolics are released from the plant part through various mechanisms such as leachation from above ground parts, root exudation, volatilization or microbial degradation. These allelochemicals besides imparting the plant allelopathic property also regulate the biotic communities of soil and alter the physical and chemical properties of soil (Nardi et al., 2000). Many studies suggest that allelopathy may contribute to the ability of particular alien species to become dominant in the native plant communities (Abdul-Wahab and Rice, 1967; Vaughn and Berhow, 1999; Ridenour and Callaway, 2001). Several aggressive weeds exhibit the phenomenon of allelopathy as a mechanism of interference which provides them competitive advantage over other plants.

El-Ghareeb (1991) studied the allelopathic effect of the invasive plant *Tribulus terrestris* on surrounding vegetation in an abandoned field of Kuwait. His study demonstrated that besides the growth inhibitory effect of plant on other plants, the soil moisture and concentration of N, P and K were significantly higher in *T. terrestris* site. In the present study too the amounts of available nutrients were significantly more in weed-invaded soils as compared to the weed free soils. Abundant evidences support the idea that higher resource availability increases the susceptibility to invasion of plant commu-

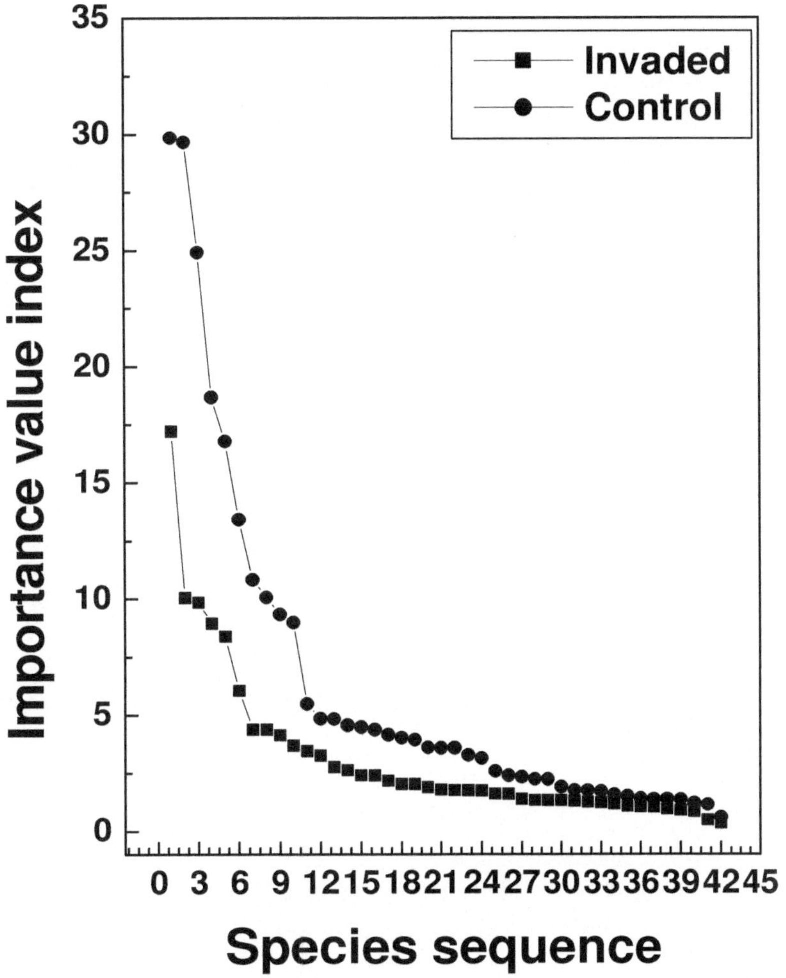

Figure 7. Distribution of common species in control and *Ageratum* invaded sites (decreasing order of Importance Value Index).

Table 3. Comparison of selected physico-chemical properties of soil collected from areas invaded by *A. conyzoides* or free from it (control).

S.N.	Parameters	Control	Invaded
1	Phenolics (ìg/100 g soil)	12.33 ± 1.05	20.88 ± 1.53**
2	pH	6.78 ± 0.01	7.09 ± 0.01**
3	EC (ìS)	129.66 ± 1.72	188.47 ± 3.35**
4	Organic Carbon (%)	0.49 ± 0.02	0.96 ± 0.06**
5	Organic Matter (%)	0.84 ± 0.03	1.66 ± 0.11**
6	N (kg/ha)	92.09 ± 3.40	213.40 ± 7.26**
7	P (ppm)	64.53 ± 3.18	122.95 ± 2.05**
8	K (ppm)	88.39 ± 4.08	142.35 ± 3.09**
9	Na (ppm)	39.29 ± 3.55	52.59 ± 4.06**
10	Ca (g/100g)	4.17± 0.28	6.67 ± 0.28**
11	Mg (g/100g)	2.17 ± 0.29	3.17 ± 0.28**
12	Cl (g/100g)	3.40 ± 0.20	5.05 ± 0.20**

** means significant from control at $P < 0.01$ after applying student *t*- test.

nities (Burke and Grim, 1996; Maron and Connor, 1996).

Further, the absorption of phenolics - the allelopathic compounds- by soil particles and their microbial breakdown and may account for the outcome of present observations (Dalton, 1999; Huang et al., 1999; Wardle et al., 1998) which are further affected by various soil factors such as soil texture, organic carbon and organic matter etc. (Kobayashi, 2004).

The reason why some invasive plants are so successful in new environments may be that they bring novel mechanisms of interactions with the recipient community. However, Dietz et al. (1996) concluded that factors other than allelopathy might be operating in nature that favours rapid establishment and persistence of dense stands of alien species.

REFERENCES

Abdul-Wahab AS , Rice EL (1967). Plant inhibition by Johnson grass and its possible significance in old-field succession. Bull Torr. Bot. Club, 94: 486-497.

AOAC (1960). Association of Official Agricultural Chemist. Methods and Analysis. AOAC, Washington, DC

Balokhra JM (1999). The Wonderland of Himachal Pradesh. Rev. and Enl. 4th Ed. H.G. Publications, New Delhi.

Barbour MG, Burk JH, Pitts WD, Gilliam FS, Schwartz MW (1999). Terrestrial Plant Ecology. 3rd Edition. Addison Wesely Longman Inc., Menlo Park, California.

Black CA (1973). Soil Plant Relationships. 2nd Ed. Scientific Publishers, Jodhpur.

Borgmann KL, Rodewald AD (2005). Forest restoration in urbanizing landscapes: Interactions between land uses and exotic shrubs. Restor. Ecol., 13(2): 334-340.

Bower CA , Gschwend FB (1952). Exchangeable cation analysis of saline and alkali soils. Soil Sci., 73: 251-261.

Burke MJW , Grime JP (1996). An experimental study of plant community invasibility. Ecol., 77: 776-790.

Cowie RH (1998*a*). Patterns of introduction of non-indigenous non marine snails and slugs in the Hawaiian Islands. Biodivers. Conserv., 7: 349-368.

Cowie RH (1998*b*). Homogenisation of Pacific Island snails. World Conservation; Quarterly Bulletin IUCN-The World Conservatin Union, 4/97-1/98: 18.

Dalton BR (1999). The occurrence and behavior of plant phenolic acids in soil environment and their potential involvement in allelochemical interference interactions: Methodological limitations in establishing conclusive proof of allelopathy. In: Principles and Practices in Plant Ecology: Allelochemical Interactions (Eds. Inderjit S, Dakshini, KMM , Foy CL) CRC Press, Boca Raton, FL, pp: 57-74.

Dietz H, Steinlein T, Winterhalter P , Ullmann I (1996). Role of allelopathy as a possible factor associated with the rising of dominance of *Bunias orientalis* L. (Brassicaceae) in some native plant assemblages. J. Chem. Ecol., 22: 1797-1811.

Dogra KS (2008). Impact of Some Invasive Species on the Structure and Composition of Natural Vegetation of Himachal Pradesh. Ph.D. thesis, Panjab University, Chandigarh, India.

Dogra KS, Kohli RK, Sood SK (2009). An assessment and impact of three invasive species in the Shivalik hills of Himachal Pradesh, India. Inter. J. Biodivers. Conserv. 1(1): 004 - 0010.

El-Ghareeb RM (1991). Suppression of annuals by *Tribulus terrestris* in an abandoned field in the sandy desert of Kuwait. J. Veg. Sci., 2: 147-154.

Holdgate MW (1986). Summary and Conclusions: Characteristics and Consequences of Biological Invasions. Philosophical Transactions of the Royle Society, London.

Huang PM, Wang MC, Wang MK (1999). Catalytic transformation of phenolic compounds in the soil. In: Principles and Practices in Plant Ecology*:* Allelochemical Interactions (Eds. Inderjit S, Dakshini KMM , Foy CL). CRC Press, Boca Raton, FL, pp: 287-306.

Kercher SM, Zedler JB (2004). Multiple disturbances accelerate invasion of reed canary grass (*Phalaris arundinacea* L.) in a mesocosm study. Oecologia, 138: 455-464.

Kinzig AP, Pacala SW, Tilman D (2001). The Functional Consequences of Bioodiversity. Princeton University Press, Princeton, NJ.

Kobayashi K (2004). Factors affecting phytotoxic activity of allelochemicals in soil. Weed Biol. Manage., 4: 1-7.

Kohli RK, Batish DR, Singh HP, Dogra KS (2006). Status, invasiveness and environmental threats of three tropical American invasive weeds (*Parthenium hysterophorus* L., *Ageratum conyzoides* L., *Lantana camara* L.) in India. Biol. Invasions, 8: 1501-1510.

Ludwig JA, Reynolds JF (1988). Statistical Ecology- A Primer on Methods and Computing. John Wiley and Sons, New York, pp: 337.

Maron JL, Connor PG (1996). A native nitrogen-fixing shrub facilitates weed invasion. Oecologia, 105: 302-312.

McKinney ML, Lockwood JL (1999). Biotic homogenization: A few winners replacing many losers in the next mass extinction. Trends Ecol. Evol., 14: 450-453.

Mishra R (1968). Ecology Work Book. Oxford and IBH Company, New Delhi, India, pp. 244.

Nardi S, Concheri G, Pizzeghello D, Sturaro A, Rella R , Parvoli G (2000). Soil organic matter mobilization by root exudates. Chemosphere, 5: 653-658.

National Focal Point for APFISN, India (2005). Stocktaking of National Forest Invasive Species Activitis, India (India Country Report 101005). New Delhi, India: Ministry of Environment and Forests.

Ohlemuller R, Walker S, Wilson JB (2006). Local vs regional factors as determinants of the invisibility of indigenous forest fragments by alien plant species. Oikos, 112: 493-501.

Olson SR, Cole CV, Watanabe FS, Dean LA (1954). Estimation of available phosphorus in soil by extraction with sodium bicarbonate. USDA Circular 939: 1-19.

Pimentel D, Lach L, Zuniga R , Morrison D (2000). Environmental and economic costs of non indigenous species in the United States. Biosciences 50: 53-65.

Ridenour WM, Callaway RM (2001). The relative importance of allelopathy in interference: The effects of an invasive weed on a native bunchgrass. Oecologia, 126: 444-450.

Schooler SS, McEvoy PB, Coombs EM (2006). Negative per capita effects of purple loosestrife and reed canary grass on plant diversity of wetland communities. Diversity Distrib., 12: 351-363.

Singh JS, Singh SP (1992). Forests of Himalaya, Structure, Functioning and Impact of Man. Gyanodaya Prakashan, Nainital, India.

Swain T, Hills WE (1959) The phenolic constituents of *Prunus domestica* I. The qualitative analysis of constituents. J. Sci. Food Agric., 10: 63-68.

Vaughn SF, Berhow MA (1999). Allelochemicals isolated from tissues of the invasive weed garlic mustard (*Alliaria petiolata*). J. Chem. Ecol., 25: 2495-2504.

Vitousek PM, Mooney HA, Lubchenco J, Melillo JM (1997). Human domination of earth's ecosystems. Science, 277: 494-499.

Walkley A, Black IA (1934). An examination of the Ditjareff method for determining soil organic matter and a proposed modification of chromic acid titration method. Soil Sci., 37: 28-38.

Wardle DA, Barker GM, Bonner KI, Nicholson KS (1998). Can comparative approaches based on plant ecophysiological traits predict the nature of biotic interactions and individual plant species effects. J. Ecol. 86: 405-420.

Wilcove DS, Rothstein D, Dobow J, Phillips A, Losos E (1998). Quantifying threats to imperiled species in the US. Biosciences, 48: 607-615.

Willis KJ, Whittaker RJ (2002). Species diversity-scale matters. Science, 295: 1245-1248.

Zhao S, Fang J (2006). Patterns of species richness for vascular plants in China's nature reserves. Diversity Distrib., 12: 364-372.

Multivariate analysis of *Harpagophytum* DC. ex Meisn (Pedaliaceae) based on fruit characters

M. Muzila*, M. P. Setshogo and S. W. Mpoloka

University of Botswana Biological Sciences, Private Bag UB 00704 Gaborone Botswana.

***Harpagophytum* is bitypic and native to Southern Africa. Its two species are *Harpagophytum procumbens* and *H. zeyheri*. *H. procumbens* is medicinal. A reliable method of identifying the species is through its fruit. However, distinguishing between *H. procumbens* and *H. zeyheri* can be difficult because of the various morphotypes. Hence, possibilities of introgression are hypothesized. The objective of this study was to test for interspecific introgression between the two species. Diagnostic characters of the fruit were subjected to multivariate analysis. Discriminant function analysis was used to identify and classify the 21 specimen types. Cluster analysis was used to test for possible appurtenance of individual fruit specimens to either the parental species (*H. procumbens* or *H. zeyheri*) or to the hybrid (*H. procumbens* X *H. zeyheri*). The study inferred the existence of hybridisation (introgression) between the two species. The hybrids can be characterised by fruit length, fruit width, arm width, arm length, and the number of seed rows. In the hybrids, the number of seed rows per loculus comes in various combinations (for example 3,3 and 3,2). And this was found to be quite important in identifying the hybrids. However, it was difficult to determine the direction of gene flow, thus, we recommend molecular analysis of the hybrids.**

Key words: Harpagophytum, fruit, introgression, multivariate.

INTRODUCTION

Harpagophytum DC. ex Meisn (Pedaliaceae) is a bitypic genus, with both species being native to Southern Africa. The two species are *Harpagophytum procumbens* and *H. zeyheri*. *H. procumbens*. The two species are perennial herbs with several prostrate annual stems that emanate from a succulent tuberous taproot. For an elaborate taxonomy of the genus see Ihlenfeldt (1988). This study focused on the biology of the fruit as it is perceived to play a major role in the taxonomy of the genus. The fruit of *Harpagophytum* is a woody capsule that is imperfectly dehiscent along its longitudinal length (Figure 1). The fruit is laterally compressed, with two obtuse protuberances on each face, armed with two rows of curved arms along the edges. Each arm bears re-curved spines. Inside the fruit are numerous seeds that are stacked in rows of 2's or 4's in each loculus, depending on the species. The two (2) rows are a diagnostic character for *H. zeyheri,* while the four (4) rows are a diagnostic character for *H. procumbens* (Ihlenfeldt, 1988).

Harpagophytum fruit capsules are morphologically variable and therefore present several taxonomic problems. It is not clear whether the various morphotypes exhibited by the species are an indication of hybridisation within the genus or they are just ecotypes of the same species. Hybridization between species often forms hybrid swams (Grant, 1971). Hybrid swams occur as a result of gene exchange between nuclear DNA or cyto-plasmic DNA (that is cpDNA or mtDNA) between species and is known as introgression. Introgression can be infraspecific, interspecific or intergeneric (Reiseberg and Brunsfeld, 1998). Infraspecific introgression involves the formation of morphotypes within the same species, interspecific introgression forms hybrids between different species of the same genus, while intergeneric

*Corresponding author. E-mail: muzilam@mopipi.ub.bw.

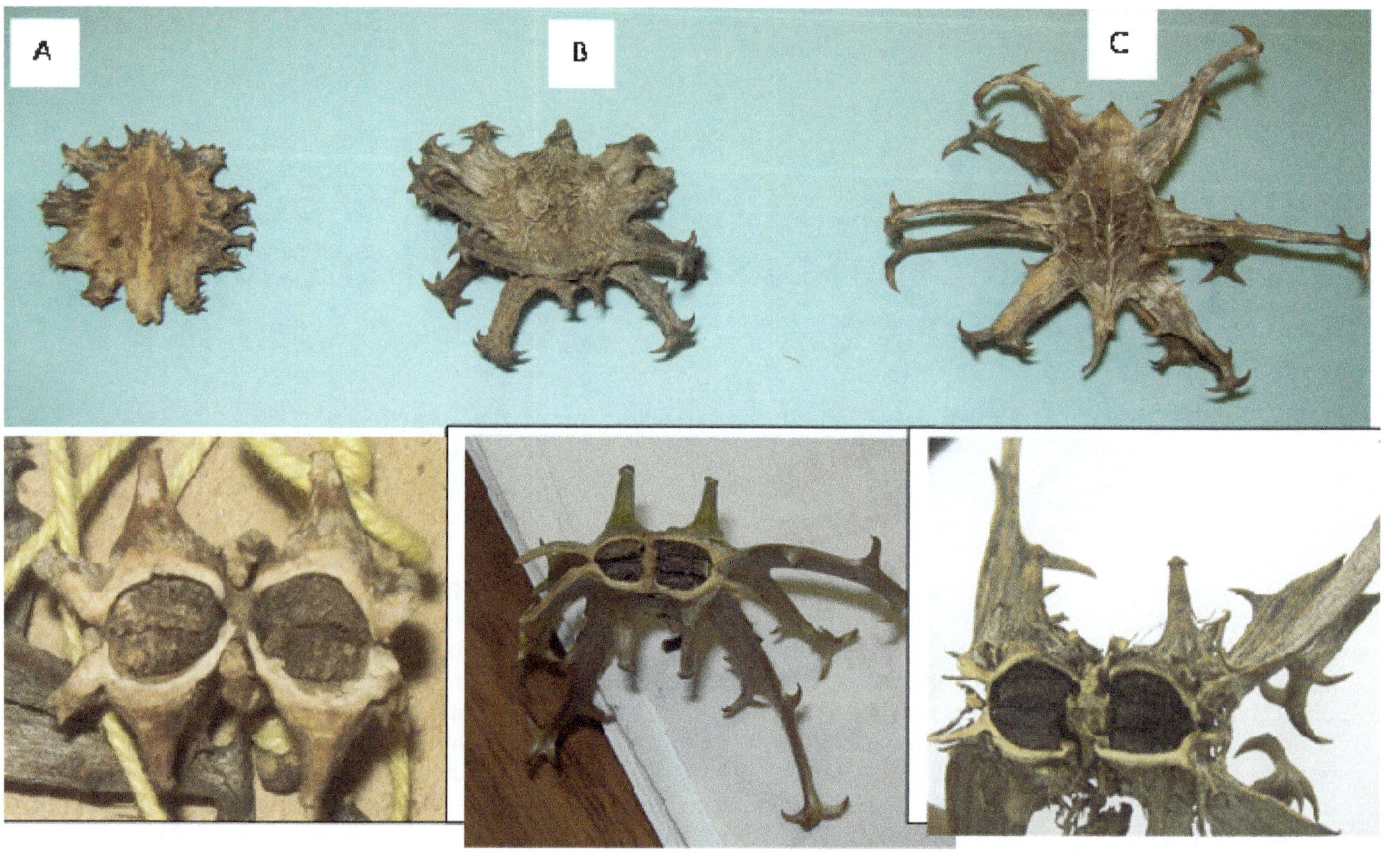

Figure 1. *Harpagophytum* fruits showing the three different forms: A) H. *zeyheri*; B) Hybrid; C) *H. procumbens*; D) two seed rows in *H. zeyheri*; E) three seed rows in one of the hybrids; F) four seed rows in *H. procumbens.*

introgression is between species of different genera (Reiseberg and Brunsfeld, 1998). Interspecific and intergeneric introgression can also lead to speciation, which is the stabilization of a particular hybrid to become a recognized species (Grant, 1971).

The exchange of genetic material in introgression can be very complex and can lead to a situation where introgression is detected in the morphology of the species but not in the nuclear DNA (Reiseberg and Brunsfeld, 1998). In other situations genetic exchange can be detected in the nuclear DNA and not in the cytoplasmic DNA or vice-versa. Therefore, all studies of introgression (that is using morphology or DNA) are important because through these the direction of gene transfer and the age of the hybrid swarm can be determined. The impact of introgression on plant diversity has been a subject of much debate (Reiseberg and Brunsfeld, 1998; Arnold and Martin, 2009). Introgression can be an evolutionary dead end (Grant, 1971), thereby reducing possibilities of genetic gain in plant breeding programs by scientists (Rangel et al., 2008) or it can reinforce the survival of the species (Reiseberg and Brunsfeld, 1998, Lanta et al., 2003, Arnold and Martin, 2009). The evolutionary dead ends are faced with extinction (Ellstrand, 1999) as they may find it difficult to adapt to changing environmental conditions. On the other hand, introgression can strengthen the genetic diversity of the species as in, for example, adaptive radiation (Grant, 1971). An analysis of gene flow is crucial in understanding speciation events and maintenance of species integrity (Curtu et al., 2007).

The main objective of the study, therefore, was to determine whether the various morphological patterns exhibited by *Harpagophytum* fruits are an indication of hybridization (introgression) between the two species or just ecotypes (Jordan, 2010) of the same species.

MATERIALS AND METHODS

Nomenclature

Diagnostic keys of Ihlenfeldt (1988) were used to distinguish between the types of *Harpagophytum* and their hybrids.

Sampling sites

Specimens were collected from six different areas of southern Botswana and deposited at UCBG herbarium in Gaborone, Botswana. The sites included the following (Table 4):

Table 1. Factor Loadings (Varimax normalized) to indicate principal components that contributed to taxa delimitation. Only factor loadings greater than 0.70 are indicated.

Characters	Factor 1	Factor 2	Factor 3
Arm length	0.74		
Arm width		0.85	
Column height	0.82		
Seed rows		0.85	
Fruit length	0.95		
Fruit width	0.77		
Fruit circumference			0.86

Sympatric zones

1) Kumakwane
2) Malotwana
3) Mmamashia
4) Oodi

Pure type zones

1) Bella - Bella)
2) Sekoma

Sampling technique

The sampling technique was systematic qualitative. Some 105 fruit specimens were collected, which comprised five (5) fruit specimens collected from 21 plant types (with the five fruits from each plant treated as duplicates of the voucher specimen). Fruits were chosen for further analysis because they play a great role in the taxonomy of the genus, and they are relatively uniform when compared to plastic traits like vegetative features.

Multivariate analysis

Morphometric analyses were conducted on each of the five replicate fruit specimens and the mean value recorded. The plant parts analyzed were as follows:

1) Length of the longest arm
2) The widest part of the longest arm (that is arm width)
3) Fruit specimen width
4) Fruit specimen length
5) The number of seed rows
6) Height of column that enclosed the seed rows
7) Fruit circumference.

Firstly, a one - way ANOVA model in STATISTICA (Anonym, 1995) was used to test for significant differences among the collected fruit specimens. Then a discriminant function analysis was used to test the hypothesis that there are three groups of *Harpagophytum*, namely *H. procumbens*, *H. zeyheri* and their hybrids. Under discriminant function analysis the scatter plot, classification matrix, box and whisker plot analysis were employed. The classification matrix was used to check for reliability of the scatterplot groupings (that is testing for confidence in the results), while the box and whisker tested for details of how the three groups differed. And finally, a cluster analysis was carried out using un-weighted pair – group averaging (UPGMA) in order to check for possible appurtenance of individual fruit specimen to either the parental/type species (*H. procumbens* or *H. zeyheri*) or to the expected hybrid (*H. procumbens* X *H. zeyheri*).

RESULTS

General morphology and number of seed rows

Based on fruit type, various morphotypes of *Harpagophytum* were observed. Features that characterize the morphotypes include arm length, arm width, fruit width, fruit length and the number of seed rows (Table 1). The number of seed rows - as a character - plays a significant role in the taxonomy of *Harpagophytum* as it immediately identifies the hybrids since the number of seed rows for *Harpagophytum* are four (4) and two (2) on each seed column for *H. procumbens* and *H. zeyheri* respectively (Ihlenfeldt, 1988). However, for the hybrids, the numbers of seed rows come in various combinations (Figure 1 and Table 2). For instance three hybrid specimens showed three (3) seed rows in one of the seed loculus and two (2) seed rows on the other (that is 3.2), while the other two hybrid specimens showed three (3) seed rows on both seed loculus (that is 3.3). Hence, upon noticing the importance of seed rows in the taxonomy of the genus, we placed a higher weight on the character prior to multivariate analysis (apriori weighting). However, the character was standardized, prior to analysis, and therefore excluded from the box and whisker analysis.

Discriminant function analysis

Scatter plot

According to the ordination scatterplot, three groups of *Harpagophytum* exist in southern Botswana (Figure 2). The groups are *H. procumbens*, *H. zeyheri* and hybrid(s) of the two species (*H. zeyheri* X *H. procumbens*). Root 1 separated the hybrids from the pure *H. procumbens* and the pure *H. zeyheri* and by so doing recognising them as a group. Root 2 allocated the hybrids to two groups as follows:

1) Those biased towards the *H. zeyheri* pure line
2) Those biased towards the *H. procumbens* pure line.

Classification matrix

Reliability on the classification of specimens into three groups was tested using the classification matrix. According to the classification matrix (Table 3), classification of the specimens into *H. procumbens, H. zeyheri* and their hybrids was perfectly precise (that is 100%

Table 2. Character parameters for the various voucher specimens used in the study.

Voucher number	Species[Ψ] identity	Arm length (mm)	Arm width (mm)	Seed rows* P < 0.00		Seed column height (mm)	Fruit length (mm)	Fruit width (mm)	Fruit circumference (mm)	Signif. p
				Seed column1	Seed column2					
OODI L3	Hybrid	7.8 ± 2.32	6.7 ± 0.76	2	2	3.40 ± 0.50	33.90 ± 3.44	16.60 ± 4.28	130.21 ± 3.17	0.00
LSHBL1	Hybrid	18.44 ± 4.32	9.6 ± 0.83	3	3	5.90 ± 0.50	36.20 ± 3.44	19.40 ± 4.03	82.50 ± 3.17	0.00
SKL1	*H. procumbens*	30.5 ± 6.58	9.1 ± 0.82	4	4	5.20 ± 1.12	47.80 ± 6.76	20.30 ± 4.28	90.01 ± 24.28	0.00
KKMWL5	Hybrid	21.6 ± 4.91	12.6 ± 0.9	3	3	4.30 ± 0.50	36.00 ± 3.44	21.10 ± 4.28	148.30 ± 3.17	0.00
KKWNL1	Hybrid	64.6 ± 12.98	3.2 ± 0.68	2	2	6.38 ± 0.50	58.74 ± 3.44	37.78 ± 4.28	122.34 ± 3.17	0.00
BBL5	Hybrid	7.52 ± 2.27	2.66 ± 0.67	2	2	2.28 ± 0.50	21.50 ± 3.44	16.38 ± 4.28	51.10 ± 3.17	0.00
KKMWL2	*H. procumbens*	43.3 ± 8.98	7.57 ± 0.78	2	2	6.89 ± 1.12	40.81 ± 6.76	24.03 ± 4.03	87.53 ± 24.28	0.00
LSBL3	Hybrid	21.11 ± 4.82	11.87 ± 0.88	2	2	5.35 ± 0.50	43.35 ± 3.44	32.21 ± 4.28	93.29 ± 3.17	0.00
BBL1	*H. zeyheri*	7.21 ± 2.21	4.99 ± 0.72	2	2	4.05 ± 0.77	32.90 ± 1.35	15.86 ± 2.99	108.10 ± 16.93	0.00
OODIL4	Hybrid	5.9 ± 1.96	3.53 ± 0.69	3	2	3.13 ± 0.50	30.02 ± 3.44	16.24 ± 4.28	69.54 ± 3.17	0.00
SKL2	*H. procumbens*	79.62 ± 15.8	9.9 ± 0.84	4	4	5.90 ± 1.12	57.01 ± 6.76	22.66 ± 4.03	11.30 ± 24.28	0.00
LSBL2	Hybrid	17.3 ± 4.1	6.48 ± 0.76	3	2	5.10 ± 0.50	51.38 ± 3.44	31.00 ± 4.28	97.52 ± 3.17	0.00
MMSL2	Hybrid	40.78 ± 8.51	13.1 ± 0.91	3	2	5.78 ± 0.50	51.58 ± 3.44	22.62 ± 4.28	113.76 ± 3.17	0.00
BBL2	Hybrid	4.2 ± 1.64	6.59 ± 0.76	2	2	4.92 ± 0.50	42.46 ± 3.44	25.54 ± 4.28	95.60 ± 3.17	0.00
OODI L2	Hybrid	12.1 ± 3.13	15.6 ± 0.97	2	2	4.90 ± 0.50	29.30 ± 3.44	19.40 ± 4.28	107.90 ± 3.17	0.00
MMSL1	Hybrid	44.28 ± 9.17	11 ± 0.86	2	2	6.49 ± 0.50	57.01 ± 3.44	22.28 ± 4.28	113.76 ± 3.17	0.00
SKL3	*H. procumbens*	64.4 ± 12.94	9.38 ± 0.82	4	4	10.30 ± 1.12	72.12 ± 6.76	27.50 ± 4.03	167. 52 ± 24.28	0.00
KKMWL4	Hybrid	3.5 ± 1.51	7 ± 0.77	2	2	9.32 ± 0.50	58.74 ± 3.44	37.78 ± 4.28	122.34 ± 3.17	0.00
BBL4	*H. zeyheri*	0.6 ± 0.97	4.74 ± 0.72	2	2	7.90 ± 0.77	38.50 ± 1.35	25.50 ± 2.99	90.30 ± 16.93	0.00
BBL3	*H. zeyheri*	5.68 ± 1.92	3.2 ± 0.68	2	2	6.00 ± 0.77	34.94 ± 1.35	19.90 ± 2.99	95.10 ± 16.93	0.00
BBL6	*H. zeyheri*	4.21 ± 1.65	2.2 ± 0.66	2	2	5.92 ± 0.77	32.56 ± 1.35	18.24 ± 2.99	83.14 ± 16.93	0.33

[Ψ] Hybrid refers to *H. zeyheri* X *H. procumbens.*
± represents standard error.
* Standard error not calculated since the character is qualitative as opposed to quantitative.

correct).

Box and whisker

The three groups revealed in the scatter plot graph (Figure 2) were further analyzed using box and whisker plots (Figure 3). Box and whisker plots provided morphometric details of how fruit length (Figure 3A), fruit width (Figure 3B), arm width (Figure 3C) and arm length (Figure 3D) were distinguished between the three groups. According to the box and whisker plots, *H. procumbens* generally posseses elongated fruits, while *H. zeyheri* has shorter fruits and the hybrids are intermediates of the two species (Figure 3A). The same pattern is repeated for arm width (Figure 3C) and arm length (Figure 3D). For fruit width, however, the hybrids posses the widest fruits followed by *H. procumbens* (Figures 1B and 3C).

Cluster analysis

Hierarchical tree

The hierarchical tree also recognized the presence

Table 3. Classification Matrix to indicate precision of classification

Rows: Observed classifications
Columns: Predicted classifications

Taxa	Percent correct	*H. zeyheri*	*H. procumbens*	Hybrid
H. zeyheri	100	4	0	0
H .procumbens	100	0	4	0
Hybrid	100	0	0	13
Total	100	4	4	13

Table 4. Collection sites of vouchers specimens used in the study.

Voucher Number	Species	Herbarium*	Collection site
OODI L3	Hybrid	UCBG	**Oodi** – 24° 28' 25.98" S 26° 02' 30.22" E
LSHBL1	Hybrid	UCBG	**Malotwana -** 24° 17' 31.93" S 26° 08' 15.64" E
SKL1	*H. procumbens*	UCBG	**Sekoma** -24° 30' 15.82" S 23° 02' 30.22" E
KKMWL5	Hybrid	UCBG	**Kumakwane** -24° 41' 42.13 S" 26° 02' 16.99" E.
KKWNL1	Hybrid	UCBG	**Kumakwane** -24° 41' 42.13" S 26° 02 16.99" E.
BBL5	Hybrid	UCBG	**Bella - Bella** -24° 30' 00.95" S 26° 02 16.99 E"
KKMWL2	*H. procumbens*	UCBG	**Kumakwane** -24° 41' 42.13" S 26° 02' 16.99" E.
LSBL3	Hybrid	UCBG	**Malotwana -** 24° 17' 31.93" S 26° 08' 15.64" E
BBL1	*H. zeyheri*	UCBG	**Bella - Bella** -24° 30' 00.95" S 26° 02 16.99 E"
OODIL4	Hybrid	UCBG	**Oodi** – 24° 28' 25.98" S 26° 02' 30.22" E
SKL2	*H. procumbens*	UCBG	**Sekoma** -24° 30' 15.82" S 23° 02' 30.22" E
LSBL2	Hybrid	UCBG	**Malotwana -** 24° 17' 31.93" S 26° 08' 15.64" E
MMSL2	Hybrid	UCBG	**Mmamashia -** 24° 30' 59.80" S 25° 58' 51.91" E

Table 4. Cont.

BBL2	Hybrid	UCBG	**Bella - Bella** -24° 30' 00.95 S" 26° 02' 16.99 E"
OODI L2	Hybrid	UCBG	**Oodi** – 24° 28' 25.98" S 26°02 30.22" E
MMSL1	Hybrid	UCBG	**Mmamashia** - 24° 30' 59.80" S 25° 58' 51.91" E
SKL3	*H. procumbens*	UCBG	**Sekoma** -24° 30' 15.82" S 23° 02' 30.22" E
KKMWL4	Hybrid	UCBG	**Kumakwane** -24° 41' 42.13" S 26° 02' 16.99" E.
BBL4	*H. zeyheri*	UCBG	**Bella – Bella** -24° 30' 00.95" S 26° 02' 16.99" E
BBL3	*H. zeyheri*	UCBG	**Bella - Bella** -24° 30' 00.95" S 26° 02' 16.99" E
BBL6	*H. zeyheri*	UCBG	**Bella - Bella** -24° 30' 00.95" S 26° 02' 16.99 "E

UCBG* - University College of Botswana Gaborone

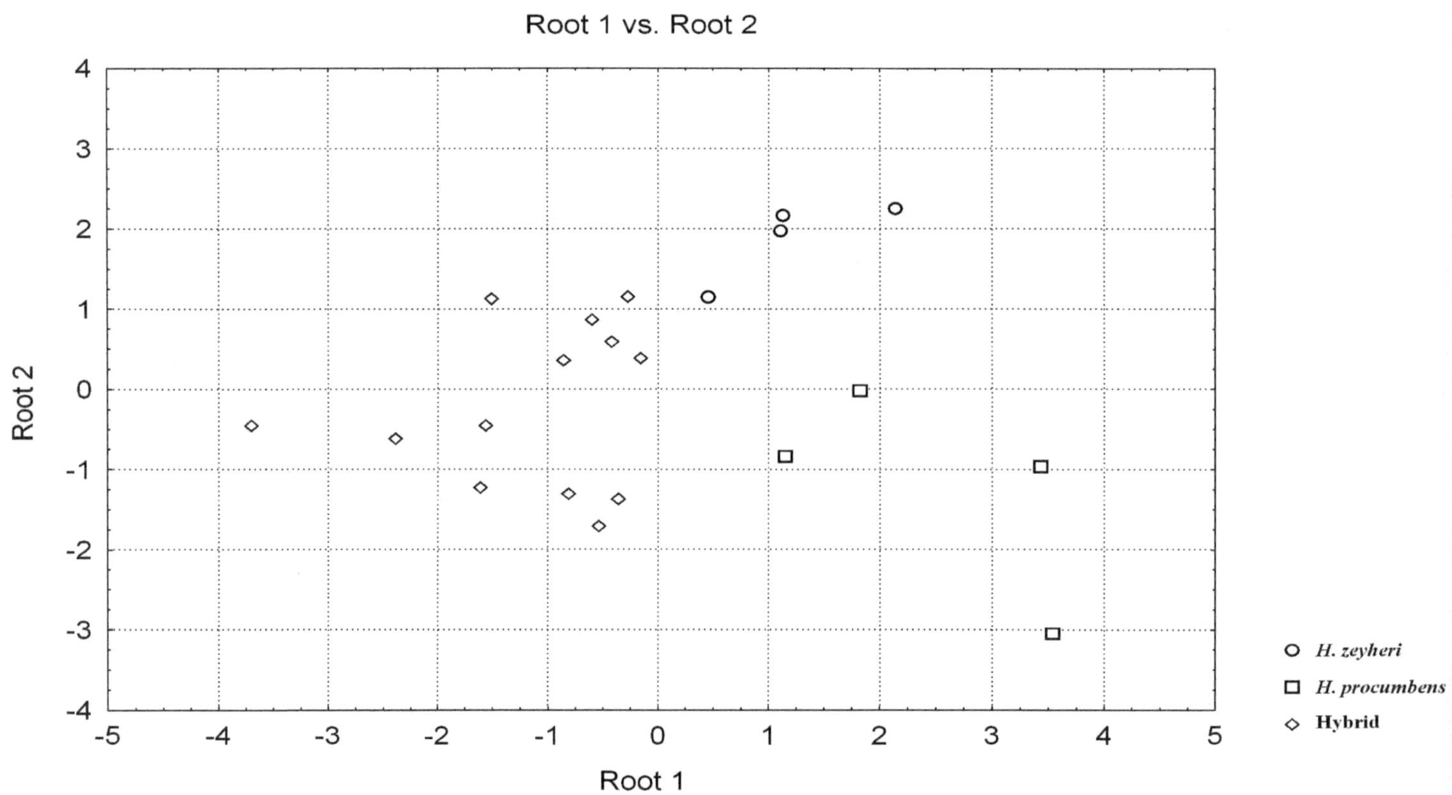

Figure 2. Scatter plot to indicate the three groups of *Harpagophytum*.

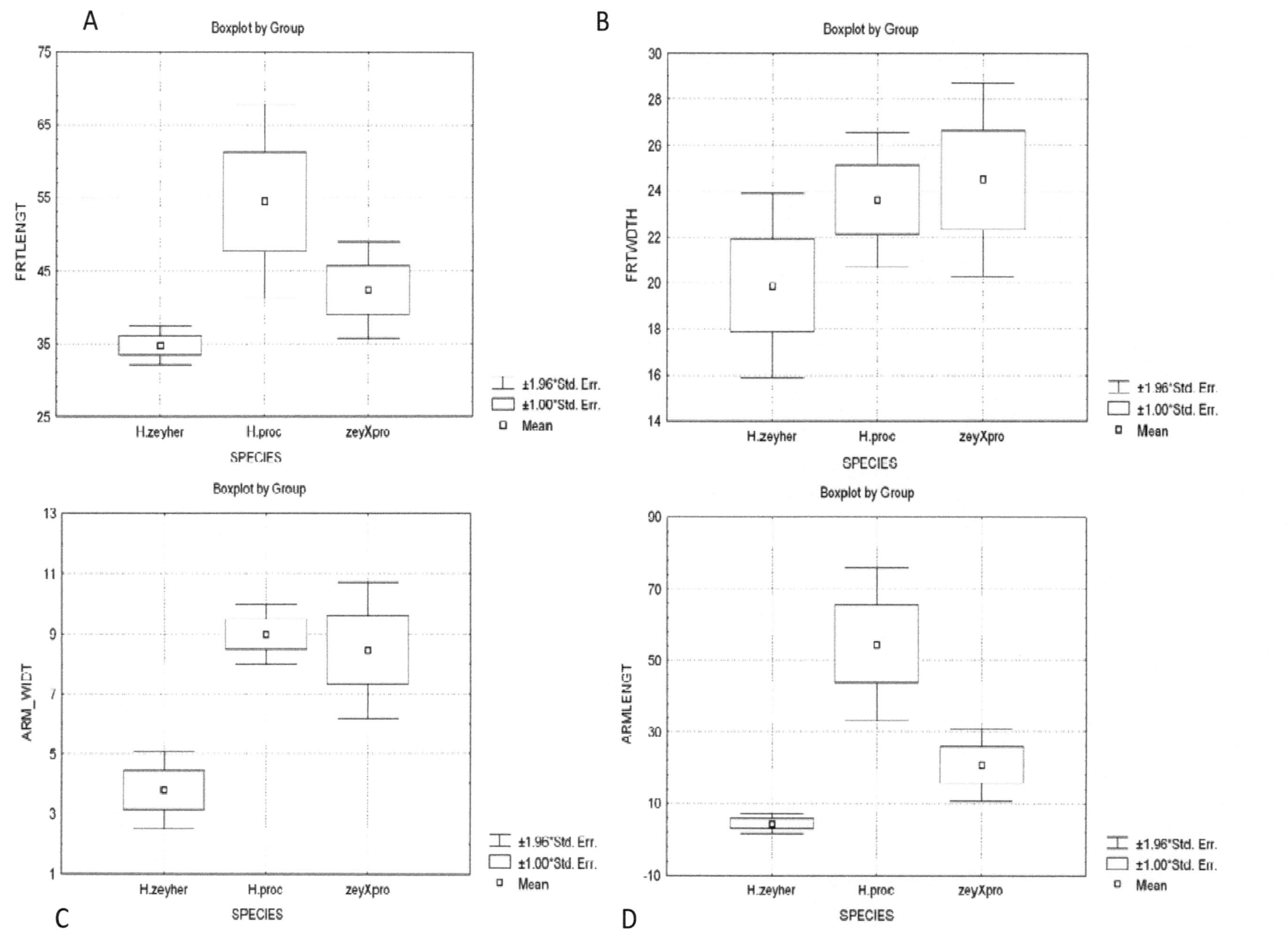

Figure 3. Box and whisker plots for the three *Harpagophytum* species: A) Fruit length; B) Fruit width; C) Arm width; D) Arm length.

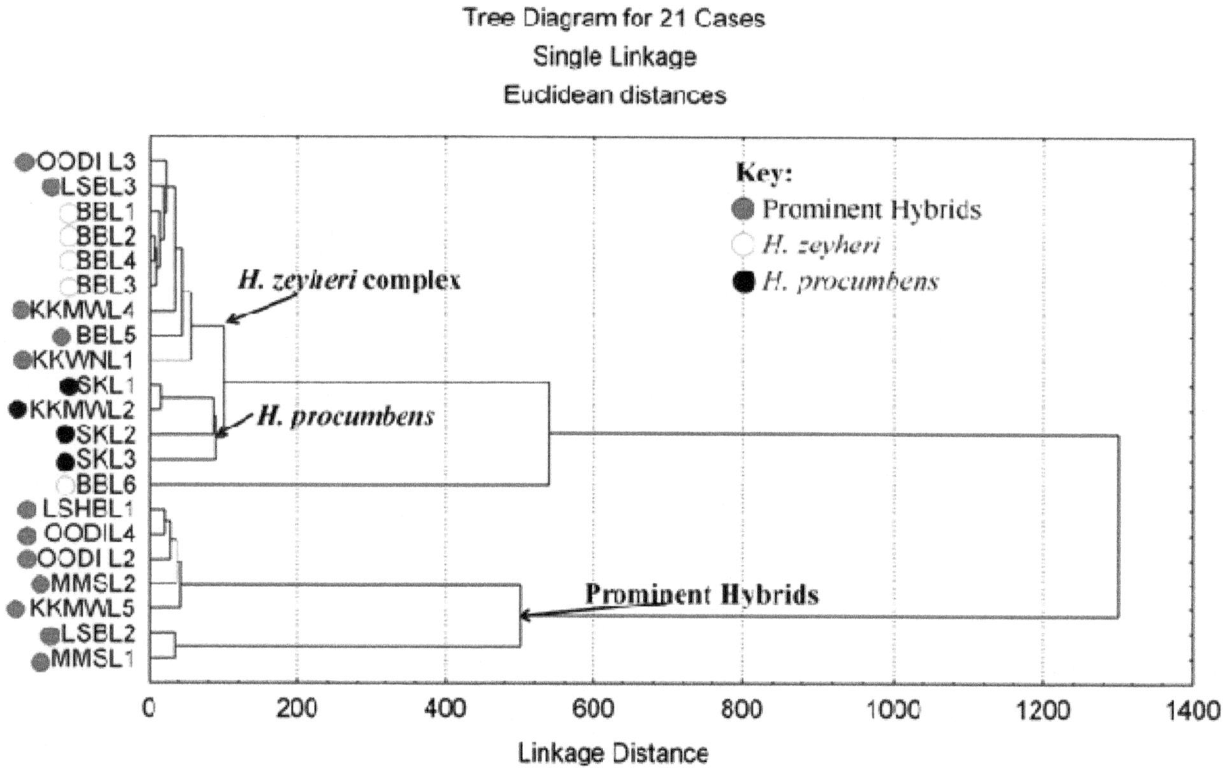

Figure 4. Tree diagram to show how the different *Harpagophytum* voucher specimens clustered.

of three groups in the study namely the *H. zeyheri* group, the *H. procumbens* group and the hybrids (Figure 1).

DISCUSSION

General morphology and number of seed rows

Number of seed rows

The three (3) seed rows in *Harpagophytum* have never been recorded in literature and are associated with speceis that we identified as hybrids. Therefore, we concluded that the three (3) seed rows are a direct evidence of introgression in *Harpagphytum*. From the literature (for example Hartmann and Ihlenfeldt, 1967; 1988) it is known that *H. zeyheri* is characterized by two seed rows on both of its seed loculus, while *H. procumbens* is characterized by four seed rows on both of its seed loculus. Hence, any deviation from that could be an indication of a new undescribed species or a hybridization event. The possibility of having a new species was ruled out because combinations in the number of seed rows per loculus have not yet stabilized (Table 2). A stabilized character is constant within different specimens of the same species and if the character was produced because of hybridization, its stabilization marks a speciation event (Walter et al., 1984).

Scatter plot and the hierachical tree

The hybrid complex (Figure 4), as identified by the hierachical tree, represents those hybrids with intermediate features (that is mid way blends). Such specimens are very difficult to classify and as such one can never tell whether the specimen is a *H. zeyheri* or a *H. procumbens*. All specimens with three seed loculus in at least one of the seed loculus fall into this category. The *H. zeyheri* complex contains those hybrids whose characters are biased towards the *H. zeyheri* species. On the other hand, Root 2 (Figure 2) in the scatter plot was well represented by specimen MMSL2, which looks like *H. procumbens* on the outside, but with the 3.2 seed row combination.

Box and whisker

The box and whisker analysis provided a quick guide to the identification of *Harpagophytum* and their hybrids. The pure lines fall within quantitative boundaries of their taxonomic descriptions as described by Ihlenfeldt and Hartmann (1970) and Ihlenfeldt (1988). And, from the results, we concluded the following to be some of the diagnostic features for the hybrids:

1) Fruits that are shorter (that is length) than those of *H. procumbens* but longer than those of *H. zeyheri*

2) Fruits that are generally wider (that is fruit width) than those of *H. procumbens* and *H. zeyheri*
3) Fruit arms are thinner (that is arm width) and shorter (that is arm length) than those of *H. procumbens* but wider and longer than those of *H. zeyheri.*

Conclusions

Morphometric analysis of fruits of *Harpagophytum* species revealed the possibility of gene flow between *H. procumbens* and *H. zeyheri.* But at this point it is difficult to determine the direction of gene flow and the frequency of hybridization. Diagnostic characters that distinguish pure lines from hybrids are the number of seed rows, fruit size (that is fruit length and fruit width) and fruit arms (that is arm width and arm length).

RECOMMENDATIONS

There is a need to do chromosomal and molecular work to get a full understanding of the *Harpagophytum* hybrid swam. The 5SDNA gene family, because it has multiple tandem repeats, should be used to determine the percentage of parental contribution by either species. Random Fragment length Polymorphism (RFLP) markers should be used to detail the frequency of occurrence and extent of introgression. In addition, cladistic analysis should be used to test for congruence since incongruence between the morphological and cpDNA data indicates the existence of localized introgression (Ferguson and Jansen, 2002).

REFERENCES

Anonym (1995). Statistica for windows volumes I and II, 2nd edition. Statsoft, U.S.A.

Arnold ML, Martin NH (2009). Adaptation by introgression. J. Biol., 8: 82.

Curtu AL, Gailing O, Finkeldey R (2007). Evidence for hybridization and introgression within a species – rich oak (Quercus spp.) community. *BMC.* Evol. Biol., 7: 1 -15.

Ellastrand CN, Prentice, HC, Hancock FJ (1999). Gene flow and introgression from domesticated plants into their wild relatives. Ann. Rev. Ecol. Syst., 30: 539 – 563.

Ferguson C, Jansen RK (2002). A Chloroplast DNA Phylogeny of Eastern *Phlox* (Polemoniaceae): Implications of Congruence and Incongruence with the *ITS* Phylogeny. Am. J. Bot., 89(8): 1324 – 1335.

Grant V (1971). Plant speciation. New York, NY, USA: Columbia University Press. *Harpagophytum procumbens* (Devil's Claw), Pedaliaceae. S. Afr. J. Bot. (In Press).

Ihlenfeldt HD (1988) Pedaliaceae. *F.Z* .8(3): 86 – 113.

Ihlenfeldt HD, Hartmann HEK (1970). *Harpagophytum. Mitt. Staatsinst. Allg.* Bot. Hamburg., 13: 15 – 69.

Jordaan A (2010). Seed coat development, anatomy and scanning electron microscopy of

Lanta V, Havrenek P, Ondrej V (2003). Morphometry Analysis and Seed Germination of *Amaranthus cruentus, A. retroflexux* and their Hybrid (*A.* X *turicensis*). Plant Soil Environ., 49 (8): 364–369.

Rangel PN, Brondani, RPV, Rangeland, PHN, Brondani C (2008). Agronomic and molecular characterization of introgression lines from the interspecific cross *Oryza sativa* (BG90-2) x *Oryza glumaepatula* (RS-16). Genet. Mol. Res., 7(1): 184-195.

Reiseberg LH, Brunsfeld SJ (1998). *Molecular Evidence and Plant Introgression.* In: Soltis DE, Solits PS Doyle JJ (eds). Molecular Systematics of Plants II: DNA Sequencing. Boston Kluwer, New York.

Walter GH, Hulley PE, Craig JFK (1984). Speciation, adaptation and interspecific competition. Oikos, 43: 246 -248.

6

The impact, uses, and ecological role of agrestals in two selected agroecosystems of Eastern India

Saurav Dwari and Amal Kumar Mondal*

Plant Taxonomy, Biosystematics and Molecular Taxonomy Laboratory, Department of Botany and Forestry, Vidyasagar University, Midnapore-721 102, West Bengal, India.

Agricultural fields contain some small plants (weeds) which are known as agrestals. Generally, weeds are perceived as unwanted intruders in agro-ecosystems that compete for resources, reduce yields, and force the use of human labor and technology to prevent crop losses; but not all weeds are undesirable. Some weeds act as valuable agro-ecosystem components. They serve as nutritious foods, and important sources of fodder and medicine. Certain weeds may limit insect damage to crops. These beneficial effects indicate that weeds are not just agricultural pests, but can also play beneficial roles in the human society. So, the main objective of the study is to know about the beneficial agrestals of the district and their roles. Detailed field surveys were made in different blocks of the district to collect data about agrestals and their usefulness. These fields contain various medicinal plants such as *Solanum nigrum* L., *Eclipta prostrata* L. *Oxalis corniculata* L., *Rungia pectinata* L. etc., edible plant *Amaranthus viridis* L., ornamental plant *Grangea maderaspatana* (L.) Poir, and aromatic plant *Trachyspermum copticum* L. It is clear from the study that these two fields of the district contain some valuable agrestals which are used by the locals.

Key words: Agrestals, agro-ecosystems, weeds, food, fodder, medicine, insects, society.

INTRODUCTION

The common definition of weeds is that they are plants in the wrong place (Harlan and Wet, 1965). Weeds are a permanent constraint to crop productivity in agriculture and they compete for nutrients, space, light, and exert a lot of harmful effects by reducing the quality, as well as quantity of the crop, if the weed populations are left un-controlled (Kavaliauskaite and Bobinas, 2006). Literally, agrestal means agricultural weed (Rao et al., 2008). Weeds are not always undesirable. In many systems, particularly in the tropics, weeds are useful elements in agricultural systems. Certain weeds have been shown to increase the profitability of a system (Showler and Reagan, 1991). However, not all weeds have a negative effect on the crop, and it is important to consider the individual system when assessing the impact and losses due to weeds in order to determine the ideal treatment (Dutta and Banerjee, 1978). At low densities, weeds often do not affect the yield (Altieri, 1988). The effect of the weed must be assessed. In many areas, weeds have both positive and negative effects. It is often possible to leave weeds with the crop for a certain period of time, the period threshold before using control methods (Dwari and Mondal, 2011).

*Corresponding author. E-mail: amalcaebotvu@gmail.com.

MATERIALS AND METHODS

Detailed field surveys of (*Momordica charantia* L. and *Cucurbita maxima* L.) were made in different places of Howrah District (Figure 1) following blocks during 2010 to 2011: Udaynarayanpur, Amta I and II, 22.34°N latitude and 88.0°E longitude; Bagnan I and II, 22.47°N latitude and 87.97°E longitude; Uluberia I and II, 22.47°N latitude and 88.11°E longitude; Bali-jagacha, 22.65°N latitude and 88.34°E longitude; Domjur, 22.64°N latitude and 88.22°E longitude; Panchla, 22.54°N latitude and 88.14°E longitude; Sankrail, 22.58°N latitude and 88.24°E longitude. The two fields of this district were selected (including all blocks) and thoroughly surveyed. The documentation of agrestals was done during April 2009 to August 2011. The study sites were divided into 2 major groups, that is, one

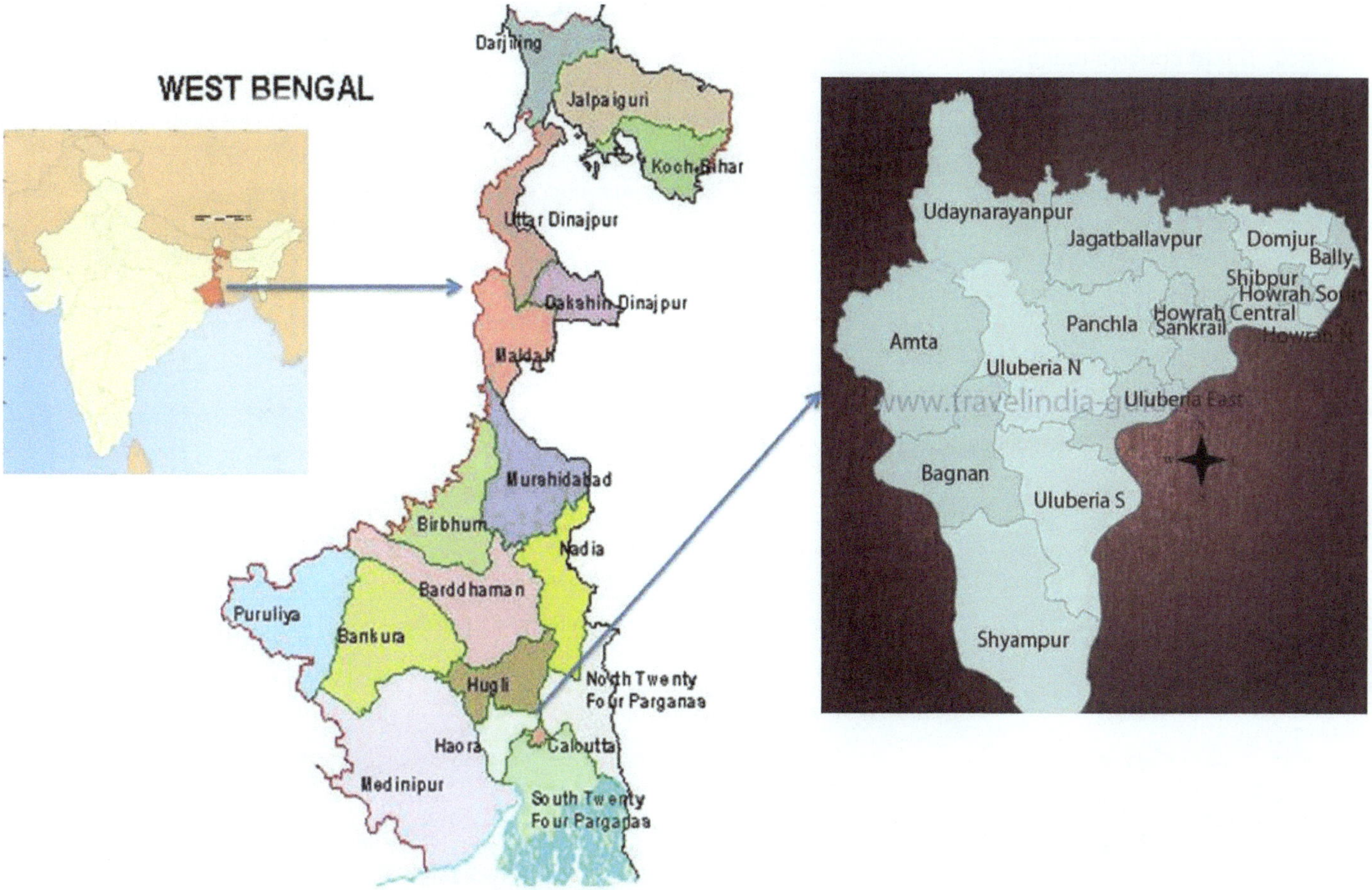

Figure 1. Howrah district, West Bengal of India (study area).

is agriculturally rich Damodar River associated group and another group was less agriculturally rich. Prior to start of the survey, some random sample areas were selected, which major portion of the area covered. Different data about the usefulness of agrestal were documented from field survey and resourceful persons (farmers and land owners). Collected voucher specimens were deposited in the herbarium of Department of the Botany and Forestry, Vidyasagar University. Weeds were identified with the help of herbarium specimen of different weeds with theS help of the experts.

RESULTS

At the end of the detailed study throughout the whole district, it was found that these two fields of this district contain many useful weeds, which can be used directly for human welfare bio-resource or is already used by local peoples. Weeds can be used directly as medicinal, food, fodder, ornamental and other human uses. Out of 23 sp. of agrestals collected from the fields, 16 species possess economic importance in various ways (Table 1). The different uses of these agrestals are medicinal, edible, fodder, ornamental and aromatic which are enumerated as follows:

Medicinal plants

Dicots

1) *Oxalis corniculata* L. (Figure 2)
Local name: Amrul
Family: Oxalidaceae
Medicinal uses: Leaves- Used in piles, anaemia, opacities of the cornea and warts (Chetty et al., 2011).

2) *Solanum nigram* L. (Figure 3)
Local name: Kakmachi
Family: Solanaceae
Medicinal uses: Whole plants used in asthma, bronchitis, diarrhea, heart and kidney disease. Root used in urinary troubles. Leaf is used as a laxative (Joy et al., 1998).

3) *Boerhavia diffusa* L. (Figure 4)
Local name: Punarnova
Family: Nyctaginaceae
Medicinal uses: Whole plants used in anaemia, constipation, arthritis, spurse, diabetes and jaundice (Chetty et al., 2011).

Table 1. Collected and reported plant species of the study area.

No.	Plant species name	Local name	Family	Importance	Occurrence
1	*Oxalis corniculata* L.	Amrul	Oxalidaceae	Medicinal and Edible	CMF
2	*Solanum nigram* L.	Kakmachi	Solanaceae	Medicinal	CMF
3	*Boerhavia diffusa* L.	Punarnova	Nyctaginaceae	Medicinal	CMF
4	*Rungia pectinata* (L.) Nees.	Chotosibjata	Acanthaceae	Medicinal	MCF
5	*Eclipta prostrata* L.	Keshute	Asteraceae	Medicinal	CMF
6	*Grangea maderaspatana* (L.) Poir.	Namuti	Asteraceae	Medicinal and ornamental	MCF
7	*Gnaphalium purpureum* L.	Kalpahibon	Asteraceae	Edible and ornamental	CMF
8	*Amaranthus viridis* L.	Ban notey	Amaranthaceae	Edible	CMF
9	*Alternanthera philoxeroides* Griseb.	Danta	Amaranthaceae	Fodder	BOF
10	*Euphorbia hirta* L.	Bara dudhe	Euphorbiaceae	Medicinal	CMF
11	*Croton bonplandianum* Baill.	Churchuri	Euphorbiaceae	Medicinal	MCF
12	*Portulaca grandiflora* Hook.	Bara lonia	Portulaceae	Ornamental	MCF
13	*Cleome viscosa* L.	Hurhuria	Capparidaceae	Ornamental	MCF
14	*Trachyspermum copticum* L.	Ban Ajwan	Apiaceae	Aromatic	MCF
15	*Dactylotenium aegypticum* (L.) Beauv.	Makra	Poaceae	Medicinal	MCF
16	*Cynodon dactylon* (L.) Pers.	*Durba*	Poaceae	Medicinal and fodder	BOF

CMF: *Cucurbita maxima* L. field, MCF: *Momordica charantia* L. field and BOF: Present in both field.

Figure 2. Flowering twig of *Oxalis corniculata L.* (Oxalidaceae).

Figure 3. Twig of *Solanum nigram* L. (Solanaceae) with fruits.

4) *Rungia pectinata* (L) Nees. (Figure 5)
Local name: Chotosibjata
Family: Acanthaceae
Medicinal uses: Leaves - Used to cooling, aperient, given to children suffering to small pox; roots - Used to febrifuge (Paria, 2005).

5) *Eclipta prostrata* (L.) (Figure 6)
Local name: Keshute
Family: Asteraceae
Medicinal uses: Leaves - used in jaundice, fever and help to promote hair growth (Joy et al., 1998).

Figure 4. Flowering twig of *Boerhavia repens* L. (Nyctaginaceae).

Figure 5. Patch of *Rungia pectinata* (L) Nees. (Acanthaceae).

Figure 6. Vegetative of *Eclipta prostata* L. (Asteraceae).

Figure 7. Flowering twig of *Grangea maderaspatana* (L) Poir. (Asteraceae).

6) *Grangea maderaspatana* (L.) Poir. (Figure 7)
Local name: Namuti
Family: Asteraceae
Medicinal uses: Leaves: Infusion of leaves considered stomachic, antispasmodic and used in irregular menstruation (Chetty et al., 2011).

7) *Euphorbia hirta* L. (Figure 8)
Local name: Bara dudhe
Family: Euphorbiaceae
Medicinal uses: Plant juice used in dysentery; Milk used in removing warts (Paria, 2005).

8) *Croton bonplandianum* Baill. (Figure 9)
Local name: Lankeswari
Family: Euphorbiaceae
Medicinal uses: Leaves - Leaf paste is applied for skin

Figure 8. Flowering twig of *Euphorbia hirta* L. (Euphorbiaceae).

Figure 9. Vegetative twig of *Croton bonplandianum* Baill. (Euphorbiaceae).

diseases and wounds and also used to stop bleeding (Bapuji and Ratnam, 2009).

Monocots

1) *Dactylotenium aegyptium* (L) Beauv. (Figure 10)
Local name: Makra
Family: Poaceae
Medicinal uses: Parched grains are administered to women who suffer from stomach ache after child birth (Paria, 2005).

2) *Cynodon dactylon* (L.) Pers. (Figure 11)
Local name: Durba
Family: Poaceae
Medicinal uses: Plant juice is astringent, diuretic, used in dropsy; hysteria, epilepsia etc; Root-decoction is diuretic (Paria, 2005).

Edible plants

1) *Amaranthus viridis* L. (Figure 12)
Local name: Ban notey
Family: Amaranthaceae
Uses: Whole plants without root used as vegetable by local peoples.

2) *Gnaphalium purpureum* L. (Figure 13)
Family: Asteraceae
Uses: Young leaves cooked by local peoples.

3) *Oxalis corniculata* L.
Local name: Amrul
Family: Oxalidaceae
Uses: Whole plants without root used after boiled by local peoples.

Fodder plants

1) *Cynodon dactylon* (L.) Pers. (Patil and Janagoudar, 1993)
Local name: Durba
Family: Poaceae

2) *Alternanthera philoxeroides* Griseb. (Figure 14)
Local name: Danta
Family: Amaranthaceae

Ornamental plants

1) *Portulaca grandiflora* Hook. (Figure 15)
Family: Portulacaceae

2) *Cleome viscosa* L. (Figure 16)
Family: Capparidaceae

3) *Grangea maderaspatana* (L.) Poir.
Family: Asteraceae

4) *Gnaphalium purpureum* L.
Family: Asteraceae

Figure 10. Flowering twig of *Dactylotaneum aegypticum* (L) Beauv. (Poaceae).

Figure 11. Vegetative of *Cynodon Dactylon* (L) Pers. (Poaceae).

Figure 12. Flowering twig of *Amaranthus viridis* L. (Amaranthaceae).

Figure 13. Flowering twig of *Gnaphalium purpureum* L. (Asteraceae).

Aromatic plants

1) *Trachyspermum copticum* L. (Figure 17)

Family: Apiaceae
The strong aroma is enhanced by toasting or frying and goes well with potatoes or fish. Legumes (lentils, beans)

Figure 14. Patches of *Alternanthera philoxeroides* Griseb. (Amaranthaceae).

Figure 15. Flowering patches of *Portulaca grandiflora* Hook. (Portulacaceae).

Figure 16. Flowering twig of *Cleome viscosa* L. (Capparidaceae).

Figure 17. Flowering twig of *Trachyspermum copticum* L. (Apiaceae).

are, however, the most important field of application. In India, where these vegetables are popular, since they provide a source of protein to the many vegetarians, they are commonly flavored with a perfumed butter frequently containing ajwain. It has a characteristic aromatic smell and pungent taste, and is widely used as a spice in

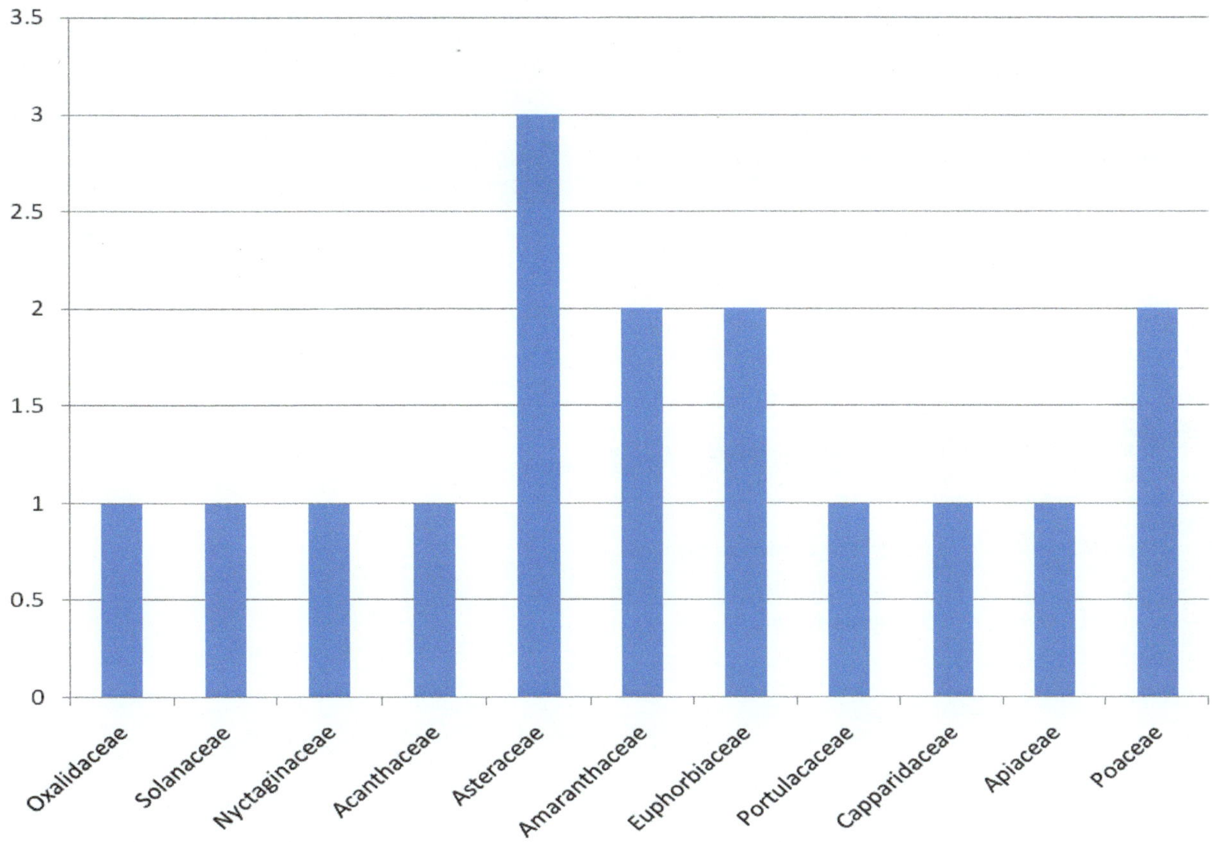

Figure 18. Number of weed species in families.

curries. Its seeds are used in small quantities for flavouring numerous foods, as preservatives (Kaur and Arora, 2010).

ACKNOWLEDGEMENT

I want to convey my gratitude to the honorable guide Dr. Amal Kumar Mondal, Associate professor of Botany, Department of Botany and Forestry, Vidyasagar University, West Bengal, India and Dr. Sanjukta Mondal (Parui), Associate Professor, WBES, Department of Zoology, Lady Brabourne College, Kolkata-700017, West Bengal, India, for their consistence help during my study period. I would like to express my thanks to all senior, junior research scholars and friends for their continuous help. I am also thankful to my parents for their support me constantly.

DISCUSSION

The whole survey thus reveals that these two fields contain many useful agrestals, which can be used as bio-resource for human welfare and some are already used by local peoples. The present study is the first that documented beneficial agrestals of these fields from this district. In West Bengal, one work by Dutta and Banerjee in 1978 was done from Hooghly and undivided Midnapore district on rice field, and they documented 124 economic important plants (Dutta and Banerjee, 1978). Local drug collectors of this district can use these fields for collection of various medicinal (10 species) plants. People can collect these edible (3 species), fodder (2 species) and ornamental plants (4 species) for their preliminary uses. Documented plants are represented in Table 1 to understand the overall structure and function of this study area at a glance and also to get an accurate and precise information of the reported plant species. Among three food plants of these fields, uses of *A. viridis* L. and *O. corniculata* L. are very high in that district. In the fields of this district, the families Asteraceae and Poaceae are the dominant families in dicots and monocots, respectively (Figure 18). In this district, it was found that people always try to make some new ornamental plants from wild plants. *G. maderaspatana* (L.) Poir. and *G. purpureum* L. are two wild plants used as ornamental plants in this district.

T. copticum L. is a very important aromatic plant also found in the *Momordica charantia* L. field. So from this study, it is clear that in low density these weeds are valuable and up to a certain level there is no requirement to remove these weeds for ecological sustainability.

REFERENCES

Altieri MA (1988). The impact, uses, and ecological role of weeds in agroecosystems. In: Altieri MA and Liebman M (ed.), Weed Management in Agroecosystems: Ecological Approaches. CRC Press, Boca Raton, Florida, United States. pp.1-6.

Bapuji JL, Ratnam SV (2009). Traditional uses of some medicinal plants by tribals of Gangaraju Madugula Mandal of Visakhapatnam district, Andhra Pradesh. Ethnobotanical Leaflets 13: 388-98.

Chetty KM, Sivaji K, Rao KT (2011). Flowering plants of Chittor district Andhra Pradesh, India. Students offset printers, Tirupati pp. 54:175-285.

Dutta SC, Banerjee AK (1978). Useful weeds of West Bengal rice fields. Econ. Bot. 32:297-310.

Dwari S, Mondal AK (2011). Studies on agrestal diversity in the sugarcane field of Howrah district, West Bengal, India: use as an important bioresource for human welfare. Int. J. Biodiver. Conserv. 3(13):686-704.

Harlan JR, de Wet JMJ (1965). Some thoughts about weeds. Econ. Bot. 145(1):16-24.

Joy PP, Thomas J, Mathew S, Skaria BP (1998). Medicinal plants. Kerala Agricultural University, Kerala, India pp.10-189.

Kaur GJ, Arora DS (2010). Bioactive potential of *Anethum graveolens*, *Foeniculum vulgare* and *Trachyspermum ammi* belonging to the family Umbelliferae - Current status. J. Med. Plants. Res. 4(2):087-094.

Kavaliauskaite D, Bobinas C (2006). Determination of weed competition critical period in red beet. Agron. Res. 4: 217-220.

Paria DN (2005). Medicinal plants resources of South West Bengal. Directorate of Forests, Government of West Bengal, Kolkata pp.1-198.

Patil VC, Janagoudar BS (1993). Chemical control of *Cynodon dactylon* (L) Pers. Volume 3. Proc. Int. Symp. Indian Society Weed Sci. Hissar 20:238-239.

Rao RP, Lakshimaiah A, Rajeswaramma PM, Reddy MS, Sadasivaiah B, Basha SK (2008). Agrestal of Southern Andrapradesh region. J. Econ. Taxon. Bot. 32(3):694-709.

Showler AT, Reagan TE (1991). Effects of sugarcane borer, weed, and nematode control strategies in Louisiana sugarcane. Environ. Entomol. 20:358-370.

Heritability of pre-weaning growth performance traits in Mengali sheep in (Balochistan) Pakistan

Muhammad Masood Tariq[2], Masroor Ahmad Bajwa[2], Ferhat Abbas[2], Abdul Waheed[1]*, Farhat Abbas Bokhari[2] and Majed Rafiq[2]

[1]Department of Animal Breeding and Genetics, University of Agriculture, Faisalabad, Pakistan.
[2]Centre for Advanced studies in Vaccinology and Biotechnology University of Balochistan, Quetta, Pakistan.

Genetic parameters were estimated for weights at birth (BW), 30 days (MW), 120 days weaning weight (WW), and pre-weaning average daily gain (PRADG) of four flocks of Mengali sheep maintained at the Experimental Station CASVAB, Quetta, (ESC), Mastung, Noshki and Quetta over a period of 5 years from 2005 to 2009. Records on 2750 lambs descended from 581 ewes and 56 rams were included in the analysis. Variance components were estimated fitting animal model using restricted maximum likelihood (REML) procedure. Genetic parameters were computed by pos-processing of the variance components. The heritability estimates for BW, MW, WW, and PRADG were 0.39 ± 0.06; 0.125 ± 0.02; 0.177 ± 0.03 and 0.23 ± 0.05, respectively. BW was highly heritable while other growth traits were found moderately heritable, showing larger proportional of environmental variances. In general, heritability estimates were moderate in early growth traits of Mengali sheep. Hence it was suggested that improvement can be achieved by mass selection.

Key words: Genetic parameters, Mengali sheep, heritability estimates.

INTRODUCTION

Mengali sheep is an important sheep breed of Balochistan which is black or brown with white patches on the body (Khan et al., 2007; Kakar and Ahmad, 2004) but poorly documented and not recognized in the census up till 2006. The animals are well adapted to the local conditions of most of the Districts of Quetta, Khuzdar, Chaghi, Kalat, Mastung, Awaran and Kharan of Balochistan Province. Source of origin of Mengali sheep is still unknown. This sheep breed is mostly raised by native Baloch tribe "Mengal" (main tribe of Chaghi area). The breed is therefore known as Mengali of the region for adoptability and profitability. Since animals of this breed have been isolated from other sheep breeds, they were expected to have developed some distinctive characteristics. Information on this breed is scanty and its current population status is not known. No evaluation or improvement program has been undertaken for this breed before. Sheep raising in Pakistan (especially in Balochistan) and neighboring countries is mostly kept by local pastoralists on extensive production system. In such a system, output is lower than in an intensive system. Within local area natural selection of breeds is an appropriate method for genetic improvement in the traditional low input production systems of small ruminants (Khan et al., 2007; Akhtar et al., 2008; Ali, 2008).

Performance traits of farm animals are determined not only by an animal's genetic potential for growth but also by maternal genetic and permanent and temporary environmental effects. Hence, to achieve optimum genetic progress in a selection program, both direct and maternal genetic components should be taken into account, especially if there is not an antagonistic relationship between them (Meyer, 1992; Snowder and Van Veilk, 2003)

Numerous studies have been conducted to estimate the heritability of performance traits of different sheep breeds reported by several researchers; Yazdi et al. (1997) in Baluchi sheep found low estimates for BW. Shaat et al. (2004) who studied the genetic trends for

*Corresponding author. E-mail: tariqkianiraja@yahoo.com.

Lamb weights in flocks of Egyptian Rahmani and Ossimi sheep observed moderate estimates for BW. Refik et al. (2009) who studied the genetic and non-genetic influences on the body weights of Turkish Merino were evaluated using REML reported low estimates for BW however, moderate for WW. Borg et al. (2009) also analyzed the data of Targhee flock used to estimate genetic and environmental relationships between lamb birth weight and adult ewe weight, condition score, and prolificacy and estimates for BW and WW found were low. In Pakistan, few workers have also done their research on the same lines: Qureshi et al. (1992) did in Kajli; Babar (1994) in Lohi. Akhtar (1996) in Hissardale sheep used paternal half sib correlation method (PHSC) while, Hussain (2006) in Thalli sheep had used restricted maximum likelihood (REML) to estimate the heritability estimates. Especially in Balochistan Province, 48% of sheep populations of Pakistan exist but scarce research has been carried out to explore their real genetic potentials. Only some efforts have been done by Sharif (2001) in Balochi and Bibrik sheep breeds, and Ali (2008) on some of the performance traits of Karakul sheep. Such heritability estimates of performance traits are not yet available for Mengali sheep. Therefore the present study was conducted in order to estimate heritability of some performance traits (birth weight to pre-weaning weight) of economic importance in Mengali sheep breed by using multivariate animal models.

MATERIALS AND METHODS

The data used in this study were scored from 2005 to 2009 on 2750 lambs obtained from 581 dams and 56 sires of Mengali sheep. Mengali flocks were kept in four stations at three different locations (Experimental Station CASVAB, UoB, Quetta (ESC), Killi Hassni, Quetta; Khadkucha, Mastung and Peer Wala, Noshki). The traits included in the present study were weights at birth (BW), 30 days (MW), 120 days, Weaning weight (WW), and Pre-weaning average daily gain (PRADG). Animal ID, sire ID, dam ID and lamb IDs were checked. Pedigree and other pertaining information were obtained from the record registers maintained at these farms. Data were checked for inconsistencies.

Animal model was fitted using REML procedure. ASREML Version 2.0 (Gilmour et al., 2007) was used for computing genetic parameters like heritability, genetic and phenotypic correlations. The statistical model fitted for estimation of heritability was as follows:

$$Y_{ijklm} = \mu + H_i + P_j + A_k + a_{ijkl} + e_{ijkl}$$

Where

μ = overall mean

H_i = effect of ith flock (1-4)

P_j = effect of jth parity group (1-2nd and later)

A_k = age of dam

a_{ijkl} = random animal effect

e_{ijkl} = random error associated with each observation, assumed as normally distributed with mean zero and variance σ_e^2

For computing genetic parameters, an analysis of variance was performed prior to fitting the statistical model. Only significant fixed effects were included in the final model.

RESULTS AND DISCUSSION

Overall means at birth weight for Mengali lambs (n = 2750) were observed in male (3.61 ± 0.36), female (3.48 ± 0.37) and combined sexes (3.54 ± 0.36 kg). The lambs born during spring were heavier (3.68 ± 0.35 kg) than the lambs born during autumn season (3.49 ± 0.33 kg). Similarly, single born lambs were (3.72 ± 0.29 kg) heavier than twin born lambs (3.45 ± 0.34 kg). Results also revealed that male lambs were heavier than female. Average for WW was 20.33 ± 3.37; 18.04 ± 3.35 and 19.16 ± 3.63 for male, female and combined sexes respectively of Mengali lambs. The data also revealed that the spring born lambs weighted at WW were heavier (19.95 ± 3.39 kg) as compared to autumn born lambs (18.54 ± 3.30 kg). Similarly, single born lambs were (20.15 ± 3.10 kg) heavier than twin born lambs (18.45 ± 3.00 kg).

The findings of the effect of sex and SOB on body weights obtained in present study are in agreement with that observed by Refiq et al. (2009) and Mokhtari and Rashidi (2010) who reported that SOB (season of birth) and sex had significant effect on BW and WW. Birth weight of lambs born in spring season was significantly heavier than autumn born lambs. However, non significant effect of SOB on birth weight was observed by Naikare et al. (1987). Deccani and Hussain (2006) findings in Thalli lambs were contrary to the present study.

The analysis of variance for the birth weight of Mengali lambs is presented in Table 1. Results have shown that birth weight of lambs varied significantly due to location, year of birth (YOB), SOB, TOB type of birth) and sex ($P < 0.05$). However, the difference in birth weight of lambs due to YOB*SOB and TOB*sex was non- significant (Table1).

The results of WW revealed that the weight at weaning varied significantly due to location, YOB, SOB, TOB, and sex ($P < 0.05$) but the differences in WW of lambs due to YOB*SOB and TOB*sex were non- significant. The single born and male lambs were heavier than the twin female lambs groups (Table 2). The significant effect of YOB, TOB, SOB and sex on BW and WW of lambs as obtained in the present study was in line with the findings of many researchers (Babar et al., 1994; Akhtar et al., 1996; Shah and Khan, 2004; Hussain, 2006; Refiq et al., 2009 and Mokhtari and Rashidi, 2010).

Heritability estimates and their standard errors are shown in Table 3. Heritability estimates were 0.39±0.06 (BW), 0.125 ± 0.02 (MW), 0.177 ± 0.03 (WW), and (PRADG), 0.23 ± 0.05. Generally, estimates of heritability were low to moderate. MW and BW had the lowest (0.125 ± 0.02) and highest (0.39 ± 0.06) estimates, respectively. Heritability estimates for BW were similar to the estimates reported by Kaul and Tomar (1982)

Table 1. Analysis of variance for birth weight (BW) in Mengali sheep.

Source of variation	d.f.	Mean squares	F	Significant (Sig.)
Location of flock	3	2.698	38.144	0.000
YOB	4	0.434	6.133	0.000
SOB	1	0.899	12.716	0.000
TOB	1	143.369	2026.869	0.000
Sex	1	3.497	49.432	0.000
YOB * SOB	4	3.162	0.447	0.775
TOB * Sex	1	0.158	2.228	0.136
Error	2363	7.073		
Total	2379	2.698		

(Sig.) = Significant ($P < 0.05$).

Table 2. Analysis of variance for 120 days (Weaning Weight) (WW).

Source of variation	d.f.	Mean squares	F	(Sig.)
Location	3	353.488	78.856	0.000
YOB	4	26.510	5.914	0.000
SOB	1	52.204	11.646	0.001
TOB	1	2130.786	475.333	0.000
SEX	1	1198.966	267.463	0.000
YEAR * SOB	4	0.612	0.137	0.969
TOB * SEX	1	0.759	0.169	0.681
Error	2362	4.483		
Total	2378			

(Sig.) = Significant ($P < 0.05$).

Table 3. Heritability estimates ±SE of various performance traits of Mengali sheep breed.

Trait	Heritability estimate ±SE[1]	Trend of estimate#
BW	0.39 ± 0.06	High
MW	0.125 ± 0.02	Low
WW	0.177 ± 0.03	Low
PRADG	0.23 ± 0.05	Moderate

[1]SE = Standard error; # = according to Refik et al. (2009).

in Corriedale sheep (0.20 ± 0.10), Chaudhry and Shah (1985) in Awassi (0.34), Bathaei (1994) in Mehraban Iranian (0.35), Snyman et al. (1997) in Afrino (0.22), Jara et al. (1998) in Corriedale lambs (0.32), Mcmanus and Miranda (1998) in Bergamasca sheep of Brazil (0.32) and in Bannur (0.29 ± 0.20), Pollott et al. (1998) in Turkish Awassi lambs (0.44 ± 0.09), Al-Shorepy (2001) in crossbred lambs (0.32) and Ali (2008) in Karakul (0.30).

The estimates of the present study were lower than those obtained by many researchers: Chaudhry and Shah (1985) in Hissardale (0.94). They were lower than those of Mahmoud (2000) in Barki (0.81), Rahmani (0.66) and Ebangi et al. (0.61) in Fulbe sheep (2001).

The estimates were higher than those reported by Yazdi et al. (1997) in Baluchi (0.14), Cloete et al. (1998) in Dohne Merino (0.04), Hussain (2006) in Thalli (0.07 ± 0.02), Refik et al. (2009) in Turkish Merino (0.14 ± 0.02), and Borg (2009) in Targhee (0.19).

Heritability estimates for WW were in agreement with the findings of many workers: Alipour and Edriss (1997) in Fars Karakul (0.11), Risam et al. (1999) in Corriedale (0.13 ± 0.04), Bedhiaf et al. (2000) in Barbary (0.001), Qureshi and Ghaffar (2002) in Kajli (0.13 ± 0.02), Babar et al. (2003) in Lohi sheep (0.147 ± 0.034), Shahroudi et al. (2002) in Kermani (0.19 ± 0.08), Simm et al. (2002) in Suffolk (0.17),

Hussain (2006) in Thalli (0.04 ± 0.11) and Borg (2009) in Targhee (0.12).

Estimates of the present study were lower than those of Chaudhry and Shah (1985) in Awassi (0.82), Kacchi (0.52), Lohi (0.55) and Hissardale (0.89), Ali (2008) in Karakul (0.30), and Refik, (2009) in Turkish Merino (0.29 ± 001). Heritability estimates for PRADG were similar to the estimates documented by Dikshit (1982) in Bannur (0.36 ± 0.23), Alrawi et al. (1982) in Awassi (0.29), Mahmoud (2000) in Barki (0.37) and in Rahmani (0.21), Shah and Khan, (2004) in Lohi (0.17 ± 0.05) and Refik et al. (2009) in Turkish Merino (0.29 ± 0.02). The present estimates were higher than those of the study reported by Bohra et al. (1980) in Malpura (0.06), Atkins (1984) in Hill (0.06 ± 0.03). Alipour and Edriss (1997) in Fars Karakul (0.068), Kominakis et al. (1999) in Boutsico (0.05), Gut et al. (2001) in White-headed Mutton (0.0042) and Hussain (2006) in Thalli (0.076 ± 0.013).

Estimates were lower than those of the study documented by Raman et al. (1981) Stavropol Merino (0.68 ± 0.31), Shrestha et al. (1985) in Suffolk (0.47 ± 0.04), Shrestha et al. (1986) in Dorset (0.50 ± 0.04) and Lopez et al. (1991) in Ile-de-France (0.61).

The moderate estimates of heritability obtained in the present study suggested that most of the observed variation in pre-weaning daily gain in this flock was due to environment factors.

Conclusion

Most of the traits were low to moderate heritable except BW that was high. The variation in the findings of present study from other researchers might be due to data set, breed, managemental, feeding, environmental conditions in which the various flocks were raised and due to method of estimation. Larger proportion of environmental variation was responsible for low heritabilities. Conclusively, it is suggested that mass selection based on current breeding practices in Balochistan Province can be an effective way to achieve higher productive performance. This information would be helpful in developing future breeding strategy and for the conservation of genetic resource.

REFERENCES

Akhtar P (1996). Genetic and phenotypic parameters of some performance traits of Hissardale sheep in Pakistan. Ph D Dissertation. Department of Animal Breeding and Genetics, Univ. Agric. Faisalabad, Pakistan.

Akhtar P, Ali S, Hussain A, Mirza MA, Mustafa MI, Sultan JI (2008). Heritability estimates of post-weaning performance traits in Hissardale sheep in Pakistan. Turk. J. Vet. Sci., 32(4): 275-279.

Ali M (2008). Effect of genetic and environment factors on performance traits of Karakul sheep. M Sc. Thesis. Department of Livestock Management, Sindh Agriculture University, Tandojam.

Alipour S, Edriss MA (1997). Estimation of genetic and environmental factors affecting some of the economical traits in Fars Karakul (Gray) breed of sheep. J. Agric. Sciences - Islamic Azad Univ. 3(I 1/12): 69-77.

Alrawi AA, Badawi FS, Said SI, Farog MS (1982). Genetic and Phenotypic parameters estimate for growth traits in Awassi sheep. Indian J. Anim. Sci.

Al-Shorepy SA (2001). Genetic parameters for growth traits of a local breed of sheep in the United Arab Emirates. J. Agric. Sci., 137(3): 365-371.

Atkins KD (1984). Genetic parameters of body weight in Hill sheep. Presented in winter meeting of British Society of Anim. Prod. 26-28 March, (Anim. parameters for Breed. Abst., 52: 5269.

Babar ME (1994). Genetic and Phenotypic parameters of some performance characteristics of Lohi Sheep. Ph D Dissertation. Deptt. of Animal Breeding and Genetics, Univ. Agric., Faisalabad, Pakistan.

Babar ME, Ahmad Z, Hussain A (2003). Heritability estimates of birth and weaning weight in Lohi sheep. Pak. J. Agric. Sci., 40(314): 214-216.

Bathaei SS (1994). Genetic parameters for pre-weaning growth traits of Mehraban Iranian fat-tailed sheep. World Rev. Anim. Prod. 29(1): 4, 7, 10, 61-65.

Bedhiaf S, Bouix J, Clement V, Bibe B, Francois D (2000). Choice importance of analysis model for the estimation of genetic parameters of growth of meat sheep in Tunisia. 7emes Rencontres autour des Recherches sur lcs Ruminants. Paris, France, 6-7 December, pp. 169-172.

Bohra SDJ, Nagarcenkar R, Sharma KNS (1980). Genetic and phenotypic preweaning body weights and gains in Malpura sheep. Indian Vet. J., 57(2): 126-129.

Borg RC, Notter DR, Kott RW (2009). Phenotypic and genetic associations between lamb growth traits and adult ewe body weights in western range sheep. J. Anim. Sci., 87: 3506-3514.

Chaudhry MZ, Shah SK (1985). Heritability and correlation of birth weight, weaning weight and 12 months weight in Lohi, Awassi, Hissardale and Kacchi sheep. Pak. Vet. J., 5(2): 67-71.

Cloete SWP, Scholtz AJ, Aucamp BB (1998 b). Environmental effects, heritability estimates and genetic trends in a Western Cape Dohne Merino nucleus flock. South African J. Anim. Sci., 28(3/4): 185-195.

Dikshit RK (1982). Genetic and non-genetic studies on pre-weaning traits in Bannur lambs. Thesis Abst. Haryana Agric. Univ., 8(4): 334.

Ebangi AL, Njoya A, Ngo-Tama AC, Awa DN, Mbah DA (2001). Genetic and phenotypic parameters of birth weight traits in Fulbe sheep in Cameroon. Revue. d'Elevage et de Medecine Veterinaire des Pays Tropicaux, 54(2): 147-151.

Gilmour AR, Gogel BJ, Cullis BR, Welham SJ, Thompson R (2007). Asreml user guide Release 2.0. VSN International Ltd. 5 The Water's House, Water House Street Hemel, Hempstead, HP1 1 ES, UK.

Gut A, Slosarz P, Szwaczkowski T (2001). Genetic trends of some performance traits in White-headed Mutton sheep. Czech J. Anim. Sci., 46(8): 363-369, (CAB Abst. 1212-1819).

Hussain A (2006). Genetic evaluation of Thalli sheep in Pakistan. Ph D Dissertation. Department of Animal Breeding and Genetics, Univ. Agric. Faisalabad, Pakistan.

Jara A, Montaldo H, Barria N (1998). Direct and maternal genetic effects for birth, weaning and 14-month weights of Corriedale breed in Magallanes. Proc. 6th World Cong, Genetics Applied to Livestock, Armidale, Australia; 11-16 January, 24: 181-184.

Kakar MA, Ahmed M (2004). An over-view of livestock in Balochistan. Livestock and Dairy Development Department Balochistan, Quetta.

Kaul KL, Tomar NS (1982). Genetic studies on Russian Merino and Stavropol sheep in India. I. Birth weight and weaning weight. Indian Vet., J. 59(11): 874-877. (Anim. Breed. Abst., 51: 3672, 1983).

Khan SA, Khan MA, Khan SA, Mehmoud S (2007). Genetic

resources and diversity in Pakistani sheep. Inter. J. Agric. Biol. 1560-5830/2007/09-6-941-944.
Kominakis A, Rogdakis E, Koutsotolis K, Mantzios A (1999). Genetic parameters of various growth traits of Boutsico ewe lambs. Pitheo and No 175; re & 26: 67-85. (CAB Abst. 1105-2651).
Lopez DT, Albardonezo GD, Rico M, Espejo D, Mateos I (1991). Genetic parameters of daily gain and conformation score in early maturing sheep breeds. Parameters Genetics de la ganacia diania de peso y la conformancion en ragzas ovinas precoces, 87A: 242-247.
Mahmoud A (2000). Heritability, genetic and phenotypic correlations of preweaning growth traits of Rahmany and Barki lambs. Alexandria J. Agric. Res., 45: 1.
McManus CM, Miranda RM (1998). Genetic analysis of production characteristics of grade Bergamasca sheep reared in the Brasilia region. Proc. 6`h World Cong., Genetics Applied to Livestock, Armidale, Australia; 1 1-16 January, 24: 226-230.
Meyer K (1992). Variance components due to direct and maternal effects for growth traits of Australian beef cattle. Livestock Prod. Sci., 31: 179-204.
Mokhtari MS, Rashidi A (2010). Genetic trend estimation for body weights of Kermani sheep at different ages using multivariate animal models. Small Rumin. Res., 88(1): 23-26.
Naikare BD, Mali SL, Jagtap DZ, Belhe ND (1987). Non-genetic factors affecting birth weight of Deccani, Merino x Deccani and Dorset x Deccani half-bred lambs. Indian J. Anim. Sci., 54(6): 587-589.
Pollott GE, Gursoy O, Kirk K (1998). The Genetics of meat and milk production in Turkish Awassi sheep. Proc. 6th World Cong., Genetics Applied to Livestock, Armidale, Australia; 11-16 January, 24: 177-180.
Qureshi MA, Ghaffar A (2002). Performance of Kajli sheep in Pakistan: 2. Influence of environment on reproduction traits. Proceed. 7th World Congress on Genetics Applied to Livestock Prod., Montpellier, France, August Session 8, pp. 1-4.
Qureshi MA, Khan MA, Din GM, Rehman A (1992). Genetic studies on Awassi sheep in Pakistan. 2. Phenotypic, genetic and environmental correlations among birth weight and some productive traits. Gomal Univ. J. Res., 11(1): 79-88.
Raman KS, Mani V, Arumugham MP (1981). Production performance of Stavropol Merinos; weaning weight. Indian Vet. J., 58(11): 893.
Refik A, Ceyhan A, Ozder M, Sezenler T (2009). Genetic and Non-Genetic Parameter Estimates for Growth Traits in Turkish Merino Lambs. J. Anim. Vet. Adv., 8(9): 1729-1734.
Risam KS, Mir MY, Kirmani MA (1999). Genetic parameters of birth and weaning weights in a closed flock of Corriedale sheep. Indian J. Small Rumin., 5(1): 1-3.
Shaat I, Galal S, Mansour H (2004). Genetic trends for lamb weights in flocks of Egyptian Rahmani and Ossimi sheep. Small Rumin. Res., 51(1): 23-28.
Shah MH, Khan FU (2004). Establishment of a Nucleus Flock of Highest Genetic Merit for Breeding, Production and Propagation. Annual Report (2003-2004). Livestock Production Research Institute Bahadurnagar, Okara, Pakistan.
Shahroudi EE, Bahrini MR, Doulk VD, Mesgaran MD (2002). The factor affecting some economical traits Kermani sheep. Iranian J. Agric. Sci., 33(3): 395-402.
Sharif M (2001). Performance evaluation of economic traits of Balochi and Bibrik sheep of Balochistan. M.Sc. Thesis. Department of Livestock Management, Sindh Agriculture University, Tandojam.
Shrestha JNB, Vesely JA, Chesnais JP (1985). Genetic and phenotypic parameters for daily gain and b y weights in Suffolk lambs. Canadian J. Anim. Sci., 65(3): 575-582.
Shrestha JNB, Vesely JA, Chesnais JP, Cuthbertson DC (1986). Genetic and phenotypic parameters for daily gains and body weights in Dorset lambs. Canadian J. Anim. Sci., 66(1): 289-292.
Simm G, Lewis RM, Grundy B, Dingwall WS (2002). Responses to selection for lean growth in sheep. Anim. Sci., 74(1): 39-50.
Snowder GD, Van Veilk LD (2003). Estimates for genetic parameters and selection strategies to improve the economic efficiency of post-weaning growth in lamb. J. Anim. Sci., 81: 2704-2713.
Snyman MA, Erasmus GJ, van Wyk JB (1995). Non-genetic factors influencing growth and fleece traits in Afrino sheep. South African J. Anim. Sci., 25(3): 70-74.
Yazdi MH, Engstrm G, Nasholm A, Johansson K, Jorjani H, Liljedahl LE (1997). Genetic parameters for lamb weight at different ages and wool production in Baluchi sheep. J. Anim. Sci., 65: 247-255.

Standardization of herbal medicines - A review

Kunle, Oluyemisi Folashade[1]*, Egharevba, Henry Omoregie[1] and Ahmadu, Peter Ochogu[2]

[1]Department of Medicinal Plant Research and Traditional Medicine, National Institute for Pharmaceutical Research and Development (NIPRD), Idu Industrial Layout Idu, PMB 21 Garki, Abuja, Nigeria.
[2]Tianjin University of Traditional Chinese Medicine, Tianjin, China.

There is increasing awareness and general acceptability of the use of herbal drugs in today's medical practice. Although, most of these applications are unorthodox, it is however a known fact that over 80% of the world population depends on herbal medicines and product for healthy living. This rise in the use of herbal product has also given rise to various forms of abuse and adulteration of the products leading to consumers' and manufacturers' disappointment and in some instances fatal consequences. The challenge is innumerable and enormous, making the global herbal market unsafe. This review seeks to enlighten stakeholders in herbal medicine on the need to establish quality parameters for collection, handling, processing and production of herbal medicine as well as employ such parameters in ensuring the safety of the global herbal market. The processes of good quality assurance and standardization of herbal medicines and products were also discussed.

Key words: Herbal medicine, standardization, quality control.

INTRODUCTION

The use of herbs as medicine is the oldest form of healthcare known to humanity and has been used in all cultures throughout history (Barnes et al., 2007). Early humans recognized their dependence on nature for a healthy life and since that time humanity has depended on the diversity of plant resources for food, clothing, shelter, and medicine to cure myriads of ailments. Led by instinct, taste, and experience, primitive men and women treated illness by using plants, animal parts, and minerals that were not part of their usual diet. Primitive people learned by trial and error to distinguish useful plants with beneficial effects from those that were toxic or inactive, and also which combinations or processing methods had to be used to gain consistent and optimal results. Even in ancient cultures, tribal people methodically collected information on herbs and developed well-defined herbal pharmacopeias. Physical evidence of the use of herbal remedies some sixty thousand years ago has been found in a burial site of a Neanderthal man uncovered in 1960 in a cave in northern Iraq (Solecki, 1975).

Indeed, well into the twentieth century, much of the pharmacopeia of scientific medicine was derived from the herbal lore of native people. The knowledge of plant-based drugs developed gradually and was passed on, thus, laying the foundation for many systems of traditional medicine all over the world. In some communities herbal medicine is still a central part of their medical system.

Medicinal plants are widely distributed throughout the world but most abundantly in tropical countries. It is estimated that about 25% of all modern medicines are directly or indirectly derived from higher plants (WHO, 2005, 2002a,b, 1999a,b, 1998a,b, 1990, 1981, 1979; De Smet, 1995; Duke and Martinez, 1994; Majno, 1975; Ackerknecht, 1973). Thus, herbal medicine has led to the discovery of a number of new drugs, and non-drug substances.

HERBAL MEDICINE

An herb is a plant or part of a plant valued for its medicinal, aromatic, or savoury qualities. Herbs can be

*Corresponding author. E-mail: yemisikunle@yahoo.co.uk.

viewed as biosynthetic chemical laboratories, producing a number of chemical compounds. Herbal remedies or medicines consist of portions of plants or unpurified plant extracts containing several constituents, which often work together synergistically. Herbal medicine or herbalism is the use of herbs or herbal products for their therapeutic or medicinal value. They may come from any part of the plant but are most commonly made from leaves, roots, bark seeds, and flowers. They are eaten, swallowed, drunk, inhaled, or applied topically to the skin Herbal products often contain a variety of naturally-occuring biochemicals from plants, many of which contribute to the plant's medicinal benefits. Chemicals known to have medicinal benefits are referred to as "active ingredients" or "active principles" and their presence depends on a number of factors including the plant species, the time and season of harvest, the type of soil, the way the herb is prepared, etc.

During the past decade, there has been increasing public interest and acceptance of natural therapies in both developing and developed countries. Due to poverty and limited access to modern medicine, about 80% of the world's population, especially in the developing countries uses herbal medicine as their source of primary healthcare (Bodeker et al, 2005; Mukherjee, 2002; Farnsworth et al., 1985; Bisset, 1994). In these communities, traditional medical practice is often viewed as an integral part of their culture. In the West, people are attracted to herbal therapies for many reasons, the most important reason being that, like our ancestors, it is believed they will help us live healthier lives. Herbal medicines are often viewed as a balanced and moderate approach to healing. Individuals who use them as home remedies and over-the-counter drugs spend billions of dollars on herbal products. As such, they represent a substantial proportion of the global drug market (WHO, 2005, 2002a, 1999a and b, 1990; Blumenthal, 2000; Blumenthal et al., 1998; Roberts and Tyler, 1997; Farnsworth et al., 1985).

To achieve the desired benefit from herbal preparations, an individual must take the required dose over a certain length of time. Although it is generally believed that most herbal preparations are safe for consumption, some herbs like most biologically active substances could be toxic with undesirable side effects (Bisset, 1994).

The variability of the constituents in herbs or herbal preparations due to genetic, cultural and environmental factors has made the use of herbal medicines more challenging than it would necessarily have been. For instance, the availability and quality of the raw materials are frequently problematic, the active principles are diverse and may be unknown, and quality of different batches of preparation may be difficult to control and ascertain. In most countries, herbal products are launched into the market without proper scientific evaluation, and without any mandatory safety and toxicological studies. There is no effective machinery to regulate manufacturing practices and quality standards. Consumers can buy herbal products without a prescription and might not recognize the potential hazards in an inferior product. A well-defined and constant composition of the drug is therefore, one of the most important prerequisites for the production of a quality drug. Given the nature of products of plant origin, which are not usually constant and are dependent on and influenced by many factors, ensuring consistent quality of products is vital for the survival and success of the industry (Bauer, 1998).

QUALITY CONTROL AND STANDARDIZATION OF HERBAL MEDICINES – CONCEPT AND SCOPE

Generally, all medicines, whether they are synthetic or of plant origin, should fulfill the basic requirements of being safe and effective (EMEA, 2005; WHO, 2002c, 1998c, 1996, 1991a,b, 1990, 1988). The term "herbal drugs" denotes plants or plant parts that have been converted into phytopharmaceuticals by means of simple processes involving harvesting, drying, and storage (EMEA, 1998). Hence they are capable of variation. This variability is also caused by differences in growth, geographical location, and time of harvesting.

Standardization of herbal medicines is the process of prescribing a set of standards or inherent characteristics, constant parameters, definitive qualitative and quantitative values that carry an assurance of quality, efficacy, safety and reproducibility. It is the process of developing and agreeing upon technical standards. Specific standards are worked out by experimentation and observations, which would lead to the process of prescribing a set of characteristics exhibited by the particular herbal medicine. Hence standardization is a tool in the quality control process.

Several problems not applicable to synthetic drugs often influence the quality of herbal drugs. For instance:

1. Herbal drugs are usually mixtures of many constituents.
2. The active principle(s) is (are), in most cases unknown.
3. Selective analytical methods or reference compounds may not be available commercially.
4. Plant materials are chemically and naturally variable.
5. Chemo-varieties and chemo cultivars exist.
6. The source and quality of the raw material are variable.

The methods of harvesting, drying, storage, transportation, and processing (for example, mode of extraction and polarity of the extracting solvent, instability of constituents, etc.) also affect herbal quality.

At present no official standards are available for herbal preparations. Those manufacturers, who are currently doing some testing for their formulations, have their own parameters, many of which are very preliminary in nature. Presently it is very difficult to identify the presences of all the ingredients as claimed in a formulation. Hence the first important task is to evolve such parameter by which the presence of the entire ingredient can be identified, various chromatographic and spectrophotometric methods and evaluation of physicochemical properties can be tried to evolve pattern for identifying the presence of different ingredient. Wherever possible these methods can be applied for quantitative estimation of bioactive group of compounds like alkaloids, flavonoids, polyphenolic components or estimation of particular compound (Wani, 2007)

The need for standardization – Producers' and consumers' perspective

In the global perspective, there is a shift towards the use of medicine of herbal origin, as the dangers and the shortcoming of modern medicine are getting more apparent.

It is the cardinal responsibility of the regulatory authorities to ensure that consumers get the medication, which guarantees purity, safety, potency and efficacy. The regulatory authorities rigidly follow various standards of quality prescribed for raw materials and finished products in pharmacopoeias, formularies and manufacturing operation through statutory imposed good manufacturing practices. These procedures logically would apply to all types of medication whether included in modern system of medicine or one of the traditional systems.

Though herbal products have become increasingly popular throughout the world, one of the impediments in its acceptance is the lack of standard quality control profile. The quality of herbal medicine that is, the profile of the constituents in the final product has implications in efficacy and safety. However, due to the complex nature and inherent variability of the constituents of plant-based drugs, it is difficult to establish quality control parameter though modern analytical technique are expected to help in circumventing this problem. Furthermore, the constituents responsible for the claimed therapeutic effects are frequently unknown or only partly explained. This is further complicated by the use of combination of herbal ingredients as being used in traditional practice. It is common to have as many as five different herbal ingredients in one product. Thus batch to batch variation starts from the collection of raw material itself in the absence of any reference standard for identification. These variations multiply during storage and further processing. Hence for herbal drugs and products, standardization should encompass the entire field of study from cultivation of medicinal plant to its clinical application.

Plant materials and herbal remedies derived from them represent substantial portion of global market and in this respect internationally recognized guidelines for their quality assessment and quality control are necessary.

Standardization and quality control of herbal crude drugs – Processes and procedures

According to WHO (1996a and b, 1992), standardization and quality control of herbals is the process involved in the physicochemical evaluation of crude drug covering aspects, such as selection and handling of crude material, safety, efficacy and stability assessment of finished product, documentation of safety and risk based on experience, provision of product information to consumer and product promotion. Attention is normally paid to such quality indices such as:

1. Macro and microscopic examination: For Identification of right variety and search of adulterants.
2. Foreign organic matter: This involves removal of matter other than source plant to get the drug in pure form.
3. Ash values: These are criteria to judge the identity and purity of crude drug – Total ash, sulphated ash, water soluble ash and acid insoluble ash etc.
4. Moisture content: Checking moisture content helps reduce errors in the estimation of the actual weight of drug material. Low moisture suggests better stability against degradation of product.
5. Extractive values: These are indicative weights of the extractable chemical constituents of crude drug under different solvents environment.
6. Crude fibre: This helps to determine the woody material component, and it is a criterion for judging purity.
7. Qualitative chemical evaluation: This covers identification and characterization of crude drug with respect to phytochemical constituent. It employs different analytical technique to detect and isolate the active constituents. Phytochemical screening techniques involve botanical identification, extraction with suitable solvents, purification, and characterization of the active constituents of pharmaceutical importance.
8. Chromatographic examination: Include identification of crude drug based on the use of major chemical constituents as markers.
9. Quantitative chemical evaluation: To estimate the amount of the major classes of constituents.
10. Toxicological studies: This helps to determine the pesticide residues, potentially toxic elements, safety studies in animals like LD50 and Microbial assay to establish the absence or presence of potentially harmful microorganisms.

The processes mentioned above involves wide array of scientific investigations, which include physical, chemical and biological evaluation employing various analytical methods and tools. The specific aims of such investigation in assuring herbal quality are as varied as the processes employed.

Physical evaluation

Each monograph contains detailed botanical, macroscopic and microscopic descriptions of the physical characteristics of each plant that can be used to ensure both identity and purity. Each description is accompanied by detailed illustrations and photographic images which provide visual documentation of accurately identified material.

Microscopic evaluation

Full and accurate characterization of plant material requires a thorough physical examination. Microscopic analyses of plants are invaluable for assuring the identity of the material and as an initial screening test for impurities.

Chemical evaluation

This covers screening, isolation, identification and purification of the chemical components. Chemical analysis of the drug is done to assess the potency of vegetable material in terms of its active principles. The chemical screening or tests may include colour reaction test, which help to determine the identity of the drug substance and possible adulteration.

Biological evaluation

Pharmacological activity of certain drugs has been applied to evaluate and standardize them. The assays on living animal and on their intact or isolated organs can indicate the strength of the drug or their preparations. These assays are known as Biological assays or Bioassay.

Purity determination

Each monograph includes standards for purity and other qualitative indices already mentioned above.

Analytical methods

Critical to compliance with any monograph standard is the need for appropriate analytical methods for determining identity, quality, and relative potency. There are a plethora of analytical methods available. However, it is often difficult to know which is the most appropriate to use, but critical among know analytical tools in monograph standardization is chromatography.

Chromatography

Chromatography is the science which studies the separation of molecules based on differences in their structure and/or composition. In general, chromatography involves moving a preparation of the materials to be separated, “the "test preparation”, over a stationary support. The molecules in the test preparation will have different interactions with the stationary support leading to separation of similar molecules. Test molecules which display tighter interactions with the support will tend to move more slowly through the support than those molecules with weaker interactions. In this way, different types of molecules can be separated from each other as they move over the support material. Chromatographic separations can be carried out using a variety of supports, including immobilized silica on glass plates (thin layer chromatography), very sensitive High Performance Thin Layer Chromatography (HPTLC), volatile gases (gas chromatography), paper (paper chromatography), and liquids which may incorporate hydrophilic, insoluble molecules (liquid chromatography).
High performance thin layer chromatography (HPTLC) is a valuable quality assessment tool for the evaluation of botanical materials. It allows for the analysis of a broad number of compounds both efficiently and cost effectively. Additionally, numerous samples can be run in a single analysis thereby dramatically reducing analytical time. With HPTLC, the same analysis can be viewed collectively in different wavelengths of light thereby providing a more complete profile of the plant than is typically observed with more specific type of analysis.

Quantitative analysis

The most appropriate quantitative analytical method with accompanying chromatograms is desirable. The primary goal of the methods is to provide validated methods to be used to quantify the compounds most correlated with pharmacological activity or qualitative markers (Wani, 2007).

Control of starting material

Control of the starting materials is essential in order to ensure reproducible quality of herbal medicinal products (De Smet, 2004; Gaedcke and Steinhoff, 2003; WHO, 2002b; Phillipson, 1993). The following points are to be considered in the control of starting materials:

Authentication and reproducibility of herbal ingredients

The problems associated with unregulated herbal products highlight the major public health issues that can arise when their herbal ingredients have not been authenticated correctly. Herbal ingredients must be accurately identified by macroscopic and microscopic comparison with authentic material or accurate descriptions of authentic herbs (Houghton, 1998). It is essential that herbal ingredients are referred to by their binomial Latin names of genus and species; only permitted synonyms should be used. Even when correctly authenticated, it is important to realise that different batches of the same herbal ingredient may differ in quality due to a number of factors such as:

1. Inter- or intra-species variation: The variation in constituents is mostly genetically controlled and may be related to the country of origin.
2. Environmental factors: The quality of a herbal ingredient can be affected by environmental factor like climate, altitude and other conditions under which it was cultivated.
3. Time of harvesting: For some herbs the optimum time of harvesting should be specified as it is known that the concentrations of constituents in a plant can vary during the growing cycle or even during the course of a day.
4. Plant part used: Active constituents usually vary between plant parts and it is not uncommon for a herbal ingredient to be adulterated with parts of the plant not normally utilised. In addition, plant material that has been previously subjected to extraction and is therefore 'exhausted' is sometimes used as adulterants to increase the weight of a batch of herbal ingredient.
5. Post-harvesting factors: Storage conditions and processing treatments can greatly affect the quality of a herbal ingredient. Inappropriate storage after harvesting can result in microbial contamination, and processes such as drying may result in a loss of thermo-labile active constituents.

Adulteration/substitution

There are instances when herbal remedies have been adulterated with other plant material and conventional medicines. Reports of herbal products devoid of known active constituents have reinforced the need for adequate quality control of herbal remedies.

Identity and purity

In order to try to ensure the quality of licensed herbal medicines, it is essential not only to establish the botanical identity of a herbal ingredient but also to ensure batch-to-batch reproducibility. Thus, in addition to macroscopic and microscopic evaluation, identity tests are necessary. Such tests include simple chemical tests, e.g. colour or precipitation and chromatographic tests. Thin-layer chromatography is commonly used for identification purposes but for herbal ingredients containing volatile oils, a gas–liquid chromatographic test may be used. Although the aim of such tests may be to confirm the presence of active principles, it is frequently the case that the nature of the active principle has not been established. In such instances chemical and chromatographic tests help to provide batch-to-batch comparability and the chromatogram may be used as a 'fingerprint' for the herbal ingredient by demonstrating the profile of some common plant constituents such as flavonoids, alkaloids and terpenes.

Identity and purity ask the most important question "is the herb the one it should be?" In answering this, a lot of quality determinants are critically examined. Such determinant as purity and chemical constituents are very important. To prove identity and purity, criteria such as type of preparation, sensory properties, physical constants, adulteration, contaminants, moisture, ash content and solvent residues have to be checked. Identity can be achieved by macro- and microscopical examinations. Voucher specimens are reliable reference sources. Outbreaks of diseases among plants may result in changes to the physical appearance of the plant and lead to incorrect identification (De Smet, 1999). At times an incorrect botanical quality with respect to the labelling can be a problem. For example, in the 1990's, a South American product labelled as "Paraguay Tea" was associated with an outbreak of anticholinergic poisoning in New York. Subsequent chemical analysis revealed the presence of a class of constituents that was different from the metabolites normally found in the plant from which Paraguay tea is made.

Assaying for those herbal ingredients with known active principles is another method of ensuring product's identity and purity. An assay should be established in order to set the criterion for the minimum accepted percentage of active substances. Such assays should, wherever possible, be specific for individual chemical substances, and high-pressure liquid chromatography and gas–liquid chromatography are the methods of choice. Where such assays have not been established, then non-specific classical methods such as titration or colorimetric assays may be used to determine the total content of a group of closely related compounds.

Purity is closely linked with the safe use of drugs and deals with factors such as values, contaminants (e.g. foreign matter in the form of other herbs), and heavy metals. However, due to the application of improved analytical methods, modern purity evaluation also includes microbial contamination, aflatoxins, radioactivity, and pesticide residues. Analytical methods such as photometric analysis, thin layer chromatography (TLC), high performance liquid chromatography (HPLC), and gas chromatography (GC) can be employed in order to establish the constant composition of herbal preparations. Depending on whether the active principles of the preparation are known or unknown, different concepts such as "normalization versus standardization" have to be applied in order to establish relevant criteria for uniformity.

Content assay is the most difficult area to perform in quality control since in most herbal drugs the active constituents are not known. Sometimes markers can be used. In all other cases, where no active constituent or marker can be defined for the herbal drug, the percentage extractable matter with a solvent may be used as a form of assay, an approach often seen in pharmacopeias. The choice of the extracting solvent depends on the nature of the compounds involved, and might be deduced from the traditional uses. For example, when an herbal drug is used to make a tea, the hot water extractable matter, expressed as milligrams per gram of air-dried material, may serve this purpose (WHO, 1998b, 1996b). A special form of assay is the determination of essential oils by steam distillation. When the active constituents (for example, sennosides in Senna) or markers (for example, alky amides in Echinacea) are known, a vast array of modern chemical analytical methods such as ultraviolet/visible spectroscopy (UV/VIS), TLC, HPLC, GC, mass spectrometry (MS), or a combination of GC and MS (GCMS), can be employed (Watson, 1999).

Good agricultural/Manufacturing practices

Quality control and the standardization of herbal medicines also involve several other steps like source and quality of raw materials, good agricultural practices and good manufacturing practices. These practices play a pivotal role in guaranteeing the quality and stability of herbal preparations (WHO, 2004, 2003, 2000, 1992, 1988b; EMEA, 2002; Blumenthal et al., 1998; Roberts and Tyler, 1997). The quality of a plant product is determined by the prevailing conditions during growth, and accepted Good Agricultural Practices (GAP) can control this. These include seed selection, growth conditions, fertilizers application, harvesting, drying and storage. In fact, GAP procedures are integral part of quality control.

Factors such as the use of fresh plants, age and part of plant collected, period, time and method of collection, temperature of processing, exposure to light, availability of water, nutrients, drying, packing, transportation of raw material and storage, can greatly affect the quality, and hence the therapeutic value of herbal medicines. Apart from these criteria, factors such as the method of extraction, contamination with microorganisms, heavy metals, and pesticides can alter the quality, safety, and efficacy of herbal drugs. Using cultivated plants under controlled conditions instead of those collected from the wild can minimize most of these factors (Eskinazi et al., 1999; Blumenthal et al., 1998; Bauer, 1998). Sometimes, the active principles are destroyed by enzymic processes that continue for long periods from collection to marketing, resulting in a variation of composition. Thus, proper standardization and quality control of both the raw material and the herbal preparations should be conducted.

Contaminants of herbal ingredients

Herbal ingredients of high quality should be free from insects, animal matter and excreta. It is usually not possible to remove completely all contaminants, hence specifications should be set in order to limit them:

1. Ash values: Incineration of a herbal ingredient produces ash which constitutes inorganic matter. Treatment of the ash with hydrochloric acid results in acid-insoluble ash which consists mainly of silica and may be used to act as a measure of soil present. Limits may be set for ash and acid-insoluble ash of herbal ingredients.
2. Foreign organic matter: It is not possible to collect a herbal ingredient without small amounts of related parts of plant or other plants. Standards should be set in order to limit the percentage of such unwanted plant contaminants.
3. Microbial contamination: Aerobic bacteria and fungi are normally present in plant material and may increase due to faulty growing, harvesting, storage or processing. Herbal ingredients, particularly those with high starch content, may be prone to increased microbial growth. Pathogenic organisms including Enterobacter, Enterococcus, Clostridium, Pseudomonas, Shigella and Streptococcus have been shown to contaminate herbal ingredients. It is essential that limits be set for microbial contamination and the European Pharmacopoeia now gives non-mandatory guidance on acceptable limits (Barnes et al., 2007).
4. Pesticides: Herbal ingredients, particularly those grown as cultivated crops, may be contaminated by DDT (dichlorodiphenyltrichloroethane) or other chlorinated hydrocarbons, organophosphates, carbamates or polychlorinated biphenyls. Limit tests are necessary for acceptable levels of pesticide contamination of herbal ingredients. The European Pharmacopoeia includes details of test methods together with mandatory limits for 34 potential pesticide residues (Barnes et al., 2007).
5. Fumigants: Ethylene oxide, methyl bromide and phosphine have been used to control pests which contaminate herbal ingredients. The use of ethylene oxide as a fumigant with herbal drugs is no longer permitted in Europe (Barnes et al., 2007).
6. Toxic metals: Lead, cadmium, mercury, thallium and arsenic have been shown to be contaminants of some

herbal ingredients. Limit tests for such toxic metals are essential for herbal ingredients.

7. Radioactive contamination: There are many sources of ionization radiation, including radionuclides, occurring in the environment. Hence, a certain degree of exposure is inevitable. (AOAC, 2005; WHO, 2000; De Smet, 1992).

8. Other contaminants: As standards increase for the quality of herbal ingredients it is possible that tests to limit other contaminants such as endotoxins and mycotoxins will be utilized to ensure high quality for medicinal purposes (Barnes et al., 2007).

Standardization of herbal medicines

This involves adjusting the herbal drug preparation to a defined content of a constituent or a group of substances with known therapeutic activity by adding excipients or by mixing herbal drugs or herbal drug preparations. Botanical extracts made directly from crude plant material show substantial variation in composition, quality, and therapeutic effects. Standardized extracts are high-quality extracts containing consistent levels of specified compounds, and they are subjected to rigorous quality controls during all phases of the growing, harvesting, and manufacturing processes. No regulatory definition exists for standardization of dietary supplements. As a result, the term "standardization" may mean many different things. Some manufacturers use the term standardization incorrectly to refer to uniform manufacturing practices, but following a recipe is not sufficient for a product to be called standardized. Therefore, the presence of the word "standardized" on a supplement label does not necessarily indicate product quality.

When the active principles are unknown, marker substances should be established for analytical purposes and standardization. Marker substances are chemically defined constituents of a herbal drug that are important for the quality of the finished product. Ideally, the chemical markers chosen would also be the compounds that are responsible for the pharmacological effects in the body. There are two types of standardization. In the first category, "true" standardization, a definite phytochemical or group of constituents is known to have activity. Ginkgo with its 26% ginkgo flavones and 6% terpenes is a classic example. These products are highly concentrated and no longer represent the whole herb, and are now considered as phytopharmaceuticals. In many cases they are vastly more effective than the whole herb. However the process may result in the loss of efficacy and the potential for adverse effects and herb–drug interactions may increase. The other type of standardization is based on the guarantee of the manufacturers for the presence of a certain percentage of marker compounds which are not indicators of therapeutic activity or quality of the herb.

In the case of herbal drug preparations, the production and primary processing of the medicinal plant or herbal drug has a direct influence on the quality of the active pharmaceutical ingredients (APIs). Due to the inherent complexity of naturally growing medicinal plants and the limited availability of simple analytical techniques to identify and characterize the active constituents solely by chemical or biological means, there is a need for an adequate quality assurance system. This assurance is also required during cultivation, harvesting, primary processing, handling, storage, packaging, and distribution. Deterioration and contamination through adulteration, especially microbial contamination, can occur at any one of these stages. It is extremely important to establish good agricultural, harvesting, and manufacturing practices for herbal starting materials in order to minimize these undesirable factors. In this regard producers, processors, and traders of medicinal plants or herbal drugs have an obligation and a role to play. The manufacturers and suppliers of herbal products should adhere to quality control standards and good manufacturing practices. Currently, only a few manufacturers adhere to complete quality control and good manufacturing procedures including microscopic, physical, chemical, and biological analysis. Organizations, such as, National Agency for Food and Drugs Administration and Control (NAFDAC) help safeguard Nigerians' health, and Health Canada help safeguard Canadians' health by carrying out premarket reviews of all drugs before they are authorized for sale. The products available in the market are analyzed regularly to ensure that they are free of unsafe ingredients and that the products actually contain the ingredients indicated on the labels.

The potency and quality of an individual herbal product may be unclear because of lack of regulation. It is obvious that for a given plant product its quality will also be determined by the prevailing conditions during the growth cycle of the plant. Therefore, for cultivated plants the good agricultural practice (GAP) system has been introduced, under which each step, including seed selection, growing conditions, use of fertilizers, and optimization of harvest time, harvesting, and drying, has to adhere to a set of criteria. It is likely that GAP procedures will become an integral part of quality control in the near future.

CRITICAL FACTORS AFFECTING THE QUALITY CONTROL OF HERBAL DRUGS

Microscopic evaluation

Quality control of herbal drugs has traditionally been based on the appearance and today microscopic evaluation is indispensable in the initial identification of herbs, as well as, in identifying small fragments of crude or powdered herbs, and detection of foreign matter and adulterants. A primary visual evaluation, which seldom needs more than a simple magnifying lens, can be used to ensure that the plant is of the required species, and that the right part of the plant is being used. At other times, microscopic analysis is needed to determine the correct species and/or that the correct part of the species is present. For instance, pollen morphology may be used in the case of flowers to identify the species, and the presence of certain microscopic structures such as leaf stomata can be used to identify the plant part used. Although this may seem obvious, it is of prime importance, especially when different parts of the same plant are to be used for different treatments. Stinging nettle (Urtica urens) is a classic example where the aerial parts are used to treat rheumatism, while the roots are applied for benign prostate hyperplasia (AOAC, 2005).

Foreign matter

Herbal drugs should be made from the stated part of the plant and be devoid of other parts of the same plant or other plants. They should be entirely free from moulds or insec, including excreta and visible contaminant such as sand and stones, poisonous and harmful foreign matter and chemical residues. Animal matters such as insects and "invisible" microbial contaminants, which can produce toxins, are also among the potential contaminants of herbal medicines (WHO, 2004, 2003; EMEA, 2002). Macroscopic examination can easily be employed to determine the presence of foreign matter, although, microscopy is indispensable in certain special cases (for example, starch deliberately added to "dilute" the plant material). Furthermore, when foreign matter consists, for example, of a chemical residue, TLC is often needed to detect the contaminants (AOAC, 2005; WHO, 1999a, 1998a).

Ash content

To determine ash content, the plant material is burnt and the residual ash is measured as total and acid-insoluble ash. Total ash is the measure of the total amount of material left after burning and includes ash derived from the part of the plant itself and acid-insoluble ash. The latter is the residue obtained after boiling the total ash with dilute hydrochloric acid, and burning the remaining insoluble matter. The second procedure measures the amount of silica present, especially in the form of sand and siliceous earth (AOAC, 2005).

Heavy metals

Contamination by toxic metals can either be accidental or intentional. Contamination by heavy metals such as mercury, lead, copper, cadmium, and arsenic in herbal remedies can be attributed to many causes, including environmental pollution, and can pose clinically relevant dangers for the health of the user and should therefore be limited (AOAC, 2005; WHO, 1998c; De Smet, 1992). The potential intake of the toxic metal can be estimated on the basis of the level of its presence in the product and the recommended or estimated dosage of the product. This potential exposure can then be put into a toxicological perspective by comparison with the so-called Provisional Tolerable Weekly Intake values (PTWI) for toxic metals, which have been established by the Food and Agriculture Organization of the World Health Organization (FAO-WHO) (De Smet, 1999; WHO, 1981, 1979). A simple, straightforward determination of heavy metals can be found in many pharmacopoeias and is based on colour reactions with special reagents such as thioacetamide or diethyldithiocarbamate, and the amount present is estimated by comparison with a standard (WHO, 1988a). Instrumental analyses have to be employed when the metals are present in trace quantities, in admixture, or when the analyses have to be quantitative. Generally, the main methods commonly used are atomic absorption spectrophotometry (AAS), inductively coupled plasma (ICP) and neutron activation analysis (NAA) (Watson, 1999).

Microbial contaminants and aflatoxins

Medicinal plants may be associated with a broad variety of microbial contaminants, represented by bacteria, fungi, and viruses. Inevitably, this microbiological background depends on several environmental factors and exerts an important impact on the overall quality of herbal products and preparations. Risk assessment of the microbial load of medicinal plants has therefore become an important subject in the establishment of modern Hazard Analysis and Critical Control Point (HACCP) schemes.

Herbal drugs normally carry a number of bacteria and molds, often originating in the soil. Poor methods of harvesting, cleaning, drying, handling, and storage may also cause additional contamination, as may be the case with Escherichia coli or Salmonella spp. while a large range of bacteria and fungi are from naturally occurring microflora, aerobic spore-forming bacteria that frequently predominate. Laboratory procedures investigating microbial contaminations are laid down in the well-known pharmacopeias, as well as, in the WHO guidelines (WHO, 2000, 1998a). Limit values can also be found in the sources mentioned. Generally, a complete procedure consists of determining the total aerobic microbial count, the total fungal count, and the total Entero-bacteriaceae count, together with tests for the presence of Escherichia coli, Staphylococcus aureus, Shigella, and Pseudomonas aeruginosa and Salmonella spp. The European Pharmacopoeia also specifies that *E. coli* and *Salmonella* spp. should be absent from herbal preparations.

Materials of vegetable origin tend to show much higher levels of microbial contamination than synthetic products and the requirements for microbial contamination in the European Pharmacopoeia allow higher levels of microbial contamination in herbal remedies than in synthetic pharmaceuticals. The allowed contamination level may also depend on the method of processing of the drug. For example, higher contamination levels are permitted if the final herbal preparation involves boiling with water. The presence of fungi should be carefully investigated and/or monitored, since some common species produce toxins, especially aflatoxins. Aflatoxins in herbal drugs can be dangerous to health even if they are absorbed in minute amounts (WHO, 2000). Aflatoxin-producing fungi sometimes build up during storage (De Smet, 1992). Procedures for the determination of aflatoxin contamination in herbal drugs are published by the WHO (2000). After a thorough clean-up procedure, TLC is used for confirmation. In addition to the risk of bacterial and viral contamination, herbal remedies may also be contaminated with microbial toxins, and as such, bacterial endotoxins and mycotoxins, at times may also be an issue (Sinha,1998; De Smet et al., 1997; De Smet, 1992). There is evidence that medicinal plants from some countries may be contaminated with toxigenic fungi (Aspergillus, Fusarium). Certain plant constituents are susceptible to chemical transformation by contaminating micro-organisms. Withering leads to enhanced enzymic activity, transforming some of the constituents to other metabolites not initially found in the herb. These newly formed constituent(s) along with the molds such as Penicillium nigricans and P. jensi may then have adverse effects (De Smet, 1992).

Pesticide residues

Even though there are no serious reports of toxicity due to the presence of pesticides and fumigants, it is important that herbs and herbal products are free of these chemicals or at least are controlled for the absence of unsafe levels (De Smet, 1992). Herbal drugs are liable to contain pesticide residues, which accumulate from agricultural practices, such as spraying, treatment of soils during cultivation, and administering of fumigants during storage. However, it may be desirable to test herbal drugs for broad groups in general, rather than for individual pesticides. Many pesticides contain chlorine in the molecule, which, for example, can be measured by analysis of total organic chlorine. In an analogous way, insecticides containing phosphate can be detected by measuring total organic phosphorus.

Samples of herbal material are extracted by a standard procedure, impurities are removed by partition and/or adsorption, and individual pesticides are measured by GC, MS, or GC-MS. Some simple procedures have been published by the WHO and the European Pharamacopoeia has laid down general limits for pesticide residues in medicine (WHO, 1996a, 1998a, 2000; De Smet, 1999; AOAC, 2005)

Radioactive contamination

Dangerous contamination, however, may be the consequence of a nuclear accident. The WHO, in close cooperation with several other international organizations, has developed guidelines in the event of a wide sspread contamination by radionuclides resulting from major nuclear accidents. These publications emphasize that the health risk, in general, due to radioactive contamination from naturally occurring radio nuclides is not a real concern, but those arising from major nuclear accidents such as the nuclear accident in Chernobyl and Fukushima may be serious and depend on the specific radionuclide, the level of contamination, and the quantity of the contaminant consumed. Taking into account the quantity of herbal medicine normally consumed by an individual, is unlikely to be a health risk. Therefore, at present, no limits are proposed for radioactive contamination (AOAC, 2005; WHO, 2000; De Smet, 1992).

Analytical methods

Published monographs in a pharmacopeia are the most practical approach for quality control of herbal drugs and there are many available (EMEA, 2005; WHO, 1998a,b, 1996a, 1998a, 1981). When pharmacopoeia monographs are unavailable, development and validation of analytical procedures have to be carried out by the manufacturer. The best strategy is to follow closely the pharmacopoeia definitions of identity, purity, and content or assay. Valuable sources for general analytical procedures are included in the pharmacopeias, in guidelines published by the WHO (AOAC, 2005; WHO, 2000). Additional information, especially on chromatographic and/or spectroscopic methods can be found in the general scientific literature. The plant or plant extract can be evaluated by various biological methods to determine pharmacological activity, potency, and toxicity.

A simple chromatographic technique such as TLC may provide valuable additional information to establish the identity of the plant material. This is especially important for those species that contain different active constituents. Qualitative and quantitative information can be gathered concerning the presence or absence of metabolites or breakdown of products (AOAC, 2005). TLC fingerprinting is of key importance for herbal drugs made up of essential oils, resins, and gums, which are complex mixtures of constituents that no longer have any organic structure. It is a powerful and relatively rapid solution to distinguish between chemical classes, where macroscopy and microscopy may fail. Chromatograms of essential oils, for example, are widely published in the scientific literature, and can be of invaluable help in identification. The instruments for UV-Visible determinations are easy to operate, and validation procedures are straightforward but at the same time precise. Although measurements are made rapidly, sample preparation can be time consuming and works well only for less complex samples, and those compounds with absorbance in the UV-Visible region. HPLC is the preferred method for quantitative analysis of more complex mixtures. Though the separation of volatile components such as essential and fatty oils can be achieved with HPLC, it is best performed by GC or GC-MS. The quantitative determination of constituents has been made easy by recent developments in analytical instrumentation. Recent advances in the isolation, purification, and structure elucidation of naturally occurring metabolites have made it possible to establish appropriate strategies for the determination and analysis of quality and the process of standardization of herbal preparations. Classification of plants and organisms by their chemical constituents is referred to as chemotaxonomy.

TLC, HPLC, GC, quantitative TLC (QTLC), and high-performance TLC (HPTLC) can determine the homogeneity of a plant extract. Over-pressured layer chromatography (OPLC), infrared and UV-Visible spectrometry, MS, GC, liquid chromatography (LC) used alone, or in combinations such as GC-MS and LC-MS, and nuclear magnetic resonance (NMR), electrophoretic techniques, especially by hyphenated chromatographic techniques, are powerful tools, often used for standardization and to control the quality of both the raw material and the finished product. The results from these sophisticated techniques provide a chemical fingerprint as to the nature of chemicals or impurities present in the plant or extract (WHO, 2002c). Based on the concept of photo equivalence, the chromatographic fingerprints of herbal medicines can be used to address the issue of quality control. Methods based on information theory, similarity estimation, chemical pattern recognition, spectral correlative chromatograms (SCC), multivariate resolution, the combination of chromatographic fingerprints and chemometric evaluation for evaluating fingerprints are all powerful tools for quality control of herbal products.

Validation

The validation of herbal products is a major public health concern both in developed and resource-poor countries, where fakers selling adulterated herbal medicines are common. In this regard, there is no control by the government agencies, despite the existence of certain guidelines in some individual countries and those outlined by the WHO. If the herbal products are marketed as therapeutic agents, and irrespective of whether the products really have any positive effects to cure and reduce the severity of the disease, it is necessary to ensure scientific validation and periodic monitoring of the quality and efficacy by drug control administrators. It is feasible that the introduction of scientific validation would control the production of impure or adulterated herbal products and would eventually ensure their rational use. This could also lead to the regulation of the industry so that only qualified physicians and health providers are allowed to prescribe the medication.

Several of the principal pharmacopoeias contain monographs outlining standards for herbal drugs. The major advantage of an official monograph published in a pharmacopoeia is that standards are defined and available, and that the analytical procedures used are fully validated. This is of major importance, since validation can be a rather time-consuming process.

By definition, validation is the process of proving that an analytical method is acceptable for its intended purpose for pharmaceutical methods. Guidelines from the United States Pharmacopeia (USPC, 1994 to 2001), the International Conference on Harmonization (ICH), and the US Food and Drug Administration (FDA) provide a framework for performing such validations. Generally, validation investigations must include studies on specificity, linearity, accuracy, precision, range, detection, and quantitative limits, depending on whether the analytical method used is qualitative or quantitative (De Smet et al., 1997). Also, of utmost importance is the availability of standards. For macroscopic and microscopic procedures in general this means that reliable reference samples of the plant must be available. A defined botanical source (e.g. voucher specimens) will normally solve this problem. Standards for chromatographic procedures are less easy to obtain. Characteristic plant constituents, either active or markers, are seldom available commercially. Sometimes an LC-MS approach can be referred to as a mode of characterization. Going one step further, after isolation of such a compound, elucidations to prove its definite structure will not be easy. The method often employed is to use readily available compounds that behave similarly in the chosen chromatographic systems, and to calculate retention values and/or times towards these compounds as a standard. Qualitative chemical examination is designed to detect and isolate the active ingredients. TLC and HPLC are the main analytical techniques commonly used. In cases when active ingredients are not known or too complex, the quality of plant extracts can be assessed by a "fingerprint" chromatogram (De Smet et al., 1997).

Labelling of herbal products

The quality of consumer information about the product is

as important as the finished herbal product. Warnings on the packet or label will help to reduce the risk of inappropriate uses and adverse reactions (De Smet et al., 1997). The primary source of information on herbal products is the product label. Currently, there is no organization or government body that certifies herb or a supplement as being labelled correctly. It has been found that herbal remedy labels often cannot be trusted to reveal what is in the container. Studies of herbal products have shown that consumers have less than a 50% chance of actually getting what is listed on the label, and published analyses of herbal supplements have found significant differences between what is listed on the label and what is in the bottle. The word "standardized" on a product label is no guarantee of higher product quality, since there is no legal definition of the word "standardized." Consumers are often left on their own to decide what is safe and effective for them and the lack of consistent labelling on herbal products can be a source of consumer frustration. Certain information such as "the product has been manufactured according to Pharmacopoeia standards," listing of active ingredients and amounts, directions such as serving quantity (dosage) and frequency of intake of the drug, must be in the label.

CONCLUSIONS

Plant materials are used throughout the developed and developing world as home remedies, in over-the-counter drug products, and as raw material for the pharmaceutical industry, and they represent a substantial proportion of the global drug market. Therefore, it is essential to establish internationally recognized guidelines for assessing their quality. Certain herbs have become popular over the years, but the general public, medical practitioners and the media still have a poor understanding of safe and effective use of herbal medicine. Evidence is emerging on the dangers of indiscriminate use of some of these herbs. As in most situations, the truth lies hidden under the media hype, poorly understood science, an exaggerated claim. The need for standardization of herbals is now very essential given the global acceptance of herbal products as remedies for various diseases and ailments.

The deployment of modern analytical tools in testing the various quality parameters for an effective quality control herbal product cannot be over emphasized. The assurance of the safety and efficacy of a herbal drug requires monitoring of the quality of the product from collection through processing to the finished packaged product. It is recommended that various government agencies should follow a more universal approach to herbal quality by adopting the WHO guidelines and also developing monographs using the various quality parameters outlined above. This will strengthen the regulatory process and minimize quality breach.

REFERENCES

Ackerknecht EH (1973) Therapeutics: from the Primitives to the Twentieth Century. Hafner Press, New York.

AOAC (2005). Official Methods of Analysis of AOAC International, 18th edn. AOAC International, Gaithersburg, MD.

Barnes J, Anderson LA, Phillipson JD (2007). Herbal medicine. 3rd Edition, Pharmaceutical Press, London. pp 1-23.

Bauer R (1998). Quality criteria and standardization of phytopharmaceuticals: Can acceptable drug standards be achieved? Drug Inform. J., 32: 101–110.

Bisset NG (1994). Herbal Drugs and Phytopharmaceuticals. CRC Press, Boca Raton, FL.

Blumenthal M (2000) Ephedra update: industry coalition asks FDA to adopt national labeling guidelines on ephedra; offers co-operative research with NIH. Herbal Gram., 50: 64-65.

Blumenthal M, Brusse WR, Goldberg A, Gruenwald J, Hall T, Riggins CW, Rister RS (1998). The Complete German Commission E Monographs. Therapeutic Guide to Herbal Medicines. The American Botanical Council, Austin, TX.

Bodeker C, Bodeker G, Ong CK, Grundy CK, Burford G, Shein K (2005). WHO Global Atlas of Traditional, Complementary and Alternative Medicine. World Health Organization, Geneva.

De Smet PAGM (2004). Health risks of herbal remedies: An update. Clinical Pharmacology Therapeutics, 76(1): 1–17.

De Smet, PAGM (1995). Health risks of herbal remedies. Drug Safety, 13: 81-93.

De Smet PAGM, Keller K, Hansel R, Chandler RF (1992). Aristolochia species In: Adverse Effects of Herbal Drugs, Springer-Verlag, Heidelberg. 1.

De Smet PAGM, Keller K, Hansel R, Chandler RF (1997). Adverse Effects of Herbal Drugs, (eds), Springer-Verlag, Heidelberg. 129(3): 137-145.

De Smet PAGM (1999). Overview of herbal quality control. Drug Inform. J., 33: 717-724.

Duke JA, Martinez RV (1994). Handbook of Ethnobotanicals (Peru). CRC Press, Boca Raton, FL.

EMEA (1998). Quality of Herbal Medicinal Products. Guidelines. European Agency for the Evaluation of Medicinal Products (EMEA), London.

EMEA (2002). Points to Consider on Good Agricultural and Collection Practice for Starting Materials of Herbal Origin. EMEA/HMPWP/31/99 Review. European Agency for the Evaluation of Medicinal Products (EMEA), London.

EMEA (2005). Guidelines on Quality of Herbal Medicinal Products/Traditional Medicinal Products, EMEA/CVMP/814OO Review. European Agency for the Evaluation of Medicinal Products (EMEA), London.

Eskinazi D, Blumenthal M, Farnsworth N, Riggins CW (1999). Botanical Medicine: Efficacy, Quality, Assurance, and Regulation. Mary Ann Liebert, New York.

Farnsworth NR, Akerele O, Bingel AS, Soejarto DD, Guo Z (1985). World Health Organ., 63: 965.

Gaedcke F, Steinhoff B (2003) Quality assurance of herbal medicinal products. In: Herbal Medicinal Products. Stuttgart: medpharm GmbH Scientific Publishers. pp. 37-66, 81-88.

Houghton P (1998). Establishing identification criteria for botanicals. Drug Inform. J., 32: 461-469.

Majno GM (1975). Healing Hand: Man and Wound in the Ancient World. Harvard University Press, Cambridge, MA.

Mukherjee PW (2002). Quality Control of Herbal Drugs: An Approach to Evaluation of Botanicals. Business Horizons Publishers, New Delhi, India.

Phillipson JD (1993). Quality assurance of medicinal plants. In: Franz C et al. First World Congress on medicinal and aromatic plants for human welfare, WOCMAP, quality, phytochemistry, industrial aspects, economic aspects. Acta Horticult., 333: 117-122.

Roberts JE, Tyler VE (1997). Tyler's Herbs of Choice. The Therapeutic Use of Phytomedicinals. The Haworth Press, New York.

Solecki RS (1975). Standardized product as well as the quality of the consumer information on the herbal remedy. hanidar IV. Science, 190: 880.

Wani MS (2007). Herbal medicine and its standardization. Pharma. info., 1: 6.
Watson DG (1999). Pharmaceutical Analysis. Churchill Livingstone, Edinburgh.
WHO (1979). The International Pharmacopeia, General Methods of Analysis, 3rd edn. World Health Organization, Geneva. 1.
WHO (1981). The International Pharmacopeia, Quality Specifications, 3rd edn.World Health Organization, Geneva. 2.
WHO (1988a). The International Pharmacopeia,Quality Specifications for Pharmaceutical Substances, Excipients, and Dosage Forms, 3rd edn. World Health Organization, Geneva. 3.
WHO (1988b). Quality Control Methods for Medicinal Plant Materials. World Health Organization, Geneva.
WHO (1990). The Use of Essential Drugs. Eighth report of the WHO Expert committee. World Health Organization, Geneva.
WHO (1992). Quality Control Methods for Medicinal Plant Materials. World Health Organization, Geneva.
WHO (1996a). Quality Assurance of Pharmaceuticals: A Compendum of Guidelines and Related Materials, Good Manufacturing Practices and Inspection. World Health Organization, Geneva. 2.
WHO (1996b). Guidelines for the Assessment of Herbal Medicines. WHO Technical Report Series, World Health Organization, Geneva. 863.
WHO (1998a). Quality Control Methods for Medicinal Plant Materials, World Health Organization, Geneva.
WHO (1998b). Guidelines for the Appropriate Use of Herbal Medicines. WHO Regional Publications, Western Pacific Series. WHO Regional office for the Western Pacific, Manila. 3.
WHO (1998c). Basic Tests for Drugs, Pharmaceutical Substances, Medicinal Plant Materials and Dosage Forms. World Health Organization, Geneva.
WHO (1999a). Quality Control Methods for Medicinal Plant Materials. World Health Organization, Geneva.
WHO (1999b). WHO Monographs on Selected Medicinal Plants, World Health Organization, Geneva. 1.
WHO (2000). General Guidelines for Methodologies on Research and Evaluation of Traditional Medicine. World Health Organization, Geneva.
WHO (2000). The WHO Recommended Classification of Pesticides by Hazard and Guidelines to Classification 2000–2002 (WHO/PCS/01.5). International Programme on Chemical Safety, World Health Organization, Geneva.
WHO (2002a). WHO Monographs on Selected Medicinal Plants, World Health Organization, Geneva. 2.
WHO (2002b). Ancient Remedies, New Disease. World Health Organization, Geneva.
WHO (2002c). General Guidelines for Methodologies on Research and Evaluation of Traditional Medicine. World Health Organization, Geneva.
WHO (2003). WHO Guidelines on Good Agricultural and Collection Practices (GACP). World Health Organization, Geneva.
WHO (2004). Guidelines on Good Agricultural and Collection Practices (GACP) for Medicinal Plants. World Health Organization, Geneva.
WHO (2005). WHO Global Atlas of Traditional, Complementary and Alternative Medicine. World Health Organization, Geneva. 1and 2.

In vitro study of effects of growth hormones on sporophyte development of *Cyathea spinulosa*

Jagadish Parajuli[1] and Sanu Devi Joshi[2,3]

[1]WWF Nepal, Kathmandu, Nepal.
[2]Nepal Academy of Science and Technology (NAST), Nepal.
[3]Central Department of Botany, Tribhuvan University, Nepal.

***Cyathea spinulosa* Wall. ex Hook. is a type genus of Cyatheaceae family. It is listed in endangered category of IUCN Red Data Book and also placed in Convention on International Trade of Endangered Species of Flora and Fauna (CITES) Appendix II. The spore germination, gametophyte morphogenesis and sporophyte rooting of *Cyathea* in response to 2.0% sucrose and three major growth hormones: Indole–3 acetic acid (IAA), indole–3 butyric acid (IBA) and naphthalene acetic acid (NAA) was studied. The study was carried out at the tissue culture laboratory of Central Department of Botany, Tribhuvan University, Nepal. The spore of *Cyathea* was reared on Knop's Basal Medium (KBM) with 2.0% sucrose after surface sterilization. The sporophyte was sub-cultured on medium with growth hormones IAA, IBA and NAA, each containing 1, 2 and 3 ppm concentrations. The study suggested that *Cyathea* exhibited bipolar germination and the gametophyte development was Adiantum type. The mature gametophyte was observed as heart shaped. Sex organs were not observed up to 24 weeks which indicates the development of apogamous sporophyte. The growth hormones affect the development of leaves and roots. The sporophyte containing three leaves and two roots was the best result obtained in KBM containing 1 ppm IAA.**

Key words: Cyatheaceae, *Cyathea spinulosa*, morphogenesis, Knop's medium, growth hormones, apogamous sporogophyte.

INTRODUCTION

Cyathea spinulosa Wall. ex Hook. is a genus of Cyatheaceae family (Leptosporangiate of Pteridophyta), and well known for its excellent beauty for outdoor decoration. This plant prefers warm and humid atmosphere for its luxuriant growth along the streams side at 335 to 2000 m altitudes in East and Central Nepal (Gurung, 1991). It has great afforestation value due to well adoptability to push the rhizome into barren soil, act as colonal of disturbed soil especially in water logged areas, shade beneath the fronds help in the establishment of other grass, the root system efficiently binds top soil, can be easily propagated by rhizome cutting and splitting (Gurung, 1992). It is economically important and the wood log is much in demand in orchid cultivation and for starch extraction (Khare et al., 2005). It was listed in endangered category of IUCN Red Data Book in 1998.

It is also included in the appendix II of Convention on International Trade in Endangered Species of Wild Fauna and Flora (CITES, 2013).

It has been possible to induce apogamous sporophytes and aposporous gametophyte without the intervention of fertilization and sporogenesis by suitably manipulating the cultural condition *in vitro*. Nutritional studies on fern have highlighted their response to cultural condition (Bir et al., 1982). Some of the studies carried out on tree fern were variation in performance across a successional mosaic (Arens, 2001), effect of sterilization and storage conditions on the viability of spores (Simabukuro et al., 1998), relationships of Papuasian Cyatheaceae (Raubeson et al., 1995), spore germination and gametophyte development (Rechenmacher et al., 2010), evidence for Gondwanan vicariance and limited transoceanic dispersal (Korall and Pryer, 2014), effects of light, temperature and pH on spore germination and early gametophytic development (Du et al., 2009), morphology of the gametophytes and young sporephytes (Huang et al., 2001). Some of the studies in which fern gametophytes have been used include investigation of growth correlations, regeneration and hormonal effects (Albaum, 1938a, b), the induction of sex organ formation (Naf, 1956, 1958), the effect of the inhibition of protein synthesis on growth (Hotta and Osawa, 1958), induction of apogamous sporophyte development (Whittier and Steeves, 1960, 1962; Mehra and Sulklyan, 1969; Kato, 1970), effect of sucrose on heteroblastic leaf development (Sussex and Cluttter, 1960; Caponetti, 1970), effect of storage temperature (Ballesteros et al., 2011) and many more.

The objectives of this research were to observe the effect of 2% sugar concentration on the gametophyte morphogenesis in the spore germination and to investigate the effect of different growth hormones on *in vitro* rooting, and micropropagation of *C. spinulosa*. *In vitro* studies of threathened species- *C. spinulosa*, so far has not been done in Nepal. Realzing the importance, it is very necessary to study *Cyathea in vitro*.

MATERIALS AND METHODS

Spores of *C. spinulosa* were collected from Godavari Botanical Gardens, Kathmandu, Nepal for micropropagation. Fresh spores were taken from May to October and dry spores were taken on remaining months.

Surface sterilization, spore culutre and preparation of explants

The mature fronds of *C. spinulosa* with spores on their ventral surface were washed serially with diluted detergent in running tap water for 30 to 45 min, and with distilled water for 3 to 4 times. The pinna was cut into small pieces and surface sterilized with 1% sodium hypochlorite solution for 12 min and finally rinsed with sterile distilled water for 5 to 6 times inside the laminar airflow chamber. Spores were taken in the sterilized Petri dish with the help of surgical blade pressing one side of the frond with the forceps. Then, spores were inoculated on 10 different cultrue tube containing slanting Knob's basal medium (KBM, 1884) containing 2% sucrose. Joshi et al. (2002) studied the effect of sucrose on spore germination of *C. spinulosa*. They found 91% as the highest percentage of spore germination in KBM with 2% sucrose. Similar protocol had been adopted in the study.

For the sub-culture of sporophytes, sporophytes were cut down into small pieces in Petri dish and inoculated in KBM containing different concentration (1, 2 and 3 ppm) of growth hormones–Indole–3 acetic acid (IAA), indole–3 butyric acid (IBA) and naphthalene acetic acid (NAA) by the help of forceps (Table 1). All the cultures were maintained at 25°C (±2) and eight-hour light daily. The responses were observed after one week of culture. Various stages of germinating spores and prothalli were observed under microscope. The slides were prepared by mounting in glycerine. Each culture was subcultured after three months on fresh media. The observations were taken in one week interval.

Acclimatization

The sporophytes were observed after 24 weeks of spore inoculation. After several sub-culture, sporophytes having developed roots were grown in soil containing soil, compost and sand in 2:1:1 ratio and high humidity was maintained covering by plastics. After few weeks the plastic cover were removed and the plants were planted in pots.

RESULTS

Spore germination and gametophytic morphogenesis

A spore, consisting of a single haploid cell, has simple requirements for differentiation. It is significant that no nutritional or hormonal gradients are imposed on the spore by the adjoining cell (Rasid, 1999). The spores did not germinate as soon as they were shed. They remained dormant for a period of time ranging from 3 to 4 weeks to 10 to 12 weeks. The rate of spore generation with respect to different period of year when spore collected is represented in Figure 1. The dormancy was found to be dependent on temperature of spore collection, maturity of spores and degree of hydration. At the lower temperature during winter, the spore took up to 12 weeks to germinate while during summer spore germinated within 3 to 4 weeks.

Swelling, expansion and greening of a spore (Figure 2) and rupture of exine were the basic pre-germination changes. That change was followed by the appearance of small protuberance like structure almost all over the spore surface. With the opening of the exine, the protoplasmic mass of the spores with intine appeared in the form of small protuberance, that is, protonemal filament. The germination of spore was bipolar resulting in the formation of small rhizoidal cell and a larger protonemal initial cell. A rhizoid was seen to be given off from the rhizoidal initial cell. The protonemal filament ultimately developed into prothalli after 8 weeks of spore inoculation (Figure 3) which contained few numbers of rhizoids.

Table 1. Culture of sporophyte on KBM containing various concentration of IAA, IBA and NAA.

Weeks of observation	IAA concentration			IBA concentration			NAA concentration		
	1 ppm	2 ppm	3 ppm	1 ppm	2 ppm	3 ppm	1 ppm	2 ppm	3 ppm
6 weeks	2 leaves + 2 roots	1 leaf + no root	2 leaves + no root	2 leaves + no root	1 leaf + no root	4 leaves + no root	2 leaves + no root	1 leaves + no root	2 leaves + 1 root
12 weeks	3 leaves + 2 roots	2 leaves + no root	3 leaves +1 root	Yellower and dead	2 leaves + no root	4 leaves + no root	3 leaves + no root	2 leaves + no root	3 leaves + 1 root
18 weeks	3 leaves + 2 roots	2 leaves + no root	3 leaves + 1 root	-	4 leaves + no root	6 leaves + no root	4 leaves + no root	3 leaves + no root	4 leaves + 1 root

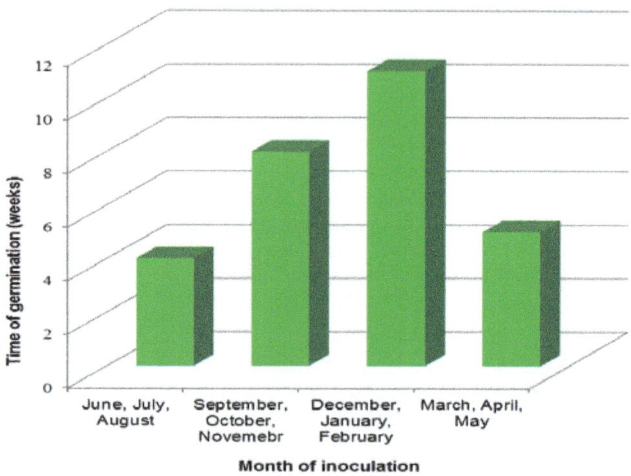

Figure 1. Effect of spore collection date on spore germination.

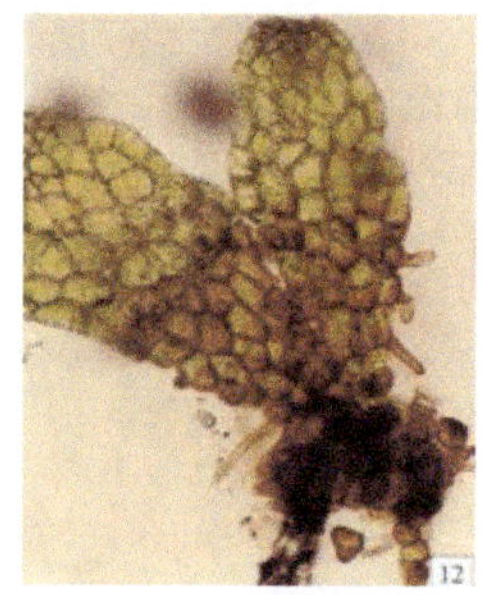

Figure 4. Gametophyte showing meristematic region at apical end and non-germinated spores below gametophye after 12 weeks (100x).

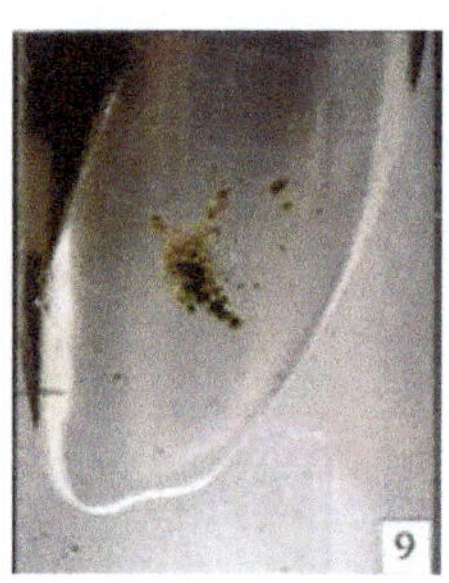

Figure 2. Sweeling, expansion and greening of spore.

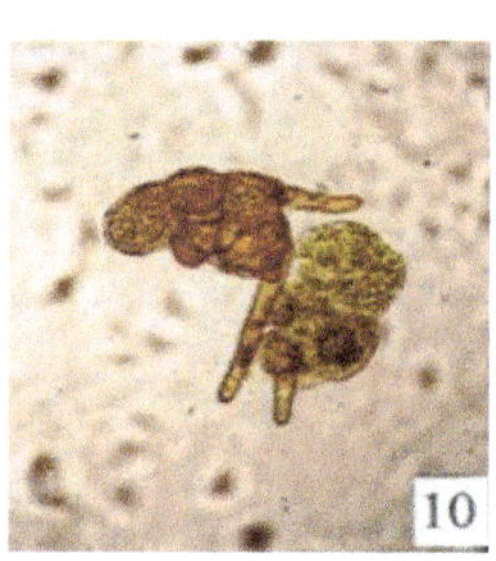

Figure 3. 8-Celled prathallus after 8 weeks (400x).

Structure of prothallus/gametophyte

Spatulate gametophytes with clearly visible meristematic cells at the apical end were observed after 12 weeks (Figure 4). At that stage, rhizoids were shorter than the length of the gametophyte. Numbers of rhizoids were more towards the distal end while no rhizoids were observed towards apical end. After 14 weeks, apical notch began to form at the anterior end below that laid the growing apex, consisting of meristematic cells. After 16 weeks, mature heart-shaped gametophytes were found. The growing apex was present in a deep notch at its anterior end, consisting of meristematic cells (Figure 5). The gametophytes were dorsoventrally flattened and numerous delicate rhizoids ascended from under surface. The rhizoids were unicellular, hyaline in colour, and double-layered wall. There were a single nucleus and a large number of chloroplasts in each cell and there were no intercellular spaces between the cells. Gametophytes of different age (20 and 22 weeks) grown on KBM + 2% sucrose are shown respectively in Figures 6 and 7. During entire developmental stages, the sex organs, that is, both antheridia and archegonia were not observed. This clearly indicated the development of apogamous sporophyte.

After three successive subcultures on KBM with 2% sucrose, the sporophytes initiation was observed only after 24 weeks of spore inoculation (Figure 8). After 26 weeks of culture, both gametophyte and apogamous sporophyte were observed in same medium (Figure 9).

Figure 5. Mature heart-shaped gametophyte clearly showing notched meristematic region and chloroplast affter 16 weeks (100x).

Figure 6. Gametophytes inside culture tube after 20 weeks.

Figure 7. Gametophytes inside culture tube after 22 weeks.

Figure 8. Development of sporophytes start after 24 weeks.

Figure 9. 26 weeks old culture showing both gametophyte and apogamous sporophyte.

Figure 10. 30 weeks old apogamous sporophyte.

After 30 weeks, the well-developed sporophyte was obtained (Figure 10). At that stage, roots were very few in number, unbranched and feebly developed. Rhizome were small, leaves were small and bipinnate or bifid.

Figure 11. Sporophyte on KBM + 1 ppm IAA after 12 weeks.

Figure 12. Sporophyte on KBM + 1 ppm IAA after 18 weeks.

Figure 13. Sporophyte on KBM + 2 ppm IAA after 6 weeks.

Figure 14. Sporophyte on KBM + 3 ppm IAA after 12 weeks.

Culture of sporophyte

The sporophyte explants were obtained from the *in vitro* grown spore on KBM with 2% sucrose. They were sub-cultured on KBM containing various concentrations of IAA, IBA and NAA.

On KBM with 1 ppm IAA, there was small protuberance like gametophyte on lower part of sporophyte with two roots observed after four weeks of culture. After six weeks, there was sporophyte with two leaves and two roots. One leaf was added and altogether formed three leaves and two roots after 12 weeks (Figure 11). This was sub-cultured on new media with same concentration of hormone. After 18 weeks of subculture, sporophyte with 3 leaves and 2 roots was observed (Figure 12). On KBM with 2 ppm IAA, there was only one leaf, root failed to form after six weeks (Figure 13). After 12 weeks, one leaf was added and two-leaved sporophyte was formed. After 18 weeks, still there was only two-leaved sporophyte, roots were not developed. On KBM with 3 ppm IAA, sporophyte only with three leaves was observed after six weeks. After 12 weeks, the sporophyte with three leaves and one root was developed (Figure 14). After 18 weeks the leaf and root number were same as 12 weeks and no further development was observed.

The sporophyte cultured on KBM supplemented with 1 ppm IBA showed two new leaves and no root after six weeks of culture (Figure 15). This remained as such for 10 weeks and after that ultimately became yellower and died. On KBM with 2 ppm IBA, after six weeks there was sporophyte with one leaf only and one gametophyte like body, but root was absent (Figure 16). After 12 weeks there was sporophyte with two leaves and no root (Figure 17). The gametophyte like body became brown and dead. After 18 weeks, sporophyte with four leaves and no root was observed (Figure 18). The sporophyte cultured on KBM supplemented with IBA 3 ppm showed four new leaves and no root after six weeks (Figure 19). Up to 12 weeks, the sporophyte remained as above. While after 18 weeks, two leaves were added and finally sporophyte with

Figure 15. Sporophyte on KBM + 1 ppm IBA after 6 weeks.

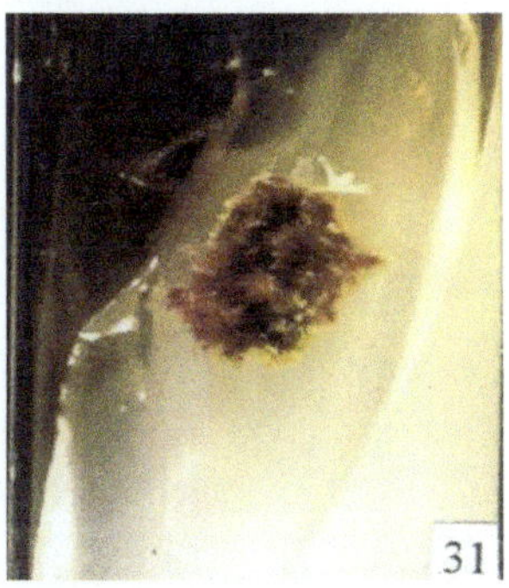

Figure 16. Sporophyte on KBM + 2 ppm IBA after 6 weeks.

Figure 17. Sporophyte on KBM + 2 ppm IBA after 12 weeks.

Figure 18. Sporophyte on KBM + 2 ppm IBA after 18 week.

Figure 19. Sporophyte on KBM + 3 ppm IBA after 6 weeks.

Figure 20. Sporophyte on KBM + 3 ppm IBA after 18 weeks.

with six leaves and no root was observed (Figure 20).

On KBM with 1 ppm NAA, the sporophyte with two leaves and no root was observed after six weeks (Figure 21). There was appearance of three leaves and no root after 12 weeks. After 18 weeks, there was sporophyte with four leaves and no root (Figure 22). When sporophyte was cultured on KBM with 2 ppm NAA, there was sporophyte with one leaf and no root after six weeks. After 12 weeks, two leaves sporophyte appeared where roots were to develop (Figure 23). After 18 weeks, sporophyte with three leaves were observed, root was not developed. On KBM supplemented with 3 ppm NAA, the sporophyte having two leaves and one root was observed after six weeks (Figure 24). After 12 weeks, three leaved and one-rooted sporophyte was found (Figure 25). The sporophyte with four leaves and one root was observed after 18 weeks (Figure 26).

When sporophytes were cultured on KBM with 2% sucrose, there were significant number of leaves but roots failed to develop. Roots were not observed up to 18 weeks (Figure 27).

Figure 21. Sporophyte on KBM + 1ppm NAA after 6 weeks

Figure 22. Sporophyte on KBM + 1 ppm NAA after 18 weeks.

Figure 23. Sporophyte on KBM + 2 ppm NAA after 12 weeks.

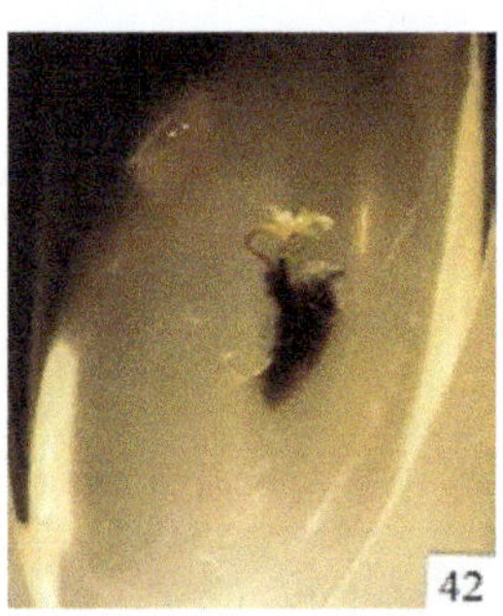

Figure 24. Sporophyte on KBM + 3 ppm NAA after 6 weeks.

Figure 25. Sporophyte on KBM + 3 ppm NAA after 12 weeks.

Figure 26. Sporophyte on KBM + 3 ppm NAA after 18 weeks.

Acclimatization

The rooted sporophytes were then transferred to soil and were kept at high humidity to avoid desiccation of the delicate fern fronds by covering with plastic from the upper side. The sporophyte with green leaves remained up to one week (Figure 28).

DISCUSSION

Sugar is an important factor for the spore germination. Kato (1973) observed that carbon compounds have a potent influence on protonema growth and secondary rhizoid formation. Parajuli and Joshi (2013) found 86% of spore germination of *Colysis latiloba* in 1% sucrocse. Similarly, in *Microsorium alternifolium* and *Microsorium scolopendria,* maximum germination was reported on media containing 1.0% sucrose and glucose while in *Melanophidium punctatum* and *Anisocampium cumingianum,* highest percentage of spore germination was reported in 0.25 and 1.5% sucrose respectively (Joshi, 1977). The findings of Miller and Miller (1961) showed that 1% sucrose for spore culture of *Onoclea sensibilis* growing under low intensities of light increased the growth rate of gametophyte. Whittier (1964, 1965) also found that vegetative growth of prothallus was accelerated by sucrose in *Cryptomium falcatum*, *Cheilanthes tomentosa* and *Cheilanthes farinose*.

Figure 27. Sporophyte on KBM after 12 weeks.

Figure 28. Acclimatization of rooted sporophyte.

The spore of ferns shows various germination pattern viz. bipolar, tripolar and amorphous germination (Rashid, 1999). Nayar and Kaur (1968) have considered bipolar pattern as most advanced where first division results in a small rhizoidal cell and large prothallial cell. *C. spinulosa* also exhibited bipolar germination in the study. The gametophyte of *C. spinulosa* was heart-shaped. Prothalli of homosporous fern follow a definite pattern of development and attain a characteristic adult form. Seven different patterns recognized (Nayar and Kaur, 1969) are: Osmuda-type, Marattia-type, Adiantum type, Drynaria-type, Kaulinia-type, Ceratopteris-type and Aspidum-type. The *C. spinulosa* gametophyte development was Adiantum-type.

Presence of sugar, carbon source for vegetative growth under sterile condition in the medium has been reported as one of the causal agents in the induction of apogamy (Kato, 1970; Mehra and Sulkalyan, 1969; Whittier 1964a, b). In the present work, sucrose at the concentration of 2% induced apogamous sporophytes in *C. spinulosa*. Similarly, Whittier and Steeves (1960, 1962) induced apogamy in *Pteridium aquilinum* by using 4.0% sucrose. They obtained greater response of apogamy at 5.0% sucrose. In *Ampelopteris* also 4.0% sucrose or 5.0% sucrose with 1 to 2 mg/l 2, 4-D was found optional for apogamous soprophyte initiation (Sulkalyan and Mehra, 1971). Earlier investigation of induced apogamy was done on soil grown gametophyte. Sulklyan and Mehra (1977) induced apogamy in the haploid and diploid gametophyte of *Nephrolepsis cordifolia* on Kundson's medium supplemented with 0.5 to 2% sucrose. Chopra and Loyal (1979) induced apogamy in four-months old culture of *Ceratopteris pteroides* containing 3-5% sucrose whereas the perennial gametophytes of *A. lepotophylla* exhibited apogamy in half year old cultures raised on sucrose free medium. Kawakami et al. (1995) has in-duced apogamous sporophyte in *Pteris multifida* asepti-cally in the dark but dark grown culture did not form apo-gamous sporophyte even when supplied with sugar (Whittier and Steeves, 1962). While, Elmore and Whittier (1974) has demonstrated the involvement of ethylene, a natural product of gametophyte metabolism as a true cause of apogamous bud induction.

In the present study, when the sporophytes of *C. spinulosa* were cultured on the KBM with different concentration of auxin, that is, IAA, IBA and NAA, best rooting was observed in KBM + IAA (1 ppm), KBM + IAA (3 ppm) and KMB + NAA (3 ppm), while no root was observed in IBA. Arockiasany et al. (2000) also found the regenerated shoots of *Pterocarpus santalinus* L. rooted on quarter strength of MS medium with IAA (1 mg/l). But IBA at 10 mg l^{-1} had a significant effect, both on rooting percentage and rooting system morphology of *Acacia tortilis raddiana* microcuttings, at NAA (1 to 5 mg l^{-1}) number of roots were low but strong (Sune et al., 2001). Copes and Madal (2000) has found 1 μM NAA treatment induced significantly greater rooting than the control in Douglas- fir cuttings. He also found 24.6 mM IBA produced greater rooting percentage. He further concluded that rooting decreased with both too little and too much auxin. Asami et al. (2000) found the rate of rooting from the adventitious shoots cultured in MS medium containing 0.1 mg l^{-1} NAA was higher than cultured in MS medium without NAA in the Japanese butterbur. Bonomo et al. (2013) noted that supplementing media with BAP affected the longiitudinal growth of the filamentous gametophytes, and the proliferation of ramification from early stage.

Conclusion

The result of the study suggested that *C. spinulosa* exhibited bipolar germination and the gametophyte development was Adiantum type. The mature gametophyte of *C. spinulosa* was observed as heart shaped. Sex organs were not observed up to 24 weeks which indicates the development of apogamous sporophyte. The sporophyte containing three leaves and two roots was the best result obtained in KBM containing 1 ppm IAA.

The knowledge on germination, morphogenesis and sporophyte development are very important to undersand the entire life cycle of *C. spinulosa* and to analyse the effect of sugar and various growth hormones. Moreover, these findings provide information to support the cultivation, management of conservation of this endangered species.

Conflict of Interests

The author(s) have not declared any conflict of interests.

ACKNOWLEDGEMENT

The authors gratefully acknowledge Central Department of Botany, Tribhuvan University (TU), Nepal for the lab facility for the study. Thanks are due to Dr. Bijaya Pant for valuable suggestion and Mr. Santosh Mani Nepal for encouragement. They are grateful to Dr. Maan Rokaya for review of this paper. They are also indebted to Dr. Sangeeta Rajbhandari, Mr. Sanjeev Rai, Mr. Sajan Subedi and Mr. Samik Mishra for their cooperation in photography.

REFERENCES

Albaum HG (1938a). Normal growth, regeneration and adventitious outgrowth formation in fern prothalia. Amer. J. Bot. 25:37-44.

Albaum HG (1938b). Inhibitions due to growth hormones in fern prothallia and sporophytes. Amer. J. Bot. 25:124-133.

Arens NC (2001). Variations in performance of the tree fern *Cyathea caracasana* (Cyatheaceae) across a successional mosaic in Andean cloud forest. Amer. J. Bot. 88(3):545-551.

Arockiasany S, Ignacimathus S, Melchias G (2000). Influence of growth regulations and explants types on *in vitro* shoot propagation and rooting of red sandal wood (*Pterocarpus santalinus* L.). Ind. J. Exp. Bio. 38(12):1270-1273.

Asami I, Ogawa RI, Inozuka M, Syumiya A (2000). Culture condition using cell tray for rooting of the adventitious shoots induced from petiole segments in Japanese butterbur. Research Bulletin of the Aichi-Ken Agricultural Research Center. 32(12):53-60.

Ballesteros D, Estrelles E, Walters C, Lbars AM (2011). Effect of storage temperature on green spore longevity for the ferns *Equisetum ramosissimum* and *Osmunda regalis*. CryoLetters. 32(2):89-98.

Bir SS, Goyal P (1982). Effect of ascorbic acid on fern gametophytes, spore germination and 2-D growth in *Pteris vittata* Lian. In Verma, S.C. Contemporary Trends in Plant Science: Kalyani publishers, New Delhi.

Bonomo MC, Martinez OM, Tanco ME, Cardonzo R, Aviles Z (2013). Spores germination and gametophytes of *Alsophila odonelliana* (Cyatheaceae) in different sterile media. Int. J. Exp. Bot. 82:119-126.

Caponetti JD (1970). Morphogentic studies on excised leaves of *Osmunda cinnamemer* L. Histological Studies of leaf development in sterile culture. Can. J. Bot. 48(6):1005-1016.

Chopra HK, Loyal DS (1979). Growth pattern and life habit of the fern gametophyte as factors in induced apogamy. In Bir SS Recent researches in Plant Science: Kalyani Publishes, New Delhi.

Convention on International Trade in Endangered Species of Wild Fauna and Flora (CITES) (2013). Appendices I, II and III.

Copes DL, Mandal NL (2000). Effects of IBA and NAA treatments on rooting Dougles-fir stem cuttings: New forests 20(3):249-257.

Du H, Li Y, Li D, Dai S, Jiang C, Shi L (2009). Effect of light, temperature and pH on spore germination and early gametophytic development of *Alsophila metteniana*. Bio. Sci. 17(2):182-187.

Elmore HW, Whittier DP (1974). Ethylene and carbohydrate requirement for apogamous bud induction in *Pteridium* gametophytes. Can. J. Bot. 52(9):2089-2096.

Gurung VL (1991). Ferns the beauty of Nepalese flora. Sahayogi Press, Kathmandu, Nepal.

Gurung VL (1992). *Cyathea spinulsosa* Wall. ex Hook: a Potential afforestation species. Ban Ko Jankari. 3:3.

Huang YM, Chiou WL, Lee PH (2001). Morphology of the gametophytes and young sporophytes of Cyatheaceae native to Taiwan. Taiwania. 46(3):274-283.

Joshi SD (1977). Morphogenetic and cytological studies in some liverworts and ferns. Unpublished Ph D thesis. Karnatak University, India.

Joshi SD, Pant B, Acharya N (2002). *In vitro* studies of spore germination and gametophytes morphogenesis in *Cyathea spinulosa*. Nuta J. 1(1):9-16.

Kato Y (1970). Physiological and morphogenetic studies of fern gametophytes and sporophytes formation in the dark cultured gametophyte of *Pteris vittata*. Bot. Gaz. 131(3):205-210.

Kato Y (1973). Control of rhizoid formation in the fern gametophytes. Cytologia. 38:117-124.

Kawakami SM, Ito M, Kawakami S (1995). Apogamous sporophyte formation in a fern *Pteris multiflora*. J. Plant Res. 108(1090):181-184.

Khare PB, Behera SK, Srivastava R, Shukla SP (2005). Studies on reproductive biology of a threatened tree fern, *Cyathea spinulosa* Wall. ex Hook. Curr. Sci. 89(1): 173-177.

Korall P, Pryer KM (2014). Global biogeography of scaly tree ferns (Cyatheaceae): evidence for Gondwanan vicariance and limited transoceanic dispersal. J. Biogeogr. 41:402-413.

Mehra PN, Sulklyan DS (1969). *In vitro* studies on apogamy and controlled differentiation of rhizome segments of the fern *Ampelopteris prolifera* (Fetz) Copel. Bot. J. Linn. Soc. 62(4):431-443.

Miller JH (1968). Fern gametophytes as an experimental material. Bot. Rev. 34:361-440.

Miller JH, Miller PM (1961). The effect of different light conditions and sucrose on the growth and development of the gametophyte of the fern *Onoclea sensibilis*. Amer. J. Bot. 48:154-159.

Naf U (1956). The demonstration of a factor concerned with the initiation of antheridia in Polypodiaceous fern. Growth. 20:91-105.

Naf U (1958). On the physiology of antherida formation in the bracken fern (*Pterdium aqulinium* (L) Kuhn.). Physiol. Plant. 11:728-746.

Nayar BK, Chandra P (1963). Observation on the morphology of the gametophyte of *Cyclosorus*. J. Ind. Bot. Soc. 42:392-400.

Nayar BK, Kaur S (1968). Spore germination in homosporous fern. J. Palynol. 4:1-14.

Nayar BK, Kaur S (1969). Types of prothallial development in homosporus ferns. Phytomorphology. 19:179-188.

Parajuli J, Joshi SD (2013). *In vitro* Morphogenesis of *Colysis latiloba* (Ching.) Ching. Int. J. Sci. Tech. Res. 2(3):247-253.

Rashid A (1999). An introduction to Pteridophyta. Vikas Publishing House, India.

Raubeson LA, Attwood DK, Stein DB (1995). The relationships of Papuasian Cyatheaceae to new world tree ferns. Amer. Fern J. 85(4):328-340.

Rechenmacher C, Schmitt JL, Droste A (2010). Spore germination and gametophyte development of *Cyathea atrovirens* (Langsd. & Fisch.) Domin (Cyatheaceae) under different pH conditions. Braz. J. Bio. 70(4) :1155-1160.

Simabukuro AE (1998). The effect of sterilization and storage conditions on the viability of the spores of *Cyathea delgadii*. Amer. Fern J. 88(2):72-80.

Sulklyan DS, Mehra PN (1971). Effect of auxin in growth and differentiation of the gametophytes of the fern *Ampelopteris prolifera* (Retz) Copel. Int. Symp. 17(23):5-6.

Sulklyan DS, Mehra PN (1977). *In vitro* morphogenetic studies in *Nephrolepis cordifolia*. Phytomorphology. 27:396-407.

Sussex IM, Clutter EG (1960). A study of the effect of externally supplied sucrose on the morphology of excise fern leaves *in vitro*. Phytomorphology. 10:87-99.

Whittier DP (1964a). The effect of sucrose on apogamy in *Cryptomium falcatum*. Amer Fern. J. 54:20-25.

Whittier DP (1964b). The influence of cultural condition on the induction of apogamy in *Pteridium* gametophytes. Am. J. Bot. 51:730-736.

Whittier DP, Steeves TA (1960). The induction of apogamy in bracken fern. Can. J. Bot. 38: 925-930.

Whittier DP, Steeves TA (1962). Further studies on induced apogamy in ferns. Can. J. Bot. 40:1525-1531.

Vegetation structure and composition in Guru Ghasidas Vishwavidyalaya in Central India

D. K. Patel

Department of Rural Technology, Guru Ghasidas Vishwavidyalaya, A Central University, Bilaspur, 495009, Chhattisgarh, India. E-mail: dplantscience@yahoo.co.in.

Diversity of plants includes their presence and habit in specific ecological area. Plants' adaptability, genetic makeup and related environment influence their composition. This present study aims to assess vegetation on campus. A total number of 273 plant species belonging to 84 families were recorded, in which Fabaceae registered as the largest family with 38 plant species. Herbs showed their maximum presence (65.93%) followed by trees (19.05%), herb/climbers (6.96%) and shrubs (8.06%). Rich presence of phanerophytes showed tropical moist humid climatic condition. Conservation practices and sustainable management strategies are essential steps for protection of natural resources, especially for the diverse vegetational group. Recorded major families of vegetation composition were Fabaceae (42.85%), Asteraceae (21.42%), Apocynaceae, Solanaceae (11.90%), Amaranthaceae and Poaceae (10.71%), while other families showed intermediate number of species. The campus is rich with diverse plant species, and by obtaining proper information about it, prime efforts can be made to conserve it.

Key words: Diversity, habit, plant species, university campus.

INTRODUCTION

Information of plant diversity is needed for the study of dynamic nature of vegetation under specific eco-environmental situation. Ecology, which is the study of diversity of species, has been used for the analysis of pattern, causes of extinction and management practices (Huston, 1994). Awareness and understanding of environment as well as biological diversity is increasing daily due to its rapid and significant role in maintaining ecosystem. Most vegetational diversity and dynamics found in tropical moist area are due to availability of better environment to the plant species. Guru Ghasidas Vishwavidyalaya GGV-Bilaspur (CV) is a central university in Central India of around 700 acres area. The site has rich diversity of plant species.

Biological resources provided by nature are not only the source of ecosystem balance, but also useful to social communities (Heywood, 1995). Vegetation of a specified ecosystem leads to and regulates the process of sustainability of the ecosystem. Bio-species may vary due to climatic condition as well as the tolerance capacity of the species in their natural habitat followed by their mode of propagation in next generation.

Species diversity and ecosystem fluctuation is regulated by human activities, deforestation, environmental changes, etc. Local climate of the area leads to the presence of the plant species followed by several environmental parameters, namely, temperature, pH, moisture, soil, biological pressure, etc. So, it becomes urgent to protect the biodiversity using effective methods.

Vegetational diversity of an area is influenced by local environmental and geographical situation and rainfall (Arora, 1995). Effect of climate change on the flora of mountains is noticed by Stanisci et al. (2005). Land management of ecosystem is a remarkable and key part of conservation of biodiversity. Odland and Birks (1999) reported that the population of vascular plants decreases with increase in altitude.

Tropical moist area includes rich species diversity. Many studies have been carried out on plant diversity of different parts of the world (Shimizu, 1991; Hussain et al., 2000; Aparajita et al., 2002; Banda et al., 2006; Kumar et al., 2006; De, 2007; Stanisci et al., 2005). Plant species respond to their environment (Harrison et al., 2001). Plant species are in high risk due to variable morphological structures (Daehler, 1998). Wilby (2007) focused on the problems of management of aquatic plants.

The aim of this study is to record the plant diversity on GGV campus for exploration of the variation in plant species based on field observations. It is a prime attempt to assess and have inventory knowledge of the plant species on GGV-Bilaspur campus (CG). Current study will be helpful for further studies in the related field.

MATERIALS AND METHODS

The present research was conducted during September 2009 to October 2010. Multiple random field observation was carried out for assessment of plants diversity (Phyto-sociological analysis) distributed on the campus. Bilaspur (CG) is a major city of Chhattisgarh State and GGV is a central university (area around 700 acres). The area is situated between 21°47' and 23°8' north latitudes and 81°14' and 83°15' east latitudes. It has an average elevation of 264 m (866 ft) near the banks of the rain-fed Arpa River with black-sandy soil (Figure 1). The climate of the area is tropical. It is hot and humid, because of its proximity to the Tropic of Cancer and depending on the monsoons for rains. There are medium rains in the monsoon season. Its summer is very hot with temperature between 30 and 47°C and between 5 and 25°C in winter.

Identification of plants was done by experts, literature, flora, and herbarium (Haines flora) and the plants were documented by following their botanical name, family, habits, and propagation method of the individual plants.

RESULTS

A total number of 273 plant species belonging to varied families (84) with different habits were recorded (Table 1). They were the major vegetation in the university campus.

Out of these plants species, herbs (180), herb/climber (19), shrubs (22), and trees (52) were noticed. Maximum plant species were recorded for Fabaceae family, whereas 39 families include single species in the university campus. Observed plant species with their family, habit, and propagation methods are listed in Table 1. Table 2 shows the family wise distributions of plants and in Table 3, the number of different habits is given. Table 4 shows plant species range belonging to their family, and in Table 5, propagation methods of different plant species are listed.

DISCUSSION

On the basis of the present findings, it is concluded that the university campus is enriched with various plants of different habits, and the knowledge about the plant species is essential for assessing them, though further strategy is needed to conserve them.

Plant species diversity (Shrubs) in Central Himalayan Region was recorded by Rikhari et al. (1997) and Ram et al. (2004). Pant and Samant (2007) reported that high biodiversity is due to variation in habitats and related environmental situations. Conserving the genetic diversity of plants is beneficial for current and also for coming generations (Olowokudejo, 1987). Carrying capacity of each ecosystem has a significant impact on long term basis and it sustains the utilization of available natural resources. Proper monitoring of the ecosystem is required for protecting the plant species.

This study indicates their rich diversity, followed by various habits (herbs, shrubs, and trees) due to suitable climatic condition as well as their survival capacity in the university campus. In the campus, herbaceous plants are mostly seen as they cover a larger part of the area (180 species of 65.93%), and are closely followed by trees (52 species of 19.05%), shrub (22 species of 8.06%) and herb/climber (19 species of 6.96%), the least of them all (Figure 2). The range base distribution related to their families was also investigated. The maximum number of plant species was recorded for a range of 0 to 5, with the plant species having a total of 67 families, whereas the single family Fabaceae which has the largest plant species has a range number of 38.

The most frequent plant species families were found in order of Fabaceae (38) > Asteraceae (18) > Euphorbiaceae, Lemnaceae (11) > Apocynaceae, Solanaceae (10) > Amaranthaceae, Poaceae (9) > Agavaceae, Araceae, Malvaceae (7) > Liliaceae, Moringaceae, Verbenaceae, Zingiberaceae (6) > Asclepiadaceae, Convolvulaceae (5) > Acanthaceae, Myrtaceae, Rubiaceae, Rutaceae (4) > Annonaceae, Comminaceae, Cucurbitaceae, Nyctaginaceae, Oxalidaceae, Polygonaceae (3) > Anacardiaceae, Apiaceae, Bixaceae, Combrataceae, Cypraceae, Hydrocharitaceae, Lytharaceae, Oleaceae, Palmae, Piperaceae, Rhmanaceae, Rosaceae, Scrophulariaceae, Vitaceae (2). The rest of the families listed in Table 2 includes single plant species.

This study also focused on the regeneration pattern of individual plant species as they propagate in new season/year for maintaining their existence in the campus. 67.40% of the studied plants show their propagation by seeds. 0.73 to 11.72% of the plants were propagated through seed/stem cutting, seed/tuber, rhizome, corm, bulb, bud, and tuber. Minimum percentage (0.38) of propagation by leaf was shown by *Kalanchoe pinnata* (Lam.) Pers.

Within the recorded mixed plant population, 11 species of the plants were aquatic in nature, recorded from water bodies in the campus. The aquatic plants are floating (*Azolla pinnata* R. Br., *Ipomoea aquitica* Forssk., *Lemna minor* Linn, *Nymphoea nouchali* Burm. F., *Pistia stratiotes* Linn, and *Trapa bispinosa* Roxb.) and submerged

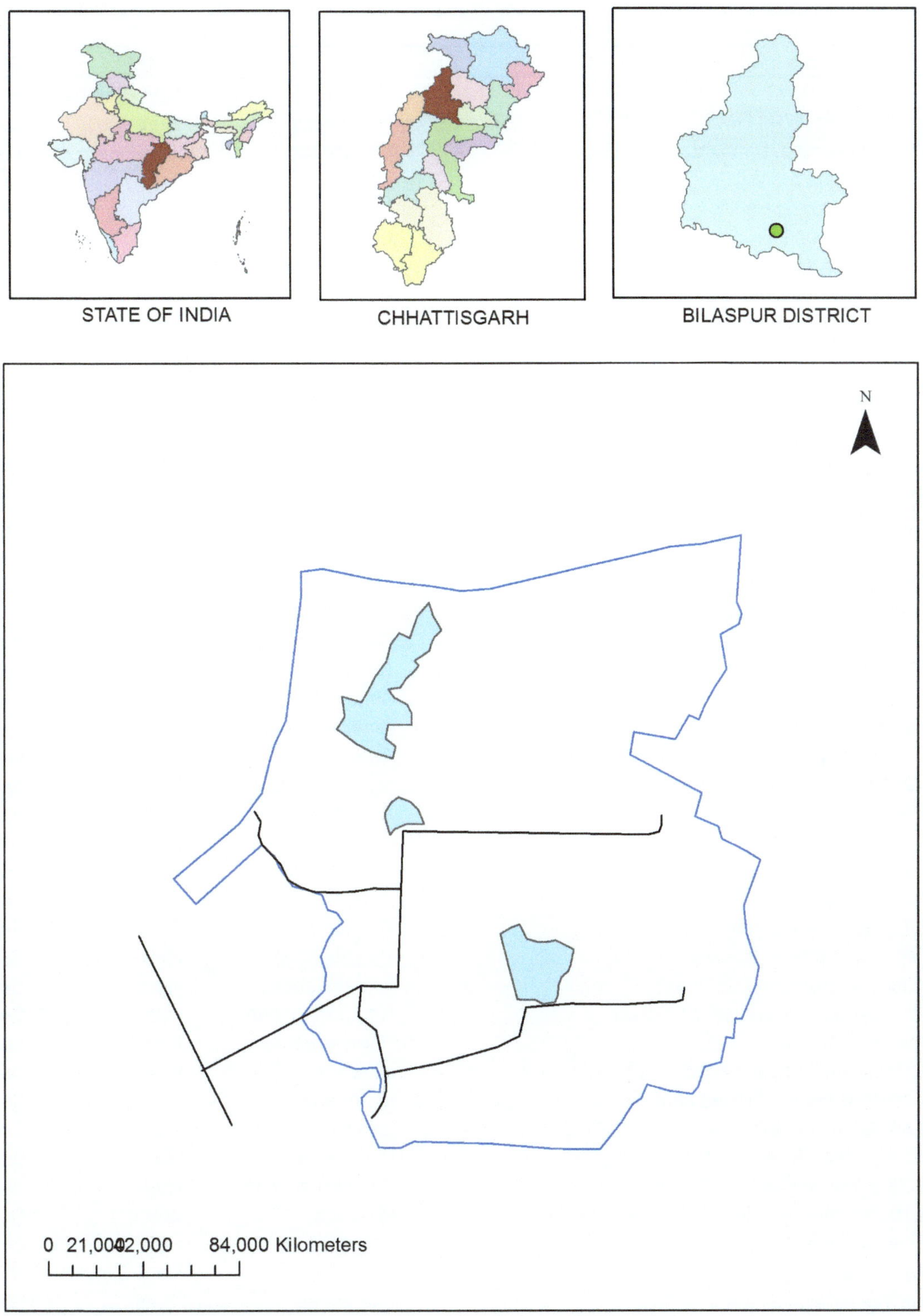

Figure 1. Location of university campus.

(*Ceratophyllum demersum* Linn., *Hydrilla verticillata* (L. F.) Royle., *Marsilea quadrifolla* Linn., *Vallisneria spiralis* Linn., and *Utricularia bifida* Linn.) in nature.

One species of Bryophyta (*Riccia fluitans* Linn.), one Pteridophyta (*Pteridium aquilinum* (L.) Kuhn.), two Gymnosperm (*Araucaria columaris* J. R. Forsk. Hook. and *Thuja oxidentalis* Linn.), one Epiphyte (*Vanda tessellate* (Roxb.) were also recorded during the course

Table 1. Diversity of plants in the G.G.V. (A Central University) Campus, Bilaspur, Chhattisgarh in Central India.

S/N	Botanical name	Family	Habit	Propagation
1	*Abrus precatorius* Linn.	Fabaceae	Herb/Climber	Seed
2	*Abutilon indicum* (L) Sw.	Malvaceae	Herb	Seed
3	*Acacia catechu* (L.F.) Willd.	Fabaceae	Tree	Seed
4	*Acacia nilotica* (L.) Willd.	Fabaceae	Tree	Seed
5	*Acacia pycnantha* Benth.	Fabaceae	Tree	Seed
6	*Acalypha indica* Linn.	Euphorbiaceae	Herb	Seed
7	*Achyranthes aspera* Linn.	Amaranthaceae	Herb	Seed
8	*Acorus calamus* Linn.	Araceae	Herb	Rhizome
9	*Adhatoda vasica* Linn.	Acanthaceae	Shrub	Stem cutting
10	*Aegle marmelos* (L.) Corr.	Rutaceae	Tree	Seed
11	*Aeschynomene indica*	Fabaceae	Herb	Seed
12	*Agave sesalana* Perr.	Agavaceae	Herb	Rhizome
13	*Ageratum conyzoiodes* Linn.	Asteraceae	Herb	Seed
14	*Ailanthus altissima* (Mill) Swingle.	Simaroubaceae	Tree	Seed
15	*Albizia lebbeck* Benth.	Fabaceae	Tree	Seed
16	*Allium cepa* Linn.	Liliaceae	Herb	Bulb
17	*Allium sativum* Linn.	Liliaceae	Herb	Bulb
18	*Aloe barbadensis* Mill	Liliaceae	Herb	Bud
19	*Alstonia scholaris* L.R.Br.	Apocynaceae	Herb	Seed
20	*Alternanthera philoxeroides Griseb.*	Amaranthaceae	Herb	Seed
21	*Alternanthera tenella* Moq.	Amaranthaceae	Herb	Seed
22	*Amaranthus caudatus* Linn.	Amaranthaceae	Herb	Seed
23	*Amaranthus spinosus* Linn.	Amaranthaceae	Herb	Seed
24	*Amorphophallus bubius* (Roxb.) Blume.	Araceae	Herb	Corm
25	*Anacardium occidentale* Linn.	Anacardiaceae	Tree	Seed
26	*Anacyclus pyrethrum* Linn.	Asteraceae	Herb	Seed
27	*Ananas comosus* Linn.	Bromeliaceae	Herb	Bud
28	*Andrographis paniculata* Nees.	Acanthaseae	Herb	Seed
29	*Annona squamosa* Linn.	Annonaceae	Tree	Seed
30	*Anthocephalus cadamba* Roxb.	Rubiaceae	Tree	Seed
31	*Antigonon leptopus* Hook & Arn.	Polygonaceae	Herb/Climber	Bulb
32	*Araucaria columaris* J. R. Forsk. Hook.	Araucariaceae	Herb	Seed
33	*Areva lanata* (Linn.) Juss. Ex Schult.	Amaranthaceae	Herb	Seed
34	*Argemone Mexicana* Linn.	Papaveraceae	Herb	Seed
35	*Asparagus racemosus* Willd.	Liliaceae	Herb	Seed/Tuber
36	*Atrocarpus heterophyllus* Lam.	Moraceae	Tree	Seed
37	*Azadirachta inaica* A. Juss.	Meliaceae	Tree	Seed
38	*Azolla pinnata* R. Br.	Salviniaceae	Herb	Bud
39	*Bacopa monnieri* Linn.	Scrophulariaceae	Herb	Stem cutting
40	*Bambusa bambos* Aka.	Poaceae	Shrub	Seed
41	*Barleria prionitis* Linn.	Acanthaceae	Herb	Seed
42	*Basella alba* Linn.	Basellaceae	Herb/Climber	Seed/Stem cutting
43	*Bauhinia variegata* Linn.	Fabaceae	Tree	Seed
44	*Biophytum sensitivum* (L.) DC.	Oxalidaceae	Herb	Seed
45	*Bixa orellana* Linn.	Bixaceae	Shrub	Seed
46	*Blumea lacera* (Burm. F.) DC.	Asteraceae	Herb	Seed
47	*Boerhaavia diffusa* Linn.	Nyctaginaceae	Herb	Seed
48	*Bombax ceiba* Linn.	Bombacaceae	Tree	Seed
49	*Bougainvillaea glabra* Choisy	Nyctaginaceae	Herb/Climber	Stem cutting
50	*Brassica compestris* Linn.	Cruciferae	Herb	Seed
51	*Butea monosperma* (Lamk.) Taub.	Fabaceae	Tree	Seed

Table 1. Continued.

52	*Caesalpinia crista* Linn.	Fabaceae	Shrub	Seed
53	*Caesalpinia pulchirima* Linn.	Fabaceae	Shrub	Seed
54	*Cajanus cajan* (L.) Millsp.	Fabaceae	Herb	Seed
55	*Caladium bicolour* Vent.	Araceae	Herb	Bulb
56	*Calistimon lanciolate* (Sm.) Sweet.	Myrtaceae	Shrub	Stem cutting
57	*Calotropis gigantean* (L.) Ait.F.	Asclepiadaceae	Shrub	Seed
58	*Calotropis procera* Aiton.	Asclepiadaceae	Shrub	Seed
59	*Canna indica* Linn.	Cannaceae	Herb	Rhizome
60	*Capsicum annuum* Linn.	Solanaceae	Herb	Seed
61	*Carica papaya* Linn.	Caracaceae	Herb	Seed
62	*Carissa carandas* Linn.	Apocynaceae	Shrub	Seed
63	*Cassia alata* Linn.	Fabaceae	Herb	Seed
64	*Cassia angustifolia* Vahl.	Fabaceae	Herb	Seed
65	*Cassia fistula* Linn.	Fabaceae	Tree	Seed
66	*Cassia occidentalis* Linn.	Fabaceae	Herb	Seed
67	*Cassia siamea* Lam.	Fabaceae	Tree	Seed
68	*Cassia tora* Linn.	Fabaceae	Herb	Seed
69	*Catharanthus roseus* (L.) G. Don.	Apocynaceae	Herb	Seed
70	*Ceiba pentandra* (L.) Gaertn.	Bombacaceae	Tree	Seed
71	*Celosia argentea* Linn.	Amaranthaceae	Herb	Seed
72	*Centella asiatica* (L.) Urban.	Apiaceae	Herb	Stem cutting
73	*Ceratophyllum demersum* Linn.	Ceratophyllaceae	Herb	Bud
74	*Cestrum nocturnum* Linn.	Solanaceae	Herb	Stem cutting
75	*Chenopodium album* Linn.	Chenopodiaceae	Herb	Seed
76	*Chlorophytum borivilianum* San. & Fer.	Liliaceae	Herb	Tuber
77	*Chrysanthemum indicum* Linn.	Asteraceae	Herb	Stem cutting
78	*Cicer arietinum* Linn.	Fabaceae	Herb	Seed
79	*Cissus quadrangualaris* Linn.	Vitaceae	Herb	Stem cutting
80	*Citrus lemon* (L.) Burm.F.	Rutaceae	Tree	Seed
81	*Cleome viscosa* Linn.	Cleomaceae	Herb	Seed
82	*Cleosia cristata* Linn.	Amaranthaceae	Herb	Seed
83	*Clerodendrum serratum* (L.) Moon.	Verbenaceae	Herb	Seed
84	*Clitoria ternatea* Linn.	Fabaceae	Herb/Climber	Seed
85	*Cocus nucifera* Linn.	Palmae	Tree	Seed
86	*Coleus forskholli* (Willd.) Briq.	Lamiaceae	Herb	Bulb
87	*Coleus blumei* Benth.	Lamiaceae	Herb	Seed/Stem cutting
88	*Colocasia esculenta* (L) Schott.	Araceae	Herb	Corm
89	*Convolvulus arvensis* Linn.	Convolvulaceae	Herb	Seed
90	*Coriandrum sativum* Linn.	Apiaceae	Herb	Seed
91	*Costus speciosus* (J. Konig) Sm.	Zingiberaceae	Herb	Seed
92	*Crinum latifolium* Linn.	Liliaceae	Herb	Bulb
93	*Crotalaria juncea* Linn.	Fabaceae	Herb	Seed
94	*Croton sparciflorus* (Morung.)	Euphorbiaceae	Herb	Seed
95	*Cucumis melo* Linn.	Cucurbitaceae	Herb	Seed
96	*Curculigo orchioides* Gaerth.	Hypoxidaceae	Herb	Rhizome
97	*Curcuma angustifolia* Roxb.	Zinziberaceae	Herb	Rhizome
98	*Curcuma aromatica* Linn.	Zingiberaceae	Herb	Rhizome
99	*Curcuma longa* Linn.	Zinziberaceae	Herb	Rhizome
100	*Cuscuta reflexa* Roxb.	Convolvulaceae	Herb/Climber	Seed
101	*Cymbopogon flexuosus* (Nees ex Steu) Wat.	Poaceae	Herb	Seed
102	*Cymbopogon martinii* (Roxb) Wats.	Poaceae	Herb	Seed
103	*Cynodon dactylon* Linn.	Poaceae	Herb	Seed/Stem cutting

Table 1. Continued.

104	*Cyperus iria* Linn.	Cyperaceae	Herb	Seed
105	*Cyperus rotundus* Linn.	Cyperaceae	Herb	Rhizome
106	*Dahelia coccinea* Cav.	Asteraceae	Herb	Bulb
107	*Dalbergia sissoo* Roxb.	Fabaceae	Tree	Seed
108	*Datura innoxia* Mill.	Solanaceae	Herb	Seed
109	*Datura metel* Linn.	Solanaceae	Herb	Seed
110	*Datura stramonium* Linn.	Solanaceae	Herb	Seed
111	*Deffenbachia bowmannii* Carriere.	Araceae	Herb	Stem cutting
112	*Delonix regia* (Boj. Ex Hook) Raf.	Fabaceae	Herb	Seed
113	*Dendrocalamus strictus* (Roxb) Nees.	Poaceae	Shrub	Seed
114	*Desmodium triflorum* Linn.	Fabaceae	Herb	Seed
115	*Desmodium motorium* (Houtt.)H.Ohashi.	Fabaceae	Herb	Seed
116	*Dioscoria bulbifera* Linn.	Dioscoriaceae	Herb	Tuber
117	*Diplocyclos palmatus* (L.) C. Jeffery.	Cucurbitaceae	Herb/Climber	Seed
118	*Dolichos lublub* Linn.	Fabaceae	Herb/Climber	Seed
119	*Dracaena dermemensis* Cvs.Engl.	Agavaceae	Herb	Stem cutting
120	*Dracaena fragrans* (L.) Ker. Gawl.	Agavaceae	Herb	Stem cutting
121	*Duranta erecta* Linn.	Verbenaceae	Herb	Seed/Stem cutting
122	*Echinochloa colona* Linn.	Poaceae	Herb	Seed
123	*Eclipta alba* (L.)Hassk.	Asteraceae	Herb	Seed
124	*Eclipta prostrata* Linn.	Asteraceae	Herb	Seed
125	*Emblica officinalis* Gaertn.	Euphorbiaceae	Tree	Seed
126	*Emilia sonchifolia* (L.) DC. exWeight.	Asteraceae	Herb	Seed
127	*Epipremnum aureum* (L.) Engl.	Araceae	Herb/Climber	Seed
128	*Eucalyptus globules* Labill.	Myrtaceae	Tree	Seed
129	*Euphorbia heterophylla* Linn.	Euphorbiaceae	Herb	Seed
130	*Euphorbia hirta* Linn.	Euphorbiaceae	Herb	Seed
131	*Euphorbia tirucalli* Linn.	Euphorbiaceae	Herb	Seed
132	*Evolvulus alsinoides* Linn.	Convolvulaceae	Herb	Seed
133	*Ficus bengalensis* Linn.	Moraceae	Tree	Seed
134	*Ficus pumila* Linn.	Moraceae	Herb/ Climber	Seed
135	*Ficus glomerata* Roxb.	Moraceae	Tree	Seed
136	*Ficus religiosa* Linn.	Moraceae	Tree	Seed
137	*Funaria hygrometrica* Hedw.	Funariaceae	Herb	Bud
138	*Galinsoga parviflora* Cav.	Asteraceae	Herb	Seed
139	*Garbera lanuginose* (DC) Schu.	Asteraceae	Herb	Seed
140	*Gmelina arboria* (Roxb.)	Verbenaceae	Tree	Seed
141	*Gossypium hirsutum* Linn.	Malvaceae	Herb	Seed
142	*Gymnema sylvestre* (Retz) R. Br.	Asclepiadaceae	Herb/ Climber	Stem cutting
143	*Hedychium coronarium* J. Koenig.	Zingiberaceae	Herb	Seed
144	*Hedyotis corymbosa* (L.) Lam.	Rubiaceae	Herb	Seed
145	*Helicterus isora* Linn.	Sterculiaceae	Tree	Seed/Stem cutting
146	*Hemidesmus indicus* (Linn.)R. Br.	Asclepiadaceae	Herb/Climber	Stem cutting
147	*Hibiscus cannabinus* Linn.	Malvaceae	Herb	Seed
148	*Hibiscus rosa- sinensis* Linn.	Malvaceae	Shrub	Stem cutting
149	*Hydrilla verticillata* (L. F.)Royle.	Hydrocharitaceae	Herb	Bud
150	*Hygrophylla spinosa* T.	Acanthaceae	Herb	Seed
151	*Hyophorbe lagenicaulis* (L. Bailey) H. E. Moore.	Palmae	Tree	Seed
152	*Hyptis suaveolens* (L.) Poit.	Lamiaceae	Herb	Seed
153	*Impatiens balsamina* Linn.	Balsaminaceae	Herb	Seed
154	*Ipomoea aquitica* Forssk.	Convolvulaceae	Herb	Seed/Stem cutting
155	*Ipomoea fistulosa* Mart.	Convolvulaceae	Herb	Seed/Stem cutting

Table 1. Continued.

156	*Ixora coccinea* Linn.	Rubiaceae	Herb	Stem cutting
157	*Jasminum sambac* (L.) Aiton.	Oleaceae	Herb	Stem cutting
158	*Jatropha curcus* Linn.	Euphorbiaceae	Shrub	Seed/Stem cutting
159	*Jatropha gossypifolia* Linn.	Euphorbiaceae	Herb	Seed/Stem cutting
160	*Kalanchoe pinnata* (Lam.) Pers.	Crassulaceae	Herb	Leaf
161	*Lagestroemia indica* (L.) Pers.	Lythraceae	Tree	Seed
162	*Lantana camara* Linn.	Verbenaceae	Shrub	Seed
163	*Lawsonia inermis* Linn.	Lythraceae	Shrub	Seed
164	*Lemna minor* Linn.	Lemnaceae	Herb	Bud
165	*Leucas aspera* Willd.	Lamiaceae	Herb	Seed
166	*Leucena lucocephala* (Lam.) de Wit.	Fabaceae	Tree	Seed
167	*Linnum usitatissium* Linn.	Liniaceae	Herb	Seed
168	*Madhuca indica* Gmel.	Sapotaceae	Tree	Seed
169	*Mangifera indica* Linn.	Anacardiaceae	Tree	Seed
170	*Marsilea quadrifolla* Linn.	Marsileaceae	Herb	Seed/Stem cutting
171	*Martynia annua* Linn.	Martyniaceae	Shrub	Seed
172	*Mentha piperata* Linn.	Lamiaceae	Herb	Seed
173	*Mentha arvensis* Linn.	Lamiaceae	Herb	Seed
174	*Mimosa pudica* Linn.	Fabaceae	Herb	Seed
175	*Mirabilis jalapa* Linn.	Nyctaginaceae	Herb	Seed/Tuber
176	*Mitragyna parvifolia* (Roxb) Korth.	Rubiaceae	Tree	Seed
177	*Momordica chrantia* Descourt.	Cucurbitaceae	Herb/Climber	Seed
178	*Moras alba* Linn.	Moraceae	Tree	Seed/Stem cutting
179	*Moringa oleifera* Lamk.	Moringaceae	Tree	Seed
180	*Mucuna pruiens* (L.) DC.	Fabaceae	Herb/Climber	Seed
181	*Muehlenbeckia platyclada* (F. Muell.) Meisn.	Polygonaceae	Herb	Seed
182	*Murraya koenigii* Spreng.	Rutaceae	Tree	Seed
183	*Murraya paniculata* (L.) Jack.	Rutaceae	Tree	Seed
184	*Musa paradiscia* Linn.	Araceae	Herb	Seed
185	*Nerium indicum* F. Le. Makino.	Apocynaceae	Herb	Seed/Stem cutting
186	*Nyctanthes arbour-tristis* Linn.	Oleaceae	Shrub	Seed/Stem cutting
187	*Nymphoea nouchali* Burm. F.	Nymphaceae	Herb	Rhizome
188	*Ocimum basilicum* Linn.	Lamiaceae	Herb	Seed
189	*Ocimum canum* Sims.	Lamiaceae	Herb	Seed
190	*Ocimum gratissimum* Linn.	Lamiaceae	Herb	Seed
191	*Ocimum sanctum* Linn.	Lamiaceae	Herb	Seed
192	*Opuntia stricta* Haw.	Cactaceae	Shrub	Stem cutting
193	*Opuntia elatior* Mill.	Cactaceae	Herb	Stem cutting
194	*Oryza sativa* Linn.	Poaceae	Herb	Seed
195	*Oxalia reticulata* Linn.	Oxalidaceae	Herb	Seed
196	*Oxalis corniculata* Linn.	Oxalidaceae	Herb	Seed
197	*Parthenium hysterophorus* Linn.	Asteraceae	Herb	Seed
198	*Passiflora foetida* Linn.	Passifloraceae	Herb/Climber	Stem cutting
200	*Peltaforum pterocarpum* (D.C.) K. Heyne.	Fabaceae	Tree	Seed
201	*Pergularia daemia* Forsk.	Asclepiadaceae	Herb	Seed
202	*Phyllanthus amarus* Linn.	Euphorbiaceae	Herb	Seed
203	*Phyllanthus niruri* Linn.	Euphorbiaceae	Herb	Seed
204	*Physalis minima* Linn.	Solanaceae	Herb	Seed
205	*Piper betel* Linn.	Piperaceae	Herb/Climber	Stem cutting
206	*Piper longam* Linn.	Piperaceae	Herb	Stem cutting
207	*Pistia stratiotes* Linn.	Araceae	Herb	Bud
208	*Pithecellobium dulce* (Roxb.) Benth.	Fabaceae	Tree	Seed

Table 1. Continued.

209	*Plumbago zeylanica* Linn.	Plumbaginaceae	Herb	Seed/Stem cutting
210	*Plumeria rubra* Linn.	Apocynaceae	Tree	Stem cutting
211	*Polianthes tuberosa* Linn.	Agavaceae	Herb	Seed
212	*Polyalthia longiflora* Sonn.	Annonaceae	Herb	Seed
213	*Polygonum barbetum* Linn.	Polygonaceae	Herb	Seed
214	*Pongamia pinnata* (L.) Merr.	Fabaceae	Tree	Seed
215	*Populus deltoids* Bartr.	Salicaceae	Tree	Seed
216	*Portulaca oleracea* Linn.	Portulaceae	Herb	Stem cutting
217	*Prunus amygdalus* Batsch.	Rosaceae	Tree	Seed
218	*Psidium guajava* Linn.	Myrtaceae	Tree	Seed
219	*Psium sativum* Linn.	Fabaceae	Herb	Seed
220	*Pteridium aquilinum* (L.) Kuhn.	Dennstaedtiaceae	Herb	Bud
221	*Pterocarpus santalinus* L.F.	Fabaceae	Tree	Seed
222	*Punica granatum* Linn.	Punicaceae	Tree	Seed
223	*Quisqualis indica* Linn.	Combretaceae	Herb/Climber	Stem cutting
224	*Rauvolfia tetraphylla* Linn.	Apocynaceae	Herb	Seed
225	*Rauvolfia serpentina* Benth.ex Kurz.	Apocynaceae	Herb	Seed
226	*Riccia fluitans* Linn.	Ricciaceae	Herb	Bud
227	*Ricinus communis* Linn.	Euphorbiaceae	Herb	Seed
228	*Rosa indica* Linn.	Rosaceae	Shrub	Stem cutting
229	*Salicornia virginica* Fern & Brack.	Amaranthaceae	Herb	Seed
230	*Sansevieria roxburghiana* Schult. & Schult. F.	Agavaceae	Herb	Rhizome
231	*Scoparia dulcis* Linn.	Scrophulariaceae	Herb	Seed
232	*Sensiveria cylindrical* Bojer.	Agavaceae	Herb	Rhizome
233	*Sensiveria trifaciata* Prain.	Agavaceae	Herb	Rhizome
234	*Sesamum indicum* Linn.	Pedaliaceae	Herb	Seed
235	*Sesbania sesban* (Jacq) W. Wight.	Fabaceae	Herb	Seed
236	*Sida cordifolia* Linn.	Malvaceae	Herb	Seed
237	*Sida acuta* Burm. F.	Malvaceae	Herb	Seed
238	*Solanum indicum* Linn.	Solanaceae	Herb	Seed
239	*Solanum nigrum* Linn.	Solanaceae	Herb	Seed
240	*Solanum xanthocarpum* Linn.	Solanaceae	Herb	Seed
241	*Sonchus asper* Linn.	Asteraceae	Herb	Seed
242	*Sphaeranthus indicus* Linn.	Asteraceae	Herb	Seed
243	*Stevia rebaudiana* (Bert.) Bertoni.	Asteraceae	Herb	Stem cutting
244	*Syzygium cumini* (L.) Skeels.	Myrtaceae	Tree	Seed
245	*Tabernimontana coronaria* Br.	Apocynaceae	Shrub	Stem cutting
246	*Tabernimontana diverticata* (L.) R. Br. ex Roem & Schutt.	Apocynaceae	Shrub	Stem cutting
247	*Tagetes erecta* Linn.	Asteraceae	Herb	Seed
248	*Tagetes paluta* Linn.	Asteraceae	Herb	Seed
249	*Tamarindus indica* Linn.	Fabaceae	Tree	Seed
250	*Tectona grandis* Linn.	Verbenaceae	Tree	Seed
251	*Tephrosia purpurea* (L.) Pers.	Fabaceae	Herb	Seed
252	*Terminalia arjuna* Roxb.	Combretaceae	Tree	Seed
253	*Thuja oxidentalis* Linn.	Cupressaceae	Herb	Stem cutting
254	*Tinospora cordifolia* (Willd.) Miers.	Menispermaceae	Herb	Stem cutting
255	*Tradeschantia fluminensis* Vell.	Commelinaceae	Herb	Bud
256	*Tradischantia pallida* (Rose D. R.) Hunt.	Commelinaceae	Herb	Bud
257	*Tradischantia spathacea* Swartz	Commelinaceae	Herb	Bud
258	*Trapa bispinosa* Roxb.	Trapaceae	Herb	Stem cutting
259	*Tridax procumbens* Linn.	Asteraceae	Herb	Seed
260	*Triticum aestivum* Linn.	Poaceae	Herb	Seed

Table 1. Continued.

261	*Typha angustifolia* Linn.	Typhaceae	Herb	Seed
262	*Urena lobata* Linn.	Malvaceae	Shrub	Seed
263	*Urtica parviflora* Roxb.	Urticaceae	Herb	Seed
264	*Utricularia bifida* Linn.	Utriculariaceae	Herb	Seed
265	*Vallisneria spiralis* Linn.	Hydrocharitaceae	Herb	Bud
266	*Vanda tessellate* (Roxb.)	Orchidaceae	Herb/Climber	Seed/Stem cutting
267	*Vitex negundo* Linn.	Verbenaceae	Shrub	Seed/Stem cutting
268	*Vitis vinifera* Linn.	Vitaceae	Herb/Climber	Stem cutting
269	*Withania somnifera* Dunal.	Solanaceae	Herb	Seed
270	*Xanthium strumarium* Linn.	Asteraceae	Herb	Seed
271	*Zea mays* Linn.	Poaceae	Herb	Seed
272	*Zinziber officinale* Rose.	Zingiberaceae	Herb	Rhizome
273	*Ziziphus jujuba* Miller.	Rhamnaceae	Tree	Seed
274	*Ziziphus mauritiana* Lam.	Rhamnaceae	Tree	Seed

Table 2. Plant species distribution according to their families.

S/N	Family	Number of species
1	Acanthaceae	4
2	Agavaceae	7
3	Amaranthaceae	9
4	Anacardiaceae	2
5	Annonaceae	3
6	Apiaceae	2
7	Apocynaceae	10
8	Araceae	7
9	Araucariaceae	1
10	Asclepiadaceae	5
11	Asteraceae	18
12	Balsaminaceae	1
13	Basellaceae	1
14	Bixaceae	1
15	Bombacaceae	2
16	Bromeliaceae	1
17	Cactaceae	2
18	Cannaceae	1
19	Caracaceae	1
20	Ceratophyllaceae	1
21	Chenopodiaceae	1
22	Cleomaceae	1
23	Combretaceae	2
24	Commelinaceae	3
25	Convolvulaceae	6
26	Crassulaceae	1
27	Cruciferae	1
28	Cucurbitaceae	3
29	Cupressaceae	1
30	Cyperceae	2
31	Dennstaedtiaceae	1
32	Dioscoriaceae	1

Table 2. Continued.

33	Euphorbiaceae	12
34	Fabaceae	36
35	Funariaceae	1
36	Hydrocharitaceae	2
37	Hypoxidaceae	1
38	Lamiaceae	11
39	Lemnaceae	1
40	Liliaceae	6
41	Linaceae	1
42	Lythraceae	2
43	Malvaceae	7
44	Marsiliaceae	1
45	Martyniaceae	1
46	Meliaceae	1
47	Menispermaceae	1
48	Moraceae	6
49	Moringaceae	1
50	Myrtaceae	4
51	Nyctaginaceae	3
52	Nymphaceae	1
53	Oleaceae	2
54	Orchidaceae	1
55	Oxalidaceae	3
56	Palmae	2
57	Papaveraceae	1
58	Passifloraceae	1
59	Pedaliaceae	1
60	Piperaceae	2
61	Plumbaginaceae	1
62	Poaceae	9
63	Polygonaceae	3
64	Portulaceae	1
65	Punicaceae	1
66	Rhamnaceae	2
67	Ricciaceae	1
68	Rosaceae	2
69	Rubiaceae	4
70	Rutaceae	4
71	Salicaceae	1
72	Salviniaceae	1
73	Sapotaceae	1
74	Scorphulariaceae	2
75	Simaroubaceae	1
76	Solanaceae	10
77	Sterculiaceae	1
78	Trapaceae	1
79	Typhaceae	1
80	Utricaceae	1
81	Utriculariaceae	1
82	Verbenaceae	6
83	Vitaceae	2
84	Zingiberaceae	6

Table 3. Distribution of plants as per their habit.

S/N	Habit	Number of plant species	Distribution (%)
1	Herb	180	65.93
2	Herb/Climber	19	6.96
3	Shrub	22	8.06
4	Tree	52	19.05
5	Total	273	100.00

Table 4. Range of distributed plant species with their family.

S/N	Plant species number range	Number of belonging family
1	0 - 5	69
2	6 - 10	11
3	11 - 15	02
4	16 - 20	01
5	20 - 40	01
	Total family	84

Table 5. Propagation methods of different plants.

S/N	Method	Number of plant species	Percentage
1	Seed	184	67.40
2	Seed/Stem cutting	016	05.86
3	Stem cutting	032	11.72
4	Seed/Tuber	002	0.73
5	Rhizome	013	4.76
6	Corm	002	0.73
7	Bulb	008	2.93
8	Leaf	001	0.38
9	Bud	013	4.76
10	Tuber	002	0.73
	Total	273	100.00

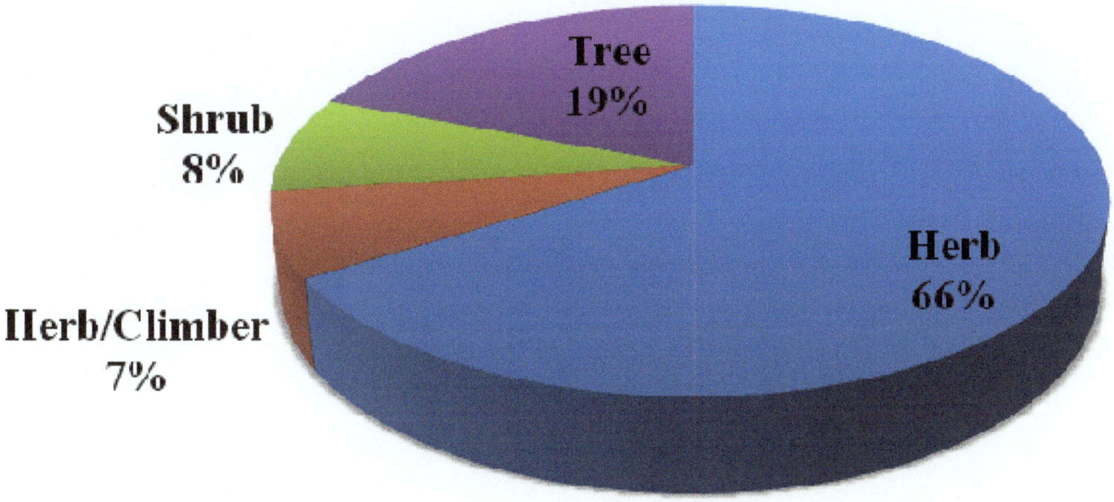

Figure 2. Habit of plants in percentage.

of the research.

REFERENCES

Aparajita H, Rawat GS, Tiwari AK (2002). Population structure of the corridor forest between Rajaji and CCorbert National Parks, Uttaranchal, India. Indian J. For. 25:310-318.

Arora RK (1995). Himalayan Resources, Diversity and Conservation, in U. Dhar(ed.), Himalayan Biodiversity, Gyanodaya Prakashan, Nainital. 2:39-55.

Banda T, Schwartz MW, Caro T (2006). Woody vegetation structure and composition along a rotection gradient in a miombo ecosystem of western Tanzania. For. Ecol. Manag. 230:179-185.

Daehler CC (1998). The taxonomic distribution of invasive angiosperm plants: ecological insights and comparison to agricultural weeds. Biol. Conserv. 84:167-180

De A (2007). Patterns of plant species diversity in the forest corridor of Rajaji- Corbert National Parks, Uttaranchal, India. Curr. Sci. 92(1):90-93.

Harrison PA, Berry PM, Dawson TE (eds) (2001). Climate change and nature conservation in Britain and Ireland: Modelling natural resources response to climate change (The MONARCH Project). UKCIP Technical report, Oxford.

Heywood VH (1995). Global Biodiversity assessment, UNEP, Combridge University Press, pp. 1140.

Hussain F, Iqbal I, Durrani MJ (2000). Vegetation studies on Ghalegay Hills, District Swat, Pakistan. Pak. J. Plant Sci. 6(1-2):1-10.

Huston MA (1994). Biological diversity: the co-exiting of species in changing landscape. Cambridge University Press, Cambridge.

Kumar A, Marcot BG, Saxena A (2006). Tree species diversity and distribution patterns in tropical forests of Garo Hills. Curr. Sci. 91(10):11-25.

Odland A, Birks HJB (1999). The altitudinal gradient of vascular plant richness in Aurland, Western Norway. Ecography 22:548-566

Olowokudejo JD (1987). Medicinal plants used as vermifuges in Nigeria and their conservation. J. Econ.Taxonomic. Bot. 9:459-466.

Pant S, Samant SS (2007). Assessment of Plant diversity and prioritization communities for conservation in Mornaula. Appl. Ecol. Environ. Res. 5(2):151-166.

Ram J, Kumar A, Bhatt J (2004). Plant diversity in six forest types of Uttaranchal, Central Himalaya, India. Curr. Sci. 86:975-978.

Rikhari HC, Adhikari BS, Rawat YS (1997). Woody species composition of temperate forests along an elevational gradient in Indian central Himalaya. J. Trop. For. Sci. 10(2):197-211.

Shimizu Y (1991). Forest types and vegetation zones of Yunnan China. J. Fac. Sci., Tokyo. Univ. Sec. III. Bot. 15:1-71.

Stanisci A, Pelino G, Blasi C (2005). Vascular plant diversity and climate change in the alpine belt of the central Apennines (Italy). Biodivers. Conserv. 14:1301-1318.

Wilby N (2007). Managing invasive aquatic plants: Problems and prospects. Aquatic Conservation: Marine Fresh Water Ecosyst. 17:659-665.

11

Conservation of folk healing practices and commercial medicinal plants with special reference to Nagaland

Rama Shankar[1] and Ramesh Babu Devalla[2]

[1]Ayurveda Regional Research Institute (CCRAS), Itanagar- Arunachal Pradesh, India-791111.
[2]Central Council for Research in Ayurveda and Siddha, 61-65, Institutional Area, Opp. D- Block, Janakpuri, New Delhi-110058, India.

Nagaland is one of the states of North Eastern India which is bordered in the East by Myanmar, Assam in the West, Manipur in the South, and Arunachal Pradesh and partly Assam in the North. It lies between 93.5 to 94.70 N longitude and 25.5 to 27.5 E latitude. The state as a whole is a hilly tract and occupying an area of 16,527 sq. km. The state consists of seven districts and 16 tribes and their sub-tribes. The state is rich in biodiversity as a source of medicinal plants as well as, the traditional method of healing practices and health management. The present paper deals with the details of studies on conservation of traditional healing practices of tribal people of Nagaland. Traditional practices in Nagaland are practiced by some of the herbalists who use the practice of treatment through single drug or in compound formulation. Due to modernization of the society these practices are gradually getting reduced which needs to be conserved through promoting of the traditional healers by taking them in practice through scientific organizations after passing through various scientific validation practices.

Key words: Conservation, folk medicinal plants, Nagaland, *Curculigo orchioides, Zanthoxylum alatum*

INTRODUCTION

Nagaland is one of the states of North Eastern India which is bordered in the East by Myanmar, Assam in the West Manipur in the South and Arunachal Pradesh and partly Assam in the North. It lies between 93.5 to 94.70° N longitude and 25.5 to 27.5° E latitude. The state as a whole is a hilly tract and occupies an area of 16,527 sq. km. The state consists of seven districts, 16 tribes and their sub-tribes. The people of the state are Naga tribes Angami, Aos, Sema, Lotha, Rengma Chakheang, Sangtam, Ronyak, Phom and Chang. They use Nagamese as local language and use Hindi, English and other languages in dialect. Main rivers flowing in the state are Dhansiri, Doyan, Dikhu, Milak, Zungki, and Tizu. The highest peak in the state is Saramati located at an altitude of 3877 MASL. Average rainfall in the state is 250 cm.

The districts under study of exploration of medicinal plants and their traditional healing practices that is, Kohima, Mokokchung, Phek and Wokha are inhabited by Angami with its sub tribes Chakroma, Tegima, Chakrima, Memi and Kezomi in Kohima, Lotha in Wokha and Ao with sub tribes Chongli, Monsa and Changki in Mokokchung district. Traditional system of healing practice is highly practiced in the state by using locally available herbs. Most of the plants used in traditional healing system are belonging to Ayurveda; Siddha and Unani system of medicine in India however, the method of use is somewhat different. Due to modernization and development of health practices these traditional knowledge is gradually depleting from its inhabitants. Some of the valuable medicinal plants are also depleting from its Natural habitat due to over exploitation and taking into commercial trade only from its wild sources.

Some of the medicinal plants like *Quercus infectoria*

*Corresponding author. E-mail: rshankar58@gmail.com

Figure 1. Map of Nagaland showing areas under studies (Marked with xxx).

are exploited for use as fire wood. Attempt has been made to develop enthusiasm amongst the educated persons towards conservation of medicinal plants in their natural habitat through in situ cultivation.

MATERIALS AND METHODS

The study was conducted with an aim to explore highly economic medicinal wealth along with traditional method of treatment of areas taken under study in the districts of Mokokchung, Kohima Phek and Wokha as shown in Figure 1. With the implementation of previous experience of medicinal plants exploration in other states authors initially has collected the medicinal plants from different areas of the districts undertaken in the study. Details of field data were mentioned in the field book and passport data sheets. Meanwhile, identification of some traditional practitioners was also made whose assistance was taken by gathering information on the method of use of herbal medicines and local herbs. Though problems were identified for exposing the information but based on previous experience of ayurvedic medicinal plants exploration and uses that some has exposed them to with the hope that their knowledge is also getting refreshed with experienced worker in the field of medicinal plants. Gradually the identity and methodology was noted and compared with the methodology of use in other systems of herbal medicine.

The collected plant materials were dried under blotting sheets and poisoned with mercuric chloride in 90% alcohol. Dried and poisoned voucher plants were mounted on herbarium sheets and the field information were written on the herbarium sheets. Identity of the collected materials was confirmed with the authentic voucher specimens kept in the herbarium and museum of Ayurveda Regional Research Institute, Itanagar. Local people were advised towards cultivation of locally available medicinal plants in their natural habitat. Species selected for such conservation are *Abroma augusta, Acorus calamus, Asparagus racemosus, Costus speciosus, Curcuma caesia, Curculigo orchioides, Eclipta prostrata, Oroxylum indicum, Quercus infectoria, Solanum nigrum, Taxus wallichiana, Thalictrum foliolosum, Zanthoxylum acanthopodium.* To serve the purpose of conservation a good germ-plasm having viable seeds, quality planting material has been established. For the development of germ plasm bank attempted efforts has been made for germination of seeds, and vegetative cuttings, etc.

OBSERVATION AND RESULTS

Records of the plant materials were compared with the available literature on the medicinal plants of Nagaland. The information gathered during field excursion in the concern three districts was totally new for the areas of study. The observation of the study was given in Table 1 by describing the botanical name, family of the plants, the names used in Indian System, local names of plants if available. The use and method of use has given in separate column.

Enumeration of folk medicinal plants

Stable cultivation measures for conservation of high demand medicinal plants like *Abroma augusta, A. calamus, A. racemosus, C. speciosus, C. buchananii, C. caesia, C. orchioides, E. prostrata, O. sanctum, O. indicum, Q. infectoria, S. nigrum, Rubia cordifolia, T. foliolosum, Z. acanthopodium,* and *Z. alatum* has been studied and accordingly, viable seeds were germinated and planted in case of *A. racemosus, C. orchoides, E. prostrata, Ocimum sanctum, O. indicum, Q. infectoria, R. cordifolia, S. nigrum, T. wallichiana, T. foliolosum* and *Zanthoxylum* species whereas rhizome, stem or root pieces of *A. racemosus, C. speciosus, C. caesia* and *R. cordifolia* were selected for cultivation. Stem pieces with two internodes are suitable for giving rise to vegetative propagules of *R. cordifolia.* Out of traditional medicines certain plants fall under threaten due to acute pressure on their habitat through different developmental activities and unsystemic collection of well recognized medicinal plants like *C. speciosus, C. caesia, O. indicum* and *T. wallichiana*. Besides, Traditional healers along with local people are also developing interest towards cultivation of some high demand medicinal plants like *Aloe barbadensis, Houttynia cordata* in their own field. There is the need of refreshing their knowledge for systemic collection and motivating towards cultivation of high demand threatened medicinal plants.

DISCUSSION

Nagaland comprising of 8 districts have been medico botanically explored and special attention was made with an attention of assessment of potentiality of medicinal plans described in established system of medicine with an attempt towards conservation of commercially viable medicinal plants. During exploration, traditionally used medicines and medicinal plants by different communities falling under explored areas of Nagaland that is, Kohima, Mokokchung, Phek and Wokha districts were also explored. Though attempt was made by several workers in the past viz., Jamir, 1989, 1990, 1991, 1997; Jamir and Rao, 1990a, 1990b; Rao and Jamir, 1980a, 1980b. But the efforts were only up to the limit of folk claims and not the status of comparative status of folklore claims and established medicinal plants in world. Attempt towards selection of commercially viable medicinal plants in different pockets of Nagaland is the first attempt. Local people were advised towards cultivation of locally available medicinal plants in their natural habitat. Species selected for such conservation are *A. augusta, A. calamus, A. racemosus, C. speciosus, C. buchananii, C. caesia, Curculigo orchioides, E. prostrata, O. sanctum, O. indicum, Q. infectoria, S. nigrum, R. cordifolia, T. wallichiana, T. foliolosum, Z. acanthopodium,* and *Z. alatum* as shown in Figure 2.

The areas taken under study through exploration of medicinal plants and their traditional uses are Kohima, Mokokchung, Phek and Wokha are inhabited by Angami with its sub tribes Chakroma, Tegima, Chakrima, Memi and Kezomi in Kohima, Lotha in Wokha and Ao with sub-tribes Chongli, Monsa and Changki in Mokochung district. All tribal communities are using herbs in its own ways for the treatment of their ailments. During the course of study, it was observed that different ethnic groups in the area have some specific knowledge of use of plants for treatment of ailments either by single drug or combination of plant parts. At several extents it was also observed that some ritual spirits are also used for treatment of ailments.

On systematic positioning of the plants used in the system of treatment, it was found that distribution of plant was common in the families of Acanthaceae, Amaranthaceae, Asclepiadaceae, Caesalpiniaceae, Euphorbiaceae, Malvaceae and Zingiberaceae. Though, certain difficulties were observed at the time of developing interaction with the healers however, they came forward for sharing their method of treatment. Traditional healers are some specific persons only in the villages who are dedicated to the job of treatment. Use of

Table 1. Enumeration of Folk medicinal plants.

Botanical name	Family	Vernacular name	Use
Abroma augusta L.f.	Sterculiaceae	Pishach Karpas/ Ulat Kambal	Root decoction taken orally in leucorrhoea.
Achyranhes bidentata Bl.	Amaranthaceae	Apamarga	Root used in vase pain and rheumatic diseases.
Acorus calamus L.	Araceae	Vacha	Rhizome paste used in asthma.
Ageratum adenophora (Spr.) King. & Rob.	Asteraceae	Sahadevi/ Chinapo	Plant juice and paste used to stop bleeding
A. Conyzoides L.	Asteraceae	Sahadevi	Plant juice and paste used to stop bleeding.
Albizia lebbek (L.) Willd.	Mimosaceae	Sirisa	Decoction of stem bark used in cough.
A. macrophylla L.	Mimosaceae	Sirisa Bheda	Decoction of stem bark used in cough.
Aloe barbadensis Mill.	Liliaceae	Ghritakumari	Leaf pulp is applied on face as face cleaner, leaf juice taken orally in case of loss of appetite. Pulp applied on forehead in headache.
Allium sativum L.	Liliaceae	Ziva	Bulb used to remove in cough
Alnus nepalensis Don	Betulaceae	Udis	Root used in diarrhea, dysentery.
Alpinea galangal (L.) Willd.	Zingiberaceae	Kulanjan	Rhizome paste is applied on joint pain and sprain.
Alstonia scholaris (L.) Br.	Apocynaceae	Saptaparna	Powder of stem bark taken to cure malaria.
Amaranthus spinosus L.	Amaranhaceae	Kateli Chaulai	Plant useful as blood purifier.
Artemisia nilagarica (Cl.) Pamp.	Asteraceae	Damanak; Ditibati (Angamis)	Leaf fume is used as mosquito repellant.
Artocarpus heterophylla Lamk	Moraceae	Panas; Piongtong (Ao)	Fruit is eaten as nutritious diet and as tonic.
Asparagus racemosus Willd.	Liliaceae	Satavari	Root pounded into paste and taken orally in jaundice and enhancing appetite. Leaf eaten as vegetable.
Bauhinea purpurea L.	Caesalpiniaceae	Kovidar / weiben-tong(Ao)	Tender leaves used as vegetable with concept to improve digestion.
B. variegata L.	Caesalpiniaceae	Kanchanar	Stem bark used for throat disorders, tender leaf used as vegetable.
Belamcanda chinensis DC.	Iridaceae	Surajakanti	Decoction of rhizome used in tonsillitis as gargling agent.
Bombax ceiba L.	Bombaxaceae	Salmili/ Alitong	Paste of root bark is applied on joint and other fracture.
Brassica juncea (L.) Czern	Brassicaceae	Lahi sag	Boiled leaf is used as vegetable.
Breynia patens Benth.	Euphorbiaceae	Bahupraja, Kambhi	Plant is used in stomach disorders as astringent, stem juice in eye diseases.
Buddleja asiatica Lour.	Loganiaceae	Bana	Root used in malarial fever.
Callicarpa arborea Roxb.	Verbenaceae	Priyangu; Kacheton (Ao)	Fruit, stem and leaf used in skin disorders.
Cassia alata L.	Caesalpiniaceae	Dadmardan	Leaf juice/ paste used in skin diseases.
Cassia fistula L.	Caesalpiniaceae	Aragvadha	Fruit pulp used in jaundice and liver diseases.
C. occidentalis L.	Caesalpiniaceae	Kasamarda	Leaf decoction used in liver disorder and skin diseases.
C. tora L.	Caesalpiniaceae	Chakramarda	Leaf decoction used orally and paste is applied on skin diseases.
Castanopsis indica DC.	Fagaceae	Kozutong	Roasted seeds are eaten
Casytha filiformis L.	Lauraceae	Akasbel	Skin diseases, hair fall antifungal.
Centella asiatica (L.) Urban	Apiaceae	Mandookparni; Insenkorklatong	Plant used as appetizer and in case of worm infestation.

Table 1. Contd.

Chenopodium album L.	Chenopodiaceae	Vastuka	Leaf cooked and taken as vegetable for digestion skin diseases and urinary complaints.
Chromolaena odorata (L.) King & Robin.	Asteraceae	Tsurknapao	Leaf juice applied on skin diseases and also to stop bleeding.
Cissampelos pareira L.	Menispermaceae	Ambastha; Patha	Root used in fever, tonic.
Citrus limon Burm.f.	Rutaceae	Nimbu	Fruit juice is taken in vomiting and indigestion.
Citrus reticulatus L.	Rutaceae	Santara /Kamla	Fruit juice is nutritious juice.
Clerodendrum colebrookianum Walp.	Verbenaceae	Nephaphu; Orematong (Ao)	Leaf eaten as vegetable, to control hypertension.
C. serratum (L.) Moon	Verbenaceae	Bharangi	Leaf decoction used in fever and malaria for 7 days.
Coffea bengalensis Heyne ex Roem.	Rubiaceae	Coffee	Fruit powder used as stimulant, beverage.
Coix lachryma jobi L.	Poaceae	Gavedhuka	Root paste used as tonic.
Costus speciosus (Koem) Sm.	Araceae	Kebuk / Kipangpointon	Rhizome used in nervous disorder and aphrodisiac tonic. Juice of crushed rhizome is used in ear ache.
Crotalaria bialata Schr.	Fabaceae	Jhunjhuniya	Leaf rubbed on itching and ring worm.
Cryptolepis buchanani Roem. & Schult	Periplocaceae	Ishwari	Root/ stem is taken in abdominal pain, cholera and dysentery. Paste mixed with oil used in rheumatoid arthritis.
Curcuma amada Roxb.	Zingiberaceae	Amra Haridra	Rhizome used in cough and digestion.
C. aromatic Salisb.	Zingiberaceae	Bonhaldi	Rhizome used in indigestion, intestinal worms, joint pains and wound.
D. caesia Roxb.	Zingiberaceae	Narkachura	Rhizome used in cough, sprain and rheumatic pain.
Curculigo orchioids Gaertn.	Hypoxidaceae	Kali musali/ Talmuli	Root used in asthma, cough, and eye diseases.
Datura metel L.	Solanaceae	Dhatura	Leaf used in poultice in rheumatoid arthritis and joint pain.
Dendrophthoe falcata (Lf.) Etting	Loranthaceae	Vanda	Stem bark antifertile, contraceptive, brain tonic
Desmodium triflorum DC.	Fabaceae	Shalparni	Root used as tonic.
Desmostachys bipinnatus (L.) Stapf.	Poaceae	Kusha	Culm used in tooth ache with mustered oil.
Dioscorea pentaphylla L.	Dioscoreaceae	Kathalu	Root used as nutritious tonic, asthma and cough.
Drymaria cordata (L.)Willd. ex R. & S.	Caryophyllaceae	Abhijola	Plant paste used in ring worm, decoction used in cough.
Drynaria quercifolia (L.) Sm.	Polypodiaceae	Aswakatri	Rhizome used as tonic, leaf pinnules used in skin diseases.
Duabanga grandiflora(Roxb. ex DC.) Walp	Lathyraceae	Thora/ Khokan	Bark paste applied on skin diseases eczema.
Eclipta prostrate (L.) L.	Asteraceae	Bhringaraj	Used in liver disorder, hair dye.
Embelia ribes Burm. f.	Myrsinaceae	Vaibidanga	Fruit used in stomach ache, worm infestation and liver diseases, root used as tonic.
Emblica officinlis Gaertn.	Euphorbiaceae	Amlaki	Fruits used in stomach pain, eye complains, hair tonic
Engelhartia spicata Bl.	Juglandaceae	Ripekam/ Chhalbih	Plant juice used in fish poisoning.
Euphorbia hirta L.	Euphorbiaceae	Laghudugdhika	Plant paste and latex used in skin diseases.
Ficus lacor Buchh.-Ham.	Moraceae	Pakara	Bark decoction used in gynecological disorders, ulcer.
Hedychium spicatum Ham. Ex Sm.	Zingiberaceae	Sati	Rhizome used in cough and pain, skin diseases.
Hedyotis scandens Roxb.	Rubiaceae	Hylubi	Stem chewed in tooth ache.
Hibiscus abelmoschus L.	Malvaceae	Lata kasturika	Seeds used as tonic, Root used as aphrodisiac tonic.

Table 1. Contd.

H. rosa- sinensis L.	Malvaceae	Japa/ China rose	Leaf paste used in hair fall and hair dying.
Hodgsonia macrocarpa (Bl.) Cogne	Cucurbitaceae	Kagoba	Fruit juice in skin disease
Houttuynia cordata Thun.	Saururaceae	Fishkuri; Alimoli (Ao)	Root used as vegetable for improving appetite.
Justicea gendarussa Burm.	Acanthaceae	Kala vasaka	Leaf used as poultice to remove joint pain and arthritis
Kalanchoe pinnata (Lamk.) Pers.	Crassulaceae	Patharchur, Duportenga	Leaf juice used in urinary trouble, amoebic dysentery, gastric complaints. Plant paste is heated and applied on rheumatism affected part.
Leea indica (Burm. f) Merr.	Leeaceae	Kurkurjihwa,	Plant used in diarrhea, dysentery, cough.
L. macrophylla Roxb. ex Hornem.	Vitaceae	Dholasamudrika,	Dysentery, cough, Anticancer.
Leucas indica (L.) Br. ex Vatke	Verbenaceae	Dronapushpi	Leaf juice used in sinusitis and other nasal trouble. Tender leaf is eaten as vegetable.
Lablab purpureus (L.) Sw.	Fabaceae	Sem	Seed tonic and high nutritious food.
Lygodium japonicum (Thunb.) Sw.	Lygodiaceae Marattiaceae	Kalijhant	Plant paste applied on eczema and ring worm and wound, tender frond edible.
Maesa chisia Don	Myrsinaceae	Kozalongritong	Fruit edible, stem bark used in eye disease.
M. indica Willd.	Myrsinaceae	Naukling/ Changchang	Root and fruits useful in women diseases. Leaves heated on fire and rubbed on skin diseases.
Majorana hortensis Moench.	Lamiaceae	Maru, Banatulasi	Leaf juice used in cough, fever. Plant used for worm repellant.
Mallotus phillippinensis Muel. Arg.	Euphorbiaceae	Kampillaka	Fruit powder used in worm infestation.
Manihot esculenta Crantz	Euphorbiaceae	Khamalu	Root eaten as heavy nutritious food.
Mesua ferrea L.	Clusiaceae	Mangitong	Flower is used in dysentery.
Morus alba L.	Moraceae	Tuda	Bark purgative, vermifuge, fruit eaten for digestion, host for silk worm
O. basilicum L.	Lamiaceae	Nangparangtong	Leaf used in diabetes.
O. sanctum L.	Lamiaceae	Tulasi	Leaf decoction is taken in cough asthma and fever, useful in diabetes.
Paederia scandens (Lour.) Merr.	Rubiaceae	Prasarini, Paduri lota / Tsumenuli	Tender leaf used as vegetable for digestion and gastric complaints Roots leaf and seeds eaten as food vegetable.
Pandanus fascicularis Lamk.	Pandanaceae	Ketaki; Korangtong (Ao)	Root juice is used for incense, eye diseases.
Parkia timoriana (DC.) Merr.	Fabaceae	Chinabean, Yongcha	Young fruit used as high nutritious vegetable.
Passiflora idulis Sims	Passifloraceae	Passion fruit/ Intsulasau	Fruits eaten as nutritious food. Leaf decoction used for hypertension.
Phlogacanthus thyrsiformis (Hard.) Mabb.	Acanthaceae	Titi vasaka	Leaf juice/ decoction used in cough and joint pain.
Physalis minima L.	Solanaceae	Makoi	Fruit used in liver diseases.
Plantago erosa Wall.	Plantaginaceae	Aswagola	Leaf used as vegetable, spike used in dysentery.
Pouzolzia hirta Hassk (Blume) Hassk.	Urticaceae	Oik	Tender leaf used as vegetable to improve appetite.
P. sanguinea (Bl.) Merr.	Urticaceae	Notke	Tender leaves and flowers are used as vegetable to improve loose motion.
Polygonum capitatum Buch. - Ham. ex Don	Polygonaceae	Knotweed	Plant used in China in improving kidney and urinary diseases.
P. macrophylla Don	Polygonaceae	Kukhari	Leaf used as vegetable to improve appetite.
Prunus persica Batsch.	Rosaceae	Aaru	Fruit eaten and improves digestion.

Table 1. Contd.

Pseudognaphalium luteo- album (L.) Hill.	Asteraceae	Balaraksha	Leaves used in vulnerary disorder.
Pteridium aquilina (L.) Kuhn	Pteridaceae	Broken fern	Rhizome used as insecticide with storage of grains.
Pterocarpus marsupim Roxb.	Pteroacrpaceae	Red sandal	Stem wood paste used in skin diseases
Punica granatum L.	Punicaceae	Dadim	Fruit used in loss of appetite, dysentery and diarrhea. Leaf used in skin diseases.
Rubia cordifolia L.	Rubiaceae	Manjishtha	Root used in skin diseases and improving liver functioning. Root and stem used as dye protecting skin diseases.
Rubus rugosus Sm.	Rosaceae	Raspberry / Jempnesu	Crushed fruits applied on skin diseases/ fruits eaten as appetizer.
Rungia parviflora Nees	Acanthaceae	Parpati	Plant decoction used in cough.
Schima wallichii (DC.) Korth.	Theaceae	Mengtang	Leaf paste applied on cuts and wound.
Selaginella bryopteris (L.) Bak.	Selaginellaceae	Sanjivani	Plants used as remedy for liver disorder and epilepsy.
Sida acuta Burm.	Malvaceae	Bala	Root powdered and used in paralysis for local application with oil.
Smilax ovalifolia Roxb.	Smilacaceae	Chobchini bheda	Root extract used in arthritis.
Smilax wightii DC.	Smilacaceae Liliaceae	Chobchini	Root decoction used for lactating ladysm, rheumati.
Solena heterophylla Lour *amplexicaulis* (Lamk.) Gandhi Syn. *Melothria heterophylla* (Lour.) Cogne	Cucurbitaceae	Bankundri	Root used in arthritis, diabetes, spermatorhoea, snake bite.
Solanum ferox L. Syn. S. *indicum* L.	Solanaceae	Kantakari	Plant powdered and decoction used in cough and asthma, berries eaten as vegetable and worm infestation.
S. myriacanthum Dunal	Solanaceae	Kantakari / Akolonkohu	Plant decoction mixed with Tita vasa *Phlogacanthusthyrsiflorus* leaf used in cough. Mature dried seeds are burnt and smoke is allowed to enter into mouth as contraceptive.
S. tabacum Sw.	Solanaceae	Tambaku Brahati	Plant leaf is taken as narcotic drug and removal of leeches in field.
Sonchus oleraceous L.	Asteraceae	Hirankhuri	Root used in jaundice, liver complaints.
Sphenomeris chinensis (L.) Maxon.	Pteridaceae	Geteypati	Leaf pounded with water and paste used in joint pain
Spilanthus calva DC.	Asteraceae	Akarkara	Head used as anesthetic drug in tooth ache, Plant cooked and taken in body ache and joint pain.
S. paniculata Wall. ex DC.	Asteraceae	Akarkara bheda	Head used as anesthetic drug in tooth ache, plant cooked and taken in body ache and joint pain.
Thalictrum foliolosum DC.	Ranunculaceae	Mamira	Root used in liver disorders, malaria and other fever.
Trema orientalis Bl.	Ulmaceae	Jivanti	Used in pain in limb.
Urena lobata L.	Malvaceae	Vanabhenda	Root used as tonic, liver disorder.
Urtica dioica L.	Urticaceae	Bichhu pat	Leaf eaten with concept of good appetizer.
Valeriana jatamansi Jones	Valerianaceae	Tagara	Root used as ingredient for hypertension.
Zanthoxylum acanthopodium DC.	Rutaceae	Timburu; Changpet (Ao)	Fruits used in liver disorder and eye diseases.

drugs in the methods of their treatment was in the form of fresh plants, powdered drugs or some sort of extracts. Due to developmental activities

Figure 2. Photographs of medicinal plants used in traditional medicine in Nagaland.

and modernization towards health practices, traditional knowledge of the health practices are gradually depleting amongst new generation and it needs to be conserved. For the conservation of these traditional healing practices there is need for the development of common platform for such types of traditional healers and their methodology of use to be modified for longer use of prepared medicines. During the course of gathering information, it was observed that some sort of facilitation to healers is required for the sustainability of drugs as well as, the mode of treatment. The local inhabitants use parts of plans mostly, leaf stem bark, root and floral parts. Most commonly occurring diseases in the area are gastric trouble, hypertension, and jaundice, and skin diseases, cuts and wounds urinary trouble. Use of Zingiberaceous members is very common in cough cold and sprains. Acanthacceous members for cough, cold and joint pain are common in practice. For the conservation of this traditional knowledge, societies and associations at state and district level have been established during the course of interaction. There is need for developing traditional medicines in feasible medicines for longer use through powdering and granule preparation under good standard of practice.

REFERENCES

Jamir NS (1989). Some interesting medico botany used by Ao- A Naga tribe. Proc. Research Devel. Indig. Drugs (INMMR), New Delhi. pp. 259-264.

Jamir NS (1990). Some interesting medicinal plants used by Nagas. J. Res. Edu. Ind. Med., 9(2): 81-87.

Jamir NS (1991). Studies on some medico herbs from North East India. Recent Adv. Med. Arom. Spices Crops, 1: 235-239.

Jamir NS (1997). Ethnobiology of Naga tribes in Nagaland: 1, Medicinal plants: Ethnobotany, 9: 101-104.

Jamir NS, Rao RR (1990a). Fifty new or interesting medicinal plants used by the Zeliangs of Nagaland (India) Ethnobotany, 2: 11-18.

Jamir NS, Rao RR (1990b). Ethnobotany of Ao and Angami Nagas of Nagaland. J. Econ.Taxon. Bot., 36: 14(3): 593-604.

Rao RR, Jamir NS (1982a). Ethnobotanical studies in Nagaland -I Medicinal Plants. J. Econ. Bot., 36: 176.

Rao RR, Jamir NS (1982a). Ethnobotanical studies in Nagaland-II 54. Medicinal Plants. J. Econ. Taxon. Bot., 3(1): 11-17.

12

Diversity and regeneration status of tree species at Nainital Catchment, Uttarakhand, India

Kiran Bargali , Priyanka Bisht, Abida Khan and Y.S. Rawat

Department of Botany, Kumaun University, DSB Campus, Nainital -263002, Uttarakhand, India.

The existence of a species in a forest community largely depends on its ability to regenerate under varied environmental conditions. In the present study, effect of biotic disturbance was assessed and analysed on diversity and regeneration of pre-dominant tree species of oak forests. For this, two adjacent sites were selected in mixed oak forest zone (2100 m asl), namely, undisturbed open area (inside zoo) and moderately disturbed open area (out side zoo). Phytosociological analysis of tree and shrub layer vegetation was done and density, diversity index, Importance Value Index etc were determined for both the sites. Regeneration status of tree species was analysed by developing population structures. Species richness (tree as well as shrub) was high in disturbed site (11 to 20) as compared to protected site (9 to 11), and in both the sites the species richness was maximum at hill slope. Among the tree species *Quercus floribunda* was dominant with highest value of IVI in protected site. Contrary to this, *Quercus lecotrichophora* was dominant in disturbed site at hill base and hill slope, while *Q. floribunda* was dominant at hill top site. Tree density ranged from 780 to 1000 trees ha^{-1} in protected site and from 260 to 780 trees ha^{-1} in disturbed site. In both the sites, *Q. leucotrichophora* and *Q. floribunda* have higher proportion of individual in younger girth classes (seedlings and saplings), indicating expanding type of population structure. However, conversion into higher girth classes was more prominent at protected site as compared to disturbed site. Thus, it can be concluded that oak species of Kumaun Himalayan region have ability to regenerate when anthropogenic pressures are negligible.

Key words: Basal area, density, diversity, girth class, population structure, regeneration.

INTRODUCTION

Successful regeneration of tree species might be considered to a function of three major components: (i) ability to initiate new seedlings, (ii) ability of seedlings and saplings to survive and (iii) ability of seedlings and saplings to grow (Good and Good, 1972). Measurement of these parameters provides an insight into the regeneration of species in a forest community. Several authors have predicted regeneration status of tree species based on the age and diameter structure of their population (Bhuyan et al., 2003, Khan et al., 1987).

Kumaun region of the Central Himalaya harbors rich biodiversity because of its unique and diverse climate conditions. Composition of the forest is diverse and varies from place to place because of varying topography such as plains, foothills and upper mountains (Singh, 2006). Several studies have described the vegetation of Kumaun (Osmaston, 1927; Dhar et al., 1997; Singh and Singh, 1987; Hussain et al., 2008). Species composition of major forest types of Central Himalaya have already been described by Ralhan et al. (1982), Saxena et al. (1984), Singh and Singh (1987), Tewari and Singh (1981) and Upreti et al. (1985). Various changes in the

Himalayan Forests are appearing in their structure, density, composition and regeneration due to biotic pressure on them namely, uncontrolled lopping and felling of trees for fuel wood, fodder and grazing (Bargali et al., 1998; Kumar et al., 2004). These biotic pressures play an important role in forest community dynamics (Pickett and White, 1985) and regulate the regeneration ability of a species. The human influence on biodiversity and ecosystem functioning have largely taken in the form of rapid, large and frequent changes in land and resource use, increased frequency of biotic invasion, reduction in species number, creation of stresses and the potential for changes in climatic systems. Such changes have a direct impact through habitat destruction and over- exploitation of resources such as overgrazing and indirect impact through their impact on the composition of the atmosphere and climate. In the present study, tree and shrub layer vegetation was analysed in a protected and disturbed oak forest area of Nainital town. The aim of this study was: (i) to analyse the composition and diversity of tree and shrub layer in selected study sites within and around Nainital Zoo, and (ii) to determine the current natural regeneration status of the tree species at both the study sites.

MATERIALS AND METHODS

Study site

The present study area Nainital is located between 29° 21' to 29°24' N latitude and 79° 25' to 79°29' E longitude. The study sites are located within and around The Bharat Ratna Pt. Govind Ballabh Pant High Altitude Zoo, Nainital in Uttarakhand state of India (Figure 1). The area is dominated by mixed oak forest situated in the hills of Sher- ka danda. The elevation range between 2100 and 2200 m above sea level. The zoo was established in 1984 and it spread over an area of 4.693 ha. It was opened to visitors on 1st June 1995 and is being managed by The Bharat Ratna Pt. Govind Ballabh Pant High Altitude Zoo Management Society, Nainital since 1st March, 2002. This zoo was established with the objectives to conserve and protect the wildlife and biodiversity. The disturbed forest site is situated outside the boundary of zoo at the same altitudinal gradient and it spread over an area of about 6.50 ha.

For the detailed study of tree and shrub layer composition and other vegetational parameters, the area was divided into three sub-sites, that is, hill base, hill slope and hill top along the altitudinal gradient. These three sites are located at different altitudinal range and dominated by *Q. leucotrichophora* and *Q. floribunda* trees. The soil is residual to fairly deep being derived from dolomite limestone, sand stones with Quartzite and silt.

Climate

The climate of the study area is influenced by the monsoon pattern of rainfall. The annual rainfall was 2214mm of which three fourth occurred in rainy season (mid-June to mid-September). The mean monthly maximum temperature ranged between 17.1°C (January) to 26.1°C (June), and mean minimum temperature ranged between 7.1°C (January) to 26.1°C (June). Winters are very severe at the higher elevations and snowfall takes place above 1,524 m and occasionally lowers down but melts and even above this snow seldom stays long on sunny slopes.

Approach

The number and size of the quadrats were determined by the running mean method (Kershaw 1973) and species area curve (Misra, 1968). Ten plots of 10×10 m at each site (disturbed and protected) were randomly established at hill base (4 plots), slope (3 plots) and top (3 plots), within and outside the zoo area for determination of species richness and other vegetational parameters (total 20 plots). Trees and saplings were analysed in 10×10 m, shrubs in 5×5 m and seedlings in 10, 1×1 m area within each plot (Curtis and McIntosh, 1950; Phillips, 1959).Circumference at breast height (cbh at 1.37 m from the ground) of all the trees (≥ 31.5) was measured in each plot. Tree mean basal area of a species ($C^2/4\pi$ where, C = sum of cbh value of all individuals of a tree species within each plot and π = 3.14) was multiplied by its density value to calculate total basal area of a tree species. Importance Value Index (IVI) was calculated following Phillips, 1959) as:

IVI = Relative frequency + relative density +relative dominance.

Equitability (EC) or species evenness was determined following Whittaker (1975) as:

$$EC = \frac{s}{\text{Log } 10\ ni - \log 10\ ns}$$

Where s is the number of species in the site and ni and ns are the IVI value of most and least important species, respectively.

Species diversity (H) for each vegetation layer in each site was determined by using Shannon –Weiner (Shannon and Weiner, 1963) information index as:

$$H = -\sum_{I=1}^{s} (Ni/N) \log_2 (Ni/N)$$

Where; Ni= Total number of individuals of a species, N= Total number of individuals of all species

The log 2 is converted into log 10 as:

$$H = -\Sigma (Ni/N) \times 3.322 \times \log_{10}(Ni/N)$$

Concentration of Dominance (CD) was measured by Simpson's index (Simpson 1949) as:

$$C = \sum_{i=1}^{s} (Ni/N)^2$$

Where, Ni is the total number of individuals of species and N is the total number of individuals of all species.

Population structure was used to express the regeneration status of individual tree species. Individuals > 30 cm circumference were categorized as trees, 10 to 30 cm as saplings, and < 10 cm as seedlings to determine the regeneration status of tree species in each forest site (Saxena et al., 1984).

Index of similarity between protected and disturbed site was calculated following Sorenson (1948) using species richness as:

$$\text{Index of similarity (IS)} = \frac{2C}{A + B} \times 100$$

Where C is the common species in comparison forest, A is the total number of species at site A and B is the total number of species at site B.

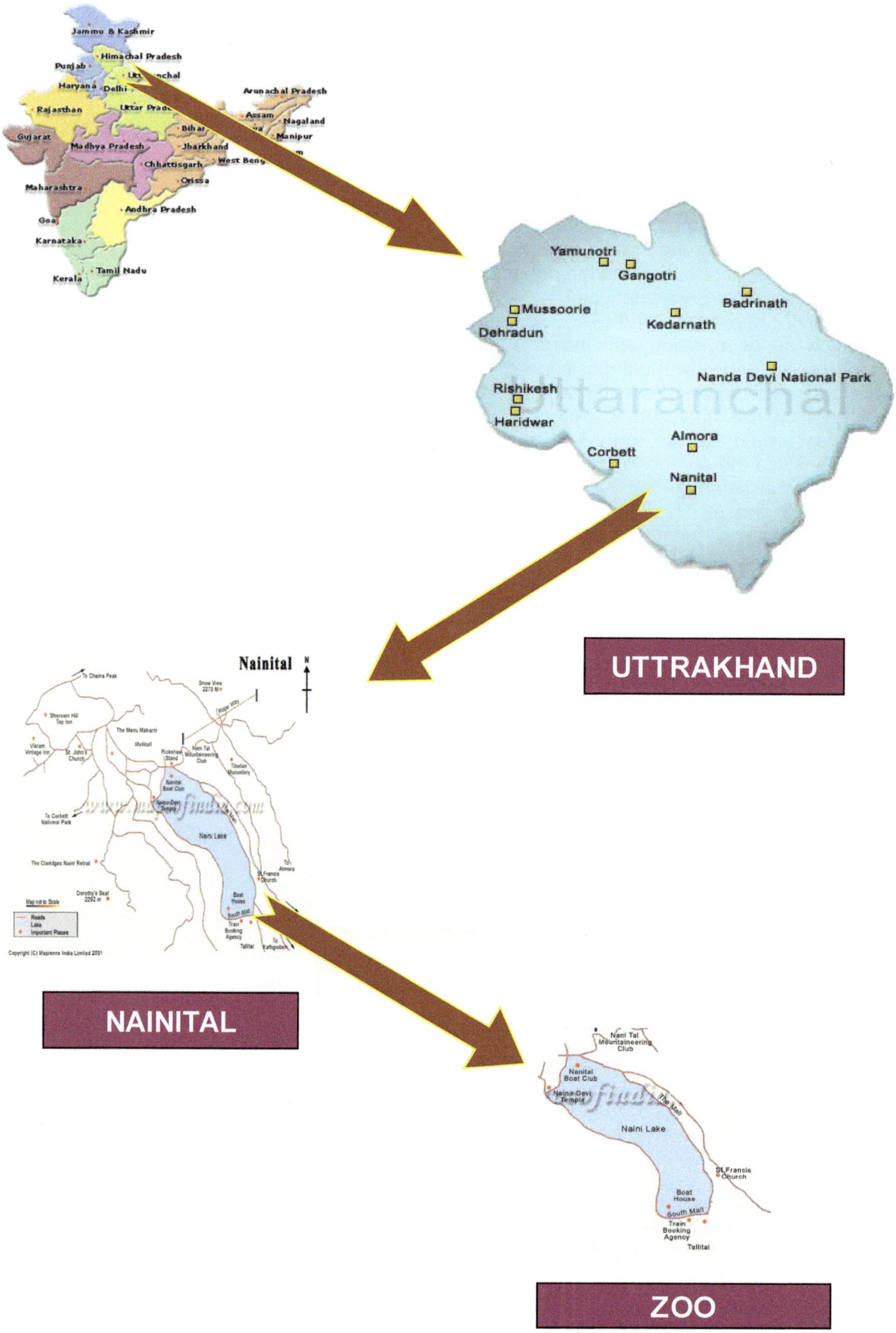

Figure 1. Map of the study area.

Table 1. Species richness and equitability in protected and disturbed forest site at hill base, hill slope and hill top.

Site	Forest feature	Forest	
		Protected site	Disturbed site
Hill base	Tree	03	03
	Shrub	07	09
	Total	10	12
	Equitability	10.11	4.39
Hill slope	Tree	02	04
	Shrub	09	16
	Total	11	20
	Equitability	4.09	7.62
Hill Top	Tree	02	04
	Shrub	07	15
	Total	09	19
	Equitability	5.33	5.19

Photo Plate 1. Protected forest site.

RESULTS

Species richness and diversity index

Total species richness (tree and shrub) was higher in disturbed forest site as compared to protected forest site (Table 1 and Photo Plate 1) and was maximum at hill slope in both the site. Most of the tree and shrub species were common at both the sites thus, resulting in 75% similarity in tree layer and 70% similarity in shrub layer. Diversity index was also higher in disturbed forest site (1.85 to 2.09) than protected site (1.00 to1.48; Table 2). Concentration of dominance also followed the same trend (Table 2).

Density of trees and shrubs

In the protected forest area, density of trees (number of tree ha^{-1}) was higher at hill slope (970 trees ha^{-1}) and it

Table 2. Diversity Index (H') and concentration of dominance (CD) for tree, sapling, seedling and shrub in protected and disturbed forest area.

Site	Parameter	Forest feature	Forest	
			Protected	Disturbed
Hill base	H'	Tree	0.44	0.52
		Sapling	0.30	0.53
		Seedling	0.45	0.51
		Shrub	0.29	0.53
		Total	1.48	2.09
	CD	Tree	0.15	0.29
		Sapling	0.13	0.38
		Seedling	0.28	0.50
		Shrub	0.05	0.03
		Total	0.61	1.20
Hill slope	H'	Tree	0.09	0.53
		Sapling	0.17	0.38
		Seedling	0.48	0.50
		Shrub	0.26	0.44
		Total	1.00	1.85
	CD	Tree	0.46	0.21
		Sapling	0.44	0.29
		Seedling	0.25	0.55
		Shrub	0.02	0.03
		Total	1.17	1.08
Hill Top	H'	Tree	0.31	0.53
		Sapling	0.41	0.38
		Seedling	0.46	0.50
		Shrub	0.34	0.47
		Total	1.52	1.88
	CD	Tree	0.35	0.15
		Sapling	0.31	0.58
		Seedling	0.46	0.24
		Shrub	0.03	0.04
		Total	1.15	1.01

decreased towards hill base (460 trees ha^{-1}) and hill top (780 trees ha^{-1}). Contrary to this, the tree density increased from hill base (260 trees ha^{-1}) to hill top in moderately disturbed forest area (780 trees ha^{-1}) (Table 3 and Photo Plate 2). Shrub species also followed the same trend in both the sites (Table 4). Among trees, *Q. floribunda* was the most dominant species (IVI 139.2-227.5) at all the three sites of protected forest area (Table 3). In the disturbed forest site, *Q. leucotrichophora* was dominant at hill base (IVI 188.6) and hill slope (IVI 164.3), while *Cedrus deodara* was dominant at hill top (IVI 128.7; Table 3). Among shrubs, *Salvia officinalis* was dominant at hill base and hill sope and *Indigofera heterantha* showed dominance at hill top of protected forest site (Table 4). In disturbed forest site, *S. officinalis* was dominant shrub at hiil base, *Hypericum oblongofolium* at hill slope and *I. heterantha* at hill top (Table 4).

Total basal area (TBA)

In protected forest area, total basal area was higher for *Q. floribunda* (9.65 to 44.28 m^2ha^{-1}) possibly due to higher density of trees (Table 5). In disturbed forest area, total basal area was higher for *Q. leucotrichophora* (3.51 to 10.96 m^2ha^{-1}) at hill base and hill slope and for *Q. floribunda* (11.4 m^2ha^{-1}) at hill top (Table 5).

Population structure

In the disturbed site, both the oak species (*Q. leucotrichophora* and *Q. floribunda*) exhibits an expanding type of population structure (Figures 2 to 4). It represents the stage with high density of trees in younger girth classes and decrease in density in higher girth classes. However, in protected site the population structure of these two oaks showed higher density in the younger girth classes and absence of trees in higher girth classes despite the fact that it has been under protection for the last 25 years. This may be due to deforestation of existing old trees of oak before protection of zoo area. Similar observation for undisturbed forest has been reported by Rao et al. (1990) and Gupta and Yadav (2005). The population structure of tree species were following three types according to the criteria given by Saxena and Singh (1984):

1. *Q. floribunda* and *Q. leucotrichophora* with higher proportion of individuals in younger girth classes as compared to higher girth classes. This type of population structure represents frequent reproduction (Knight, 1975).
2. *Cupressus torulosa* and *C. deodara* with high density in the intermediate girth classes and decreasing density towards lower and higher girth classes. According to Knight (1975) this type of population structure represents infrequent reproduction.
3. Seedlings and saplings were absent in case of *Rhododendron arboreum* and *Gravillea robusta* indicating that these species were not regenerating and may be replaced by some other tree species in future.

DISCUSSION

Our understanding of the various aspects of tree regeneration in the Himalaya is far from adequate. Unregulated man-made disturbances have greatly limited

Table 3. Density (number of trees ha^{-1}) and Importance Value Index (IVI) of tree species at different sites in protected and open forest area at hill base, hill slope and hill top.

Site	Species	Forest			
		Protected		Disturbed	
		Density	IVI	Density	IVI
Hill base	*Quercus leucotrichophora*	60	70.36	140	188.66
	Quercus floribunda	280	139.20	80	72.17
	Cupressus torulosa	120	90.40	-	-
	Rhododendron arboreum	-	-	40	39.17
	Total	460		260	
Hill slope	*Quercus leucotrichophora*	-	-	191	164.31
	Quercus floribunda	940	227.57	100	67.45
	Cupressus torulosa	-	-	100	49.01
	Cedrus deodara	-	-	25	19.17
	Gravillea robusta	30	73.91	-	-
	Total	970		325	
Hill top	*Quercus leucotrichophora*	140	88.90	50	21.81
	Quercus floribunda	640	211.08	250	82.1
	Cupressus torulosa	-	-	180	67.31
	Cedrus deodara	-	-	300	128.79
	Total	780		780	

Photo Plate 2. Open forest site.

the possibility of designing studies for assessing the response of species to a particular scale and kind of disturbances. The broadleaf trees (particularly *Q. leucotrichophora* and *Q. floribunda*) are repeatedly lopped for leaves and firewood, leading to their gradual disappearance. As a result of these activities, the forest sites have been either reduced or dramatically modified and have led to the expansion of xerophytic conditions (Singh and Singh, 1987). This protected zoo site would help to formulate planning for long- term studies as far as

Table 4. Density (number /100 m^2) of shrub species in protected and disturbed forest site at hill base, hill slope and hill top.

Species	Hill base		Hill slope		Hill top	
	Protected	Disturbed	Protected	Disturbed	Protected	Disturbed
Achyranthes aspera Linn	-	-	2.4	2.7	-	5.2
Berberis asiatica Roxb.ex. D.Don	1.8	1.8	2.4	2.3	4.0	4.2
Coriaria nepalensis Wall	-	-	-	0.9	-	3.8
Cotoneaster microphylla Wall	-	-	-	5.3	-	4.0
Desmodium elegans DC	-	0.4	-	1.1	0.7	3.1
Goldfussia dalhousiana Nees in DC	1.6	1.6	7.0	3.7	-	1.6
Hyparicum oblongifolium Choicsy	1.2	9.2	-	7.0	-	4.5
Indigofera hetrantha Wall. Ex. Brandis	1.4	12.0	4.7	5.5	3.4	9.3
Myrsine africana Wall	-	0.2	-	1.0	-	3.4
Pyracantha crenulata (Don) Roem	-	-	-	2.4	-	2.6
Rosa brunonii Lindl	-	-	2.2	0.2	0.4	0.2
Rubus ellipticus Wall	-	0.6	-	-	-	1.9
Rubus niveus wall	-	-	1.1	0.9	3.1	0.5
Salvia officinalis Linn.	13.4	10.6	8.4	3.2	1.6	
Sarcococca saligna Wall	3.4	0.8	1.0	2.4	1.6	0.7
Urtica parviflora Roxb.	2.0	-	0.9	1.0	-	-
Wickstroemia canescens Meissn	-	-	-	0.7	-	1.1

Table 5. Total basal area (m^{2}ha^{-1}) in protected and disturbed forest site at hill base, hill slope and hill top.

Site	Species	Forest	
		Protected	Disturbed
Hill base	*Quercus leucotrichophora*	35.87	10.96
	Quercus floribunda	44.28	7.90
	Cupressuss torulosa	44.60	-
	Rhododendron arboreum	-	0.61
	Total	124.75	19.47
Hill slope	*Quercus floribunda*	16.66	8.86
	Gravillea robusta	4.55	-
	Quercus leucotrichophora	-	9.13
	Cupressuss torulosa	-	5.15
	Cedrus deodara	-	1.02
	Total	21.21	24.16
Hill Top	*Quercus floribunda*	9.65	11.47
	Quercus leucotrichophora	4.85	3.51
	Cupressus torulosa	-	5.71
	Cedrus deodara	-	4.78
	Total	14.5	25.47

regeneration parameters and recovery of degraded broadleaf species of Himalayan region are concerned.

The distribution of plant species along the altitudinal transect (Hill base to Hill top) indicate that the distribution of each species is determined by its own ability to survive, grow and reproduce successfully in different environmental conditions. In protected site, a total of 17 species of plants were recorded, out of which 4 were trees and 13 were shrubs. In disturbed forest site, a total of 23 species were recorded with 5 tree and 18 shrub

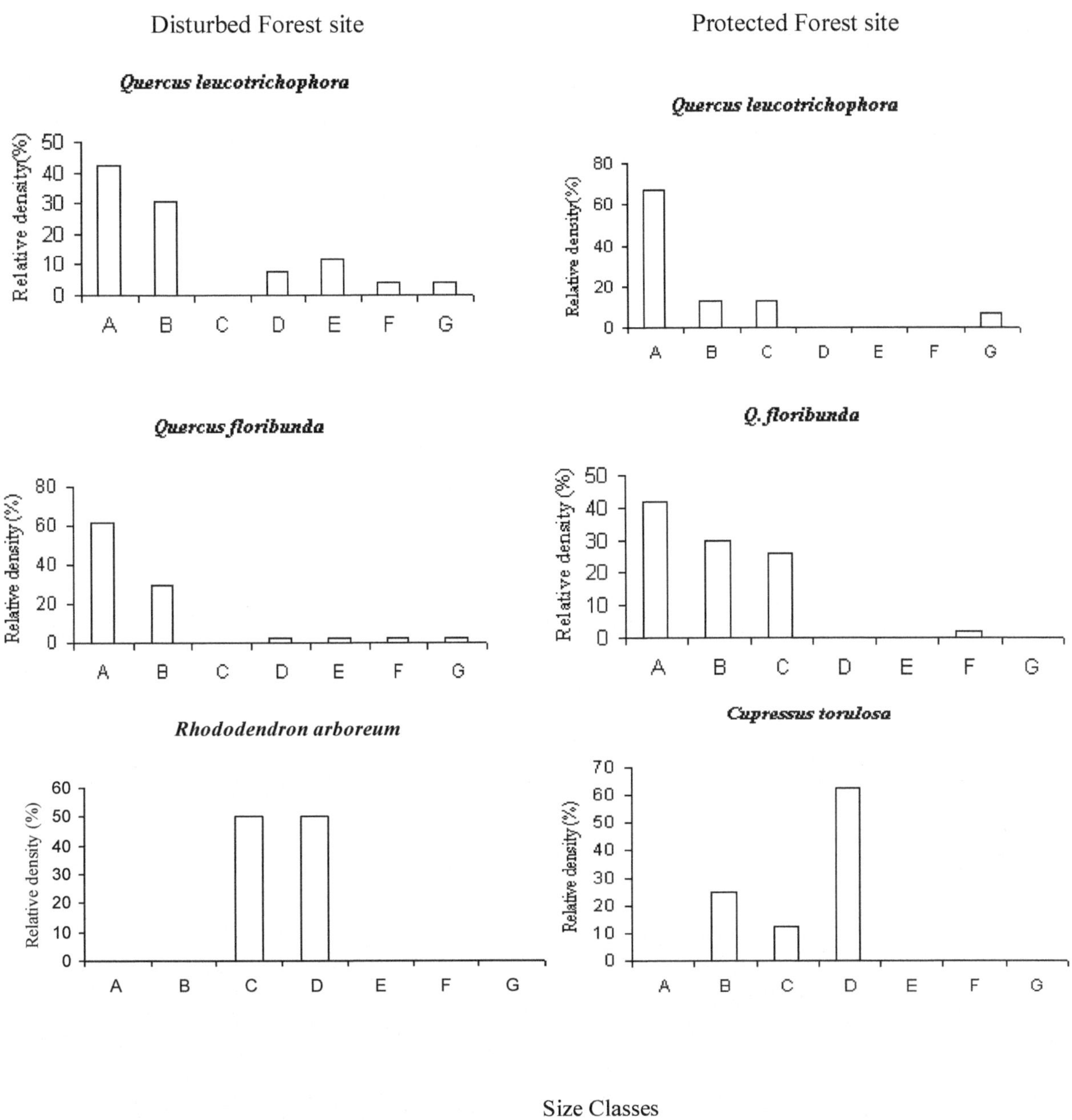

Figure 2. Population structure at Hill base of disturbed and protected forest site. A= seedlings, B= sampling, trees: C= 31-60 cm, D=61-90 cm, E=91-120 cm, F= 121-150 cm and G=>150 cm CBH.

species. These results indicate that moderate disturbance increase species richness (Singh and Singh, 1992). The distribution and species pattern in this region largely depend on the altitude and climatic variable like rainfall and temperature (Kharkwal et al., 2005). The mean basal area was greater for *Q. leucotricophora*, however, total basal area was greater for *Q. floribunda* due to high tree density particularly in protected site. *C. torulosa* and *C. deodara* were present at hill slope and hill top of disturbed forest site as they are high altitude species. The range of diversity was 0.09 to 0.55 for trees, 0.17 to 0.53 for saplings and 0.45 to 0.51 for seedlings.

Age distribution is an important population characteristic which is the consequence of both natality and mortality. The ratio of various age groups in a population determines the reproductive status of the population and indicates the future course (Odum, 1971). The data on the population structure of *Q. leucotrichophora* and *Q. floribunda* at both the site suggest that the impact of human disturbance on density of seedlings and saplings was least and was significantly high as far as density of trees of higher girth classes is concerned (Figures 2 to 4). The protected site has highest number of seedlings followed by sapling and young trees. Absence of trees

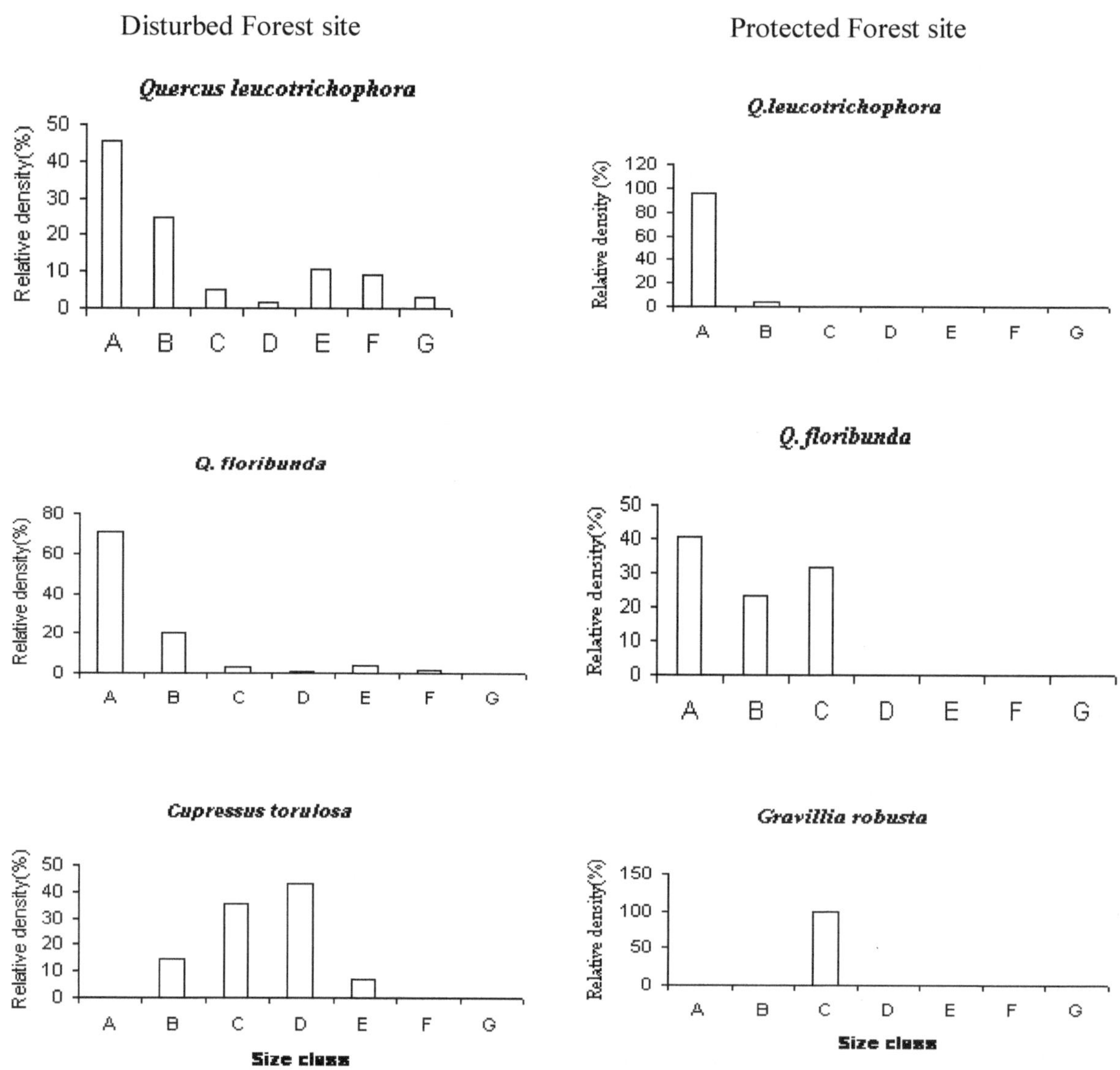

Figure 3. Population structure at Hill slope of disturbed and protected forest site. A= Seedlings, B= sampling, trees: C= 31-60 cm, D=61-90 cm, E=91-120 cm, F= 121-150 cm and G=>150 cm CBH.

in higher girth class indicates that this site represents young forest type. It may be due to deforestation of old trees in the past. A greater proportion of individual in lower size classes compared to large size classes of *Q. floribunda* indicates frequent reproduction (Singh and Singh, 1992). This site was protected by the Forest Department in the year 1984, and biotic disturbances were checked completely, resulting in greater survival of seedlings and saplings. According to Singh and Singh (1992) more individual in intermediate size classes and decreasing numbers towards the higher and lower size classes in *Q. leucotrichophora* indicates that the population is on the way of extinction. In a number of forest stands *Q. leucotrichophora* was regenerating well until recently but not now. *Q. floribunda* has been regenerating well in its own forest. Absence of trees in lower girth classes and slightly higher density in the higher girth classes in disturbed forest may be due to the presence of some old trees exhibiting the adverse effect of human disturbance. It may be suggested that partial disturbance favours the regeneration of oak species through increased micro-sites for seedling establishment and protection of seedlings from overgrazing and trampling by cattle (Gupta and Yadav, 2005). Mature trees were lopped for fuel wood and fodder by the local inhabitants, which not only reduce the vigour but also seed production and ultimately led to poor regeneration of species. The presence of some trees of *Rhododendron arboreum*, *C. torulosa* and *C. deodara* in middle girth classes indicate that these species are not regenerating well.

Forests are the important source of subsistence,

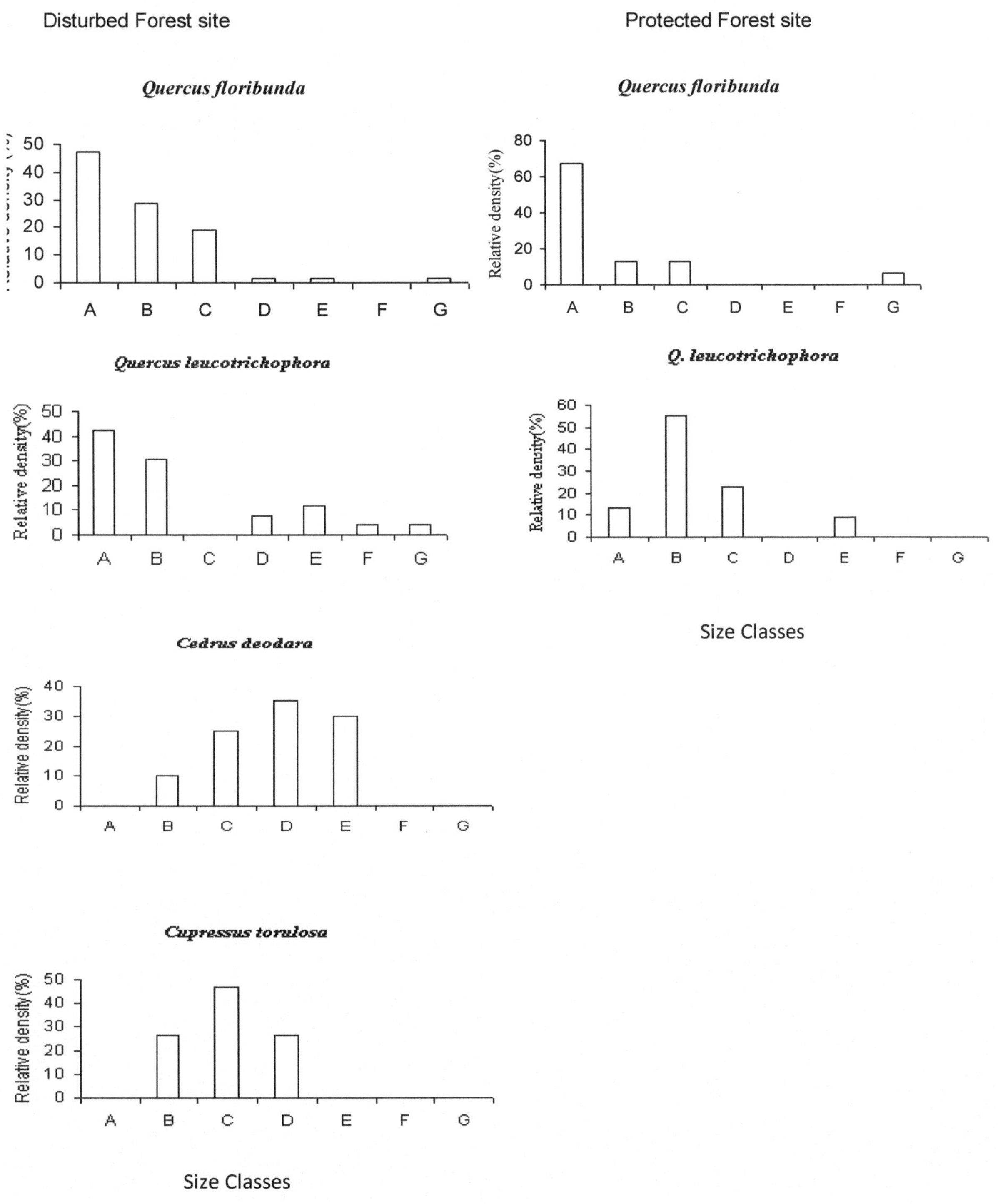

Figure 4. Population structure at Hill top of disturbed and protected forest site. A= Seedlings, B= sampling, trees: C= 31-60 cm, D=61-90 cm, E=91-120 cm, F= 121-150 cm and G=>150 cm CBH.

employment, revenue earning and raw material to various industrial uses. Above all they are most important component of life support system on the earth and provide various ecological services. The present study was conducted to analyse the impact of biotic disturbances on regeneration status of tree species as disturbances are critical elements in controlling the composition and functioning of forest ecosystems. Regeneration is a critical phase of forest management, because it maintains the desired species composition after disturbance. Excessive lopping, concomitant reduction in seed production and enhanced weevils' infestation of acorns seems to adversely affect regeneration of *Q. leucotrichophora* (Singh and Singh, 1992). Much of brown oak (*Quercus semecarpifolia*) and adjacent alpine meadows have been historically under influence of grazing of migratory livestock such as sheep and goats. The failure of brown oak to regenerate is a case of environmental semi surprise (Singh et al., 1997). The findings of the present study indicate that if anthropogenic pressure continues to operate future perpetuation of oaks (*Q. floribunda* and *Q. leucotrichophora*) in the forest may be threatened. Thus, proper conservation measures have to be formulated to protect these valuable tree species. Awareness programmes regarding the importance of forests and sustainable harvesting may help to minimize the pressure on the forests.

ACKNOWLEDGMENTS

Authors thank authorities of State Forest Department, Uttarakhand for permitting us to undertake the research work inside the Nainital Zoo. We are also thankful to Prof. Y. P. S. Pangtey, Emeritus Professor, Department of Botany, Kumaun University, Nainital for identification of plants and Head, Department of Botany for providing necessary facilities.

REFERENCES

Bargali K, Usman S, Joshi M (1998). Effect of forest covers on certain site and soil characteristics in Kumaun Himalaya. Ind. J. For. 21(3): 224-227.

Bhuyan P, Khan ML, Tripathi R S (2003). Tree diversity and population structure in undisturbed and human- impacted stands of tropical wet evergreen forests in Arunachal Pradesh, Eastern Himalaya, India. Biodivers. Conserv. 12: 1753-1773.

Curtis JT, Mclntosh RP (1950).The interrelations of certain analytical and synthetic phytosociological characters. Ecology 31: 434-455.

Dhar U, Rawal RS, Samant SS (1997). Structural diversity and representativeness of forest vegetation in a protected area of Kumaun Himalaya, India: implication for conservation. Biodivers. Conserv. 6: 1045-1062.

Good NF, Good RE (1972). Population dynamics of tree seedlings and saplings in mature Eastern hardwood forest. Bull. Torrey Bot. Club. 99: 172-178.

Gupta SK, Yadav AS (2005). Population structure of tree species in the Sariska Tiger Reserve: Effect of aspects of hill slopes and human disturbance. Bull. Nat. Inst. Ecol. 15: 35-41

Hussain MS, Sultana A, Khan JA, Khan A (2008). Species composition and community structure of forest stands in Kumaun Himalaya, Uttarakhand, India. Trop. Ecol. 49(2): 167-181.

Khan ML, Roy JPN, Tripathi RS (1987). Population structure of some tree species in disturbed and protected subtropical forests of northwest India. Acta Ecol. 247-255.

Kharkwal G,Mehrotra P, Rawat YS, Pangety YPS (2005). Phytodiversity and growth form in relation to altitudinal gradient in the Central Himalayan (Kumaun) region of India. Curr. Sci. 89(5): 873-878.

Knight DH (1975). A phytosociological analysis of species- rich tropical forest of Barno Colorado Island, Panama. Ecol. Monogr. 45: 259-284.

Kumar M, Sharma CM, Rajwar GS (2004). A study on the community structure and diversity of a sub-tropical forest of Garhwal Himalaya. Indian Forester 130(2): 207-214.

Odum EP (1971). Fundamentals of Ecology (3rd edition). W.B. Saunders Company, Philadelphia and London, p. 546.

Osmaston AE (1927). A forest flora for Kumaun. Govt. Press, United Province, Allahabad.

Phillips EA (1959). Methods of vegetation study. Henry Holt and Co. Inc. New York.

Pickett STA, White PS (1985). The Ecology of Natural Disturbances and Patch dynamics. Academic Press, New York.

Ralhan PK, Saxena AK, Singh JS (1982). Analysis of forest vegetation at and around Nainital in Kumaun Himalaya. Proc. India Natl. Sci. Acad. B48 1: 121-137.

Rao P, Barik SK, Pandey HN Tripathi RS (1990). Community composition and tree population structure in a sub-tropical broad-leaved forest along a disturbance gradient. Vegetatio 88: 151-162.

Saxena AK, Singh JS (1984). The population structure of certain Himalayan forest, associations and implications concerning their future composition. Vegetatio 50: 3-22.

Saxena AK, Singh SP Singh JS (1984). Population structures of forests of Kumaun Himalaya: Implication for management. J. Environ. Manage. 19: 307-324.

Shannon CE, Weiner W (1963). The mathematical theory of communication. University of Illinois Press, Urbana, Illinois.

Simpson EH (1949). Measurement of diversity. Nature 163: 688-692.

Singh JS, Singh SP (1987). Forest vegetation of Himalaya. Bot. Rev. 52: 80-192.

Singh JS, Singh SP (1992). Forests of Himalaya: Structure, functioning and impact of man. Gynodaya Prakashan, Nainital, India.

Singh SP, Rawat YS, Garkoti SC (1997). Failure of brown oak (*Quercus semecarpifolia*) to regenerate in Central Himalaya: A case of environmental surprise. Curr. Sci. 73(4): 371-374.

Sorenson T (1948). A method of establishing groups of equal amplitude in plant sociology based on similarity of species content. Detkong Danske Vidensk Selk bio Skr (Copenhagen), 5: 1-34.

Tewari JC Singh SP (1985). Analysis of woody vegetation in a mixed oak forest of Kumaun Himalaya. Proc. Indian Nat. Sci. Acad. 52: 332-347.

Upreti N, Tewari JC, Singh SP (1985). The oak forests of the Kumaun Himalayas (India) 1: Composition, diversity and regeneration. Mountain Res. Dev. 5(2): 163-174.

Whittaker RH (1975). Communities and Ecosystems. Macmillan Publishing Co., New York.

Efficacy of different insecticides against mushroom phorid Fly, *Megaselia halterata* (Wood) in Punjab, Pakistan

Muhammad Hussnain Babar[1]*, Muhammad Ashfaq[1], Muhammad Afzal[2], Muhammad Hamid Bashir[1] and Muhammad Asif Ali[3]

[1]Department of Agriculture Entomology, University of Agriculture, Faisalabad, Pakistan.
[2]Department of Agriculture Entomology, University College of Agriculture, University of Sargodha, Sargodha, Pakistan.
[3]Institute of Horticultural Sciences, University of Agriculture, Faisalabad, Pakistan.

Phorid fly, *Megaselia halterata* is the most destructive insect pest of mushroom causing serious threat to mushroom production and cultivation in Pakistan. Chemical control is the most effective and quicker approach used by our farming community. So there was a dire need to evaluate some commercial insecticides against this pest. The present study was therefore, conducted to evaluate six commercial insecticides (deltamethrin, spinosad, spintoram, trichlorphon, malathion and permethrin) against phorid fly for three consecutive years (2008 to 2010). The insecticides evaluation was done on the bases of three parameters viz. percent reduction in adult emergence, mushroom damage rate by phorid fly larvae and mushroom yield per unit area. All insecticides caused significant reductions in adult emergence and mushroom damage rates as compared to control. On overall bases, the average of three years data revealed that maximum reduction of adult emergence and minimum damage rate by phorid flies as well as maximum yield per plot was observed in spinosad treated plots, followed by trichlorphon and deltamethrin which were statistically at par. The results of present study showed that deltamethrin, spinosad and trichlorphon may be employed in the field for controlling mushroom phorid fly in Pakistan.

Key words: Efficacy, insecticides, *Agaricus bisporus*, phorid fly.

INTRODUCTION

Mushroom is an important crop all over the world, which is full of proteins, vitamins and minerals, exceptionally high in vitamins, especially in vitamins of group B, like, thiamin (vitamin B1), riboflavin (vitamin B2), pantothenic acid (vitamin B3) and nicotinic acid (vitamin B5) (Chang and Miles, 1991). The fresh mushrooms contain 85 to 95% water, 3% proteins, 4% carbohydrates, 0.1% fats and 1% minerals (Agrahar-Murugkar and Subbulakshmi, 2005). Edible mushrooms are a rich source of some physiological agents, of medicinal importance, that possess antitumour, anticancerous, anticholesterolic, cardiovascular, antiviral and antibacterial agents (Nayana and Janardhan, 2000; Manpreet et al., 2004; Alam et al., 2007). They are also, found to be useful against the diabetes, ulcer and lung diseases (Halpern and Miller, 2002; Wasser, 2002). Mushroom flies are continuous threat to mushroom industry in Pakistan. Damage by phorid fly directly occurs, due to the feeding of their larvae on mycelia and fruiting body or indirectly due to the excreta secreted by the larvae which alter the chemical and physical properties of the compost (Greenslade and Clift, 2004). The adults of this insect are

*Corresponding author. E-mail: babar1235@yahoo.com.

vectors of *Agaricus bisporus* pathogens, especially of the *Verticillium* spp. and cause a cosmetic damage to the fruiting bodies of the mushrooms (Rinker and Snetsinger, 1984). This pest must be controlled during early stages of mushroom production as to avoid a significant damage and yield losses (Jess and Bingham, 2004).

Pakistan enjoys diverse climate enabling great potential to grow different kinds of mushrooms but most of the growers feel reluctant while growing mushrooms due to the severe infestation of phorid flies each year at different farms which sometimes results into total failure of the crop. Due to these reasons mushrooms sometimes need to be imported from China, Korea, India and Australia.

In Pakistan, commercial cultivation of mushrooms exists on a small scale with a few private farms and no pesticide has yet been registered against this pest. The present study was therefore, designed to find out the efficacy of some commercially available pesticides against phorid fly of mushroom. It is the first study in Pakistan regarding the mushroom insect-pests control.

MATERIALS AND METHODS

The research work was carried out in the Department of Agriculture Entomology, Mushroom Culture Laboratory, Institute of Horticultural Sciences University of Agriculture, Faisalabad, Pakistan and Mushroom houses of different mushroom growers of Punjab.

Preparation of mushroom compost

Mushrooms compost was prepared, by following the method (Babar et al., 2005).

Compost material

Chicken manure = 120 kg,
Wheat straw = 300 kg
Gypsum = 10.50 kg,

300 kg of wheat straw was moisturized daily up to three days to make it soft for the microbial activity. Later on, 120 kg chicken-manure was added to the moistened wheat straw and heaped up, one meter and given 6 turnings, according to the pre-planned time schedule.

Turning schedule of the compost

0 day = making heap.
4th day = first turning.
8th day = second turning.
11th day = third turning
14th day = fourth turning + 1/2 gypsum.
17th day = fifth turning. + next 1/2 gypsum.
20th day = sixth turning (last turning).

Pasteurization

The compost was pasteurized, at 60°C, for 6 h (Noble and Gaze, 1994), in a controlled temperature chamber and cooled to ambient temperature and aerated, in order to reduce the ammonia contents.

Collection and rearing of phorid flies

The insects used in the present study were obtained from laboratory cultures of phorid flies maintained by the method reported by Richardson and Hesling (1978). The adults were collected from different mushroom farms of Punjab with the help of an aspirator which were subsequently shifted into transparent plastic pots (ca.3 litre) and transferred to the insect culture lab of Agri. Entomology Department, University of Agriculture Faisalabad. As the larvae directly feed on button mushroom (*Agaricus bisporus*) mycelium (Hussey, 1959 and White 1985) therefore, the adults were introduced in a rearing chamber (380 × 280 × 150 mm^3) containing a jar of 200 gm compost, fully colonized with mushroom-mycelia placed on a moist towel. Saccharide syrup was also provided as a carbohydrate source for flies. The flies were allowed to lay eggs for 72 h at 25 ± 1°C. After three days, the flies were expelled out from the rearing chamber. Within 21 to 30 days, a new generation was emerged under same temperature regimes. Six commercial insecticides were used in this study are tabulated as follows in table 1.

Experimental design

The experiments were carried out in RCBD (Randomized Complete Block design), at the farmer's field (mushroom houses) in three consecutive growing years during 2008 and 2010. The growth cell was used in designed to those which are employed commercially, but smaller in size (dimensions 1 m × 1 m). Each growing cellar was considered as a replication, and three replicates were used for each treatment. The experiments were repeated in three separate growing periods.

Biological assays

Pasteurized and colonized compost growing cellar were kept in the growth room. When spawn running completed then kept for 14 days at 26 ◦C and 80 to 90% relative humidity, the compost was amply spawned and then covered with 5cm thick covering of moss peat, a homogeneous mixture of black peat (90%) and calcium carbonate (10%). After casing, forty pairs (40:40) of phorid fly were introduced over the casing layer in each growing cellar covered with transparent nylon gauze dome-shaped cages and was allowed to lay eggs for 3 to 4 days. All the adult flies were subsequently removed, and when tiny larvae started to hatch then hatching date was determined by microscopic examination of small amounts of growing substrate taken randomly from growing cellars and then every plot was treated with insecticide to control the insect pest at early stage.

After casing, the bags were kept at 24 to 26°C for 6 to 7days, and then fresh air was introduced until a temperature of 16.5 ± 0.5°C and a relative humidity of 80 to 85% were maintained throughout cropping. Light trap was installed to assess the emerging adult's population. Treatments were assessed by calculating the emergent flies of the mushroom. At weekly intervals for 5 weeks, emerging flies were counted in each treatment including control. The number of mushroom damaged by the larvae was also recorded separately for each treatment.

Data analysis

All data collected were analyzed statistically for calculation of ANOVA. The means were separated by Tukey's HSD test by using

Table 1. Chemical insecticides to control the mushroom phorid.

Treatment	Common name	Trade name	Manufacturer	Dose (ml/L)
T1	Deltamethrin	Decis® 2.5EC	Bayer Crop Sciences, Germany	0.4
T2	Spinosad	Tracer® 240SC	Dow Agro Sciences, USA	0.4
T3	Spintoram	Radiant® 120SC	Ali Akbar Enterprises, Pakistan	0.2
T4	Trichlorphon	Dipterex® 80WP	Bayer Crop Sciences, Germany	6.8
T5	Malathion	Fyfanon® 57EC	Ali Akbar Enterprises, Pakistan	7.5
T6	Permethrin	Coopex® 0.5%	Ali Akbar Enterprises, Pakistan	0.25
T7	Control	-		-

Table 2. Mean percentage reductions in phorid fly (*Megaselia halterata*) population and mean percentage damaged of mushroom (*Agaricus bisporus*) caused by the larvae during three growing periods by using chemical insecticides.

Treatment	Period I		Period II		Period III	
	RAE	MDR	RAE	MDR	RAE	MDR
T1=Deltamethrin	84.75 (± 0.90)^{a}	10.50 (± 0.69)^{c}	86.01 (± 0.83)^{a}	9.40 (± 0.44)^{c}	84.59 (± 1.14)^{a}	10.37 (± 0.58)^{c}
T2= Spinosad	86.43 (± 0.85)^{a}	9.70 (± 0.67)^{c}	88.02 (± 1.07)^{a}	9.07 (± 0.33)^{c}	86.25 (± 1.13)^{a}	9.17 (± 0.30)^{c}
T3= Spintoram	83.58 (± 0.87)^{a}	11.43 (± 0.70)^{c}	84.97(± 0.93)^{ab}	10.80 (± 0.32) ^{c}	83.33 (± 1.10)^{ab}	11.07 (± 0.39)^{c}
T4= Trichlorphon	85.60 (± 0.84)^{a}	10.33 (± 0.68)^{c}	87.14 (± 1.03)^{a}	9.70 (± 0.33)^{c}	85.35 (±1.12)^{a}	9.83 (±0.33)^{c}
T5= Malathion	75.69 (± 0.89)^{c}	16.47 (± 1.03)^{b}	77.03 (± 0.85)^{c}	14.67 (±0.50)^{b}	75.50 (± 1.14)^{c}	16.47 (± 1.16)^{b}
T6= Permethrin	79.58 (± 0.88)^{b}	15.00 (± 0.67)^{b}	81.02 (± 0.74) ^{c}	13.63 (±0.76)^{b}	79.29 (±1.12)^{bc}	15.33 (± 0.87)^{b}
T7= Control	0.00^{d}	50.10 (± 0.21)^{a}	0.00^{d}	48.70 (±0.87)^{a}	0.00^{d}	48.57 (± 0.63)^{a}

Means sharing similar letters in each column are not different significantly (Tukey's HSD). RAE= Reduction in adult emergence and MDR= mushroom damage rate.

computerized statistical software SPSS® ver. 16.

The percentage reduction in adult emergence for each treatment in each of the three growing periods was calculated by using the formula (Erler et al., 2009):

% Reduction = [(B-A)/B] × 100

Where A = total number of emerging adults in the each treatments and B= total number of emerging adults in the control. Whereas the percentage damage of mushroom was determined by using the formula (Erler et al., 2009).

% Damage mushroom = (number of mushroom damaged/total number of mushroom) × 100

The mushroom yield data was expressed as Kg / M^2.

RESULTS

Effect of chemical insecticides on adult emergence

There were significant differences in RAE (Reduction in Adult Emergence) in all the three years of study (Table 2). In the first period of study there were significant differences (F = 1519.61, df 6, 14, prob. 0.000) between the treatments. Spinosad showed the highest results (86.43% ± 0.85) in RAE which was statistically at par with deltamethrin, spintoram and trichlorfon. In the second period of study, almost similar significant (F = 1424.45, df6, 14, prob. 0.000) results were obtained and also in the 3rd study period significant (F = 903.28, df 6, 14, prob. 0.000) results were obtained which showed that the four products as mentioned above were good in causing death of phorid flies. In all the three study periods two of the insecticides remained less effective viz. malathion and permethrin.

When all the treatments are compared with respect to the mean percentage reduction in adult emergence, the order of effectiveness is generally as follows: spinosad > trichlorphon > deltamethrin

Table 3. Average yield of mushroom (*A. bisporus*) obtained during three growing periods by using chemical insecticides against phorid fly (*M. halterata*).

Treatments	Yield (kg/M^2)		
	Period I	Period II	Period III
T1=Deltamethrin	7.95 (± 0.08) ab	8.10 (± 0.13) a	7.84 (± 0.08) ab
T2= Spinosad	8.30 (± 0.05) a	8.02 (± 0.08) ab	8.12 (± 0.15) a
T3= Spintoram	7.81 (± 0.09) bc	7.95 (± 0.10) ab	7.64 (± 0.07) bc
T4= Trichlorphon	8.00 (± 0.08) ab	8.12 (± 0.15) a	8.25 (± 0.09) a
T5= Malathion	7.35 (± 0.13) d	7.55 (± 0.18) c	7.20 (± 0.08) d
T6= Permethrin	7.50 (± 0.04) cd	7.65 (± 0.16) bc	7.42 (± 0.05) cd
T7= Control	4.35 (± 0.03) e	4.47 (± 0.04) d	4.54 (± 0.04) e

Means sharing similar letters in a column are not different from each other (Tukey's HSD test).

> spintoram > permethrin > malathion respectively.

Effects on mushroom damage rate by phorid fly larvae

All the chemical insecticides reduced the mushroom damage rate by the larvae of phorid fly and had significantly less mushroom damage rate compared with the control (Table 2). In terms of mean percentage damage of mushroom by the larvae, the lowest damage rates were observed (9.70%, 9.07% and 9.17%) in spinosad treatment during three growing periods successively.

During the 1st growing period there was a significant difference ($F = 429.18$, df 6, 14, prob. 0.000) among the treatments with spinosad showing very less damage rate by the larvae (9.70% ± 0.67) but again this was statistically at par with other three insecticide results viz. deltamethrin (10.50% ± 0.69), spintoram (11.43% ± 0.70) and trichlorfon (10.33% ± 0.70). When we compare results during 2nd year of study we see almost similar results as spinosad (9.07% ± 0.33) showing less damage rate as compared to the other three insecticides statistically at par namely deltamethrin (9.40% ± 0.44), spintoram (10.80% ± 0.32) and trichlorfon (9.70% ± 0.32). During 3rd year study the results were similar like spinosad (9.17% ± 0.30) showing less damage rate as compared to the other three insecticides statistically at par namely deltamethrin (10.37% ± 0.58), spintoram (11.07% ± 0.39) and trichlorfon (9.83% ± 0.33). The other two products showed higher damage rate by the larvae showing their less control during the three consecutive periods. Based on the mushroom damage rates, the order of efficacy of the treatments was as follows: spinosad > trichlorphon > deltamethrin > spintoram > permethrin > malathion > control.

Effect on mushroom yield

The yield in kg/m^2 for cumulative flushes for each individual treatment is shown in table 3. In the 1st period of study significant ($F = 304.61$, df 6, 14, prob. 0.000) differences in yield were obtained between the treatments. Maximum yield (8.30 ± 0.05 kg/m^2) was observed where spinosad was applied followed by trichlorphon, deltamethrin, spintoram, permethrin, malathion and control (8.00, 7.95, 7.81, 7.50, 7.35 and 4.35 kg/m^2 respectively). During 2nd growing period again significant ($F = 105.15$, df 6, 14, prob. 0.000) differences were observed between the treatments. Maximum yield (8.12 kg/m^2) was observed in trichlorphon treatment followed by deltamethrin, spinosad, spintoram, permethrin, malathion and control (8.10, 8.02, 7.95, 7.65, 7.55 and 4.47 kg/m^2 respectively). Similarly, during 3rd growing period significant ($F = 215.40$, df 6, 14, prob. 0.000) differences in yield were observed. Maximum yield (8.25 kg/m^2) was observed in trichlorphon treatment followed by spinosad, deltamethrin, spintoram, permethrin, malathion and control (8.12, 7.84, 7.64, 7.42, 7.20 and 4.54 kg/m^2, respectively).

DISCUSSION

Spinosad is a new chemistry insecticide with novel mode of action. It was first registred against dipterous insects in the United States in 2005 (Juliane et al., 2006). Spinosad, an environment friendly alternative to malathion (Peck and McQuate, 2000; Vargas et al. 2001).

The present study showed that in mushroom house condition spinosad was found to be most effective insecticide which had significantly reduced the adult emergence (86.43, 88.02 and 86.25%) than any of the others insecticides during three successive growing period. Similar conclusion was obtained by Sameh et al. (2009) who indicated that spinosad can reduce the olive fruit fly infestation up to 86.3%. The results of present study also match with previous studies (Kirsten et al., 2005; Braham et al., 2007; El-Aw et al., 2008; Gabriele et al., 2011). While trichlorphon deltamethrin and spintoram was also found most effective chemicals to control the phorid fly of mushroom. They reduced the phorid fly population more than 80% which is also confirmed by the previous studies (Chihiya et al., 2006; Yee et al., 2007; Palumbo and Richardson, 2008; Mahmoud et al., 2009; Ahmad et al., 2010; Raga and Sato, 2011; Saeidi et al., 2011) but they used these chemicals against fruit fly which is also a dipterous insect. Malathion and permethrin were proved effective but their efficacy was less as compared with other insecticides. Similar results were obtained by Braham et al. (2007) (in Tunisia) and El-Aw et al. (2008) who reported that spinosad was most effective insecticide than malathion in controlling dipterous insect pests. The main reason of ineffective of heavy use of malathion and permethrin was created the resistance problems, previous studies (Kaufman and Rutz, 2001; Hussain et al., 2005; Fourie et al., 2006; Braham et al., 2007; Roger et al., 2009; Saeidi et al., 2011) confirmed the results.

The results of present study indicate that all the insecticides reduce the phorid fly emergence and mushroom damage rate. When taking into consideration yield values from the treatments, the lowest values were obtained from the control. The yield values also show that all treatments gave good yield but spinosad and trichlorphon were proved most effective against this pest and gave maximum yield as compared with other treatments.

Results from this study suggest that commercial insecticides suppress the populations of mushroom flies by reducing adult emergence and have potential to control phorid fly of mushroom by the chemicals. Mushroom growers are not using insecticides against this pest in Pakistan, so resistance problem is not present here. While in some mushroom growing European countries, America, Canada and Australia where maximum farmers use insecticides to control this pest have created the resistance problems against diazinon, organophosphate and pyrethroids (Smith and White, 1996; Smith, 2002; Hsu et al., 2004; Magana et al., 2007; Ahmad et al., 2010; Jamal et al., 2011; Raga and Sato, 2011).

ACKNOWLEDGEMENT

This study was supported by the Higher Education Commission (HEC) of Pakistan in the form of Indigenous Ph. D fellowship for which the author is highly thankful.

REFERENCES

Agrahar-Murugkar D, Subbulakshmi G (2005). Nutritional value of edible wild mushrooms collected from the Khasi Hills of Meghalaya. J. Food Chem., 89: 599-603.

Ahmad S, Ahmed FS, Khan RR, Nadeem MK (2010). Evaluation of insecticide resistance in two strains of fruit fly, *Bactrocera zonata* (*Tephritidae*: Diptera), with fruit dip method, Pak. Entomol. 32 (2): 163-167.

Alam N, Hossain MS, Khair A, Amin SMR, Khan A (2007). Comparative Study of Oyster mushrooms on Plasma Lipid Profile of Hypercholesterolemic Rats. Bangl. J. Mushr., 1(1): 15-22.

Babar MH, Afzal M, Bashir MH, Ali MA (2005). Relative efficacy of different acaricides (Dicofol, Abamectin and Hexythiazox) against mushroom mites (Pygmephoroidae). Pak. J. Agri. Sci., 42 (3-4): 45-47.

Braham M, Pasqualini E, Ncira N (2007). Efficacy of kaolin, spinosad and malathion against *Ceratitis capitata* in *Citrus* orchards. Bull. Insectol., 60(1): 39-47.

Chang ST, Miles PG (1991). Recent trends in World Production of Cultivated Edible Mushrooms. Mushr. J., 503: 15-18.

Chihiya J, Gadzirayi CT, Mutandwa E (2006). Effect of three different treatment levels of deltamethrin on the numbers of dung beetles in dung pats, Afr. J. Agric. Res., 1(3): 74-77.

El-Aw MAM, Draz KAA, Hashem AG, El-Gendy IR (2008). Mortality comparison among spinosad, actara, malathion and methomyl-containing baits against peach fruit fly, *Bactrocera zonata* Saunders (Diptera: Tephtitidae) under laboratory conditions. J. Appl. Sci. Res., 4(2): 216-223.

Erler F, Polat E, Demir H, Cetin H, Erdemir T (2009). Control of the mushroom phorid fly, *Megaselia halterat* (Wood), with plant extracts. Pest Manag. Sci., 65: 144-149.

Fourie LJ, Stanneck D, Horak IG (2006). Efficacy of a topically applied combination of imidacloprid and permethrin against *Stomoxys calcitrans* on dogs, Intern. J. Appl. Res. Vet. Med., 4(1): 29-33.

Gabriele L, Hack R, Smid G (2011). Efficacy of spinosad against the poultry red mite, *Dermanyssus gallinae* (Mesostigmata: Dermanyssidae), in laboratory and field trials. Zoosymposia, 6: 282-287.

Greenslade P, Clift AD (2004). Review of pest arthropods recorded from commercial mushroom farms in Australia. Austr. Mycol. 23: 77-93.

Halpern GM, Miller AH (2002). Medicinal Mushrooms. M. Evans and Company, Inc, New York.

Hsu JC, Wu WJ, Feng HT (2004). Biochemical mechanisms of malathion resistance in oriental fruit fly, *Bactrocera dorsalis* (Hendel). Pl. Protect. Bull., 46: 255-266.

Hussain R, Ashfaq M, Saleem MA, Ahmed S (2005). Toxicity of some insecticides with novel modes of action against malathion–resistant and organophosphate susceptible strains of *Tribolium castaneum* larvae, Int. J. Agri. Biol., 7(5): 768-772.

Jamal AE, Nugud AD, Abdalmagid MA, Bashir AI, Brair M, Elnaeim IH (2011). Susceptibility of *Culex quinquefasciatus* (Diptera: Culicidae) in Khartoum locality (Sudan) to malathion, temephos, lambda-cyhalothrin and permethrin insecticides, Sudanese J. Public health. 6(2): 56-62.

Jess S, Bingham JFW (2004). Biological control of sciarid and phorid pests of mushroom with predatory mites from the genus *Hypoapsis* (Acari: Hypoaspidae) and the entomopathogenic nematode *Steinernema feltiae*, Bull. Entomol. Res., 94: 159-167.

Kaufman PE, Rutz DA (2001). Susceptibility of house flies (Diptera: Muscidae) exposed to commercial insecticides on painted and unpainted plywood panels. Pest manag. Sci., 128: 174-178.

Kirsten SP, Isaacs R, Wise JC, Gut LJ (2005). Protection of fruit against infestation by apple maggot and blueberry maggot (Diptera: Tephritidae) using compounds containing spinosad. J. Econ. Entomol., 98 (2): 432-437.

Magana C, Hernandez-Crespo P, Ortego F, Castanera P (2007). Resistance to malathion in field populations of *Ceratitis capitata*. J. Econ. Entomol., 100: 1836-1843.

Mahmoud MF, El-Kady G, Osman MAE (2009). Toxicity of spinetoram to onion thrips (*Thrips tabaci*) and to green peach aphid (*Myzus persicae*) under laboratory and field conditions. Tunisian J. Pl.

Prot., 4 (2): 221-227.
Manpreet K, Giridhar S, Khanna PK (2004). *In vitro* and *in vivo* antioxidant potentials of *Pleurotus florida* in experimental animals. Mushr. Res., 13(1): 21-26.
Nayana J, Janardhanan KK (2000). Antioxidant and antitumour activity of *Pleurotus florida*. Curr. Sci., 79(7): 941-943.
Noble R, Gaze RH (1994). Controlled environment composting for mushroom cultivation: substrates based on wheat and barley straw and deep litter poultry manure. J. Agri. Sci., 123(1): 71-79.
Palumbo J, Richardson YAJ (2008). Efficacy of radiant (spinetoram) against western flower thrips in romaine lettuce, Vegetable Report. 152: 96-103.
Peck SL, McQuate GT (2000). Field test of environmentally friendly malathion replacements to suppress wild Mediterranean fruit fly (Diptera: Tephritidae) populations. J. Econ. Entomol., 93: 280-289.
Raga A, Sato ME (2011). Toxicity of neonicotinoids to *Ceratitis capitata* and *Anastrepha fraterculus* (Diptera: Tephritidae). J. Pl. Prot Res., 51(4): 413-417.
Richardson PN, Hesling JJ (1978). Laboratory rearing of the mushroom phorid fly, *Megaselia halterata* (Diptera: Phoridae). Annu. Appl. Biol., 88: 211-217.
Rinker DL, Snetsinger RJ (1984). Damage threshold to a commercial mushroom by a mushroom-infesting phorid (Diptera: Phoridae). J. Econ. Entomol., 77: 449-453.
Roger I, Vargas RI, Pinero JC, Jacome I, Revis HC, Prokopy RJ (2009). Effectiveness of malathion against different ages of melon fly (Diptera: Tephritidae). Proc. Hawaiian Entomol. Soc., 41: 15-23
Saeidi K, Nur-Azura A, Omar D, Abood F (2011). Efficacy of various insecticides on safflower fly, *Acanthiophilus helianthi* Rossi (Diptera: Tephritidae) in Kohgiloyeh and Boyerahmad province (Iran) Austr. J. Bas. Appl. Sci., 5(12): 2660-2664.
Sameh AM, Ahmed E, Mostafa M, Metwally E, Nabil MG (2009). Efficacy of spinosad, lufenuron and malathion against olive fruit fly, *Bactrocera oleae* (Gmelin) (Diptera: Tephritidae) Egypt. Acad. J. biolog. Sci., 2(2): 171-178.
Smith JE (2002). Dimilin resistance in mushroom sciarids. Mushr. J., 656: 15-19.
Smith JE, White PF (1996). Diazinon resistance in mushroom pests. HDC Project News, 36: 12-15.
Vargas RI, Peck SL, McQuate GT, Jackson CG, Stark JD, Armstrong JW (2001). Potential for area wide integrated management of Mediterranean fruit fly (Diptera: Tephritidae) with a braconid parasitoid and a novel bait spray. J. Econ. Entomol., 94: 817-825.
Wasser SP (2002). Medicinal mushrooms as a source of antitumor and immunomodulating polysaccharides. Appl. Microbiol. Biotechnol., 60: 258-274.
White PF (1985). Pests and pesticides. In the Biology and Technology of the Cultivated Mushroom (Eds. Flegg PB, Spencer DM and Wood DA) Chichester: John Wiley and Sons Ltd. pp. 279-293
Yee WL, Jack O, Nash MJ (2007). Mortality of *Rhagoletis pomonella* (Diptera: Tephritidae) exposed to field-aged spinetoram and azinphos-methyl in Washington state. Florida Entomol., 90(2): 335-342.

14

Linking population size to conservation needs of the Grey Parrot in Cameroon

Simon A. Tamungang[1], Robert A. Cheke[2], Ghislain P. Kougoum[3] and Eric S. Ntiri[4]

[1]Department of Animal Biology, University of Dschang, BP 146 Dschang, Cameroon.
[2]Natural Resources Institute, University of Greenwich at Medway, Central Avenue, Chatham Maritime, Kent ME4 4TB, UK.
[3]Institut Supérieur du Sahel, Université de Maroua, BP 46 Maroua, Cameroun.
[4]African Insect Science for Food and Health (ICIPE), P.O. Box 30772-00100, Nairobi, Kenya.

Unsustainable exploitation of the wild Grey Parrot is a complex and challenging phenomenon for governments and international organizations to tackle. The need to reverse the negative impacts of exploitation on the Grey Parrot population and to conserve it for future generations formed the basis of this study. Population data on the Grey Parrot in Cameroon were obtained using the point count method with two counting bands (r = 0 - 25 m and r = 25 m - ∞). Parrot densities ranged from 0.50 parrots/km^2 in the Littoral Region to 2.16 parrots/km^2 in the South Region, with a mean for the five regions where nearly all the parrots occur at 1.26 parrots/km^2. By multiplying these regional density estimates of forest cover within each region, a national population size range of 164376 – 251231 parrots was estimated. Regional conservation status of the Grey Parrot was described using the Cameroon classification system of wildlife threats. Emerging conservation challenges and mitigation measures for sustainable management of the bird are proposed.

Key words: Cameroon, conservation status, Grey Parrot, *Psittacus erithacus*, population size.

INTRODUCTION

The African Grey Parrot *Psittacus erithacus* attracts much interest in the international pet market (Mulliken et al., 1992; Juste, 1996; BirdLife International, 2010). Stemming from an assessment that 21% of the global population is harvested yearly, it is currently assessed as vulnerable on the IUCN Red Data list (Birdlife International, 2010; 2011). In addition, there has been a decline in the number of Grey Parrots throughout their range and this is attributed to the parrot trade and habitat loss. (McGowan, 2001; Birdlife International, 2010). The Grey Parrot's longevity, intelligence and ability to mimic human voices and other sounds make it a highly sought-after pet (Pepperberg, 1999) and trade in the birds brings in high returns to those involved (Birdlife International, 2010; Beissinger, 2001; Mulliken et al., 1992).

In the light of DNA studies, the Grey Parrot of Africa is now recognised as comprising two species (Melo and O'Ryan, 2007). The Congo Grey Parrot, *P. erithacus* of the Congo Basin occurs from east of the River Comoé in Côte d'Ivoire to Angola and the western form, the Timneh Parrot *P. timneh* is found from west of the River Comoé in Côte d'Ivoire to Sierra Leone. Although it is an internationally protected bird species, the trade in the Congo Grey Parrot thrives in Cameroon (Tamungang and Cheke, 2009) and in many other African countries, including those where *P. timneh* occurs.

This paper is on the Congo Grey Parrot which occurs in Cameroon as one of its range states. The bird will be hence referred to in this paper as the 'Grey Parrot'.

The recognition of the above mentioned debilitating factors on the natural populations of the Grey Parrot became the basis on which the Animal Committee of CITES, in their 22nd meeting in 2006, recommended the categorization of the Grey Parrot's range countries, based on the severity of negative impacts (CITES, 2006). On their list, Cameroon, Côte d'Ivoire, Guinea, Liberia and Sierra Leone were placed under the *Urgent Concern* category; the Republic of Congo, Democratic Republic of Congo and Equatorial Guinea were under the *Possible Concern* category, with most of the rest of the other range countries listed under *Least Concern*.

Furthermore, a two year moratorium, which started on 1st January 2007, was placed on countries in the *Urgent Concern* category, who were asked to carry out measures to forestall the negative situation (CITES 2006, 2007). The need for scientific data on the Grey Parrot for informed management decisions by the Cameroon government and the need to forestall the negative situation in the range states recommended by CITES stimulated this study, which aimed to determine the population size of the Grey Parrot in Cameroon and to propose sustainable ways for its conservation.

MATERIALS AND METHODS

Stratified Random Sampling (Sutherland, 2006) was used for data collection. For the purpose of easy demarcation of each stratum, the administrative regions of Cameroon, except for the three northern regions (Adamaoua, North and Extreme North) which are not part of the endemic range of the Grey Parrot (Figure 1), were chosen as strata, which were further subdivided into sub-strata using 17 protected areas, 8 Important Bird Areas (IBAs) outside protected areas (BirdLife International, 2004) and 7 other sites where the parrots were known to occur. All ecological zones and administrative regions with Grey Parrots within them were represented (Figure 1). Transects of 1 km length were randomly selected in each sample area and 10 paths were covered within them, contributing a total length of 10 km.

We used the point count method to census Grey Parrots, a method widely used to sample bird communities in tropical forests and calculate changes in bird abundance over time (Dawson, 1981; Hutto et al., 1986; Hill et al., 2005; Seavy et al., 2005; Volpato et al., 2009). Point counts involve mainly visual and auditory detections of birds with fixed or unlimited radius plots (Blondel et al., 1981; Hutto et al., 1986; Volpato et al., 2009). Their efficiency and accuracy are influenced by observer effort, which may affect the information obtained such as species abundance (Rosenstock et al., 2002; Bart and Earnst 2002; Betts et al., 2005). If well spaced, a sample series of points in an area will provide more representative data than a few transects. The point count therefore has an advantage over transects, being easier to incorporate into a formal study design.

Each point consisted of a circle made up of two counting bands or radii (Bibby et al., 1992). The first band had a radius of 25 m (r = 0 - 25 m). Any parrot that was detected outside the 25 m radius was recorded in the second band with its radius set as infinity (r = 25 m - ∞). It was assumed that as the radius decreased, the probability that Grey Parrots would be detected increased, both in terms of the number of birds detected and the numbers recorded in a series of point counts. The goal was to use a radius as large as possible, but within which detection of all parrots could be reasonably assured, in both open and dense vegetation types. Thus, the 25 m radius was a compromise between the open and closed habitats.

The study area covered the whole of the southern part of Cameroon where rainforest exists (Figure 1). Sample points were selected randomly on each path in a protected or non-protected area, to ensure that a cross-section of the major vegetation types in the geographical range of the Grey Parrot was sampled.

Accurate distance measurements between counting points is very important in this method since bird detection is associated with a certain distance. All walked distances were measured using two pedometers (Scanner Mark II) by two observers. Each pedometer was adjusted to the normal strides of the observer at the beginning of each data collection session. All members of the research team were drilled on distance estimation prior to point data collection to reduce the bias of distance estimates in the field. The distance between two adjacent counting points on the same path was 250 m, which was decided after considering the rate of parrot encounters/km and the associated vegetation types and structures. This distance was chosen to ensure that detection from different points remained statistically independent (Reynolds et al., 1980) and it represented a compromise between sample size and parrot conspicuousness in the dense tropical rainforest. A 1-km path length selected for data collection produced 4 points with 250 m between them and this was considered reasonable to enhance the rate of parrot encounters per counting session. For each study site, a minimum of 10 km was covered, thereby producing a minimum of 40 counting points per site per visit. The 32 sample sites in the whole country produced a minimum of 1280 counting points per sampling round. It is known that activity patterns of Grey Parrots change with seasonality (Tamungang et al., 2003), so we collected data in both the dry and wet seasons. Both dry and rainy season data were collected separately each/per year, with a minimum of 2560 samples of counting points per year [2560 (= 1280x 2 wet/dry season counts/year) and carried out during 4 years (2008-2011) amounts to 2560 x 4 = 10240 counting points]. In this paper, we limited our results to the mean of combined yearly parrot counts.

The duration of data recording at a point was 10 min. A longer duration was considered more likely to record birds making long movements from previously sampled points, which would invalidate the critical assumptions of the method. Shorter durations of 5 min would not have allowed sufficient time to detect parrots that were inconspicuous at counting points, so 10 min was taken as the best compromise between the longer and shorter durations.

Detections between count points were recorded as "present" (+) but these data were not used in the final calculation of population size, being only used to show the presence/absence of birds at the sampling site. Sightings of flying birds and all auditory detections were not included in the dataset, but were simply recorded as present at the site. Two or more observers made individual counts and then compared their results before recording a final total, so as to reduce counting bias. To minimize variations associated with indices of abundance, the counting of parrots was conducted at times when there was little change in conspicuousness of the birds (Dawson, 1981). Generally, Grey Parrots are known to be more active in the morning and evening than around mid-day and in the afternoon (Tamungang and Cheke, 2009; Clemmons, 2003). Data collecting periods were therefore standardized to fit within peak periods of activity of the birds (06:00-10:00 h and 14:00-18:00 h).

Datasets were analysed using relevant statistical packages (SPSS, Map Info, and Microsoft EXCEL). To calculate Grey Parrot densities and numbers for each region of the country, the following formula for point counts within and beyond a fixed radius (Bibby et al., 1992) was used:

$$\text{Density} = (\text{Log}_e\,(n/n_2) \times n/m(\pi r^2)$$

Where, n is the total number of birds counted; n_2 is the number beyond the fixed radius; n_1 is the number counted within the radius (r) so that $n = n_1 + n_2$; m is the total number of counts; r is the fixed

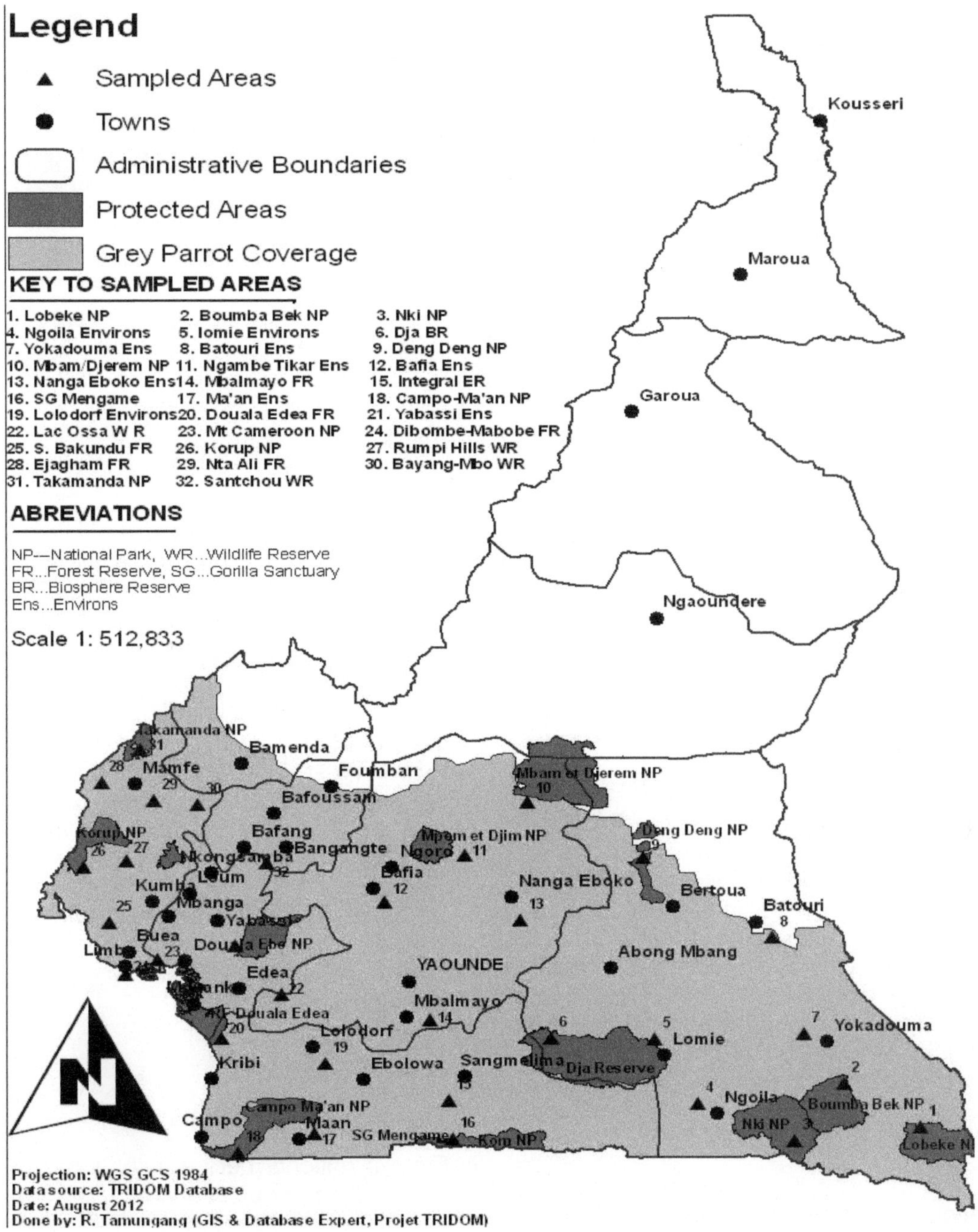

Figure 1. Original Grey Parrot coverage (range) in relation to sampled areas covered in four seasons in Cameroon during 2008-2011.

radius.

RESULTS

Grey parrot abundance

The natural range of the Grey Parrot in Cameroon falls in major parts of the South West, Littoral, South, Centre and East Regions and small parts of the North West and West Regions. This range of the parrot is relatively large as compared to the size of the country (about 25.4%). Within this range, but omitting data for the North West and West Regions, yearly densities of Grey Parrots were obtained from the mean of the estimated annual dry and rainy season densities for the study period (2008-2011) (Table 1).

The regional mean parrot densities in Cameroon ranged from 0.50 to 2.16 parrots/km^2/ (Table 1), and the mean of these means was 1.26 parrots/km^2 for the study period.

Table 1. Mean Grey Parrot (GP) annual densities ($parrots/km^2$), with standard deviations (Stdev) and upper and lower 95% confidence levels (CL), in five regions of Cameroon during 2008-2011.

Region	GP density distribution				Mean	Stdev	Lower CL (95%)	Upper CL (95%)
	2008	2009	2010	2011				
Centre	0.71	0.70	0.82	0.71	0.74	0.05	0.69	0.79
East	1.99	1.66	2.87	2.11	2.16	0.44	1.73	2.59
Littoral	0.50	0.40	0.50	0.68	0.50	0.11	0.39	0.61
South	1.33	1.56	1.98	2.00	1.72	0.28	1.44	2.00
South West	0.47	0.40	1.98	1.95	1.20	0.77	0.45	1.95

Table 2. Regional estimated Grey Parrot population sizes in Cameroon during 2008-2011.

Region	Forest area/km^2	Parrot density/km^2	Parrot population size	Lower CL (95%)	Upper CL (95%)
Centre	14058.47	0.74	10403	9700	11106
East	62559.15	2.16	135128	108227	162028
Littoral	6973.58	0.50	3487	2720	4254
South	27275.43	1.72	46914	39277	54551
South West	9893.17	1.20	11872	4452	19292

Translating densities into populations

The surface areas occupied by rainforest within the endemic range of the parrots in each of the 5 sampled regions were obtained from the Ministry of Forestry and Wildlife (MINFOF) in Cameroon (MINFOF, 2007). The rainforest habitat in Cameroon is incidentally the area covered by the Grey Parrot range. These rainforest areas were mapped relative to the survey points and protected areas already shown in Figure 1 (see Grey Parrot coverage). Similar information was also obtained on the total surface area of each region of the country. Parrot densities were obtained from the formula:

Density = Number of individual animals/surface area occupied

From this equation, we obtained the number of parrots from the densities as presented in Table 2, with the CLs calculated in relation to the amount of forest that are still surviving in each region.

The national parrot population size obtained from the sum of the regional population sizes was 207804, with a range based on the CLs of 164376–251231 parrots.

Regional population status

We adopted the Cameroonian classification system of wildlife threats (found in the 1994 Forestry and Wildlife Law) to describe the regional conservation status of the Grey Parrot (Table 3). Thus, out of the seven regions that harbour Grey Parrots in Cameroon, three regions (North West, West and Littoral) were listed as Class A, where the birds are seriously threatened, while two regions (South West and Centre) were listed as Class B, where the populations are vulnerable to threat and thus are of concern and need sustainable management activities. Two regions (East and South) were listed as Class C (Table 3).

Parrots were abundant and could be sustainably harvested. Sustainable limits of the populations need to be determined before harvesting could be done, and then only above such limits.

DISCUSSION

Threat to population size

From a national perspective, a previous estimate of the Grey Parrot population in 1998 for the country was 300,000 - 500,000 (Fotso, 1998). The current study suggests that there has been a drop to about 50% of the 1998 population in 14 years, but it is important to note that census methods used in the two studies were not the same. Generally, the high variability in regional population sizes of the parrots is indicative of the variety of factors that influence wildlife at different levels in the regions (Tamungang and Cheke, 2012; Chardonnet et al., 2002). Such factors include past history of harvesting for the pet trade, distribution and abundance of habitat resources (in quality, quantity and diversity), and human influences (demographic pressure and forest disturbance, as well as community awareness of parrot conservation (Collar and Juniper, 1992). These threat factors could be summarised into two major categories: parrot trade and habitat destruction.

Table 3. Regional conservation status of the Grey Parrot in Cameroon during 2008-2011.

Region	GP population size	Conservation status	Explanation
North West	Very low (inadequate data)	Class A	-Birds are threatened with extinction. -It is strictly forbidden to carry out any harvesting activity. -Urgent restoration programmes are needed to bring the population up to abundance level.
West	Very low (inadequate data)	Class A	
Littoral	2720-4254	Class A	
South West	4452-19292	Class B	-Birds are vulnerable to threat. - A few birds can be harvested under administratively supervised conditions, along with population development programmes.
Centre	9700-11106	Class B	
South	39277-54551	Class C	-Birds are abundant. -Legal harvesting can be sustainably carried out above the carrying capacity of the population.
East	108227-162028	Class C	

Conservation status is adapted from the Cameroonian classification system of Wildlife threats (MINFOF, 1994).

Parrot trade threat

The parrot trade is a big business in Cameroonian society involving the public and private sectors (Chupezi et al., 2006; Ngenyi, 2003). It begins with the trappers and villagers in the rural communities and extends to the middlemen who dwell in the towns and cities. The driving force behind the trade is poverty alleviation and unemployment. Major suppliers of Grey Parrots are those in the rainforest areas of Cameroon where, at the regional level, three regions (with the estimated quota in 2006: South 31%; Centre and East, 23% each) were the major sources of Grey Parrots in the country. The international trade from Cameroon is financially motivated as exporters get much higher prices per bird abroad than their local counterparts. In recent years, the Grey Parrot has been the most exploited and exported wild bird species in Cameroon. From 1981-2005, Cameroon exported 367,166 birds with a yearly average of 15,299 (Tamungang and Cheke, 2006).

From 1990 - 1996, Cameroon exported 48% of the Grey Parrots from the 25 exporting countries in Africa, thereby positioning itself as the leading exporter of the wild Grey Parrots in the world (Tamungang and Cheke, 2006). Official figures do not account for parrots that are smuggled across borders into neighbouring countries, consumed locally or that die in the process of trapping and transportation. Effectively dismantling the illegal parrot trade remains a big challenge to the Cameroon Government.

Habitat destruction threat

Another major threat to Grey Parrot population growth in Cameroon is habitat degradation and its complete loss in some places. Threats are felt by the parrot at various essential habitat-based sites used by the bird. The following major threatened sites are discussed:

Roosting sites: All the roosts identified by this study were threatened by anthropogenic pressure. These threats were predominantly through human predation on the roosting birds and /or habitat (roost) destruction, e.g. felling of roosting trees for socioeconomic reasons Similar observations were made by Dändliker (1992) in Ghana. Loss of roosting sites has many negative consequences on parrot population growth including reduction of population size.

Foraging sites: Farmers attributed the birds' habit of feeding heavily on the fruits of economic tree species, such as plum *Dacryodes edulis* and oil palm nut *Elaeis guineensis,* to reductions in their yields. Various methods were adopted for chasing them away, trapping or killing the birds including shooting with guns and catapults. This is a type of parrot-human conflict that tends to reduce bird population growth (CITES, 2011).

Nesting sites: The felling of nest trees for timber exploitation, farming and infrastructure development brought about the destruction of nests, some of which might have contained eggs or fledglings. The scarcity of nesting sites could be due to forest exploitation activities (Madindou and Mulwa, 2008). Scarcity of nest sites can lead to prevention of reproductive activities by mature females, or mature birds may migrate to neighbouring countries in search of suitable nesting sites.

Some explanation could be advanced for the seemingly high densities in the Centre and South Regions in spite of the many towns, cities and agro-industrial plantations within them. We observed relatively high levels of public sensitization to parrot conservation in Campo (in the

South) and at Ndikinimeki (in the Centre), where the bulk of the parrots roost in villages close to human houses and gain protection from the villagers. Although the parrot habitats were located around human settlements, there seemed to be a good level of sustainable co-existence between parrots and the inhabitants. The conditions of the forest habitats in and around these villages were seriously fragmented. The further away from the village, the less the fragmentation and degradation process of the forest. It was observed that the parrots travelled far from the roosts to forage and breed. Some forms of community participation in parrot conservation and human protection of parrots from poachers were well developed in Campo. The creation there of a tourism management committee and construction of camps for tourists to use while watching parrots, especially at roosts, was illustrative of this approach.

Grey Parrots frequently nest in natural cavities in *Terminalia superba* trees, popularly called *fraquet* in French, whose wood is also in high demand for house construction. Many low-income earners use '*fraquet*' in constructing houses, market centres and cheap furniture. There is intensive exploitation of this resource to meet the needs of the increasing local population or by timber exploiting firms to establish low-cost houses (locally called "*Kara-boat*") for workers. By indiscriminate felling of the trees, potential nesting sites are destroyed. Destruction of nesting sites constrains natural increases in parrot numbers because trees that produce these nesting sites occur at low densities.

Recommendations for grey parrot population development

Restructure the parrot trade to boost wild population growth

The sustainability of the parrot trade in Cameroon depends on its structure and function. A well structured and managed trade is likely to be profitable for the benefit of both the people and wild parrot populations. The current structure has many short-comings including lack of understanding of procedures to be followed by both traders and administrators, problems of information flow and lack of coordination of stakeholders to achieve common goals of the trade. It is recommended that a guideline-document should be produced identifying the challenges in the sector and outlining possible solutions. For the trade to be more functional, integrative and profitable, the following points should be considered in the guideline document:

1. Trapping season: Parrots should not be trapped during their breeding seasons. These seasons vary slightly from the South West to the East Region but the peak period is from June to October in most parts of the country. Trapping should therefore be carried out from November to April each year and trapping should be limited to fledglings.

2. Identification of trappers and traders: All legal parrot traders should identify their trappers with names, photographs and copies of identity cards. They should be issued with permits to show that they are working for the designated traders. These permits should be carried and if necessary displayed at any time the trappers or traders are in the field.

3. Verification and control of trappers: Effective means of verification and management of parrot trappers before, during and after trapping should be established. This action will go a long way to ensure that trapping rules are obeyed for the benefit of all the stakeholders in the sector.

4. Period for issue of parrot trapping permits: All permits for commercial purposes should be issued from October to December but should be valid only in January to December of the next year. This means that parrots that have not been trapped or exported within the above period cannot be carried over to the following year. This measure will ensure that yearly export quotas are not exceeded.

5. Parrot trapping in protected areas: Parrots should never be trapped in protected areas, even by valid permit holders for the trade, except for research purposes and then only under officially supervised conditions.

Encourage rural community participation in parrot conservation

Rural communities are the backbone for any meaningful wildlife conservation practice, as they are the resident custodians of the parrots that thrive in their forests. The integrated approach to parrot conservation in Cameroon cannot be successfully implemented without regarding the rural communities as major stakeholders.

1. Provide incentives to local communities: The government should provide incentives to galvanise these communities towards sustainable parrot conservation through community development projects. Assistance of rural communities in creating income-generating activities for sustenance during their annual programmes is very important. In this way they will have a sense of recognition and responsibility to be vanguards of parrot management programmes and can become strong government collaborators. Enhancing community level participation in sustainable parrot management will be a major breakthrough for wildlife management in Cameroon, since the level of grass-root wildlife conservation awareness is currently very low.

2. Communities should work with parrot trappers: The community should work with parrot trappers/traders in their region to ensure that the laws and regulations in this sector are fully implemented.

Improve policy and law enforcement measures

Wildlife law enforcement at all levels of society has remained a major challenge in Cameroon. The 1994 Forestry and Wildlife law should be revised and effectively implemented in the field. There is an urgent need for more infrastructure, equipment and personnel to deter and combat wildlife crimes in general.

Scientific and research requirements

Field research on parrots is generally a difficult and specialised domain, especially on the Grey Parrot that thrives in the humid and dense tropical forest. Scientific research in this domain will offer capacity building services, sustainable protection of wild parrots, and informed management and policy decisions.

1. A parrot GIS database be created and regularly updated in Cameroon. Uses of such a functional database are many, including population monitoring and regulation, informed management and policy decisions and trade regulation.

2. More emphasis should be placed on studies of demographic trends of parrots for long- and short-term analysis, reproduction, ecology and behaviour, radio-tracking studies to determine home range and seasonal movements, and their effects on the birds and changes over time in habitat use.

3. Habitats with substantial parrot populations should be identified and protected from socio-economic activities. Such targeted habitats include those used for nesting, roosting and feeding. An inventory of these sites should be carried out at regional levels and the village communities around them should be organised to safeguard their protection.

Encourage captive breeding of Grey Parrots in the country. Parrots from captive-bred programmes can be used as pets locally or exported, thereby reducing harvesting pressures on the population in the wild. Captive-bred parrots can be released in the wild after long periods of acclimatization to boost wild population growth rates.

ACKNOWLEDGEMENTS

Immense thanks is due to all who assisted in the project, they are too numerous to mention here. Fritz Oben analysed the data. Gilbert Mofor, Edwin Penn and Ernest Achu, all of PARROTPRO Research Group, collected field data. Estella Tamungang processed the manuscript and Rowan Martin reviewed it. This study was carried out with financial support from the Ministry of Forestry and Wildlife, Cameroon; the Loro Parque Fundación, Spain, and the British Ecological Society, UK.

REFERENCES

Bart J, Earnst SL (2002). Double sampling to estimate density and population trends in birds. *Auk.* 119:36-45.

Beissinger SR (2001). Trade of live wild birds: potential, principles and practices of sustainable use. In *Conservation of Exploited Species*:

Reynolds JD, Mace GM, Redford KH, Robinson JG (eds) (1980). Cambridge University Press, Cambridge, UK. pp.182-202.

Betts MG, Neal PP, Simon J, Nocera J (2005). Point count summary statistics differentially predict reproductive activity in bird habitat relationship studies. J. Ornithol. 146:151-159.

Bibby CJ, Burgess ND, Hill DA (1992). *Bird Census Techniques*. Academic Press, London. p.256.

Bird Life International (2000). *Threatened Birds of the World*. Lynx Edicions and BirdLife International, Barcelona & Cambridge. (*The authority on the lamentable state of the global avifauna.*)

BirdLife International (2004). State of the World's Birds: Indicators for our Changing World. An assessment from BirdLife Partners. p.75.

BirdLife International (2010). CITES and the wild bird trade. http://www.birdlife.org/action/change/cites/index.html.

BirdLife International (2011). Species factsheet: *Psittacus erithacus.*: IUCN Red List for birds. http://www.birdlife.org.

Blondel J, Ferry C, Frochet B (1981). Point counts with unlimited distance. In: Ralph C J, Scott J M, editors. Estimating numbers of terrestrial birds. Stud. Avian Biol. 6:414-420.

Chardonnet P, Ders Clers B, Fisher J, Gerhold R, Jori F, Lamarque F (2002). The value of wildlife. Sci. Tech. Rev. 21(1):15-51.

Chupezi TJ, Ndoye O, Mpele TO (2006). Commodity-chain analysis for the capture and trade in the African Grey Parrots (*Psittacus erithacus erithacus*) in Cameroon. Prepared for Policy and Distributional Equity in Natural Resource Commodity Markets: Commodity Chain Analysis as a Policy Tool Project. http://www.indiana.edu/~iascp/bali/papers/Chupezi_Tieguhong_Ndoye_Mpele.pdf. Accessed on 20 December 2012.

CITES (2006). Summary records of the twenty-second meeting of the Animals Committee, Lima, Peru, 7-13 July 2006. http://www.cites.org/eng/com/ac/22/E-AC22-summary-record.pdf. Accessed on 10 July 2011

CITES (2007). 55th Meeting of the standing committee. The Hague, Netherlands, 2 June 2007. Review of Significant trade. SCSS Doc. 17.

CITES (2011). Lessons Learnt from Non-detriment Findings. 19th meeting of the Plant Committee (PC19Inf.10), Geneva, Switzerland,

Clemmons JR (2003). Status survey of the African Grey Parrot (*Psittacus erithacus timneh*) and development of a management program in Guinea and Guinea Bissau. Unpublished report to the CITES Secretariat, Geneva, Switzerland.

Collar NJ, Juniper AT (1992). Dimensions and causes of the parrot conservation crisis. In *New World parrots in crisis: solutions from conservation biology*: Beissinger S R and Snyder N.F.R. (eds), Smithsonian Institution Press, Washington, D.C, pp.1-24.

Dändliker G (1992). The African Grey Parrot in Ghana: a population survey. A contribution to the biology of the species. A study of its commercial exploitation and management recommendations.

Dawson DG (1981). Counting birds for a relative measure (index) of density studies. Avian Biol. 6:12-16.

Fotso R (1998). Survey status of the distribution and utilization of the Grey Parrot (*Psittacus erithacus*) in Cameroon. Unpublished report to the CITES secretariat, Geneva, Switzerland.

Hill D, Fasham M, Tucker G, Shewry M, Shaw P (2005). *Handbook of Biodiversity Methods: Survey, Evaluation and Monitoring*. Cambridge University Press.

Hutto RL, Pletschet M, Hendricks P (1986). A fixed radius point count method for non-breeding season use. *Auk* 103:593-602.

Juste JB (1996). Trade in the Gray Parrot (*Psittacus erithacus*) on the Island of Principe (São Tomé and Principe, Central Africa): Initial assessment of the activity and its impact. Biol. Conserv. 76:101-104.

Madindou I, Mulwa R (2008). Some conservation aspects of the African Grey Parrot (*Psittacus erithacus*) in Kakamega Forest, Kenya: assessment of trade and habitat destruction effects. http://www.africanbirdclub.org/club/documents/Grey ParrotReportII.pdf. Accessed on 10 July 2010.

MCGowan P (2001). Status, management and conservation of the African Grey Parrot, (*Psittacus erithacus*) in Nigeria. Unpublished report to the CITES secretariat, Geneva, Switzerland.

Melo M, O'Bryan C (2007). Genetic differentiation between Príncipe Island and mainland populations of the grey parrot (*Psittacus erithacus*), and implications for conservation. Mol. Ecol. 16:1673-1685.

Ministry of Forests and Fauna (MINFOF) (2007). *Bref aperçu du secteur forestier camerounaise.* MINFOF, Yaoundé, Cameroon.

Ministry of Forestry and Wildlife (MINFOF) (1994). Law N° 94/01 of 20 January 1994 fixing Forestry, Wildlife and Fishery regimes in Cameroon . Yaoundé, Republic of Cameroon.

Mulliken TA, Broad SR, Thomsen JB (1992). The wildlife bird trade: an overview. *Perceptions, conservation and management of wild birds in trade.* TRAFFIC Int, pp 1-42. www.traffic.org/species-reports/traffic_species_birds 4.pdf. Accessed on 7 March 2010.

Ngenyi A (2003). The African Grey Parrot (*Psittacus erithacus*): status and commercial exploitation in Cameroon. *Report for WWF, Jengi SE project,* 40pp. http://cameroun-foret.com/system/files/18_13_29.pdf. Accessed on 10 April 2012.

Pepperberg IM (1999). The Alex Studies. Harvard University Press. Cambridge

Reynolds RT, SCOTT JM, Nussbaum RA (1980). A variable circular-plot method for estimating bird numbers. *Condor* 82:309-313.

Rosenstock SS, Anderson DR, Giesen KM, Leukering T, Cartes FM (2002). Land bird counting techniques: current practices and alternative. *Auk* 119:46-53.

Sutherland WJ (2006). Ecological census techniques: a handbook. 2nd ed. Cambridge University Press, UK, p.410.

Tamungang SA, Cheke RA (2006). Sustainable Trade and Conservation of Parrots in Cameroon . Annual technical and financial progress reports. Field activities reports from PARROTPRO, Cameroon to Loro Parque Fundación, Spain. p.213.

Tamungang SA, Cheke RA (2009). Ecology and conservation of the Grey Parrot in Cameroon. Annual technical and financial progress report: field activities report from PARROTPRO, Cameroon. Unpublished report to Loro Parque Fundación, Spain.

Tamungang SA, Ajayi SS (2003). Diversity of food of the Grey Parrot (*Psittacus erithacus*) in Korup National Park, Cameroon. Bulletin of the African Bird Club, 10:33-36.

Volpato GH, Lopes EV, Mendonça LB, Boçon R, Bisheimer MV, Serafini PP, Anjos LD (2009).The use of point count method for bird survey in the Atlantic forest. Zoologia 26:74-78.

Mitochondrial cytochrome b variation in populations of the cutaneous leishmaniasis vector *Phlebotomus papatasi* across eastern Tunisia

Boudabous Raja[1]*, Jaouadi K.[1], Haouas N.[1], Mezhoud H.[1], Bdira S.[2], Amor S.[2] and Babba Hamouda[1]

[1]Laboratoire de Parasitologie- Mycologie code 99UR/08-05, Faculté de Pharmacie, Département de Biologie Clinique B, 1 Rue Avicenne, 5000 Monastir Tunisia, University of Monastir-Tunisia.
[2]Service d'Hygiène de Monastir Tunisia.

Cutaneous leishmaniasis remains a serious health problem in Tunisia. It is the most extensive and prevalent form with thousands of cases reported each year since the start of a large epidemic in 1982, with 96% of cases were recorded in central and southern regions of the country. In Tunisia, *Phlebotomus papatasi* (Scopoli) (Diptera: Psychodidae) is the main vector of cutaneous leishmaniasis caused by *Leishmania major*. *P. papatasi* was collected from peridomestic animal shelters and around the houses in several provinces in eastern Tunisia. Tunisian *P. papatasi* population was used in other reports as one of the countries for investigation from one exemplar, so it does not reflect the real structure of the country. Indeed, it is important to understand the population differentiation of sand fly vectors because it can show where potential vectors can invade from, and how quickly they can be controlled after local measures is applied. In this study, we characterized the Tunisian *P. papatasi* population by using mitochondrial cytochrome b (Cyt b) of sand flies as gene marker and situate it among the Mediterranean population. Our results indicate a low level of genetic variability among Tunisian populations of *P. papatasi* and we do not see any "isolation by distance" over a large geographical scale among Tunisian provinces.

Key words: *Phlebotomus papatasi,* cytochrome b, mtDNA, Tunisia, population differentiation, sandflies.

INTRODUCTION

Cutaneous leishmaniasis caused by *Leishmania major* Yakimoff and Schokhor (Kinetoplastida: Trypanosomatidae) is prevalent in arid and semiarid zones of northern Africa, the Middle East and central Asia (Ashford, 1996). The distribution of *Phlebotomus papatasi* (Scopoli) (Diptera: Psychodidae) extends beyond that of *L. major* and its reservoir hosts to southern Europe and eastern regions of the Indian subcontinent, and throughout its range this sand fly is often abundant in domestic and peridomestic biotopes far from the colonies of its gerbil reservoir hosts (Feliciangeli, 2004). In Tunisia, investigations in the zoonotic cutaneous leishmaniasis foci in the arid region showed that *P. papatasi* is the proven vector of *L. major* (Rioux et al., 1986) and *Meriones shawi* Duvernoy is the main reservoir host (Rioux et al., 1982). *P. papatasi* was recorded in Tunisia in 1912 (Langéron, 1912) and it is considered a ubiquitous species (Rioux et al., 1986b). It is well adapted to the conditions of arid climate (Rioux et al., 1986) and its density increases with the aridity (Croset et al., 1978). Cutaneous leishmaniasis remains a serious healthy problem in Tunisia. It is the most extensive and prevalent form with thousands of cases reported each year since the start of a large epidemic in 1982 ((Ben Rachid and Ben Ismail, 2001; DSSB, 2001). Indeed, the Tunisian Ministry of Health (MMH) reported 48914 cases of cutaneous leishmaniasis caused by *L. major* between 1982 to 1994. Among them, 96% of these cases were recorded in central and southern regions of

*Corresponding author. E-mail: hamouda.babba@gnet.tn.

the country (Ben Ismail et al., 1987). The eastern cost of Tunisia is located close to the zoonotic cutaneous leishmaniasis focus of Kairouan. Thus, human movements between these two cities could facilitate the disease transmission. It is important to understand the population differentiation of sand fly vectors because it can show where potential vectors can invade from and how quickly after local control measures. Our first investigation of the population differentiation of *P. papatasi* in Tunisia was to compare populations from the center of Tunisia with natural *P. papatasi* populations from the Mediterranean basin (Syria, Turkey, Cyprus, Jordan, Israel, Palestine, Egypt, Italy and Morocco); for which a maternally-inherited gene marker was used: mitochondrial cytochrome b (Cyt b) of sand flies.

In this study, a larger fragment of sandfly Cyt b was used to see if there is any "isolation by distance" over a larger geographical scale among Tunisian provinces. The mitochondrial genome is one of the best studied of all types of DNA. It has been shown to be a good source of accessible genetic variation, and analysis of mitochondrial DNA (mtDNA) variation has been used to understand evolutionary biology both at the intraspecific and interspecific levels (Avise, 1994). Their rapid rates of evolution, clonal inheritance, and lack of recombination have made it a valuable tool for phylogeographic studies and a source of markers for species and geographic populations. This makes it a clear choice for providing markers for identifying Tunisian sand flies. In this study, a fragment of sand fly cyt b was used to do the molecular characterization of Tunisian *P. papatasi* population, situate it among the Mediterranean population and to see if there is any “isolation by distance” over a larger geographical scale among Tunisian provinces; because, the other reports included Tunisian *P. papatasi* population as one of countries for investigation from one exemplar, so it does not reflect the real structure of the country.

MATERIALS AND METHODS

Sand flies

We analyzed *P. papatasi* sand flies (n = 90) from three localities of eastern cost of Tunisia (Figure 1a). We used data of *P. papatasi* sand flies from 25 field and colony populations originating from 10 countries (Syria, Turkey, Cyprus, Jordan, Israel, Palestine, Egypt, Italy, Tunisia and Morocco) (Figure 1b) (Hamarsheh et al., 2007). Some populations like those from Palestine, Israel and Egypt were selected based on previous isoenzyme studies (Kassem et al., 1990) that suggested the presence of genetic polymorphism. Tunisian collections were carried out from 1 September to 30 October 2005 during the main season of activity of adult sand flies in Tunisia. Sand flies were collected on sticky papers and Centers for Disease Control (CDC); miniature light traps were set overnight to sample sand flies in domestic animal shelters.

All sand flies were identified by their species-specific Cyt b sequences. Most were also identified based on morphological characters of the head and abdominal terminalia which were slide mounted in Berlese fluid following dissection with sterilized forceps and micro-needles.

PCR amplification of sandfly Cyt b

Genomic DNA was extracted from the thorax, wings, legs and abdomen of individual sand flies using the QIAmp DNA Mini Kit (Qiagen, Germany). For Cyt b CB3FC (forward) (5'-3' CA(C/T)ATT CAACC(A/T)GAATGATA) was used with N1N-FA (reverse) (5'-3'GGTA(C/T)(A/T)TTGCCTCGA(A/T)TTCG(A/T)TATGA) (Esseghir et al., 1997) to amplify a 3' fragment of 545 bp which contains the 3'end of Cyt b gene (Ready et al., 1997). 1-1.5 µl of extracted DNA was added to each 50 µl of polymerase chain reaction (PCR) mix, which contained Invitrogen buffer with-out 2 mM $MgCl_2$, 100 µM of each dNTP and 40 pmol of each primer. 1.5 units of Taq DNA polymerase (Invitrogen) were always used. PCR consisted of 5 cycles of denaturation at 94°C for 30 s, annealing at 40°C for 30 s and extension at 72°C for 1 min followed by 35 similar cycles but with annealing at 44°C and a final extension at 72°C for 10 min, all on a PTC-100™ programmable thermal controller cycler (MJ Research INC). PCR products were purified before analysis using the (Promega Wizard® SV gel and PCR clean-up system according to the supplier's specifications. This step was necessary; it removed bands that interfered with interpretation of results. Samples were stored at -20°C.

Direct sequencing of PCR products

The Genetic Molecular Human Laboratory (LGMH), Medicine Faculty of Sfax-Tunisia performed automated sequencing of variants. Sequence data were aligned using the program Sequencher 4.6 Demo.

Data analysis

Sequences (442 bp without primers) were manually edited and aligned using the online version of Multalin (Corpet, 1988) (http://multalin.toulouse.inra.fr/multalin). Gaps were treated as missing data.

Phylogenetic analysis

Sequence alignment is performed using the ClustalW routine included in the MEGA version 3.1 software (Kumar et al., 2004) and checked by eye. Maximum likelihood-ratio test analyses are performed using MEGA 3.1 software. Statistics file substitution model: HKY85Gamma, shape parameter: 97.761 transition/ transversion ratio: 9.024, number of categories: 4, proportion of invariant: 0.159. Tree style: radial (by tree Dyn).

RESULTS

Sequence analysis

In all cases, PCR amplification produced one band of approximately 500 bp (including primers). All sequences are available in GenBank under accession numbers JN686468 to JN686488. The sequence analysis generated 7 haplotypes defined by 4 polymorphic sites which were parsimony informative. Not more than one different nucleotides was identified at each segregating site indicating minimal homoplasy. Two of the substitutions were A → G transitions, one was C → T transition and one was G insertion (Table 2). The most

Figure 1a. Original and extensive focus of zoonotic cutaneous leishmaniasis caused by *L. major* in Tunisia.

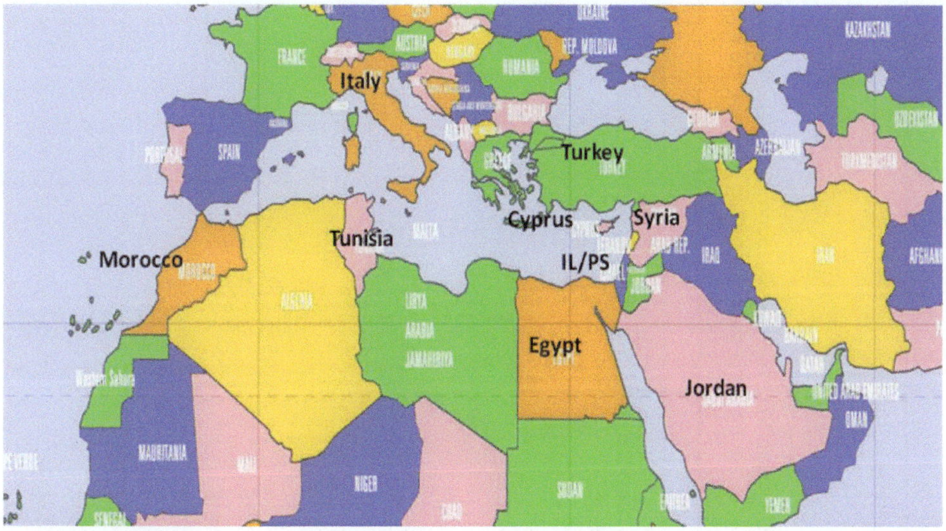

Figure 1b. Approximate geographical origins of *Phlebotomus papatasi* populations used for mtDNA analysis (IL/PS: Israel/Palestine).

Table 1. Haplotype frequencies in Tunisian *P. papatasi* population.

Locality	Code	Haplotypes (Hap)						
		Hap 1 SOS35	Hap 2 *DQ381827	Hap 3 *DQ381815	Hap 4 Pap40	Hap 5 Pap52	Hap 6 Mah54	Hap7 Mah50
Monastir (n = 30)	MON	0	20	8	1	1	0	0
Mahdia (n = 30)	Mah	11	6	11	0	0	1	1
Sousse (n = 30)	SOS	30	0	0	0	0	0	0

*Hamarsheh et al. (2007).

Table 2. Unique CB3 haplotypes found for *Phlebotomus papatasi* in this study and those by Hamarsheh et al. (2007). Alignment of variant nucleotide position for 442-bp mitochondrial cytochrome b gene. Only the polymorphic positions are shown; spaces represent identity with respect to the first reference sequence.

Parameter	Variant character position																																		
442- character CB3 haplotype	0	0	0	0	0	0	0	0	0	0	0	0	1	1	1	1	1	1	2	2	2	2	2	2	3	3	3	3	3	4	4	4	4	4	4	4
	0	0	0	1	2	6	6	7	8	9	9	9	0	0	1	1	1	7	1	2	3	4	7	8	2	2	8	9	9	0	0	0	0	1	2	3
	1	3	8	7	3	0	9	7	9	0	3	6	2	5	2	4	5	5	9	8	7	6	0	8	3	7	3	3	6	2	3	5	6	4	5	2
	C	T	A	G	T	A	C	T	G	A	T	A	C	A	-	A	A	G	T	T	T	A	C	A	G	G	T	A	C	A	A	A	A	A	-	T
Hap1															G																					
Hap2																															G					
Hap3																																				
Hap4															G																G					
Hap5															G																G					
Hap6															G																					
Hap7								C																												
DQ381815																															G					
DQ381816					C									G																						
DQ381817														G											A											
DQ381818														G																						
DQ381819				A										G			G									A		C				C	T			
DQ381820																					A															
DQ381821											C							A	C																	
DQ381822																					C						A									
DQ381823											C							A	C																	
DQ381824											C								C																	
DQ381825						G					C	G							C		T	T														
DQ381826																				C																
DQ381827																															G					

Table 2. Contd.

DQ381828									C	C							
DQ381829				T							T	G					
DQ381830				T				G			T					G	C
DQ381831				T							T						
DQ381832													T	G			
DQ381833				A	A	G											
DQ381834	A	C	T												C		
DQ381835							T							G			

frequent haplotypes were haps 1 and 2. Other shared haplotype (hap 3) was less frequent. Private haplotypes were found in individuals from two localities of Tunisia (haps 4, 5, 6 and 7) (Table 1). Sequences were compared by BLAST analysis with the last 440 bp of the mtDNA (cyt b) fragment (21 haplotypes) of widely separated collections of *P. papatasi* available in the GenBank databases (accession number DQ381815 to DQ381835) (Figure 2). Pair wise comparison between sequences from Tunisia obtained in this study and those available in GenBank were carried out and nucleotide differences (D) were calculated. This revealed a nucleotide variation ranging from 0.2 to 2% (Table 3). Haplotype 2 and hap 3 correspond to GenBank number sequence DQ381827 and DQ381815 respectively (Hamarsheh et al., 2007). For Tunisian haplotypes, except (haps 2 and 3), all the other five haplotypes were the found for the first time.

Most of the haplotypes obtained in this study revealed low nucleotide variation with those from Syria (0.23 to 0.45%) and high nucleotide difference (1.14 to 1.82%) with those from Egypt, Israel and Italy. Except some haplotypes from Palestine, Egypt and Turkey (GenBank: DQ381818, DQ381820, DQ381826, DQ381827 and DQ381835) present a medium nucleotide variability ranged between 0.23 to 0.68% (Table 3).

Phylogenetic analysis

The relationships among all haplotypes were revealed by maximum likelihood-ratio analysis. The haplotypes fall into five divergent groups; A, B, C, D and F (Figure 2), forming a well organized branching pattern. The topology of the network does not coincide with the geographical location of the populations (Table 3). Within each major clade, internal and terminal nodes can be interpreted as old and recently haplotypes respectively (Posad and Crandall, 2001). The position of the DQ381815, Haps 1, 3 and 6 haplotypes reflected an internal positioning within the network, indicating it to be the most ancestral haplotypes followed by more distantly related haplotypes. Tunisian haplotypes (Haps 2, 5 and 4) were the old haplotypes of haplotypes from Syria, Turkey and Palestine (Figure 2).

DISCUSSION

The mitochondrial genome is maternally inherited, non-recombining and highly polymorphic in insects (Avis, 1994) including sand flies (Esseghir et al., 1997). It is therefore not surprising that the cyt b sequences are good markers for characterizing *P. papatasi* populations. One or more haplotypes predominate at each geographical location with the rarest differing by single nucleotide exchanges. Differences in haplotypes shared between populations indicate that some haplotypes are more frequent than others. It has been postulated that older haplotypes have higher frequencies in the population because they have a greater probability of producing younger ones as mutational derivatives with lower frequencies (Crandall and Templeton, 1993; Smouse, 1998). It means Haps 1, 2 and 3 seems to be the older haplotypes of *P. papatasi* Tunisian populations. The tendency of some haplotypes having low, medium and high nucleotide differences or their separation into five groups (by network analysis) provides evidence of a past fragmentation event. This segregation suggests that *P. papatasi* populations from widely separated areas are substantially genetically differentiated as evidenced by differences in the frequencies of their most common haplotypes. Previous studies showed that H02 (hap 2 in this study) was derived from the widespread IRN33 (hap 3 in this study)

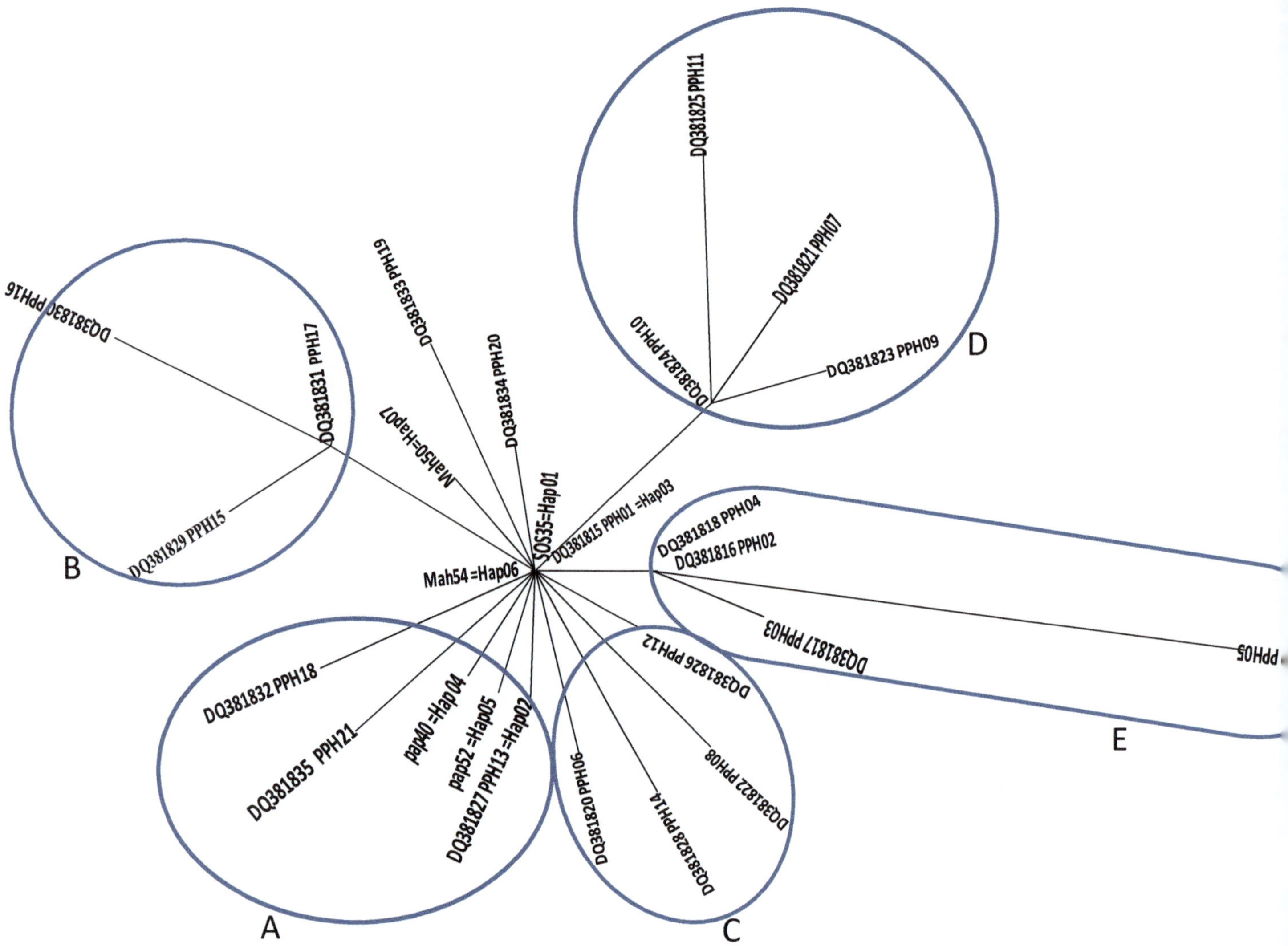

Fig. 2: Radial tree obtained by Maximum likelihood analysis for Tunisian and 21 other Mediterranean cyt b sequences of *P. papatasi*.

(Esseghir et al., 1997; Parvizi et al., 2003). It was frequent and shared between almost all populations from Syria, Turkey, Cyprus, Palestine and Tunisia, but sequence haplotype DQ381815 (hap 3 in this study) was found in some individuals from Tunisia and Cyprus (Hamarsheh et al., 2007). The haplotype H01 (hap 3 in this study) was reported to be ancestral for other haplotypes (Hamarsheh et al., 2007). The same haplotype was reported as ancestral for Iranian haplotypes. This seems to be the ancestral ones for Tunisian haplotypes as it has low nucleotide differences with all haplotypes (Table 2). By the maximum likelihood analysis of the haplotypes indicates that Haps 1 and 3 are probably the most ancestral haplotypes from which the other haplotypes which differ from Haps 1 and 3 by two or more mutational steps have evolved in radial manner (Figure 2). We found the same topology obtained by Hamarsheh et al. (2007) by median-joining network analysis.

No clear phylogeographical pattern was observed for the populations because shared haplotypes were present at the external nodes or at different nodes. These DNA fragments, particulary cytb have been regarded as good genetic markers to assess the relationships between closely related populations of *P. papatasi* in the Mediterranean region (Hamarsheh et al., 2007). Accordingly, our results indicate a low level of genetic variability among Tunisian populations of *P. papatasi* which is in line with a recent phylogenetic study reporting a genetic homogeneity among 26 populations of *P. papatasi* from 18 countries (Depaquit et al., 2008). Individual sand flies rarely migrate for more than 1000 m (Alexander, 1987; killick-kendrick, 1990). Therefore, genetic differences were expected between widely separated *P. papatasi* populations. Previous experiments carried by Kassem et al. (1993) and Ghosh et al. (1999) using isoenzyme electrophoresis of sand flies from Egypt

Table 3. Geographical origins and nucleotide differences between haplotypes of *P. papatasi* populations.

Parameter			Origine of haplotypes	Hap1	Hap2	Hap3	Hap4	Hap5	Hap6	Hap7
DQ381815	*IRN033 **PPH01	Hap 3	Syria/Tunisia	0.23	0.23	0	0.45	0.45	0.45	0.23
DQ381816			Egypt	0.68	0.68	0.45	0.91	0.91	0.91	0.91
DQ381817			Egypt	0.68	0.68	0.45	0.91	0.91	0.91	0.91
DQ381818	*H08		Egypt	0.45	0.45	0.23	0.68	0.68	0.45	0.45
DQ381819			Egypt	1.82	1.82	1.59	2.05	2.05	2.05	1.82
DQ381820			Egypt	0.23	0.45	0.23	0.45	0.45	0.23	0.23
DQ381821			Israel	0.91	0.91	0.68	1.14	1.14	1.14	0.68
DQ381822			Israel	0.68	0.68	0.45	0.91	0.91	0.68	0.68
DQ381823			Israel	1.14	1.14	0.91	1.36	1.36	1.36	1.14
DQ381824			Palestine	0.68	0.68	0.45	0.91	0.91	0.91	0.68
DQ381825			Palestine	0.45	0.45	0.91	0.68	0.68	0.68	0.45
DQ381826	*H09		Palestine	0.45	0.45	0.23	0.68	0.68	0.68	0.45
DQ381827	*H02 **PPH13	Hap 2	Palestine/Turkey	0.45	0.00	0.23	0.23	0.23	0.68	0.45
DQ381828			Jordan	0.68	0.68	0.45	0.91	0.91	0.91	0.68
DQ381829			Italy	0.91	0.91	0.68	1.14	1.14	1.14	0.91
DQ381830			Italy	1.36	1.36	1.14	1.36	1.36	1.36	1.36
DQ381831			Italy	0.68	0.68	0.45	0.91	0.91	0.91	0.68
DQ381832			Syria	0.45	0.45	0.23	0.68	0.68	0.68	0.45
DQ381833			Syria	0.68	0.68	0.45	0.91	0.91	0.91	0.68
DQ381834			Turkey	1.14	1.14	0.91	1.36	1.36	1.36	1.14
DQ381835			Turkey	0.68	0.23	0.45	0.45	0.45	0.91	0.68

**Hamarsheh et al. (2007); *Esseghir et al. (1997) and Parvizi et al. (2003).

and Jordan valley documented the presence of polymorphisms at 14 of 25 loci in Egyptian strains and 7 of 20 loci in the Jordan valley strains. Our study and those carried out by others indicate that mtDNA can be used successfully to characterize *P. papatasi* populations.

REFERENCES

Alexander JB (1987). Dispersal of Phlebotomine sand flies (Diptera: Psychodidae) in a Colombian coffee plantation. J. Med. Entomol., 24: 552-558.

Avise JC (1994). Molecular Markers, Natural History and Evolution. Chapman and Hall, New York.

Ben Ismail R, Helal H, Bach Hamba D, Ben Rachid MS (1987). Natural infestation of Phlebotomus papatasi in a focus of zoonotic cutaneous leishmaniasis in Tunisia. Bulletin of Tropical Medicine and Hygiene, 80, 613-614.

Ben Rachid MS, Ben Ismail R (1989). Parasitic diseases and environment in Tunisia. International Symposium on Environmental Sciences. INRST Editions, Tunis, pp. 17-28.

Corpet F (1988). Multiple sequence alignment with hierarchical clustering, Nucl. Acids Res., 16 (22): 10881-10890.

Crandall KA, Templeton AR (1993). Empirical tests of some predictions from coalescent theory with applications to intraspecific phylogeny reconstruction. Genetics, 134: 959–969.

Croset H, Rioux JA, Maistre M, Bayar N (1978). Sandflies from Tunisia (Diptera, Phlebotomidae). Systematic development, chorology and ethology. Annals of Human and Comparative Parasitology, 53, 711-749.

Darley U, Capaldi VM, Takamiya RA, Millett SF, Wilson MT, Malatesta F, Sarti P (1987). Reconstitution and molecular analysis of the respiratory chain, *In* Darley-Usmar VM, Rickwood D, Wilson MT (ed.), Mitochondria: a practical approach. IRL press, Oxford, England. pp. 113-126.

Depaquit J, Lienard E, Verzeaux GA, Ferté H, Bounamous A, Gantier JC, Hanafi HA, Jacobson RL, Maroli M, Moin-Vaziri V, Müller F, Ozbel Y, Svobodova M, Volf P, Léger N (2008). Molecular homogeneity in diverse geographical populations of *Phlebotomus papatasi* (Diptera, Psychodidae) inferred from ND4 mtDNA and ITS2 rDNA Epidemiological consequences. Infect. Genet. Evol., 8(2): 159-70.

DSSB (2001). Annual Report of Activities. Directorate of Basic Health Care, Health Ministry, Tunisia.

Esseghir S, Ready PD, Killick-Kendrick R, Ben-Ismail R (1997). Mitochondrial haplotypes and phylogeography of *Phlebotomus* vectors of *Leishmania major*. Insect Mol. Biol., 6: 211–225.

Feliciangeli MD (2004). Natural breeding places of Phlebotomine sandflies. Med. Vet. Entomol., 18: 71-80.

Ghosh KN, Mukhopadhyay JM, Guzman H, Tesh RB, Munstermann LE (1999). Interspecific hybridization and genetic variability of *Phlebotomus* sand flies. Med. Vet. Entomol., 13: 78–88.

Hamarsheh O, Presber W, Abdeen Z, Sawalha S, Al-Lahem A, Schönian G (2007). Genetic structure of Mediterranean populations of the sandfly Phlebotomus papatasi by mitochondrial cytochrome b haplotype analysis. Med. Vet. Entomol., 21(3): 270-7.

Kassem HA, Fryauff DJ, Shehata MG, El Sawaf BM (1993). Enzyme polymorphism and genetic variability of one colonized and several field populations of *Phlebotomus papatasi* (Diptera: Psychodidae). J. Med. Entomol., 30: 407–413.

Killick-Kendrick R (1990). Phlebotomine vectors of the leishmaniases: a review. Med. Vet. Entomol., 4: 1-24.

Kumar S, Tamura K and Nei M (2004). MEGA3: Integrated software for Molecular Evolutionary Genetics Analysis and sequence alignment. Oxford J. Briefings Bioinfor., 5(2): 150-163.

Langeron M (1921). Second parasitological mission in Tunisia. Tamerza (September-October 1919).Arch. Inst. Pastor of North Africa. 1(1). 347-382.

Parvizi P, Benlarbi M, Ready PD (2003). Mitochondrial and *Wolbachia* markers for the sand fly *Phlebotomus papatasi*: little population differentiation between peridomestic sites and gerbil burrows in Isfahan province, Iran. Med. Vet. Entomol., 17: 351–362.

Ready PD, Day JC, de Souza AA, Rangel EF, Davies CR (1997). Mitochondrial DNA characterization of populations of *Lutzomyia whitmani* (Diptera: Psychodidae) incriminated in the peri-domestic and silvatic transmission of *Leishmania* species in Brazil. Bull. Entom. Res., 87: 187-195.

Rioux JA, Perieres J, Killick-Kendrick R, Lanotte G,et Bailly M (1982). Ecology of leishmaniasis in southern France. 17. Sampling of Phlebotomus by the method of adhesive traps. Comparison with the capture technique on human bait. Ann. Parasitol. Hmm. Comp. 57, 631.

Rioux JP, Lanotte G, Petterb F, Derreure J, Akalay O, Pratlong F, Velez ID, Fikri NB, Maazoun R, Denial M, Jarry DM, Zahaf A, Ashford RW , Cadi-Soussi M, Killick-Kendrick R, Benmansour N, Moreno G, Perieres J, Guivard E, Zribi M, Kennou MF, Rispail P, Knechtli R, Serres E (1986b). The cutaneous leishmaniasis in the Western Mediterranean basin. From the enzymatic identification to the eco-epidemiological analysis. The example of three "focus" Tunisian, Moroccan and French, in: Leishmania. Taxonomy and phylogeny. Applications eco-epidemiological, Rioux JA (ed). IMEEE, Montpellier, 365-395.

Rioux JP, Lanotte G, Pratlong F (1986a). Leishmania killick n.sp. (Kinetoplastida: Trypanosomatidae), in: Taxonomy and phylogeny of Leishmania. Eco-epidemiological Applications, Rioux JA (Coll. Int. CNRS / INSERM, 1984, (ed.). IMEEE, Montpellier, 139-142.

Smouse PE (1998). To tree or not to tree. Mol. Ecol., 7: 399-412.

16

Prevalence and seasonal periodicity of endophytic coelomycetous fungi in Tamil Nadu, India

Subban Kamalraj and Johnpaul Muthumary

Centre for Advanced Studies in Botany, University of Madras, Maraimalai Campus, Guindy, Chennai-600 025, Tamil Nadu, India.

The present study reports 2,644 endophytic isolates; only 74 different morphospecies of coelomycetes have been isolated and identified from 14,400 segments of healthy tissues such as leaves (4,800), stem (4,800) and bark (4,800). Each plant tissue with 300 segments was studied in a Petri dish containing 10 segments; totally, 30 plates were studied with four seasons. Totally, 74 coelomycetes were isolated and identified to belong to 16 genera by classical taxonomy. Interestingly, the *Pestalotiopsis* spp. was found to be dominant as compared to other genera in all the collection spots in the four different seasons.

Key words: Coelomycetes, endophytes, parasites, morphospecies, colony frequency, *pestalotiopsis*, relative percentage occurrence.

INTRODUCTION

About 100,000 species of fungi have been formally described by taxonomists (Kirk et al., 2008), but the global biodiversity of the fungal kingdom is not fully understood (Mueller and Schmit, 2006). Several authors have suggested that the majority of undiscovered endophyte diversity occurs in leaves of tropical trees (Promputtha et al., 2005). Endophytes are the potential producers of novel biologically active compounds of immense value in agriculture, medicine and industry (Tan et al., 2000; Tan and Zou, 2001; Aly et al., 2010; Shankar and Krishnamurthy, 2010). Coelomycetes are anamor-phic fungi producing asexual spore (conidia) in a pycni-dium, acervulus or stroma (Sutton, 1980; Nag Raj, 1993). Currently, 1000 genera comprising 7000 species have been described (Kirk et al., 2001). The previous reports strongly indicated that endophytic coelomycetous fungi were studied for over three decades (Webber and Gibbs, 1984; Hudson, 1986, Johnson and Whitney, 1989a, b; Stone et al., 1996; Suriyanarayanan and Vijayakrishna, 2001). The diversity of endophytic coelomycetes from the tree species was subjected to biotechnological approach because of the colony frequencies accumulated over the time; the age of leaves and positively correlated needles with the fungal colonization (Bernstein and Carroll, 1977b; Petrini and Carroll, 1981; Stone, 1987; Helander et al., 1993; 1994; Carroll, 1995; Magan et al., 1996; Gaylord et al., 1996; Faeth and Hammon, 1997; Hata et al., 1998; Poteri et al., 2001; Arnold and Herre, 2003). Many reports confirmed that the endophytic fungi such as *Pestalotiopsis* spp., *Pestalotia* spp., *Phomopsis* spp., *Phyllosticta* spp., *Melanconium* spp., *Colletotrichum* spp., *Phoma* spp., *Chaetomium* spp., *Botryodiplodia* spp., *Bartalinia* spp. and *Monochaetia* spp. can produce medicinally important bioactive secondary metabolites (Firakova et al., 2007). In which, interestingly, more endophytic coelomycetous fungi have been reported (Petrini, 1986; Wang et al., 2000; Zhou et al., 2001; Yuan et al., 2006; Muthumary and Sashirekha, 2007).

The delicate balance between host and endophytic organism seems to be controlled in part by chemical factors, for example, herbicidal natural products produced by the fungus versus antifungal metabolites biosynthesized by the host plant (Tadych and White, 2009). The coelomycetous fungi are considered to have evolved from saprophytes (Zavos et al., 1988), and in turn, parasites are assumed to have given rise to endophytes (Clay, 1988). In the present study, first endophytic diversity of coelomycetes was studied in *Mangifera indica* Linn. (Anacardiaceae), *Rhododendron arboreum* Sm. (Ericaceae), *Taxodium mucronatum* Tenore. (Cupressaceae) and *Taxodium distichum* Rich. (Cupressaceae) in India.

MATERIALS AND METHODS

Study area

R. arboretum Sm., *T. distichum* Rich. and *T. mucronatum* Ten. were collected from the forests at Tamil Nadu Botanical Garden (2623 m above sea level) in Tamil Nadu National Park, Ooty, Nilgiri District of Tamil Nadu, South India (Lat.10° 38′ 11.49′′; Long. 76° 0′ 77.15′′) which is an evergreen hill forest with much canopy, and *M. indica* Linn. was collected from Botany Field Laboratory, University of Madras, Maduravoyal, Chennai, Tamil Nadu.

Isolation of endophytic fungi

The collected samples were washed thoroughly by running tap water. From 300 segments (approximately 0.5 cm^2) of each 50 leaves, stem and bark tissues from each plant were surface sterilized and placed on PDA medium poured Petri plates for the study of fungal endophytes (Fisher et al., 1993; Dobranic et al., 1995; Schulz et al., 1998). The Petri plates were incubated in a light chamber for a period of 3 to 4 weeks (Bills and Polishook, 1992) at 23 ± 2°C. The light regime was 12 h light followed by 12 h darkness (24 h/day). Throughout the experiment, two Philips "Cool white" (6500 K) florescent lamps (Philips, India) providing energy of 50 $\mu E/m^2/S$ was maintained. A Philips 100 W lamp served as the source of incandescent light. The Petri dishes were observed at regular intervals starting from the second day onwards for the fungal growth. The emerging fungal propagules were isolated, purified and maintained by subsequent subcultures (Bills 1996).

Preparation of semipermanent slides

After the sporulation of each fungus, the conidiomata of the coelomycetes were mounted, cleared and examined; the preparation was tapped vigorously to break the conidiomata and then reheated. In this way, the conidia would be released and if one is diligent, some conidiogenous cells will be evident with the developing conidia attached. The vertical sections of conidiomata were taken to confirm the nature of the fructification. Specimens with fruit bodies were sectioned with the help of sterile scalpel blade. The selected sections (10 to 15 µm) were observed under light microscope by using lactophenol mount. Lactophenol and lactophenol with cotton blue stain were used for the preparation of semipermanent slides. The slides were sealed with DPX mountant (BDH Chemical Division, Glaxo Laboratories Ltd. Mumbai, India).

Colonization frequency (CF%)

The colonization frequency for endophytic coelomycetes species in the leaf, stem and bark tissue was calculated by using the following formula (Hata and Futai 1995):

$$CF\ (\%) = \frac{\text{The number of segments colonized by an endophytic species}}{\text{Total number of segments}} \times 100$$

Relative percentage occurrence (RPO%) of each group of fungi

Relative percentage occurrence of different genera of species was calculated using the following formula:

$$RPO\ (\%) = \frac{\text{Density of colonization of one group}}{\text{Total density of colonization of all groups}} \times 100$$

RESULTS

Isolation of endophytic mycobiota

The results indicated the colony frequencies of endophytic coelomycetous species from 4 plants belonging to three families. The number of endophytic coelomycetes in barks of *T. mucronatum* and *T. distichum* were significantly higher than stem and leaves. Moreover, the highest colonization frequency was found in leaves of *M. indica* and *R. arboreum* as compared to stem and bark.

This study was undertaken for four different seasonal variations, that is, South West Monsoon (SWM), North East Monsoon (NEM), Winter Season (WS) and Summer Season (SS). The colonization frequency of leaves, stem and bark significantly increased during monsoon and decreased during summer and winter seasons (Figure 1).

Identification

The characteristics of the fungi were observed by making microslide preparations of the fruit bodies and conidia. They were identified on the basis of the structure of fruit body (conidiomata), conidiogenous cells and conidia (Sutton, 1980). The presence of appendage on the conidia which is now being recognized as an important taxonomic criterion was studied critically (Guba, 1961; Ellis, 1971; Sutton, 1973a, b, 1980; Onions et al., 1981; Nag Raj 1993) and measurement of the conidia was done. Identification of coelomycetes up to different morphospecies levels were carried out using standard manuals (Guba, 1961; Sutton, 1980; Nag Raj, 1993).

Isolation of endophytic coelomycetous mycobiota at four different seansons

South west monsoon

A total of 189 isolates, 15 different morphospecies belonging to 7 genera were isolated and identified from *M. indica*. The most frequently isolated genera were *Colletotrichum*, *Pestalotiopsis*, *Phomopsis* and *Phyllosticta*. Single species of the genera *Botryodiplodia*, *Pestalotia* and *Phoma* were also recorded (Table 1a). The

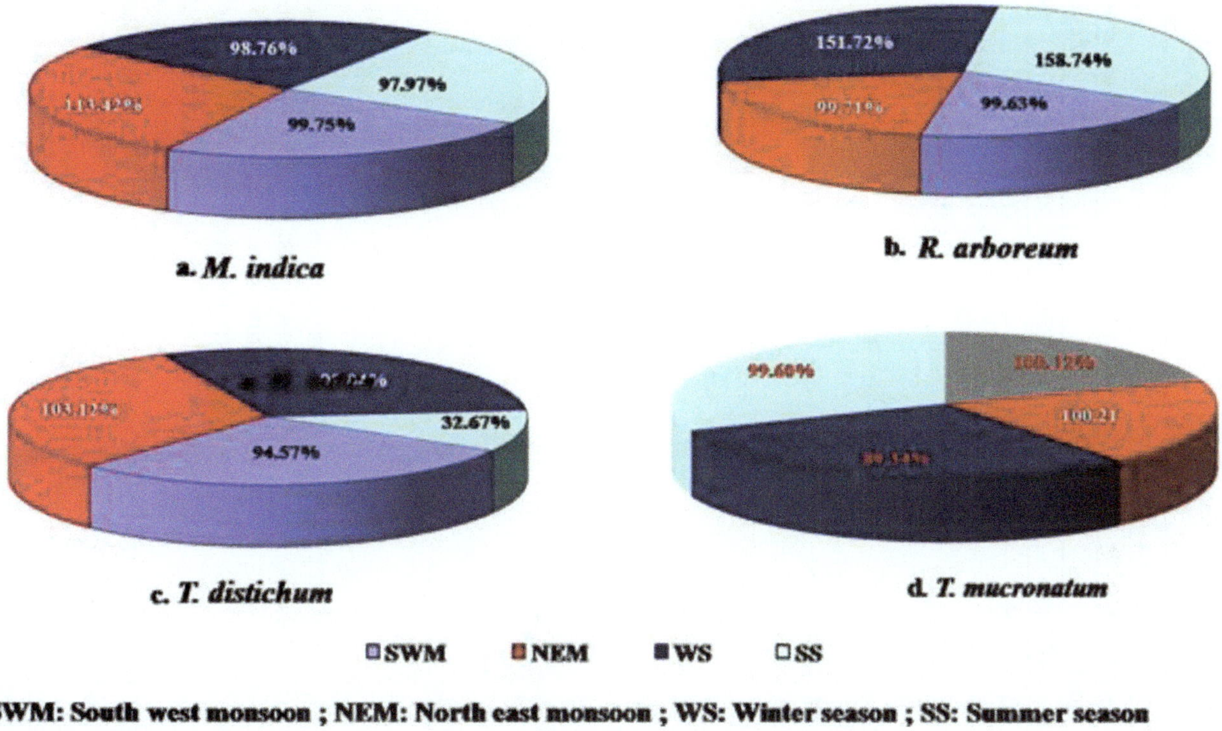

Figure 1. Total relative percentage occurrence of endophytic coelomycetous fungi on different seasons

The maximum RPO was found in *Pestalotiopsis* spp. (48.61%) followed by *Phomopsis* spp. (14.2%).

In the case of *R. arboretum*, 129 isolates existed at 16 different morphospecies belonging to 8 genera which were isolated and identified. Among them, *Colletotrichum, Pestalotiopsis, Pestalotia* and *Phomopsis* occurred predominantly. Single species of the genera *Bartalinia, Botryodiplodia, Camarosporium* and *Fusamen* were also observed (Table 1b). In this study, the maximum RPO was found in *Pestalotiopsis* spp. (47.11%) followed by *Colletotrichum* spp. (13.5%) and *Phomopsis* spp. (9.57%).

Totally, 215 isolates were recorded from *T. distichum*. Of these, 22 different morphospecies under 11 genera were isolated and identified. Among them, *Botryodiplodia*, *Colletotrichum*, *Fusicoccum*, *Monochaetia*, *Pestalotiopsis*, *Pestalotia*, *Pestalozziella*, *Phomopsis*, *Phyllosticta*, *Seimatosporium* and *Septoria* occurred predominantly. Single species of the genera *Bartalinia, Botryodiplodia, Camarosporium* and *Fusamen* spp. were also recorded (Table 1c). In this study, the maximum RPO was found in *Pestalotiopsis* spp. (37.6%) followed by *Pestalotia* spp. (9.72%) and *Seimatosporium* sp. (8.83%).

Similarly, a total of 209 isolates were recorded. Of these, 20 different morphospecies that existed in the leaves, stem and bark belonging to 10 genera were isolated and identified from *T. mucronatum*. *Colletotrichum, Pestalotiopsis, Phomopsis, Phyllosticta* and *Septoria* were isolated. Single species of the genera *Botryodiplodia, Fusamen, Monochaetia, Truncatella* and *Pestalotia* were also observed (Table 1d). The maximum RPO was found in *Pestalotiopsis* spp. (43.40%) followed by *Phyllosticta* spp. (10.62%) and *Colletotrichum* spp. (10.11%).

North east monsoon

From *M. indica*, a total of 195 isolates were obtained. Among these, 15 different morphospecies under 7 genera such as *Colletotrichum, Pestalotiopsis, Phomopsis* and *Phyllosticta* occurred predominantly. Single species of the genera *Botryodiplodia, Pestalotia* and *Phoma* were also recorded. In this study, the maximum RPO was found in *Pestalotiopsis* spp. (46.6%) followed by *Colletotrichum* spp. (17.89%).

A total of 171 isolates were recorded from *R. arboretum*. Of these, 16 different morphospecies under 8 genera were isolated and identified (Table 1b). Among them, *Colletotrichum, Pestalotiopsis, Pestalotia* and *Phomopsis* occurred predominantly. Single species of the genera *Bartalinia, Botryodiplodia, Camarosporium* and *Fusamen* were also observed. In this study, the maximum RPO was found in *Pestalotiopsis* spp. (57.47%) followed by *Colletotrichum* spp. (15.75%).

A total of 173 isolates were recorded from *T. distichum*. Of these, 22 different morphospecies that existed in the leaves, stem and bark belonging to 11 genera were isolated and identified. *Botryodiplodia*, *Colletotrichum*, *Pestalotiopsis*, *Pestalotia*, *Phomopsis* and *Phyllosticta* occurred predominantly. Single species of the genera *Seimatosporium, Septoria, Fusicoccum, Monochaetia*

Table 1a. CF and RPO of endophytic coelomycetous fungi isolated from different samples of *M. indica* during various seasons between 2006 and 2007

Endophyte	Number of segments colonized															
	South west monsoon (June - Sep)				North east monsoon (Oct - Nov)				Winter season (Dec - Feb)				Summer season (Mar - May)			
	L	S	B	RPO (%)	L	S	B	RPO (%)	L	S	B	RPO (%)	L	S	B	RPO (%)
Botryodiplodia sp. MUBL1086	0.33	0.66	0.66	2.61	0.66	0.66	0.66	3.04	0.66	0.33	0.66	2.82	0.33	0.33	0.33	2.47
Colletotrichum sp. MUBL1042	1.33	1.0	0.66		2.0	0.66	1.33		2.0	0.66	2.66		0.66	0.66	2.0	
Colletotrichum sp. MUBL1043	2.0	2.66	3.0	20.04	2.33	1.33	0.66	17.89	2.33	1.0	0.66	18.22	0.66	0.33	0.33	12.4
Colletotrichum sp. MUBL1044	-	0.33	0.66		2.0	0.33	0.66		0.33	0.33	0.33		0.33	-	-	
Colletotrichum sp. MUBL1045	0.33	0.33	0.33		0.33	-	-		-	-	0.33		-	-	-	
Pestalotiopsis sp. MUBL1000	2.0	1.0	1.66		1.33	2.0	3.0		3.33	1.33	2.0		1.0	2.0	2.66	
Pestalotiopsis sp. MUBL1001	3.0	3.33	2.0		2.33	1.33	2.66		1.0	2.0	1.66		0.33	0.66	2.0	
Pestalotiopsis sp. MUBL1002	3.0	3.66	2.66	48.61	2.0	3.0	3.66	46.6	2.33	2.66	1.33	44.5	2.0	1.33	3.0	51.5
Pestalotiopsis sp. MUBL1003	2.0	0.66	1.33		1.0	2.33	2.0		1.66	1.33	1.0		1.33	0.66	0.33	
Pestalotiopsis sp. MUBL1004	1.33	2.0	1.0		1.33	1.66	0.66		1.0	2.33	1.0		2.0	0.66	0.66	
Pestalotia sp. MUBL1062	0.33	-	-	0.52	-	-	0.33	0.50	-	-	0.33	0.56	-	-	-	-
Phomopsis sp. MUBL1070	1.0	2.0	0.66	14.25	2.33	2.0	0.66		3.0	1.33	2.0	20.5	2.33	2.0	2.0	28.3
Phomopsis sp. MUBL1071	1.33	1.66	2.33		2.33	1.0	2.0	15.87	2.66	1.0	2.0		2.0	1.0	2.0	
Phoma sp. MUBL1106	0.33	0.33	0.33	1.57	0.33	0.33	0.33	15.2	0.33	-	-	0.56	-	-	-	-
Phyllosticta sp. MUBL1078	2.0	0.66	1.0	12.15	2.66	0.66	2.0	14.32	1.33	1.66	2.0	11.9	0.33	-	0.33	3.3
Phyllosticta sp. MUBL1079	1.0	2.0	1.0		1.0	1.33	1.66		1.0	0.66	0.33		0.33	0.66	0.33	
Number of coelomycetes species recorded	15	15	15	-	15	14	15	-	14	14	15	-	13	11	12	-
Total Number of coelomycetes isolates observed	64	67	58	-	72	56	67	-	70	50	55	-	41	31	48	-

L: Leaves; S: stem; B: bark; RPO: Relative percentage occurrence; CF: colony frequency.

and *Pestalozziella* were also observed. The maximum RPO was found in *Pestalotiopsis* spp. (41.5%) followed by *Colletotrichum* spp. (13.2%) and *Phomopsis* spp. (11.55%) (Table 1c).

From *T. mucronatum*, a total of 205 isolates were recorded. Of these, 20 different morphospecies belonging to 10 genera were isolated and identified. *Colletotrichum*, *Pestalotiopsis, Phomopsis, Phyllosticta, Septoria* and *Truncatella* were recorded predominantly. Single species of the genera *Botryodiplodia, Fusamen, Monochaetia* and *Pestalotia* were also observed (Table 1d). The maximum RPO was found in *Pestalotiopsis* spp. (37.04%) followed by *Phyllosticta* spp. (10.62%) and *Colletotrichum* spp. (10.11%).

Winter season

Totally, 175 isolates were obtained from *M. indica*. Among them, 15 different morphospecies belonging to 7 genera were isolated and identified. Of these, *Colletotrichum, Pestalotiopsis, Phomopsis* and *Phyllosticta* were routinely served.

Single species of the genera Botryodiplodia *Pestalotia* and *Phoma* were also recorded (Table 1a). In this study, the maximum RPO was found in *Pestalotiopsis* spp. (44.5%) followed by *Colletotrichum* spp. (20.5%). Among the 185 isolates from *R. arboretum*, 15 different morphospecies under 7 genera were isolated and identified. In which, *Pestalotiopsis, Pestalotia* and *Phomopsis* species occurred predominantly. Single species of the genera *Bartalinia, Botryodiplodia, Camarosporium* and *Colletotrichum* were also recorded (Table 1b). In this study, the maximum RPO was found in *Pestalotiopsis* spp. (60.49%) followed by *Colletotrichum* spp. (18.35%).

From *T. distichum*, 164 isolates were recorded

Table 1b. CF and RPO of endophytic coelomycetous fungi isolated from different samples of *R. arboreum* during various seasons between 2006 and 2007.

Endophyte	Number of segments colonized															
	South west monsoon (June – Sep)				North east monsoon (Oct – Nov)				Winter season (Dec – Feb)				Summer season (Mar – May)			
	L	S	B	RPO (%)	L	S	B	RPO (%)	L	S	B	RPO (%)	L	S	B	RPO (%)
Bartalinia sp. MUBL1096	0.33	0.33	0.33	2.37	-	1.0	-	1.75	-	0.33	0.33	1.07	-	-	-	-
Botryodiplodia sp. MUBL1087	0.66	0.66	-	3.16	0.66	1.33	1.66	5.24	0.33	1.0	1.33	3.77	0.66	0.33	1.0	3.77
Colletotrichum sp. MUBL1046	0.33	1.0	0.66		0.33	3.0	1.33		-	-	-		1.33	2.0	0.33	
Colletotrichum sp. MUBL1047	0.33	0.66	2.0	13.5	2.0	-	1.0	15.75	2.66	2.0	1.0	18.35	2.66	1.0	0.66	18.93
Colletotrichum sp. MUBL1048	0.33	0.33	-		0.66	-	0.66		2.0	1.33	2.33		1.0	0.33	0.66	
Camarosporium sp. MUBL1102	-	1.33	2.0	7.99	-	0.33	2.0	4.0	0.33	-	-	0.53	0.33	-	-	0.62
Fusamen sp. MUBL1098	-	0.33	2.66	7.17	-	1.0	1.33	3.3	-	0.33	0.33	1.07	2.66	2.0	3.33	15.17
Pestalotiopsis sp. MUBL1005	0.66	1.0	3.00		1.33	2.66	2.66		2.0	4.0	3.0		2.33	1.0	0.66	
Pestalotiopsis sp. MUBL1006	-	0.33	2.33		0.66	2.0	2.33		1.0	1.33	2.0		3.33	3.0	1.33	
Pestalotiopsis sp. MUBL1007	-	0.66	2.0	47.11	2.0	2.33	1.0	57.47	3.33	2.66	2.0	60.49	2.0	0.66	1.30	53.66
Pestalotiopsis sp. MUBL1008	0.66	0.33	3.00		0.33	3.0	3.33		3.66	4.0	2.66		1.33	3.0	2.66	
Pestalotiopsis sp. MUBL1009	-	2.0	3.66		1.00	3.66	3.33		2.66	2.0	1.0		2.0	0.66	3.0	
Pestalotia sp. MUBL1063	-	0.33	1.33	8.76	-	0.66	1.0	6.40	1.0	0.66	0.66	5.90	1.33	0.16	0.33	4.59
Pestalotia sp. MUBL1064	0.33	1.0	0.66		0.33	0.33	1.33		0.66	0.33	0.33		-	0.33	0.33	
Phomopsis sp. MUBL1072	0.33	1.0	1.0	9.57	-	-	0.33	5.80	0.33	0.66	1.0	8.07	0.33	-	-	0.62
Phomopsis sp. MUBL1073	-	1.33	0.33		0.66	1.66	0.66		1.33	1.0	0.66		-	-	-	
Number of coelomycetes species recorded	09	16	14	-	11	13	15	-	16	14	14	-	12	12	12	-
Total number of coelomycetes isolates observed	12	38	75	-	30	69	72	-	64	65	56	-	64	47	47	-

L: Leaves; S: stem; B: bark; RPO: relative percentage occurrence; CF: colony frequency.

in this season. Of these, 22 different morphospecies belonging to 11 genera were isolated and identified such as *Colletotrichum*, *Fusicoccum*, *Monochaetia*, *Pestalotiopsis*, *Pestalotia*, *Phomopsis* and *Phyllosticta*. Single species of the genera *Pestalozziella*, *Seimatosporium*, *Septoria* and *Botryodiplodia* were also observed (Table 1c). The maximum RPO was found in *Pestalotiopsis* spp. (45.0%) followed by *Pestalotia* spp. (10.95%).

A total of 173 isolates were recorded. Of these, 19 different morphospecies that existed in the leaves, stem and bark belonging to 10 genera such as *Colletotrichum, Pestalotiopsis, Phomopsis, Phyllosticta, Septoria* and *Truncatella* were isolated from *T. mucronatum*. The single species of the genera *Botryodiplodia, Fusamen, Monochaetia* and *Pestalotia* species were also observed (Table 1d). The maximum RPO was found in *Pestalotiopsis* spp. (35.20%) followed by *Colletotrichum* spp. (12.10%) and *Phomopsis* spp. (10.96%).

Summer season

A total of 120 isolates were obtained. Of these, 13 different morphospecies that existed in the bark, leaves and stem belonging to 5 genera were isolated and identified from *M. indica*. Among them, *Colletotrichum, Pestalotiopsis, Phomopsis* and *Phyllosticta* occurred predominantly. Single species of the genus *Botryodiplodia* was also recorded. In this study, the maximum RPO was found in *Pestalotiopsis* spp. (51.5%) followed by *Colletotrichum* spp. (28.3%).

A total of 158 isolates were recorded. Of these, 12 different morphospecies that existed in the leaves, stem and bark belonging to 7 genera were isolated and identified in *R. arboretum*. Among them, *Colletotrichum, Pestalotiopsis, Pestalotia*

Table 1c. CF and RPO of endophytic coelomycetous fungi isolated from different samples of *T. distichum* during various seasons between 2006 and 2007.

Endophyte	Number of segments colonized															
	South west monsoon (June - Sep)				North east monsoon (Oct - Nov)				Winter season (Dec - Feb)				Summer season (Mar - May)			
	L	S	B	RPO (%)	L	S	B	RPO (%)	L	S	B	RPO (%)	L	S	B	RPO (%)
Botryodiplodia sp. MUBL1089	0.66	0.33	1.66	3.69	0.33	0.66	2.0	5.18	-	2.33	2.0	7.92	-	1.33	0.66	2.28
Colletotrichum sp. MUBL1052	0.33	1.0	0.66		0.33	1.33	0.66		-	1.66	0.66		-	0.66	0.33	
Colletotrichum sp. MUBL1053	-	0.66	0.33	9.26	-	2.66	1.0		-	2.0	1.0		0.66	0.33	1.0	
Colletotrichum sp.MUBL1054	-	2.33	1.33		-	1.0	0.66	13.2	-	1.0	0.66	12.76	0.66	0.33	1.0	5.71
Fusicoccum sp. MUBL1104	0.33	0.66	0.33	1.84	-	0.33	0.33	1.14	0.33	0.33	0.33	1.20	-	-	-	-
Monochaetia sp. MUBL1101	-	0.33	0.33	0.92	-	0.66	0.66	2.28	-	0.33	0.66	1.81	-	-	-	-
Pestalotiopsis sp. MUBL1016	-	2.66	3.0		-	1.0	2.0		-	2.33	2.0		-	2.0	2.33	
Pestalotiopsis sp. MUBL1017	-	0.66	1.33		-	1.33	2.0		-	3.0	2.0		-	1.33	1.66	
Pestalotiopsis sp. MUBL1018	-	2.33	1.33		-	0.66	2.0		-	1.0	2.0		-	1.0	0.66	
Pestalotiopsis sp. MUBL1019	-	3.0	3.33	37.6	-	1.0	2.0	41.5	-	1.0	2.0	45.0	0.33	0.66	0.33	17.5
Pestalotiopsis sp. MUBL1020	-	2.66	2.33		-	3.33	1.33		-	2.0	1.33		-	0.66	1.33	
Pestalotiopsis sp. MUBL1021	-	3.66	3.33		-	0.66	3.66		-	1.0	3.66		-	1.33	0.66	
Pestalotiopsis sp. MUBL1022	-	0.66	0.33		0.33	2.0	1.0		-	0.33	1.0		033	0.33	0.33	
Pestalotia sp. MUBL1066	0.33	0.66	2.66	9.72	-	1.66	2.33		-	0.66	2.33		-	-	-	
Pestalotia sp. MUBL1067	-	0.66	2.66		0.33	0.66	1.0	9.79	-	2.0	1.0	10.95	-	0.66	0.33	1.13
Pestalozziella sp. MUBL1105	-	3.0	0.66	5.10	-	1.0	2.33	5.77	-	0.66	0.66	2.41	-	0.33	0.33	0.75
Phomopsis sp. MUBL1076	-	1.0	1.66		-	3.0	1.0		-	-	0.33		-	-	-	
Phomopsis sp. MUBL1077	-	1.0	2.33	8.35	0.33	2.0	0.66	11.55	-	0.66	2.0	5.47	-	0.66	0.33	1.13
Phyllosticta sp. MUBL1082	-	1.33	3.0	7.88	-	0.66	0.33		-	1.0	2.33		-	-	0.33	
Phyllosticta sp. MUBL1083	-	0.66	0.66		0.33	0.66	2.0	6.33	-	0.33	0.33	7.29	-	1.33	0.33	1.90
Seimatosporium sp. MUBL1107	-	3.33	3.0	8.83	-	0.33	2.33	4.61	-	0.66	1.0	3.03	-	-	1.33	1.52
Septoria sp. MUBL1093	-	0.66	0.33	1.38	-	0.33	0.33	1.14	-	0.33	0.33	1.20	-	0.33	0.33	0.75
Number of coelomycetes species recorded	4	22	22	-	6	22	22	-	1	21	22	-	4	16	18	-
Total number of coelomycetes isolates observed	5	100	110	-	6	81	86	-	1	74	89	-	6	40	41	-

L: Leaves; S: stem; B: bark; RPO: relative percentage occurrence; CF: colony frequency.

and *Phomopsis* occurred predominantly. Single species of the genera *Botryodiplodia*, *Camarosporium* and *Fusamen* were also recorded (Table 1b). In this study, the maximum RPO was found in *Pestalotiopsis* spp. (53.66%) followed by *Colletotrichum* spp. (18.93%).A total of 87 isolates were recorded. Of these, 20 different morpho-species that existed in the leaves, stem and bark belonging to 9 genera were isolated and identified in *T. distichum*. *Colletotrichum*, *Pestalotiopsis* and *Phyllosticta* were obtained. Single species of the genera *Pestalozziella, Phomopsis, Pestalotia, Seimatosporium, Septoria* and *Botryodiplodia* were also observed (Table 1c). The maximum RPO was found in *Pestalotiopsis* spp. (17.5%) followed by *Colletotrichum* spp. (5.71%).

A total of 100 isolates were recorded in *T. mucronatum*. Of these, 20 different morpho-species that existed in the leaves, stem and bark belonging to 10 genera such as *Colletotrichum, Pestalotiopsis, Phomopsis, Phyllosticta, Septoria*

Table 1d. CF and RPO of endophytic coelomycetous fungi isolated from different samples of *T. mucronatum* during various seasons between 2006 and 2007.

Endophyte	Number of segments colonized															
	South west monsoon (June-Sep)				North east monsoon (Oct-Nov)				Winter season (Dec-Feb)				Summer season (Mar-May)			
	L	S	B	RPO (%)	L	S	B	RPO (%)	L	S	B	RPO (%)	L	S	B	RPO (%)
Botryodiplodia sp. MUBL1088	-	0.66	0.33	1.43	-	1.33	1.0	3.41	-	1.0	0.66	2.87	-	0.33	0.33	1.98%
Colletotrichum sp. MUBL1049	0.33	1.0	0.66		-	2.66	2.33		0.33	1.0	1.66		-	1.0	0.66	
Colletotrichum sp. MUBL1050	-	2.33	1.0		-	1.33	2.0		-	1.66	2.33		-	1.66	1.33	
Colletotrichum sp. MUBL1051	-	0.66	1.0	10.11	0.33	0.33	0.33	13.63	-	-	-	12.10	-	0.66	0.33	16.92%
Fusamen sp. MUBL1099	-	1.33	3.0	5.67	-	1.66	1.33	4.37	-	0.66	0.33	1.71	-	-	0.33	0.99%
Monochaetia sp. MUBL1100	-	2.0	3.33	7.72	-	2.66	2.33	7.30	-	1.0	2.33	5.77	-	1.0	0.66	4.98%
Pestalotiopsis sp. MUBL1010	0.33	2.66	3.0		-	4.0	2.33		-	2.0	1.66		0.33	2.0	1.0	
Pestalotiopsis sp. MUBL1011	-	3.66	2.0		-	2.33	1.33		-	1.0	0.66		-	1.66	0.66	
Pestalotiopsis sp. MUBL1012	-	2.66	1.66	43.40	-	1.33	2.0	37.04	-	3.33	3.0	35.20	-	2.0	1.33	49.89%
Pestalotiopsis sp. MUBL1013	-	1.0	2.33		-	2.66	1.0			2.33	2.0		-	1.0	1.33	
Pestalotiopsis sp. MUBL1014	-	3.66	2.33		-	1.33	2.0		-	1.66	2.66		-	1.0	1.33	
Pestalotiopsis sp. MUBL1015	-	2.66	2.0		-	2.66	2.33		0.33	1.33	1.0		-	1.66	1.33	
Pestalotia sp. MUBL1065	0.33	2.0	1.66	5.78	-	2.0	3.66	8.28	-	2.0	3.0	8.67	-	1.0	0.66	4.98%
Phomopsis sp. MUBL1074	-	1.0	1.66		-	3.0	1.0		-	1.0	2.3		-	0.33	0.33	
Phomopsis sp. MUBL1075	-	1.33	0.66	6.73	-	2.66	0.33	9.25	-	2.66	3.66	10.96	-	1.0	0.33	5.97%
Phyllosticta sp. MUBL1080	-	1.0	1.33		-	2.0	1.33		-	1.33	0.33		-	0.33	0.33	
Phyllosticta sp. MUBL1081	-	3.0	2.0	10.62	0.33	1.33	0.33	7.78	-	0.66	1.33	33	-	0.33	0.33	3.96%
Septoria sp. MUBL1091	-	1.0	2.0		-	1.66	1.0		-	1.0	0.33		-	1.0	0.66	
Septoria sp. MUBL1092	-	0.66	1.33	7.23	-	1.0	1.66	7.78	0.33	0.33	0.66	4.59	-	0.33	0.33	6.96%
Truncatella sp. MUBL1108	-	0.33	0.66	1.43	-	0.33	0.66	1.44	-	0.33	0.33	1.14	-	0.66	0.33	2.97%
Number of coelomycetes species recorded	3	20	20	-	2	20	20	-	3	19	19	-	1	19	20	-
Total number of coelomycetes isolates observed	3	104	102	-	2	112	91	-	3	79	91	-	1	57	42	-

L: Leaves; S: stem; B: bark; RPO: relative percentage occurrence; CF: colony frequency.

and *Truncatella* were isolated. The single species of the genera *Botryodiplodia, Fusamen, Monochaetia* and *Pestalotia* species were also observed (Table 1d). The maximum RPO was found in *Pestalotiopsis* spp. (49.89%) followed by *Colletotrichum* spp. (16.92%) and *Septoria* spp. (6.96%).

In the results, 74 endophytic coelomycetous fungi were isolated and successfully cultured in PDA medium. The identified coelomycetes were deposited at Madras University Botany Laboratory (MUBL), University of Madras, Chennai 600 025, Tamil Nadu, India. They were also maintained on PDA slants under refrigeration.

DISCUSSION

The fungal endophytes were a diverse group of organisms forming associations almost ubiquitously throughout the plant kingdom (Vega et al., 2008).

The range of associated plants and fungi may be altered as climate changes, and not necessarily in a synchronous fashion. Plants may lose or gain endophytes, with as of yet unknown impacts on defense and fitness, although generalist species may provide indirect defense in new habitats more often than not (Van and Putten, 2003).

Changes in distribution may bring plants into competition with previously established local species, making the fungal community and particularly the pathogenic role of fungus important in determining outcomes of competition with non-native invasive species (Dighton, 2003; Molinari and Knight, 2010).

The colonization frequency was used to compare diversity (Orachaipunlap, 2009). In this present study, the colonization frequency in four different seasonal variations of endophytes was recorded. However, other factors that may contribute to changes in the endophytic community include leaf age, weathering of leaf cuticle, the presence of wounds, increased exposure to propagules with time and changes in leaf physiology, leaf chemistry and also soil characteristics (Lumyong et al., 2009; Marquez et al., 2010).

In this study, the maximum isolates were observed with highest total CF% in SWM, NEM and WS, whereas, the meager amount of total CF% was observed in SS. Similarly, the higher colonization frequency and greater diversity of endophytes were observed during the monsoon and winter seasons than in summer by Thalavaipandian et al. (2011). The greater number of fungal isolates suggested that colonization by endophytes is correlated with climatic factors, such as rainfall and atmospheric humidity, which might influence the occurrence of some endophytic species (Wilson and Carroll, 1994).

In the present study, since the leaf lamina of *M. indica* and *R. arboreum* was broader as compared to the needles of gymnosperms, it has higher moisture content; hence, it anchors large number of endophytic colonization (Bernstein and Carroll, 1977b). An evaluation of the endophytic coelomycetes in *T. distichum* and *T. mucronatum* showed that since leaves are needle in shape, the lesser number of coelomycetes were found. Moreover, the diversity and ecological role of endophytes in tropical angiosperms are almost entirely unexplored (Hawksworth, 1993; Rodrigues and Petrini, 1997). The colonization frequency of endophytes in this study was within the range of many host plants studied in the tropics (Fröhlich et al., 2000; Photita et al., 2001; Suryanarayanan et al., 2003).

The frequently isolated fungi such as *Pestalotiopsis* spp., among the 74 different morphospecies belonging to 16 genera where observed in all the seasons in the four different plants. One of the most commonly found endophyte of Yew tree was *Pestalotiopsis* spp. Generally, this genus was among the most commonly isolated endophytic fungi of tropical plants (Strobel and Daisy, 2003; Cannon and Simmons, 2002). Moreover, Mahesh et al. (2005) isolated and identified 18.2% of coelomycete endophytes which includes *Pestalotiopsis* spp. and *Phoma eupyrena* from the living inner bark tissues of Neem tree. Interestingly, the present studies on the endophytic coelomycetes diversity in bark, leaf and stem of *M. indica*, revealed that *Pestalotiopsis* spp. were most dominant. Similarly, these results also supported that *Pestalotiopsis* spp. occurred most dominantly in tropical plants collected from various parts of South India and followed by the *Colletotrichum* spp. (Cannon and Simmons, 2002; Shankar and Shashikala, 2011).

Goveas et al. (2011) observed the percentage frequency of colonization of the endophytic fungi in leaves as compared to stem. Similarly, in this study, the most colony frequency was observed in leaves as compare to stem and bark in *M. indica* and *R. arboreum* than the other two plants at different climatic conditions.

Therefore, the present investigation was carried in the coelomycetes diversity for discovering and identifying novel strains. In the future, investigation should continue with these isolates obtained to recover the potential novel compound for human beings.

ACKNOWLEDGEMENT

The authors are thankful to Prof. R. Rengasamy, Director, CAS in Botany, University of Madras, Marimalai Campus, Guindy, Chennai-600 025 for providing necessary facilities. The first author (Subban Kamalraj) thank the University Grants Commission (UGC), under the scheme of Rajiv Gandhi National Fellowship (RGNF) (Award letter No. F. 14-2(SC)/2007(SA-III), Government of India, New Delhi, for financial support.

REFERENCES

Aly AH, Debbab A, Kjer J, Proksch P (2010). Fungal endophytes from higher plants: a profilic source of phytochemicals and other bioactive natural products. Fungal Diversity. 41:1-16.

Arnold AE, Herre EA (2003). Canopy cover and leaf age affect colonization by tropical fungal endophytes: ecological pattern and process in *Theobroma cacao* (Malvaceae). Mycologia. *95*:388-398.

Bernstein ME, Carroll GC (1977b). Microbial populations on Dougles Fir needle surfaces. Microb. Ecol. 4:41-52.

Bills GF (1996). Isolation and analysis of endophytic fungal communities from woody plants. In: *Endophytic Fungi in Grasses and Woody Plants*, Redlin SC, Carris LM (Eds.) APS Press: St Paul, Minnesota USA, pp. 31-65.

Bills GF, Polishook JD (1992). Recovery of endophytic fungi from *Chamaecyparis thyoides*. Sydowia. 44:1-12.

Cannon PF, Simmons CM (2002). Diversity and host preference of leaf endophytic fungi in the Iwokrama Forest Reserve. Guyana Mycologia. 94(2):210-220.

Carroll GC (1995). Forest endophytes: pattern and process. Can. J. Bot. 73:S1316-S1324.

Clay K (1988). Fungal endophytes of grasses: a defensive mutualism between plants and fungi. Ecology. 69: 10-16.

Dighton J (2003). Fungi in Ecosystem Processes. New York: Marcel Dekker.

Dobranic JK, Johnson JA, Alikhan QR (1995). Isolation of endophytic fungi from eastern larch (*Larix laricina*) leaves from New Brunswick. Canada, Can. J. Microb. 41:194-198.

Ellis MB (1971). *Dematiaceous Hyphomycetes*. CMI Kew Surrey England, p. 608.

Faeth SH, Hammon (1997). Fungal endophytes in oak trees: long-term patterns of abundance and associations with leaf miners., Ecology. 78:810-819.

Firakova S, Sturdikova M, Muckova M (2007). Bioactive secondary metabolites produced by microorganisms associated with plants. Biologia. 62:251-257.

Fisher PJ, Petrini O, Sutton BC (1993). A comparative study of fungal endophytes in leaves, xylem and bark of *Eucalyptus nitens* in Australia and England. Sydowia. 45:338-345.

Fröhlich J, Hyde KD, Petrini LE (2000). Endo-phytic fungi associated with palms. Mycol. Res. 104:1202-1212.
Gaylord ES, Preszler RW, Boecklen WJ (1996). Interactions between host plants, endophytic fungi, and a phytophagous insect in an oak (*Quercus grisea_Q-gambelii*) hybrid zone. Oecologia. 105:336-342.
Goveas SW, Madtha R, Nivas SK, D'Souza L (2011). Isolation of endophytic fungi from *Coscinium fenestratum* -a red listed endngered medicinal plant. Eurasia J. Biosci. 5:48-53.
Guba EF (1961). Monograph of *Pestalotia* and *Monochaetia*. Harvard University Press, Cambridge, Massachusetts, pp. 342.
Hata K, Futai K (1995) Endophytic fungi associated with healthy pine needles and needles infested by the pine needle gall midge, *Thecoliphosis japonensis*. Can. J. Bot. 73:384-390.
Hata K, Tsuda M, Futai K (1998). Seasonal and needle age dependent changes of the endophytic mycobiota in *Pinus thunbergii* and *Pinus densiflora* needles. Can. J. Bot. 76:245-250.
Hawksworth DL (1993). The tropical fungal biota: census, pertinence, prophylaxis and prognosis. In: Aspects of Tropical Fungal Mycology, (eds S Isaac, JC Frankland, R Watling, AJS Whalley). Cambridge University Press, Cambridge, pp. 265-293.
Helander ML, Neuvonen S, Sieber T, Petrini O (1993). Simulated acid rain affects birch leaf endophyte populations. Microb. Ecol. 26:227-234.
Helander ML, Sieber TN, Petrini O, Neuvonen S (1994). Endophytic fungi in scots pine needles: spatial variation and consequence of simulated acid rain. Can. J. Bot. 72:1108-1113.
Hudson H (1986). *Fungal Biology.* Edward Arnold London.
Johnson JA, Whitney NJ (1989a). A study of fungal endophytes of needles of balsam fir (*Abies balsamea*) and red spruce (*Picea rubens*) in New Brunswic, Canada, using culture and electron microscope techniques. Can. J. Bot. 67:3513-3516.
Johnson JA, Whitney NJ (1989b). An investigation of needle endophyte colonization patterns with respect to height and compass direction in a single crown of balsam fir (*Abies balsamea*). Can. J. Bot. 67:723-725.
Kirk PM, Cannon PF, David JC, Stalpers JA (2001). Ainsworth and Bisby's Dictionary of the Fungi, 9th (Eds.). CABI Bioscience, UK, p. 655.
Kirk PM, Cannon PF, Minter DW, Stalpers JA (2008). *Dictionary of the Fungi. 10th ed.* Wallingford: CABI. ISBN 0-85199-826-7.
Lumyong S, Techa W, Lumyong P, McKkenzie EHC, Hyde KD (2009). Endophytic fungi from *Calamus kerrianus* and *Wallichia caryotoides* (Arecaceae) at Doi Suthep- Pui National Park, Thailand. Chiang Mai Journal of Science. 36:158-167.
Magan N, Smith MK, Kirkwood IA (1996). Effects of atmospheric pollutants on phyllosphere and endophytic fungi. In: *Fungi and Environmental Change*, Frankland JC, Magan M, Gadd GM, (Eds.) Cambridge University Press, New York, pp. 90-101.
Mahesh B, Tejesvi MV, Nalini MS, Prakash HS, Kini KR, Ven Subbiah, Shetty HS (2005). Endophytic mycoflora of inner bark of *Azadirachta indica* A. Juss. Curr. Sci. 88: 1-2.
Marquez SS, Bills GF, Acuna LD, Zabalgogeazcoa I (2010). Endophytic mycobiota of leaves and roots of the grass *Holcus lanatus*. Fungal Diversity. 41: 115–123.
Molinari N, Knight C (2010). Correlated evolution of defensive and nutritional traits in native and non-native plants. Bot. J. Linn. Soc. 163 (1):1-13.
Mueller GM, Schmit JP (2006). Fungal biodiversity: what do we know? What can we predict? Biodiversity and Conservation, 16:1-5.
Muthumary J, Sashirekha S (2007). Detection of taxol, an anticancer drug from selected coelomycetous fungi. Ind. J. Sci.Technol. 1:1-10.
Nag Raj TR (1993). Coelomycetous anamorphs with appendage bearing conidia, Mycologue Publications Waterloo Ontario Canada.
Onions A, Allosopp MS, Eggins D (1981). Smith's introduction to industrial mycology, 7th (Eds.) Arnold London p. 398.
Orachaipunlap K, Roengsumran S, Sihanonth P (2009). Diversity of Endophytic Fungi Isolated From Plant Leaves of Deciduous Dipterocarp Forest in Tak Province. Kasetsart J. (Nat. Sci.), 43:182-188
Petrini O (1986). Taxonomy of endophytic fungi of aerial plant tissues. In: *Microbiology of the Phyllosphere,* Fokkema NJ, Heuvel JVD (Eds.) Cambridge University Press Cambridge MA, pp. 175-187.
Petrini O, Carroll GC (1981). Endophytic fungi in foliage of some Cupressaceae in Oregon. Can. J. Bot. 59:629-636.
Photita W, Lumyong S, Lumyong P, Hyde KD (2001). Endophytic fungi of wild banana (*Musa acuminata*) at Doi Suthep Pui National Park. Thailand, Mycological Research. 105:1508-1513.
Poteri M, Helander ML, Saikkonen K, Elamo P (2001). Effect of *Betula pendula* clone and leaf age on *Melampsoridium betulinum* rust infection in a field trial. Forest Pathol. 31:177-185.
Promputtha I, Jeewon R, Lumyong S, McKenzie EHC, Hyde KD (2005). Ribosomal DNA fingerprinting in the identification of non-sporulating endophytes from *Magnolia liliifera* (Magnoliaceae). Fungal Diversity. 20:167-186.
Rodrigues KF, Petrini O (1997). Biodiversity of endophytic fungi in tropic regions. In: Biodiversity of Tropical Microfungi. (Ed. KD Hyde), Hong Kong University Press, Hong Kong, China, pp. 57-70.
Schulz B, Guske S, Dammann U, Boyle C (1998). Endophyte-host interaction. II. Defining symbiosis of the endophyte-host interaction. Symbiosis. 25:213- 227.
Shankar NB and Shashikala J (2010). Diversity and structure of fungal endophytes in some climbers and grass species of Malnad region, Western Ghats, Southern India. Mycosphere. 1(4):265-274.
Shankar NB, Krishnamurthy YL (2010). Endophytes: the real untapped high energy biofuel resource. Current Science. 98:883.
Stone JK (1987). Initiation and development of latent infections by Rhabdocline parkeri on Douglas fir. Can. J. Bot. 65:2614-2621.
Stone JK, Sherwood MA, Carroll GC (1996). Canopy microfungi: function and diversity, Northwest Sci. 70: 37-45.
Strobel GA, Daisy B (2003). Bioprospecting for microbial endophytes and their natural products. Microbiol. Mol. Biol. R. 67: 491-502.
Suryanarayanan TS, Venkatesan G, Murali TS (2003). Endophytic fungal communities in leaves of tropical forest trees: diversity and distribution patterns. Curr. Sci. 85: 489-492.
Suryanarayanan TS, Vijaykrishna D (2001). Fungal endophytes of aerial roots of *Ficus benghalensis.* Fungal Divers. 8:155-161.
Sutton BC (1973a). Coelomycetes, In: *The fungi,* Ainsworth GC, Sparrow FK, Sussman AS (Eds.), IV pp. 513-582.
Sutton BC (1973b). Hyphomycetes from Manitoba and Saskatchewan Canada, Mycol. Pap. 132:1-143.
Sutton BC (1980). *The Coelomycetes. Fungi Imperfecti with pycnidia, acervuli and stromata,* Commonwealth Mycological Institute Kew Surrey England, p. 696.
Tadych M, White JF (2009). Endophytic microbes, Rutgers University New Brunswick, NJ, USA.
Tan RX, Meng JC, Hostettmann K (2000). Phytochemical investigation of some traditional Chinese medicines and endophyte cultures. Pharmaceutical Biology. 38:22-32.
Tan RX, Zou WX (2001). Endophytes: a rich source of functional metabolites. Natural Product Reproduction. 18: 448–459.
Thalavaipandian A, Ramesh V, Bagyalakshmi, Muthuramkumar S, Rajendran A (2011). Diversity of fungal endophytes in medicinal plants of Courtallam hills, Western Ghats, India. Mycosphere. 2(5):575-582.
Van Wim H, der Putten WH (2003). Plant defense belowground and spatiotemporal processes in natural vegetation Ecology. 84 (9):2269-2280.
Vega FE, Posada F, Aime MC, Pava-Ripoll M, Infante F, Rehner SA (2008). Entomopathogenic fungal endophytes. Biological Control. 46:72-82.
Wang J, Li G, Lu H, Zheng Z, Huang Y, Su W (2000). Taxol from *Tubercularia* sp. strain TF5, an endophytic fungus of *Taxus mairei.* FEMS Microb. Lett. 193:249-253.
Webber J, Gibbs JN (1984). Colonization of elm bark by *Phomopsis oblonga*. Trans. Br. Mycol. Soc. 82:348-352.
Wilson D, Carroll Gc (1994). Infection studies on *Discula quercina,* an endophyte of *Quercus garryana.* Mycologia. 86: 635-647.
Yuan JI, B.I. Jian-Nan, Yan Bing, Zhu Xu-Dong (2006). Taxol producing fungi: a new approach to industrial production of taxol. Chinese J. Biotechnol. 22:1-6.
Zhou DP, Sun JQ, Yu HY, Ping WX, Zheng XQ (2001). *Nodulisporium*, a genus new to China. Mycosystema. 2:277-278.

Biochemical studies on *Plantago major* L. and *Cyamopsis tetragonoloba* L.

Mohamed, I. Kobeasy[1], Osama, M. Abdel-Fatah[1], Samiha M. Abd El-Salam[2] and Zahrat El-Ola M. Mohamed[2]*

[1]Department of Biochemistry, Faculty of Agriculture, Cairo University, Giza, Egypt.
[2]Department of Crops Technology Research, Food Technology Research Institute, Agriculture Research Center, Giza, Egypt.

Plantago major **(seeds and leaves) and** ***Cyamopsis tetragonoloba*** **(guar beans) were analyzed for general chemical components including, fatty acids and amino acids. Guar beans had high contents of proteins, fats and total hydrolysable carbohydrates.** ***Plantago*** **leaves high percentage of linolenic acid was characterized by 56.19%. While,** ***P. major*** **seeds and Guar beans had high percentages of linoleic acid (25.41 and 48.99%, respectively). Essential and non essential amino acids were present in all samples and Guar beans had high amounts of glutamic, arginine, aspartic and leucine. Total phenols, total flavonoids and tannin content were the highest in** ***Plantago*** **leaves. Antioxidant activity of ethanolic, hot and cold water extracts of** ***Plantago*** **leaves and seeds and guar beans were evaluated.** ***Plantago*** **leaves extracts exhibited higher antioxidant activity than** ***plantago*** **seeds and guar beans extracts. The ethanolic, hot and cold extracts of plant induced anticancer activity with various degrees. Ethanolic extract of** ***P. major*** **leaves possessed the greatest effect on tumor cell growth (Dead 74%) followed by hot water extract of** ***P. major*** **leaves (Dead 54.6%).**

Key words: *Plantago major*, Guar beans, flavonoids, phenolic compounds antioxidant activity, antileukemia.

INTRODUCTION

Cancer is the leading cause of death in the developing and developed countries. Cancer claims over six million lives globally each year (Pezzuto, 1997).

There is a growing trend in the use of medicinal plant because of their medical effectiveness, low toxicity and the many natural anticancer agents derived from these plants (Ozaslan et al., 2007). *Plantago major* L. is a perennial plant from Plantaginaceae family.

It is introduced to the Nordic countries parallel to the introduction of the first primitive cultivated fields in the stone age nearly 4000 years ago (Jonsson, 1983). *P. major* spread by man from Europe throughout the world. The Indians named it 'White man's footprint' because it is found everywhere the Europeans had been. This adapted into the genus name *Plantago* that is from Latin *planta*, meaning sole of the foot.

It is an old medicinal plant that has been known for centuries, but it is regarded as weed by many people (Samuelsen, 2000). It is renowned as a traditional herbal medicine throughout the world (McCutcheon et al., 1995).

P. major has also been used as an anesthetic, antiviral, anti-inflammatory, astringent, anti-helmintic, analgesic, analeptic, antihistaminic, anti rheumatic, antitumor, anti-ulcer, diuretic, expectorant and hypotensive in traditional medicine (Grigorescu et al., 1973; Matev et al., 1982; Franca et al., 1996). Moreover, water soluble compounds isolated from *Plantago* spp. (especially *P. major*) have been reported to induce an immunostimulating activity on human lymphocyte proliferation (Chiang et al., 2003).

Pathological studies showed that, *P. major* L. extract induced inhibitive effect on ehrlich ascites tumor (EAT) and prevented tumor extension (Ozaslan et al., 2007). In addition, its crude extract inhibits tumor activity on the ehrlich ascites tumors (EAC) (Ozaslan et al., 2009).

Guar or cluster bean (*Cyamopsis tetragonoloba* L.) belongs to the family Leguminaceae and is grown in tropical Africa and Asia. *C. tetragonoloba* bean is commercially grown for its seeds as a source of natural

*Corresponding author. E-mail: zahoramahmoud@yahoo.com.

polysaccharide (galactomannan), commercially known as guar gum. It has a number of uses in food and other industries, such as paper, textiles, oil well drilling and pharmaceuticals (Whistler and Hymowitz, 1979).

C. tetragonoloba L. is a well-known traditional plant used in folklore medicine. It acts as an appetizer, cooling agent, digestive aid, laxative, and is useful in dyspepsia and anorexia Anti-ulcer, anti-secretory, cytoprotective, hypoglycemic, hypolipidemic and anti-hyperglycemic effects (Mukhtar et al., 2006). In addition, Guar beans are potentially high sources of additional phytochemicals (Wang and Morris, 2007). The aim of this study was to determine the main components beside biologically active compounds of *P. major* L. and *C. tetragonoloba* L. and to evaluate their antioxidant and anticancer effects.

MATERIALS AND METHODS

Chemicals and raw materials

1, 1-Diphenyl-2-picryl hydrazyl (DPPH) and quercetin were purchased from Sigma Chemical Co. (St., Louis, USA). Gallic acid, tert-butyl-4-hydroxy toluene (BHT), Folin-Ciocalteu reagent, and methanol were purchased from Merck Co. (Germany). Aerial parts of the *P. major* L. (leaves and seeds) were collected from Medicinal and Aromatic Plants Farm in Kanater, Agriculture Research Center, Egypt. Guar (*C. tetragonoloba* L.) beans were obtained from Crop Research Institute, Agriculture Research Center, Giza, Egypt.

Chemical analysis

Proximate chemical composition of crude proteins, fats, ash and crude fiber contents were determined as described in Association of Official Analytical Chemists (AOAC) methods (2005). Total hydrolysable carbohydrates were determined as glucose using phenol sulphuric acid reagent (Dubois et al., 1956).

Amino acid profiles were determined using amino acid analyzer (Beckman, 7300/G300) according to the method mentioned by AOAC (2006).

Identification of fatty acids

Fatty acid profiles were determined according to the method of (Aura et al., 1995) utilizing a gas chromatograph (HP 6890).

Preparation of plant extracts

Plant materials were air dried at room temperature and ground in a mortar. Fifty grams of each plant powder was extracted in 500 ml of ethanol by maceration (48 h). The solvent was removed under vacuum at temperature below 50°C, and then the extracts were freeze-dried.

Hot water extract of samples was prepared from plants according to the standard methods of Chang and Yeung (1988). Dried whole plants (100 g) were boiled in 1000 ml of distilled water for 1 h and filtered by gauze. The filtrate was collected and the residue was further extracted with 1000 ml of distilled water. The same procedure was repeated three times. The aqueous extract obtained from three successive extraction was mixed, concentrated *in vacuo*, and then lyophilized. The dried extract was collected and stored at 4°C until use.

Cold water samples (100 gm) were suspended in (1 L) of distilled water, stirred for 12 h at about 25°C, left to stand at the same temperature for another 12 h. The solution in each case was passed through folded muslin. The process was repeated twice. The aqueous extract obtained was mixed, concentrated *in vacuo*, and then lyophilized. The dried extract was collected and stored at 4°C until use.

Total flavonoid determination

Total flavonoids were determined using aluminum chloride colorimetric method as reported by Chang et al. (2002).

Total phenol determination

Total phenol content was estimated by Folin-Ciocalteu method of McDonald et al. (2001) and gallic acid was used a standard.

Determination of tannin content

Quantitative estimation of tannin for each sample was carried out as catechin equivalents using the vanillin–HCl/methanol (Price et al., 1978).

Identification of phenols and flavonoids by HPLC

Phenols and flavonoids were fractionated using HPLC (HP 1100), using a hypersil C_{18} reversed–phase column (250 × 4.6 mm) with 5 µm particle size. UV Detector was set at 254 nm. Flow rate was 0.3 ml/min. Two mobile phase: A (0.5 ml acetic acid/99.5 ml distilled water), mobile phase B (0.5 ml acetic acid/99.5 ml acetonitrile). The separation was performed at room temperature 25°C (Merfort et al., 1997).

Free radical scavenging activity determination

The stable 1, 1-diphenyl-2-picryl hydrazyl radical (DPPH) was used for determination of free radical-scavenging activity of the extracts (Koleva et al., 2002). Different concentrations of each herbal extract were added, at an equal volume, to ethanolic solution of DPPH (0.025 g/l). After 30 min at room temperature, the absorbance was measured at 517 nm absorption of blank sample containing the same amount of ethanol and DPPH. The experiment was repeated three times. BHT and quercetin were used as standard controls.

Inhibition (antioxidant activity) was calculated by the following formula:

$$\% \text{ Inhibition} = [(AB - AA)/ AB] \times 100$$

where
AB—absorption of blank sample (t=0 min)
AA—absorption of tested extract solution (t=30 min).

Viability of tumor cells

This study was performed on cells harvested from adult leukemia patients or healthy relatives admitted to the National Cancer Institute (NCI), Cairo University.

The cytotoxicity of each extract on Acute Myeloblastic Leukemia (AML) cells was determined by the MTT assay (Selvakumaran et al., 2003). The MTT assay is a test of metabolic competence based upon assessment of mitochondrial performance relying on the

Table 1. Chemical composition of *P. major* L. (seeds and leaves) and *C. tetragonoloba* L. beans (% on dry weight basis).

	P. major leaves	*P. major* seeds	*C. tetragonoloba*	L.S.D. at 5%
Crude Protein	$14.50^c \pm 0.06$	$15.01^b \pm 0.07$	$25.58^a \pm 0.11$	0.1639
Fat	$0.\ 9^c \pm 0.01$	$1.45^b \pm 0.02$	$2.94^a \pm 0.02$	0.0370
Ash	$16.8^a \pm 0.1$	$7.14^b \pm 0.06$	$3.6^c \pm 0.05$	0.1442
Crude fiber	$13.7^b \pm 0.13$	$26.9^a \pm 0.3$	$10.3^c \pm 0.10$	0.3929
Total hydrolysable carbohydrate	$48.89^b \pm 0.44$	$45.87^c \pm 0.36$	$56.90^a \pm 0.6$	0.9627

Values are means ± SD of three measurements. Means in the same raw with different letters are significantly different ($p < 0.05$).

conversion of yellow MTT to the purple formazan derivative by mitochondrial succinate dehydrogenase in viable cells. Mononuclear cells were separated from other blood cells by Ficoll hypaque density gradient (Pharmacia, Uppsala, Sweden). AML cells were then washed with three changes of PBS. The cell counts were adjusted to 3×10^3 cell /well and plated in 100 µl of medium/well in 96-well plates (Costar Corning Rochester, NY). After overnight incubation, extracts were in various concentrations (100, 200 µg/ml) with cytotoxicity to human normal lymphoid cell line (reported in elsewhere); 3 wells were included in each concentration. After treatment with extracts for one day, 20 µl of 5 mg/ml MTT (pH 4.7) were added per well and cultivated for another 4 h; the supernatant fluid was removed, then 100 µl DMSO were added per well and shaken for 15 min. The absorbance at 570 nm was measured with a microplate reader (Bio-Rad, Richmond, CA), using wells without cells as blanks. All experiments were performed in triplicate.

Cytoviability % = A_{570} of treated cells / A_{570} of control cells × 100.

Statistical analysis

Statistical analysis was carried out according to (Fisher, 1970) LSD (Least squares difference) test was used to compare the significant differences between means of treatment (Waller and Duncan, 1969).

RESULTS AND DISCUSSION

Chemical composition

Chemical composition of *P. major* L. and *C. tetragonoloba* L. is shown in Table 1. The results showed that, the *C. tetragonoloba* L. beans contained higher contents of proteins, fats and total hydrolysable carbohydrates compared to *P. major* L.leaves and seeds; meanwhile, it had lower values of ash and crude fibers.

In this respect Kays et al. (2006) found that the protein, fat, carbohydrate and ash contents of C. *tetragonoloba* L. beans were ranged between 22.9 - 30.6%, 2.88 - 3.45%, 50.2 - 59.9% and 3.04 - 3.53%, respectively. Al-Hafedh and Siddiqui (1998) reported that the C. *tetragonoloba* L. beans had 32.81% crude proteins, 3.18% crude fats, 4.19% ash and 10.87% crude fibers. In addition Khatta et al. (1988) examined four cultivars of C. *tetragonoloba* L. and found 24.5 - 32.9% crude proteins, 2.4 - 3.3% crude fats, 3.2 - 4.0% ash and 9.0 - 10.2% crude fibers.

Plantago seeds had higher contents of crude fibers and proteins than that of *plantago* leaves. These data are in the line with those obtained by Romero-Baranzini et al. (2006) who mentioned that, *plantago ovata* seeds contained crude proteins (17.4%), crude fats (6.7%), ash (2.7%) and carbohydrates (48.6%). It is well known that protein content differs among cultivars due to differences in genotype and environmental conditions during developing and maturation of the grains. On the other hand, leaves and seeds of *P. ovata* contained crude proteins of 21.87% and 13.12%, respectively (Romero-Baranzini et al., 2006).

Fatty acid composition

The saturated and unsaturated fatty acids of the *P. major* L. seeds and leaves and *C. tetragonoloba* L. beans are shown in Table 2.

From the tabulated data, it could be stated that *P. major* L. and *C. tetragonoloba* L. are rich in unsaturated fatty acids. *C. tetragonoloba* L.and *P. major* seeds contained high amounts of linoleic acid $C_{18:2}$ (48.99 and 25.41%, respectively), while *plantago* leaves had the highest amounts of linolenic acid $C_{18:3}$ (56.19%). These results are in agreement with those obtained by Liu et al. (2002) who reported the major fatty acids of *P. major* L. leaves were linolenic, linoleic and palmitic acid, but smaller amounts of stearic and oleic acid.

In the meantime, $C_{20:0}$ recorded 3.49 and 0.41% for *P. major* L. seeds and leaves, respectively. These results are in agreement with Samuelsen et al. (2000) who reported that arachidic acid was isolated from *P. major* L. only, and not from any other investigated *Plantago* species.

Amino acid composition

In order to obtain more information about the chemical composition of the *P. major* L. and *C. tetragonoloba*, the

Table 2. Fatty acid composition % of *P. major* L. (seeds and leaves) and *C. tetragonoloba* L. beans (% of total lipid).

Fatty acid	*P. major* L. leaves	*P. major* L. seeds	C. *tetragonoloba* L. beans
$C_{12:0}$	-	10.18	-
$C_{14:0}$	-	9.27	-
$C_{16:0}$	17.99	15.48	14.72
$C_{16:1}$	0.73	0.37	0.07
$C_{18:0}$	1.62	3.84	7.81
$C_{18:1}$	4.08	11.89	21.85
$C_{18:2}$	18.73	25.41	48.99
$C_{18:3}$	56.19	20.07	5.11
$C_{20:0}$	0.41	3.49	1.16
$C_{20:1}$	0.26	-	0.28
Unsaturated	79.98	57.74	76.31
Saturated	20.02	42.26	23.69
Unsaturated F.A/ Saturated F.A	3.22	1.37	3.99

Table 3. Amino acid composition of *P. major* L. (seeds and leaves) and *C. tetragonoloba* L. beans (g/100g sample) on dry weight basis.

Amino acids	*P. major* L. leaves	*P. major* L. seeds	C. *tetragonoloba* L. beans
Lysine	0.79±0.03	0.51±0.008	1.05±0.04
Threonine	0.64±0.01	0.48±0.009	0.75±0.01
Valine	0.92±0.03	0.70±0.01	0.81±0.03
Isoleucine	0.92±0.03	0.45±0.008	0.79±0.03
Leucine	1.30±0.05	0.77±0.03	1.31±0.05
Phenylalanine	0.80±0.03	0.55±0.01	0.89±0.03
Tyrosine	0.64±0.01	0.41±0.008	0.87±0.03
Total essential amino acid	6.01	3.87	6.47
Serine	0.62±0.02	0.59±0.01	1.07±0.04
Proline	0.64±0.03	0.62±0.03	0.71±0.04
Glycine	0.70±0.02	0.70±0.01	1.27±0.05
Alanine	0.75±0.04	0.59±0.02	0.93±0.05
Aspartic	1.85±0.03	1.05±0.04	2.22±0.04
Glutamic	1.97±0.03	2.16±0.05	4.41±0.08
Histidine	0.32±0.006	0.26±0.006	0.53±0.01
Arginine	0.70±0.02	0.72±.03	2.41±0.09
Total non-essential amino acids	7.55	6.69	13.55

Values are means ± estimated uncertainty (U).

amino acids were determined and presented in Table 3.

The results showed that, the C. *tetragonoloba* L. beans contained lightly higher total essential amino acids (6.47%) compared with *plantago* seeds (3.87%).

These results are in agreement with those obtained by Khalil (2001) who mentioned that, histidine of *P. major* L. leaves and seeds are (0.32 and 0.256%, respectively). However, it contained high amounts of other amino acids especially glutamic acid, aspartic acid, leucine, and valine. In the same line, Romero-Baranzini et al. (2006) found that *P. ovate* contained high amount of glutamic and aspartic amino acids.

Biological active compounds

Total phenol, total flavonoid and tannin contents of *P. major* L. leaves and seeds, and *C. tetragonoloba* L. beans are shown in Table 4.

The results in Table 4 indicate that *P. major* L. leaves contained the highest total phenol content (13.05. mg/g) followed by *P. major* L. seeds (7.43 mg/g) and

Table 4. Biological active compounds of *P. major* L. (seeds and leaves) and *C. tetragonoloba* L. beans (mg/g) on dry weight basis.

Constituent	*P. major* L. leaves	*P. major* L. seeds	*C. tetragonoloba* beans	L.S.D. at 5%
Total phenol (mg gallic/g)	13.05^{a}±0.10	7.43^{b}±0.07	3.76^{c}±0.04	0.1496
Total flavonoid (mg Quercetin/g)	6.41^{a} ±0.04	3.03^{b} ±.03	0.83^{c} ±0.01	0.05652
Tannins (mg Catechine /g)	5.63^{a} ±0.06	2.43^{b} ±0.03	0.76^{c} ±0.01	0.07838

Values are means ± SD of three measurements. Means in the same raw with different letters are significantly different ($p < 0.05$).

Table 5. Phenolic and falvonoide compounds of *P. major* L. (seeds and leaves) and *C. tetragonoloba* L. beans (mg/100 g).

Compounds	*P. major leaves*	*P. major seeds*	*C. tetragonoloba beans*
Gallic	4.8096	0.686	1.090
pyrogallic	-	-	-
Catechol	0.241	2.356	4.103
Catechine	15.924	3.684	1.003
Caffeic acid	8.164	9.832	0.222
Vanillic acid	2.478	1.365	-
Syrengic	4.158	3.074	1.788
Caffeine	-	-	3.158
Ellagic	132.110	-	1.055
Coumarin	2.347	0.4387	-
Ferulic acid	32.439	48.584	1.885
Colchicine	25.026	1.828	6.851
Luteolin	10.079	7.107	1.04
Quercetin	2.089	1.563	0.030
Isorhhomnetin	-	-	-
Rutin	4.323	4.071	0.338
Quercetrin	0.323	0.542	1.056
sakaruntin	-	-	-
Naringenin	0.086	0.892	0.045
Hypersoid	0.992	-	-
Chrisin	0.0224	0.013	0.043
Kaempferol	0.825	0.347	0.026

C. *tetragonoloba* L. beans (3.76 mg/g).

It is also clear that, flavonoids and tannin are present in lower levels in C. *tetragonoloba* L. beans (0.83 and 0.76 mg/g, respectively) compared to *P. major* L. leaves (6.41 and 5.63 mg/g, respectively) and *P. major* L. seeds (3.03 and 2.43 mg/g).

These results are in agreement with those obtained by Gálvez et al. (2005) who found that the flavonoids content ranged between 0.69 - 3.09% in *Plantago* species. Souri et al. (2008) reported that the phenols content of *P. major* L. seeds were 672 mg/100 g.

Also, Grubešić et al. (2005) demonstrated that the tannins content in *Plantago* species leaves was (0.56 - 2.26%). Kaushal and Bhatia (2006) outlined that guar beans contained flavonoids in the range of 0.13 to 0.23%.

Identification of phenols and flavonoid

The data of phenolic acids and falvonoid compounds of *P. major* L. seeds and leaves, and C. *tetragonoloba* L. beans are shown in Table 5.

The data indicate that the *P. major* L. (leaves and seeds) and C. *tetragonoloba* L. comprised biologically active phenolic compounds ferulic, caffeic and vanillic acids, and flavonoid compound, that is, luteolin. Ferulic acid was the *major* phenolic compound in the *P. major* L. seeds and leaves, while in *C. tetragonoloba* L. beans colchicine was the *major* phenolic compound. On the other hand, luteolin was the *major* flavonoid compound in the *P. major* L. seeds and leaves.

The aforementioned data are in line with data of

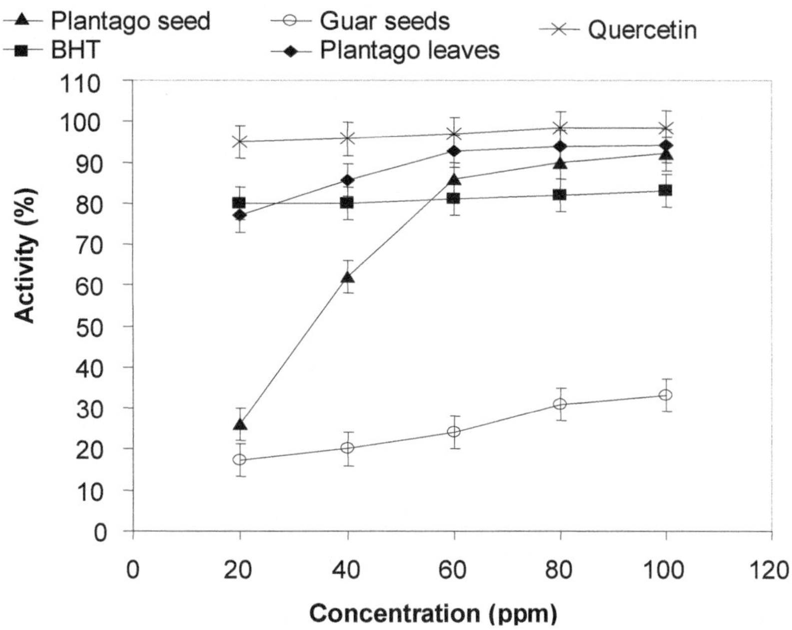

Figure 1. Antioxidant activity of ethanolic extract of *P. major* L. (seeds and leaves) and *C. tetragonoloba* L. (guar beans).

Kawashty et al. (1994) who studied the flavonoid profile of 18 species of *Plantago.* They indicated that, the flavonoid compound luteolin was the major component in most of *Plantago* species.

In this context Samuelsen et al. (2000) reported that, *Plantago* contains the following biologically active compounds (vanillic acid), flavonoid (luteolin) and phenolic compounds (caffeic and ferulic acids).

In addition, Kaushal and Bhatia (2006) reported that gallic acid as well as its derivatives, kaempferol, caffeic acid and ellagic acid were derived from *C. tetragonoloba* L. beans. Furthermore, Wang and Morris (2007) stated that C. *tetragonoloba* L. beans seem to be a good source of quercetin and kaempferol.

Antioxidant activity

The antioxidant activity of ethanolic, hot water and cold water extracts of *P. major* L. (seeds and leaves) and *C. tetragonoloba* L. (guar beans) are illustrated in Figures 1, 2 and 3, respectively.

It could be observed that the highest antioxidant activity was found in ethanolic extract of *P. major* L. leaves even at low concentration of 20 ppm, and increased with increasing the concentration up to 60 ppm which is also the most effective concentration of *P. major* L. seeds. It is worthy to mention that C. *tetragonoloba* L. beans exhibited low antioxidant activity at different concentrations of the ethanolic extract (Figure 1).

Furthermore, the data illustrated in Figures 2 and 3 clearly indicated that, either hot or cold water extract of *P. major* L. leaves exhibited the most active ones, when compared with *P. major* L. and C. *tetragonoloba* L. seeds. It is also observed that, ethanolic extracts were more active than the hot and cold water extracts of the samples under investigation. In general the activity increased with increasing the concentration of extract.

However, the radical scavenging activity of the plant extracts can be arranged in the following order: *P. major* L. leaves > *P. major* L. seeds > *C.tetragonoloba L.* beans.

Pourmorad et al. (2006) reported that antioxidant activity *P. major* L. leaves are 89.3% at 0.8 mg/ml concentration.

The results of the present study are in agreement with Gálvez et al. (2005), Souri et al. (2008) and Beara et al. (2009) who examined *Plantago* species for anti oxidant activity and reported that it could be regarded as a possible new sources of natural antioxidants.

Anti tumor

The results of the cytotoxic effect of ethanolic, hot and cold extracts from *P. major* L. leaves and seeds and *C. tetragonoloba* L. beans on AML cells are summarized in Table 6.

The results showed that all extracts possessed anticancer activity with differing degrees. A dose-dependent inhibition of cell proliferation was observed for most of the all extracts tested in this study.

It is clear that ethanolic extract of *P. major* L. leaves had the greatest effect on tumor cell growth (Dead 74% ±

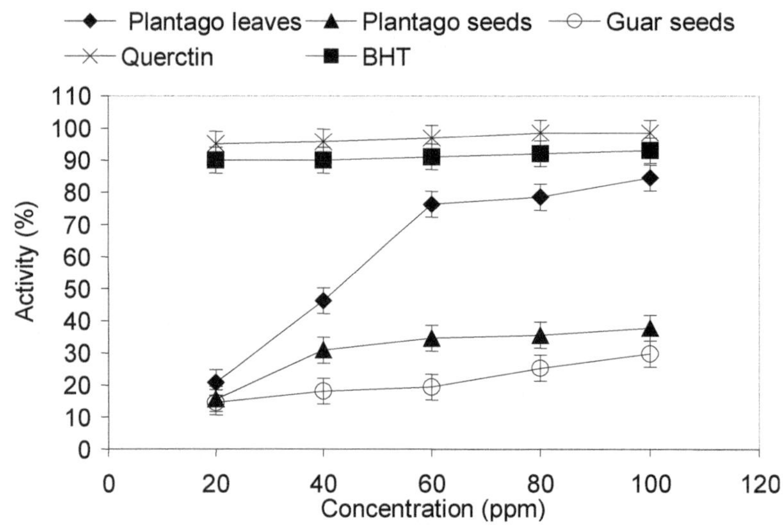

Figure 2. Antioxidant activity of hot water extract of *P. major* L. (seeds and leaves) and *C. tetragonoloba* L. (guar beans).

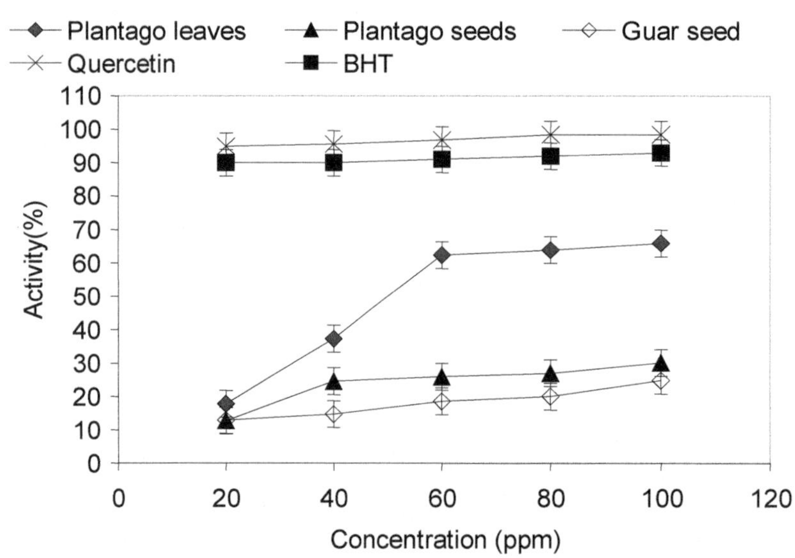

Figure 3. Antioxidant activity of cold water extract of *P. major* L. (seeds and leaves) and *C. tetragonoloba* L. (guar beans).

0.35) followed by hot water extract of *P .major* L. leaves (Dead 54.6% ± 1.21).

In this respect, Gomez et al. (2000) observed that *P. major* L. leaf extracts activate nitric oxide and TNF-α production of macrophages-mediated lymphocyte proliferation. Macrophages play an important role in modulating humoral and cellular immunity against infectious diseases and cancer.

Gálvez et al. (2003) demonstrated that *Plantago* spp. extracts have shown growth inhibitory and cytotoxic effects on melanoma cell lines and breast adenocarcinoma. It is thought that cytotoxic activity depends basically on flavonoids, flavone and luteolin.

Luteolin exhibited inhibitory effect on various human cancer cell lines, such as renal A-549, ovary SK-OV-3, melanoma SK-MEL-2, XF-498, HCT15, gastric HGC-27, breast MCF-7 and human leukemia cells (Le Bail et al., 1998; Post and Varma, 1992; Ryu et al., 1994).

Chiang et al. (2003) showed that hot water extracts of *P. major* L. and *Plantago asiatica* possessed effects of immunomodulatory activity on human mononuclear cells proliferation.

In addition, Okazaki et al. (2002) reported that fibers had a remarkable effect in preventing colorectal cancer

Table 6. The effect of extracts on National Cancer Institute (NCI) 20 (male – 65 years old) cell after incubation for 24 h.

Plant extract	Concentration (µg/ml)		
	50	75	100
	Dead %	Dead %	Dead %
C. tetragonoloba L Hot water	$14.2^{fg}\pm2.66$	$30.7^{c}\pm1.68$	$39.3^{d}\pm0.43$
C. tetragonoloba L cold water	$11.8^{g}\pm1.40$	$26.31^{d}\pm2.85$	$28.4^{h}\pm2.35$
C. tetragonoloba L ethanolic	$24.6^{bc}\pm2.58$	$31.0^{c}\pm3.56$	$33.4^{ef}\pm2.29$
P. major L. seeds Hot water	$27.5^{b}\pm1.30$	$31.0^{c}\pm2.66$	$31.6^{fg}\pm1.57$
P. major L. seeds cold water	$16.3^{ef}\pm1.17$	$22.8^{d}\pm1.40$	$29.9^{gh}\pm2.26$
P. major L. seeds ethanolic	$19.2^{de}\pm2.35$	$35.5^{b}\pm2.58$	$34.7^{e}\pm0.32$
P. major L. leaves hot water	$22.2^{cd}\pm1.92$	$37.4^{b}\pm1.30$	$48.7^{b}\pm1.23$
P. major L. leaves cold water	$26.5^{b}\pm2.57$	$39.2^{b}\pm1.17$	$43.5^{c}\pm1.20$
P. major L. leaves ethanolic	$47^{a}\pm1.76$	$62.1^{a}\pm2.35$	$74.6^{a}\pm0.35$
L.S.D. at 5%	3.4948	3.9573	2.6628

Values are means ± SD of three measurements. Means in the same row with different letters are significantly different ($p < 0.05$).

and other cancer risks.

Conclusion

This investigation shows that the *P. major* and *C. tetragonoloba* contained important biologically active compounds and *P. major* leaves had the highest total phenol, flavonoid and tannin content. In addition, ethanol, cold and hot extracts of the same plants showed antioxidant activity, but the highest antioxidant activity was found in ethanolic extract of *P. major* leaves .Also, ethanolic extract of *P. major* leaves had the greatest effect on tumor cell growth followed by hot water extract of *P. major* leaves.

REFERENCES

Al-Hafedh YS, Siddiqui AQ (1998). Evaluation of *C. tetragonoloba* seed as a protein source in Nile yilapia, *Oreochromia niloticue* (L.) practical diets. Aquacult. Res., 29: 701-708.

AOAC (2005). Official Methods of Analysis of AOAC international 18th Ed. Association of Official Analytical Chemists, Washington, D.C., pp. 1-14.

AOAC (2006) Official Methods of Analysis of AOAC international 18th Ed. Association of Official Analytical Chemists, Washington, D.C., pp. 24-37.

Aura AM, Forssel P, Mustranta A, Poutanen K (1995). Transesterification of soy lecithin by lipase and phospholipase. J. Am. Oil Chem. Soci., 72: 1375-1379.

Beara N, Lesjak MM, Jovin EĐ, Balog KJ, Mimica-Duki NM (2009). Plantain (*Plantago* L.). Species as Novel Sources of Flavonoid Antioxidants. J. Agric. Food Chem., 57(19): 9268-9273.

Chang C, Yang M, Wen H, Chern J (2002). Estimation of total flavonoid content in propolis by two complementary colorimetric methods. J. Food Drug Anal., 10: 178-182.

Chang RS, Yeung HW (1988). Inhibition of growth of human immunodeficiency virus *in vitro* by crude extracts of Chinese medical herbs. Antiviral Res., 9: 163-176.

Chiang LC, Ng LT, Chiang W, Chang MY, Lin CC (2003). Immunomodulatory activities of flavonoids, monoterpenoids, triterpenoids, iridoid glycosides and phenolic compounds of *Plantago* species. Planta Med., 69(7): 600-604.

Dubois M, Gilles KA, Hamilton J, Rebers PA, Smith F (1956). Colorimetric method for determination of related substances. Anal. Chem., 28: 250-256.

Fisher RA (1970). Statistical Method for Research Workers, Eds. Oliver and Boyed, Edinburgh, pp. 140-142.

Franca F, Lago EL, Marsden PD(1996).Plants used in the treatment of leishmanial ulcers due to Leishmania (Viannia) braziliensis in an endemic area of Bahia, Brazil. Rev. Soc. Bras. Med. Trop., 29: 229-232.

Gálvez M, Martin-Cordero C, Houghton, PJ, Ayuso MJ (2005). Antioxidant activity of methanol extracts obtained from *Plantago* species. J Agric. Food Chem., 53: 1927-1933.

Gálvez M, Martin-Cordero C, Lopez-Lazaro M, Cortes F, Ayusa MJ (2003). Cytotoxic effect of *Plantago* spp. on cancer cell lines. J. Ethnopharm., 88: 125-130.

Gomez FR, Calderon CL, Scheibel LW, Tamez-Guerra P, Rodriguez-Padilla C, Tamez- Guerra R, Weber RJ (2000). Immunoenhancing properties of *plantago major* leaf extract. Phytother. Res., 14: 617-622.

Grigorescu E, Stanescu U, Basceanu V, Aur MM (1973) Phytochemical and microbiological control of some plant species used in folk medicine. II. *Plantago lanceolata* L., *Plantago media* L., *Plantago major* L. Rev. Med. Chir. Soc. Med. Nat. Iasi., 77: 835-841.

Grubešić RJ, Vukovic J, Kremer D, Vladimir-Knězevic, S (2005) Spectrophotometric method for polyphenols analysis: Prevalidation and application on *Plantago* L. species. J. Pharmaceut. Biomed. Anal., 39: 837-842.

Jonsson S (1983). Blomsterboken. Markens Herbs, heather and trees. Technological Publishers, Oslo.

Kaushal GP, Bhatia S (2006). A study of polyphenols in the seeds and leaves of guar (Cyamopsis *tetragonoloba* L. Taub). J. Sci. Food Agric., 33: 461-470.

Kawashty SA, Gamal ED, Abdalla MF, Saleh, NA (1994). Flavonoids of *Plantago* species in Egypt. Biochem. Syst. Ecol., 22: 729-733.

Khalil MM (2001). Biochemical and technological studies on the production of isolated guar protein. Nahrung, 45(1): 21-24.

Khatta VK, Kumar N, Gupta PC (1988). Chemical composition and amino acid profile of four varieties of guar (*Cyamopsis tetragonoloba*) seed. Indian J. Anim. Nutr., 5: 326-326.

Koleva I, Van Beek TA, Linssen JPH, de Groot A, Evstatieva LN (2002). Screening of plant extracts for antioxidant activity: a comparative study on three testing methods. Phytochem. Anal., 13: 8-17.

Le Bail JC, Varnat F, Nicolas JC, Habrioux G (1998). Estrogenic and antiproliferative activities on MCF-7 human breast cancer cells by flavonoids. Cancer Lett., 130: 209-216.

Liu L, Ho P, Zhou Y, Hocart C, Zhang R (2002). Fatty acid profiles of leaves of nine edible wild plants: An Australian study. J. Food Lipids,

9: 65-69.
Matev M, Angelova I, Koichev A, Leseva M, Stefanov G (1982). Clinical trial of a *Plantago major* preparation in the treatment of chronic bronchitis., Vitr. Boles, 21: 133-137.
McCutcheon AR, Roberts TE, Gibbons E (1995). Antiviral screening of British Colombian medicinal plants. Ethnopharm., 49: 101-110.
McDonald S, Prenzler PD, Autolovich M, Robards K (2001). Phenolic content and antioxidant activity of olive extracts. Food Chem., 73: 73-84.
Merfort I, Wray V, Barakat HH, Hussen SAM, Nawwar MAM, Willuhan G (1997). Flavonol triglycerides from seeds of *Nigella sativa*. Phytochem., 46(2): 359-363.
Mukhtar HM, Ansari SH, Bhat ZA, Naved T (2006). Antihyperglycemic Activity of Cyamopsis tetragonoloba Beans on Blood Glucose Levels in Alloxan-Induced Diabetic Rats Pharm. Biol., 44(1): 10-13.
Okazaki H, Nishimune T, Matsuzaki H (2002). Increased incidence rate of colorectal tumors due to the Intake of a soluble dietary fiber in rat chemical Carcinogenesis can be suppressed by substituting partially An insoluble dietary fiber for the soluble one Int. J. Cancer, 100: 388-394.
Ozaslan M, Karagöz K, Kalender ME, Kilic IH, Sari I, Karagöz A(2007). *In vivo* Antitumoral effect of *Plantago major* L.extract on Balb/C Mouse with Ehrlich Ascites Tumor. Am. J. Chin. Med., 35(5): 841-851.
Ozaslan M, Karagoz K, Kiliç IH, Cengiz B, Kalender ME, Güldür ME, Karagoz A, Zümrütdal ME (2009). Effect of *Plantago major* sap on Ehrlich ascites tumors in mice. Afr. J. Biotechnol., 8(6): 955-959.
Pezzuto JM (1997). Plant-derived anticancer agents. Biochem. Pharmacol., 53: 121-133.
Post JFM, Varma RS 1992). Growth inhibitory effects of bioflavonoids and related compounds on human leukemic CEM-C1 and CEM-C7 cells. Cancer Lett., 67: 207-213.
Pourmorad F, Hosseinimehr SJ, Shahabimajd N (2006). Antioxidant activity, phenol and flavonoid contents of some selected Iranian medicinal plants. Afr. J. Biotechnol., 5(11): 1142-1145.
Price ML, Socoyoc SV , Butter LG (1978). A critical evaluation of the vanillin reaction as an assay for tannin in sorghum grain. Agric. Food Chem., 26: 1214-1218.
Romero-Baranzini AL, Rodriguez OG, Yanez-Farias GA, Barron-Hoyos JM, Rayas-Duarte P (2006). Chemical, Physicochemical, and Nutritional Evaluation of *Plantago* (*Plantago ovata* Forsk). Cereal Chem., 83(4): 358-362.
Ryu SY, Choi SU, Lee CO, Lee SH, Ahn JW, Zee OP (1994). Antitumor activity of some phenolic components in plants. Arch. Pharmacol. Res., 17: 42-44.
Samuelsen AB (2000). The traditional uses, chemical constituents and biological activities of *Plantago major* L. A review. J. Ethnopharm., 71: 1-21.
Selvakumaran M, Pisarcik DA, Bao R, Yeung AT, Hamilton TC (2003). Enhanced cisplatin cytotoxicity by disturbing the nucleotide excision repair pathway in ovarian cancer cell lines. Cancer Res., 63: 1311-1316.
Souri E, Amin G, Farsam H, Barazandeh TM (2008). Screening of antioxidant activity and phenolic content of 24 medicinal plant extracts. Daru., 16: 83-87.
Waller WM, Duncan DB (1969) A Bayes rule for the symmetric multiple comparison problem, J. Am. Statistical Assoc., 64: 1484-1499.
Wang ML, Morris JB (2007). Flavonoid content in seeds of guar germplasm using HPLC. Plant Genetic Resources: Characterization and Utilization, 5: 96-99.
Whistler RL, Hymowitz T (1979). Guar: Agronomy, Production, Industrial Use, and Nutrition, Purdue University Press, West Lafayette, IN, pp. 1-118

The folk biology of South American-native shrub, *Mimosa pigra* L. [Leguminosae] and its invasive success in Rwanda

Seburanga, J. L.[1]*, Kaplin, B. A.[2], Bizuru, E.[1] and Mwavu, E. N.[3]

[1]Faculty of Science, National University of Rwanda, University Avenue, Box 117, Huye, Rwanda. [2]Department of Environmental Studies, Antioch University, New England, 40Avon St. Keene, NH 0343l, USA.
[3]Department of Forestry, Biodiversity and Tourism, Makerere University, P.O. Box 7062, Kampala, Republic of Uganda.

Although, the probability of an alien plant species to become invasive in natural or semi-natural habitats is relatively low, the introduction of new taxa should always be regarded as a sensitive issue because the effects of an insertion of an invasive species are likely to be prohibitive as well as external. Ideally, plant species with an invasive potential should not be allowed to enter a country and, if they do, the next best scenario is early detection and eradication. Unfortunately, in Rwanda, despite its location in the Albertine Rift, a biodiversity hotspot with many endemic and endangered species, little is known about the status of introduced taxa. In this article, we combine the insight from library resources, herbarium records and vegetation surveys to document the status of an introduced species, *Mimosa pigra* L. The results comprise evidence that this species has reached the invasive status in Rwanda and support the plea that it be officially listed as such. We suggest that Egypt through Uganda may have been the primary source of *M. pigra* L. propagules that first infested Rwanda through the Akagera-Nyabarongo river system, earlier than Mid-19th century. We presume that under the shadow of its non-specific Kinyarwanda name 'u-mu-gey-o', it was able to spread undisturbed and invade Rwanda's central and eastern floodplains before scientists could take notice of its true identity.

Key words: Mimosa, invasive, Nyabarongo, floodplain, Umugeyo, infestation, range.

INTRODUCTION

According to Yates et al. (2004), 'plant invasions are a current threat to biodiversity conservation, second only to habitat loss and fragmentation'. Indeed, introduction of invasive species in critical ecosystems such as wetlands, croplands, waterways and high-endemism forests, although a subtle action whose harmful effect might require several years to be realized, poses insidious and potentially long-range threat to environmental security and can cost infested countries billions of dollars in lost revenue, property damage and eradication expenses (Kapustka, 2007). In financial terms, the estimated annual damage from displacement of local taxa by invasive species in the United States alone amounts to more than US$120 billion (Pimentel et al., 2005); equivalent to nearly 65 times the 2011/2012 annual budget of the Republic of Rwanda. In particular, environmental weeds, a subset of plants that invade natural vegetation, adversely affecting native biodiversity and/or ecosystem functioning, are considered to be one of the greatest threats to nature conservation (Williams and West, 2000).

Although, the probability of an alien plant species to become invasive in natural or semi-natural habitats is relatively low [approximately 0.01 in New Zealand, where only 240 out of nearly 20 000 non-indigenous plant species are considered weeds (Fowler et al., 2000)], the introduction of new taxa should always be regarded as a sensitive issue because the effects of an insertion of an invasive species are likely to be not only prohibitive (cost so high that it prevents people from taking rehabilitating action), but also external (cost not carried by those who caused the damage). For instance, a decision as simple as that of bringing in a few propagules of *Eichhornia crassipes* (Mart.) Solms [Pontederiaceae] for ornamental purposes resulted in a disastrous situation for Rwanda's eastern river and lake systems, prompting policy makers to ban its cultivation [by Order No. 51/162 of 4/5/1955] since the 1950s. This step did not successfully hinder its spread due to its incredibly high propagule pressure. Instead, the species expanded its range to become a trans-boundary concern and was officially recognized to invade the world's second largest lake, East Africa's Lake Victoria, in 1988 (Moorhouse et al., 2001). Furthermore, by the time a species is declared invasive under existing legislation, it often has little impact because the spread of the species and its associated damage to ecosystems has already begun (www.dpi.qld.gov.au, 2012.03.03).

Ideally, plant species with an invasive potential should not be allowed to enter a country and, if they do, the next best scenario is early detection and eradication (Fowler et al., 2000). Unfortunately, in Rwanda, located in the Albertine Rift, a biodiversity hotspot with many endemic and endangered species (Eilu et al., 2004; Plumptre et al., 2007; Rutagarama & Martin, 2006), little is known about the invasive status of the majority of introduced species. Among these is the south American-native *Mimosa pigra* L. [Leguminosae], which has almost unnoticeably invaded Rwanda's floodplains for decades.

M. pigra has been on the African continent for not less than 200 years, with the known earliest record date being that of 1826 in Egypt (GISP, 2007). A recent study suggests that it immigrated from South America, its ancestral area, during late Miocene through trans-Atlantic long-distance dispersal (Simon et al., 2011). The presence of *M. pigra* as a naturalized or invasive taxon in North-eastern Africa has been attested by many authors (Sheded and Hassan, 1999; Vilà et al., 1999; EIAR, 2010). In neighbouring Uganda, *M. pigra* was found to be particularly abundant in the Lake Victoria eastwardly-adjacent district of Rakai (Byenkya et al., 2004; Chapman et al., 2001). Its presence was also noted in Doho wetland, north to Lake Victoria and 60 km from Kilimandjaro Mountain (Kalema and Ssegawa, 2007). In Burundi, *M. pigra* has been reported to threaten Lake Tanganyika's riparian communities (lta.iwlearn.org, 10.07.2012). Still in Burundi, but closer to south eastern Rwanda, *M. pigra* was referred to as one of the most potential harmful species whose spread should be contained before it invades the Ruvubu National Park via the Ruvubu River (Masharabu, 2011).

Not only it is prickle-armed to form impenetrable thickets which protect it against trembling or destruction by humans and foraging animals, but also has incredible powerful reproductive machinery with a single plant being able to produce up to 9000 seeds per year. In addition, the pods containing these seeds are light and hairy enough to ease propagation by wind and water, especially during flooding periods (Lonsdale, 1993). According to the same author, even in case of wind dispersal alone, it still would achieve a spread rate of up to 18.3 m per year. However, in a natural environment, this rate can be tremendously magnified as illustrated by records in Zambia's Kafue Flats (Douthwaite and van Lavieren, 1977). With an original population size of 100 ha in 1986, *M. pigra* infestation had expanded to cover 2500 ha by 2003, achieved an average spread rate of 132 m $year^{-1}$ (Indira, 2007). In this article, we discuss the unusual distribution pattern of *M. pigra* in Rwanda, identify its primary source of infestation and elucidate its identity.

METHODS

Description of the study area

The study was carried out in Rwanda, Central Africa, between 1°03-2°50 south of the Equator and 28°52-30°54 east of the Greenwich Meridian Line. Field data collection lasted for three months, starting from Early-May 2012 to Mid-August 2012, and covered riparian and floodplain plant communities of Nyabarongo River. The area covered by this study is part of both the Albertine Rift, a biodiversity hotspot which harbors many endemic and endangered species (Eilu et al., 2004; Plumptre et al., 2007; Rutagarama and Martin, 2006), and the Nile River Basin, one of the great rivers of the world. The Ministerial Order No 008/16.01 of 13/10/2010, establishing the list of wetlands in Rwanda, classifies Nyabarongo-associated wetlands in the category of those to be fully protected and the process of giving part of them the RAMSAR site statute has already begun. Downstream Nyabarongo River, Akagera wetlands have been under special protection since 1934 when the Akagera National Park was gazetted.

Concept definition

Richardson et al. (2000), supported by Ipou et al. (2011) and Pysek et al. (2004), define an invasive species as a 'naturalized plant that produces reproductive offspring, often in very large numbers, at considerable distances from parent plants [with an approximated spread rate of more than 2 m/year], and thus have the potential to spread over a considerable area'. In Flora Europaea, a taxon is considered to be 'effectively naturalized' once it has 'established in a single station for at least 25 years, or is reported as naturalized in a number of widely separated localities' (Tutin et al., 1964). To be considered 'alien' a taxon should have moved at least 100 km from the nearest locality where it is native (Richardson et al., 2000).

Colautti and MacIsaac (2004) attempted to provide a practical reference as to when a species should be labeled invasive, suggesting that invasive species [also referred to as 'metaphytes' in opposition to 'diaphytes' or taxa established in a non-permanent way (Sobrino et al., 2002)] are 'those taxa that are non-native to the

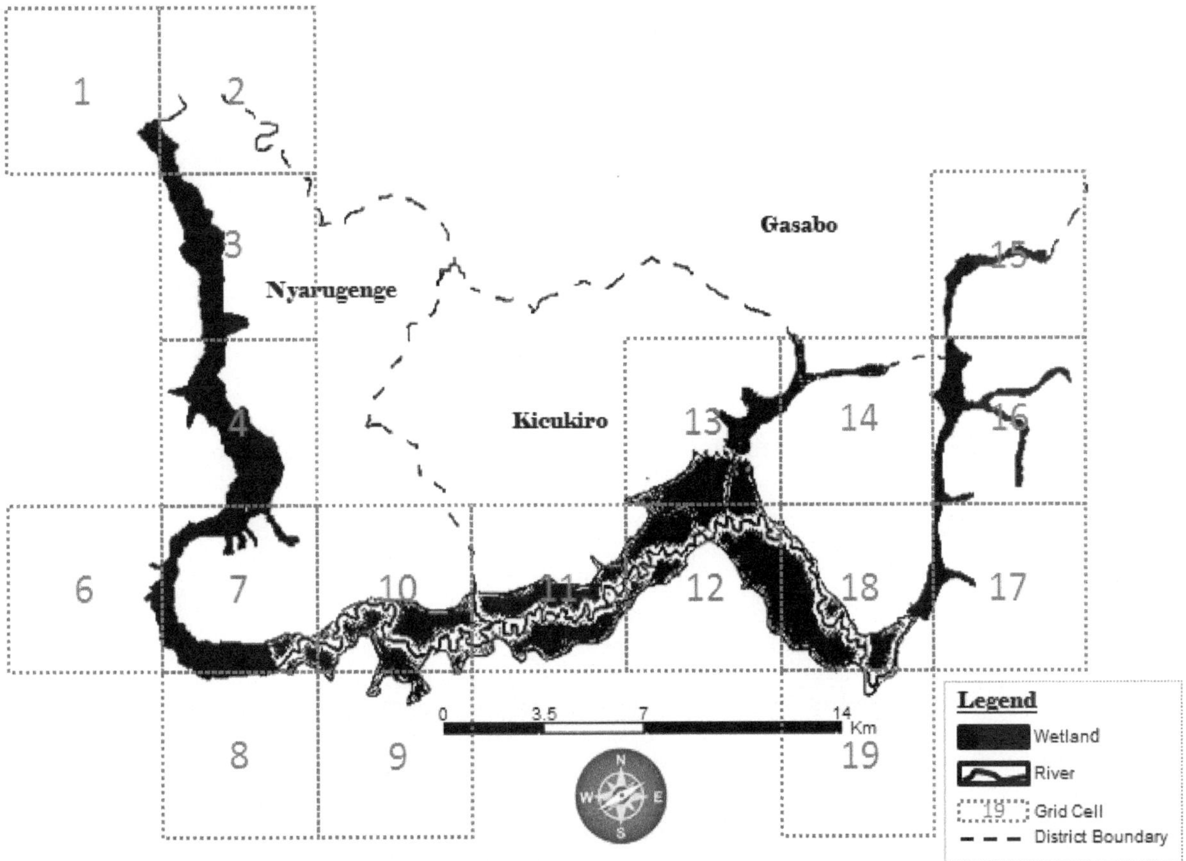

Figure 1. Localization of field data collection sites (Nyabarongo floodplains).

ecosystem under consideration and whose naturalization has resulted into either (1) widespread but rare, (2) localized but dominant or (3) widespread and dominant occurrence'. However, Simberloff (2010) still recognizes that there are no clearly defined operational criteria for a threshold above which a species can be said to have reached the status of invader in a given area.

For the purpose of this study, the following minimal conditions were applied: [1] cultivated (CV): the taxon is represented by specimens that were planted by humans; [2] casual (CAS): the taxon is represented by at least one individual plant growing wildly; [3] naturalized (NAT): the species occur by a wild population sustained over at least 2 lifecycles (3 seasons for seasonal and annual plants and 2 generations of seedlings for perennial taxa); [4] invasive (INV): the taxon is naturalized in more than 3 localities separated by at least 100 to 1000 m, the estimated routine maximum dispersal range of most plant species in the tropics (Corlett, 2009); and [5] transformer (TRF): the taxon spreads over a continuous area of more than 500 m^2 [an average-sized ecotope or smallest ecologically distinct landscape feature (Ellis et al., 2006)], with a mean distance between individuals or groups of plants of less than one time the canopy range of an individual plant of the concerned species; equivalent to a canopy cover of more than 50% (Bernez et al., 2006; Westhoff and van der Maarel, 1978).

Data acquisition

Levels of infestation

A 5.5 x 5.5 km^2 grid was overlaid over the study area's map (Figure 1). Only grid cells adjacent to the wetlands were counted. In each grid cell, Six 200 x 100 m^2 quadrats were defined and thoroughly searched using the timed-meander method (Huebner, 2007). We left the quadrat when it was entirely walked or as soon as *M. pigra* was checked off (Gulezian and Nyberg, 2010).

Source of infestation

A desk-based study was conducted during which we delved into the available literature and herbarium collections to retrieve additional information on past distribution patterns of *M. pigra* in Rwanda and identify the probable source of primary infestation. Records at National Herbarium of Rwanda and references to *M. pigra* in the works of Troupin (1966, 1971, 1978, 1982, 1983, 1985a and b) and Bloesch et al. (2009) were the main source of these secondary data. Rigorously selected online plant databases were also accessed, including the Global Invasive Species Database <www.issg.org> managed by Invasive Species Specialist Group (ISSG) of the IUCN Species Survival Commission and the Agro-forestry Tree Database <www.worldagroforestrycenter.org> of the International Centre for Research in Agroforestry.

Adoption of a name

Rwanda's folk taxonomists, upon assigning to it the Kinyarwanda name of 'u-mu-gey-o', confused the identity of *M. pigra* with that of environmentally less harmful plant species, *Acacia brevispica* Harms [Leguminosae], *Ptelorobium stellatum* (Forsk.) Brenan. [Leguminosae], *Acacia monticola* Brenan ex. Exell. [Leguminosae], the yet to be declared invasive in Rwanda *Caesalpinia decapetala*

Table 1. Historical records of *M. pigra* L. in Rwanda (Source: National Herbarium of Rwanda).

Sites of record	Author's name	Sample code	Year of record
Rusumo (Kirehe District)	Runyinya, B.	644	1977
Kanyinya, Lac Ihema floodplains (ANP)	Bouxin and Radoux	1216	1969
Plain Ndayitarirahe (Uruwita, ANP)	Troupin, G.	6957	1958
Rusumo (nearby Akagera waterfalls	Troupin, G.	15855	1976
Kamurugenze (Akagera National Park	Troupin, G.	7466	1958
Karama Station (Bugesera District)	Bouxin and Radoux	1379	1972
Bigugu, Kigali- Kanombe (MVK)	Mvukiyumwami, J.	54	1982
Kijojo (Byumba, now Nyagatare District)	Christiaensen, A.R.	802	1954
Uruwita Plain (Akagera National Park)	Bouxin and Radoux	26	1969
Uruwita Plain (Akagera National Park)	Troupin, G.	5312	1957
Mwendo (Bugesera District)	Liben, L.	868	1953
Gabiro (Gatsibo District)	Robyns, W.	9573	1938
Gabiro- Kagitumba Road, Km 18	Der Veken, P.V.	10731	1974
Near Gabiro (Gatsibo Dstrict)	Germain, R.	1138	1942
Nyabarongo (Shyorongi, Rulindo Dstrict)	Bouxin and Radoux	1758	1970
Route Kigali- Butare, Km 15 (Kamonyi)	Bouxin and Radoux	1790	1970
Icyanya, Kigali	Becquet, A.	247	1932

(Roth) Alston [Leguminosae], and the taxonomically distant *Rosa* sp. [Rosaceae]. To derive the best predictors of this group denomination, we compared its scores on 21 key characters with the scores of three of its name sakes: *C. decapetala* (Roth) Alston [Leguminosae], *Rosa* sp. [Rosaeae] and *A. brevispica* Harms [Leguminosae]. The selected characters included the morphology (presence of prickles on stems, the hairy character of fruits, sensitivity of leaves, type of flowers ...), ecology (ability for wild growth, most preferred habitats, ability to form dense thickets...), function of the plants (readiness for ornamental, medicinal, and live fencing use, source of animal fodder and possibility of soil fertility improvement through nitrogen fixation), and characteristic denomination of the species ('catclaw', in English, 'Umugeyo' and 'Bwara', in Kinyarwanda). Scoring was done in the form of binary coding system, in which '1' was assigned to a species if it tested positive for the character, and '0' assigned if it did not.

Data analysis

M. pigra range was mapped in ArcGis 9.3. ANOVA was used to determine the level of significance of observed differences between mean altitude and time of sample collection in relation to the sites of record (Akagera and Nyabarongo basins). Based on their respective scores on selected key characters and through PCA (Principle Component Analysis) and Hierarchical Clustering (UPGMA, Unweight Pair Group Method with Arithmetic Mean), predictors of association patterns of *M. pigra* and its close cousin species were identified. In particular, the question as to why *M. pigra* was given the Kinyarwanda name of 'u-mugey-o', normally used to refer to the Rwandan-Native *A. brevispica* Harms [Leguminosae], but also to the introduced *C. decapetala* (Roth) Alston [Leguminosae] and *Rosa* sp. [Roasaceae], was addressed. A predictor of this group denomination was recognized.

RESULTS

Levels of infestation

The first scientific record of *M. pigra* in Rwanda dates back to 1932, shortly after the arrival of European colonizers. However, the collections continued to be taken for decades and available specimens at the National Herbarium of Rwanda include samples deposited recently as of 1982. The majority of these samples were collected within two periods of time, separated by 10 years: 1948 to 1958 (35.3 % of samples) and 1968 to 1978 (47% of samples) (Figure 2a).

The findings of this study revealed that *M. pigra* occurs within Nyabarongo and Akagera basins, between 1°04' and 2°24' south latitude and between 1°54' and 30°54' east longitude. This range covers the entire Kigali City region, approximately two third of the Eastern Province and small areas of the Southern (Kamonyi District) and Northern Provinces (Rulindo and Gicumbi district) (Figure 2b). No record of its presence has so far been made outside this range. The majority of records were taken in Akagera National Park and its surrounding swamp system (Table 1).

The analysis of current levels of infestation showed that *M. pigra* was present in 72% of the total number sampling units, with higher frequency within the undisturbed wetlands than in the regularly plowed areas. Table 2 and Figure 3 illustrate the coordinates of key occurrence sites within Kigali city and neighboring rural districts.

Source of infestation

The optimal distribution of *M. pigra* in Rwanda ranges between 1300 and 1500 m of altitude, with more than ¾ of records having been collected in this range. A few cases of *M. pigra* striving at mid altitude (1600-1800 m) were observed within the Nyabarongo River Basin in Kicukiro, Rulindo and Kamonyi districts (Figure 4a).

The mean altitude of record was 1454 and 1377 m for records taken within floodplains of Nyabarongo and

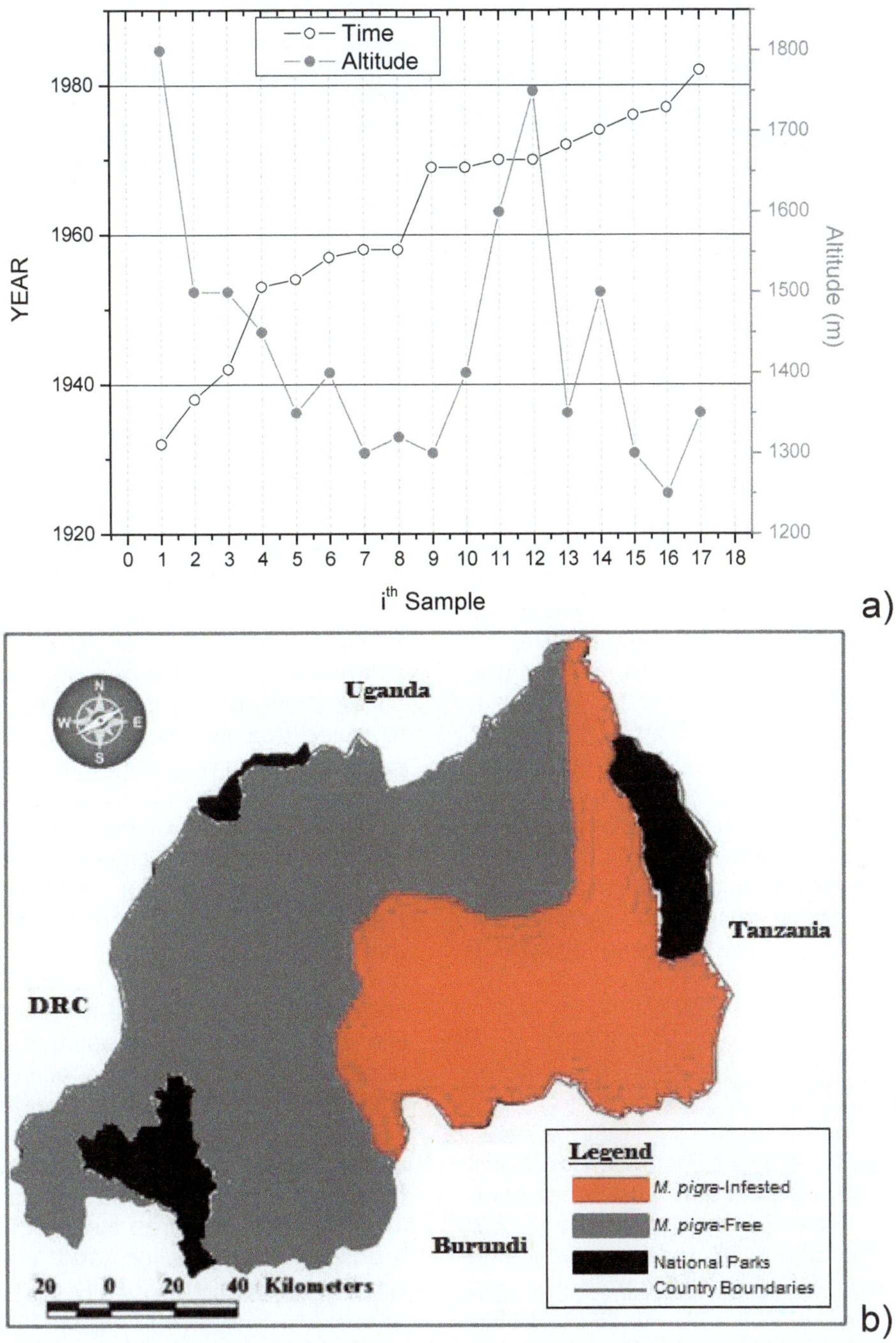

Figure 2. Past (a) and modern-day (b) distribution range of MP in Rwanda.

Akagera rivers, respectively. However, the analysis of variance showed that the apparent difference between these means was not statistically significant ($F = 1.93$, $p>0.05$) (Figure 4b).

The majority of records in the Akagera Basin were taken by the year 1960 while, in the same period, only 33% of the samples had been collected in the Nyabarongo Basin (Figure 5a). As a whole, the collections in the Basin of Akagera River seem to be close in time, starting later and finishing earlier, despite the fact that they represent more than half of the records (Figure 5b). However, the statistical test showed that means of year of record are comparable (1961 and 1963, respectively) ($F = 0.074$, $p>0.05$).

Adoption of a name

A comparison of four name sake species' scores on 21

Table 2. Key modern-day occurrence sites of *M. pigra* L. in Kigali city and surrounding rural districts.

Site	District	Latitude	Longitude	Elevation (m)	Ecosystem
Kibuza	Kamonyi	1°58'51"	29°58'05"	1393	Swamp
Mugomero	Kamonyi	2°00'37"	29°53'38"	1611	Swamp
Karambo	Rulindo	1°52'29"	29°59'26"	1377	Floodplain
Nyabugogo	Gasabo	1°55'59"	30°02'52"	1371	Swamp
Gahanga	Kicukiro	2°03'39"	30°05'11"	1345	Riparian
Isumo	Rwamagana	1°56'13"	30°16'39"	1389	Floodplain
Ruliba	Nyarugenge	1°57'28"	30°00'14"	1365	Riparian
Kidogo	Bugesera	2°09'59"	30°13'50"	1316	Riparian

key characters revealed, as it was expected, that *Rosa* sp. is significantly different from the three others (Figure 6).

In contrast, unexpectedly, *M. pigra* was found to be significantly different from both *A. brevispica* Harms and *C. decapetala* (Roth) Alston, despite the fact that the three species belong to the same family of Leguminosae. Conversely, cluster analysis further compared the 21 characters in the light of the species' scores on each and showed two main clusters (Figure 7).

One cluster includes characteristics specific to only *M. pigra*, including: (1) densely hairy pods, (2) pods clustered around an insertion point, (3) sensitive leaves, and (4) almost a riparian-restricted growth. The other cluster comprises characters *M. pigra* shares with the other three species such as (1) the presence of prickles on stems, (2) the ability to form dense thickets, (3) the readiness to be used as a hedge and (4) the rambling growth of stems.

DISCUSSION

Levels of infestation

Based on its modern-day high occurrence score (present in 72% of the total number of quadrats), and considering its historical records (17 herbarium records made in spatially wide (altitude: 1250 to1800 m; maximum distance between two record sites: 137 km) and temporally long ranges (1930s-1980s) *M. pigra* seems to have started invading Rwanda's eastern riparian and floodplains decades ago, and therefore, in this study, was assigned an invasive status in Rwanda.

The classification of *M. pigra* under the category of invasive species in Rwanda is in line with Richardson et al. (2000), Ipou et al. (2011) and Pysek et al. (2004) who define an invasive species as a 'naturalized plant that produces reproductive offspring, often in very large numbers, at considerable distances from parent plants [with an approximated spread rate of > 2 m/year], and thus have the potential to spread over a considerable area'. This was confirmed by our findings on *M. pigra* range of occurrence in Rwanda. At least 17 records of *M. pigra* presence in Rwanda were made at different locations from 1932 to 1982, some of the locations being separated by more than 100 km. Thirty years after these records were made, this study revisited *M. pigra* occurrence in the floodplains of Nyabarongo River. The results showed that *M. pigra* was present in the majority of quadrats (72 %). Key coordinates of record sites are reported in this article to help in further studies. Although estimating the spread rate of *M. pigra* in Rwanda was out of the scope of this study, the opportunistic observations revealed that *M. pigra* patch size averages at 3 to 4 m and that in many places, patches can be as close as to only 1 to 5 m away from each other.

At the beginning of this study, we had the feeling that, *M. pigra* was significantly expanding to other districts of Rwanda. However, our findings suggest that its range was restricted to Nyabarongo and Akagera Basins and that its invasion success in the last decades can be attributed more to the expansion of its 'grain' than to the change in the 'extent' of its distribution pattern. Altitude, distance to source of infestation and land use systems in upstream areas may have acted as barriers to its establishment in other regions. Instead, it seemed to be busy filling in the gaps to dominate landscapes within the already infested areas. This is in accord with the predictions of Theoharides and Dukes (2007) who suggested that after the first three stages of invasion have occurred (transport, colonization and establishment), the plant species spends significantly a longer period of time (also referred to as the 'lag time') in landscape spread process or dispersal within the region. Apparently, the overall occurrence was still too low such that when the ever first syllabus of the spermatophyte flora of Rwanda was compiled in 1971 (Troupin, 1971), this plant did appear among the 200 plant species it aimed to describe. Both in Rwanda and Burundi, it has been characterized by modest occurrence until recently when its spread was accelerated to the point of receiving the attention of environmentalists (lta.iwlearn.org, 10.07.2012). Similarly, in Zambia, unparalleled spread rates of *M. pigra* were registered in Kafue flats since 1980s following the construction of hydro-electric dams which caused disturbance in the flood regimes and expanded the range of

Figure 3. Modern-day occurrence of *M. pigra* L. in Rwanda: [1] in Kibuza wetland, Kamonyi District (upper left corner), [2] in Gahanga sector, along the Kigali-Bugesera road (upper right corner); [3] in Karambo, along the road to Ruli; and [4] in Nyabugogo wetland, along the road to Kigali-Gatuna [Photos© J. L. Seburanga, 2012].

floodplains (Sheppe, 1985). This view was shared by GISP (2007) who recognized that *M. pigra* spread in Africa has increased in the last three decades, often resulting from changes in rain fall and flooding regimes (Davis et al., 2000). During the 30 years that followed the latest specimen collection in 1982, its grain was expanded such that, in a number of sites, it is now appropriate to put it in category of invasive plant species with widespread and dominant occurrence (Colautti and MacIsaac, 2004). If this spread tendency is maintained, the species will soon fall in the category of 'transformer' species (Richardson et al., 2000).

Although, the study of dispersal ecology was not conducted in the particular case of Rwanda, altitude seems to be the best predictor of *M. pigra* spread. For instance, it was never recorded in Lake Kivu shores [on the other side of the Congo-Nile ridge] despite the fact that this region appears to be rather suitable to its establishment and that its cousin *Mimosa invisa* Mart. ex Colla was reported to occur in the Rusizi District in 1980s (Troupin, 1982). Also encouraging is to realize that it was never encountered in the biodiversity-rich afromontane region of the Albertine rift hotspot (Plumptre et al., 2007).

However, there is a room to worry if nothing is done to contain its spread. It is intriguing to note its presence in Rwanda at 1600-1750 m (Figure 4) and in neighboring Uganda at 1900-2000 m (Kalema and Ssegawa, 2007). This issue is of significance in terms of biodiversity conservation as no dedicated conservationist would wish to see *M. pigra* penetrating into the montane forest belt of the Albertine Rift (1500-2500 m) (Plumptre et al., 2007). Therefore, environmentalists in Rwanda should stay alert and do all what it takes to prevent this species from entering these forests.

Source of infestation

M. pigra occurrence appeared to be increasing as one moves downstream from Nyabarongo River (in Rwanda) to Lake Victoria (in neighboring Uganda); that is, from the central plateau (1350-1800 m of altitude) to the eastern lowlands (1250-1500 m of altitude). Considering the fact that the known earliest records of *M. pigra* in the great lakes region of East Africa dates back to 1862 in the Acholi District of Northern Uganda (GISP, 2007; plants.jstor.org, 01.09.2012), we suggest that *M. pigra*

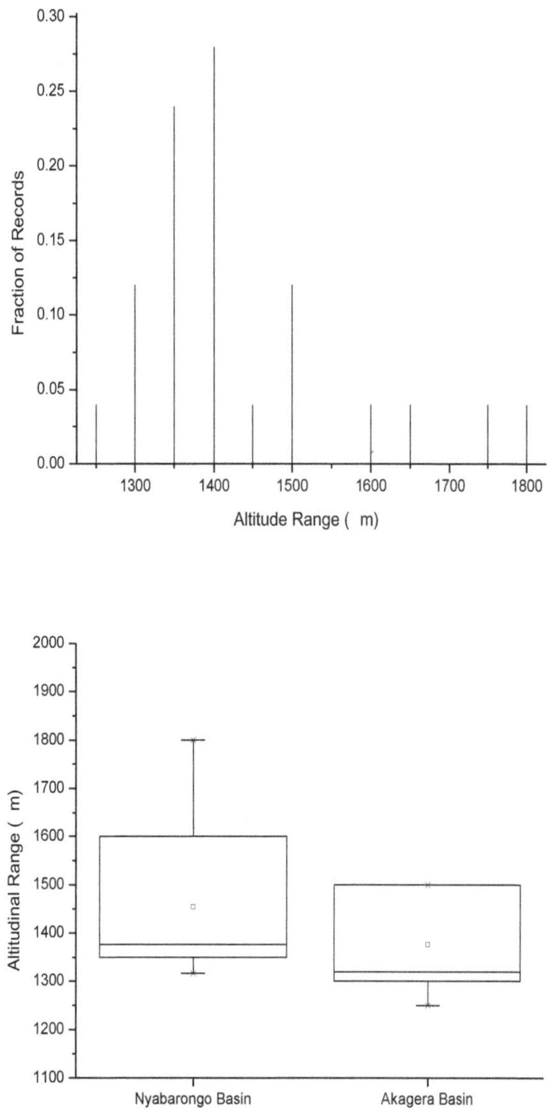

a)

b)

Figure 4. a and b: *M. pigra* distribution in Nyabarongo and Akagera basins in relation to altitude of record.

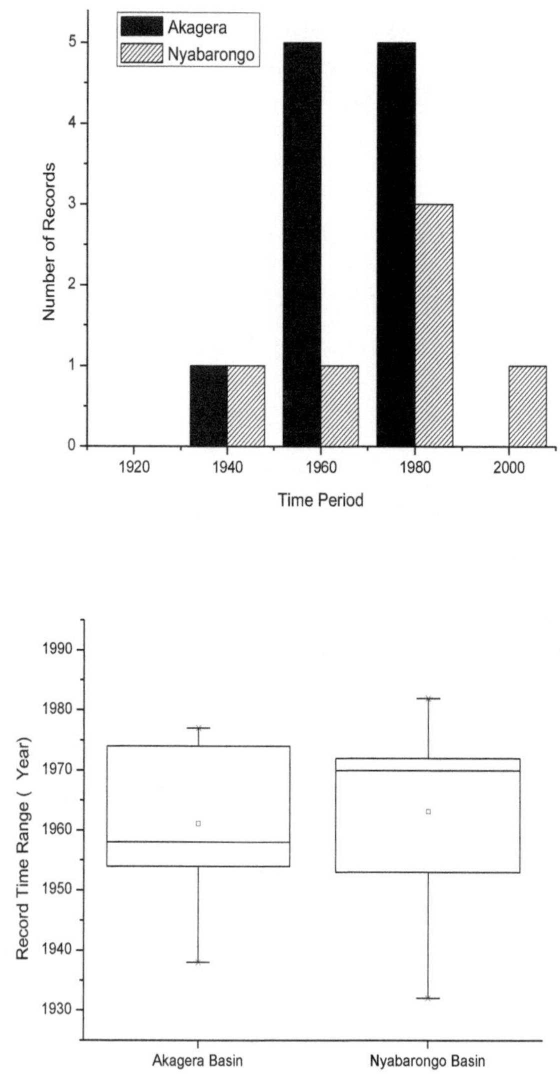

a)

b)

Figure 5. a and b: *M. pigra* distribution in Nyabarongo and Akagera basin in relation to year of record.

migrated upwardly from Egypt [where it was first recorded in Africa] through White Nile to Uganda and continued to reach Rwanda's Nyabarongo floodplains via the trans-boundary Akagera River. This would have required millennia if applied an average dispersal rate of up to 76 m year^{-1} in a wetland system (Lonsdale, 1993) and considering the 6500 km of distance traveled. This suggests three scenarios: [1] *M. pigra* native to Africa had a loose time budget to spread, [2] *M. pigra* is exotic to Africa but was introduced to Egypt earlier than 1900th century, and more frequent and long range flooding insured an incredibly higher spread rate, [3] *M. pigra* is exotic to Africa but was introduced independently to Egypt, Uganda and Rwanda. our attempt to detect a relationship between the time and place of collection of historical *M. pigra* records and its expansion pattern in Rwanda was not statistically supported (F = 1.93, p>0.05 and F = 0.074, p>0.05) and seems to be in line with the third scenario. However, based on our expert judgment and the knowledge drawn from the extensive literature review, we suggest that this may have been due to the small character of our sample size (only 17 records collected between 1932 and 1982 were taken into account because we were dealing with a time-restricted research question, and we could not enlarge this sample by adding on modern-day records).

It follows that choosing between the scenarios would not be easy as long as the debate around the African versus South American origin of *M. pigra* is not closed. However, there is strong evidence of association of *M. pigra* occurrence and Nyabarongo, Akagera and Nile Rivers'

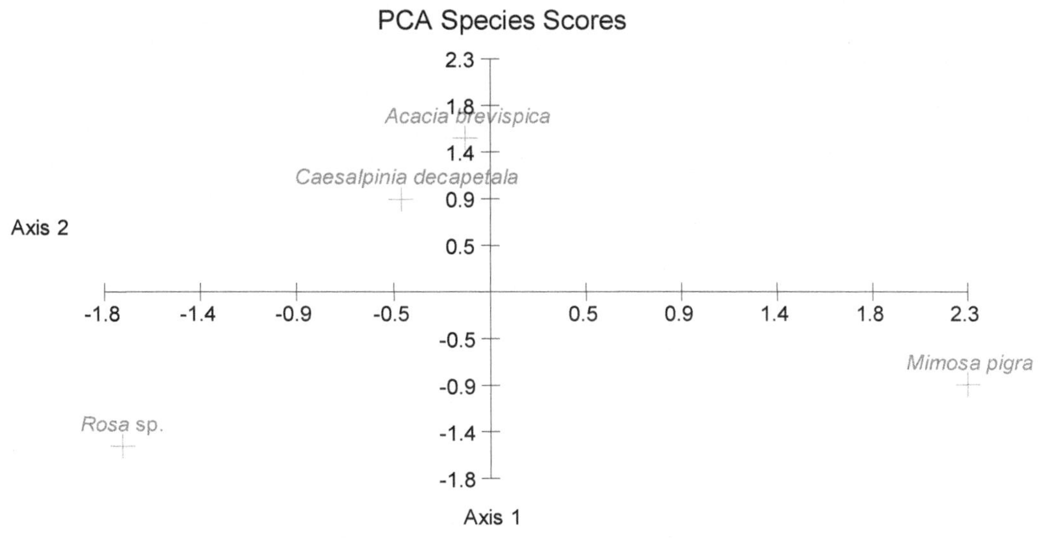

Figure 6. Species character-based PCA score chart of *M. pigra* and its three name sake plant species.

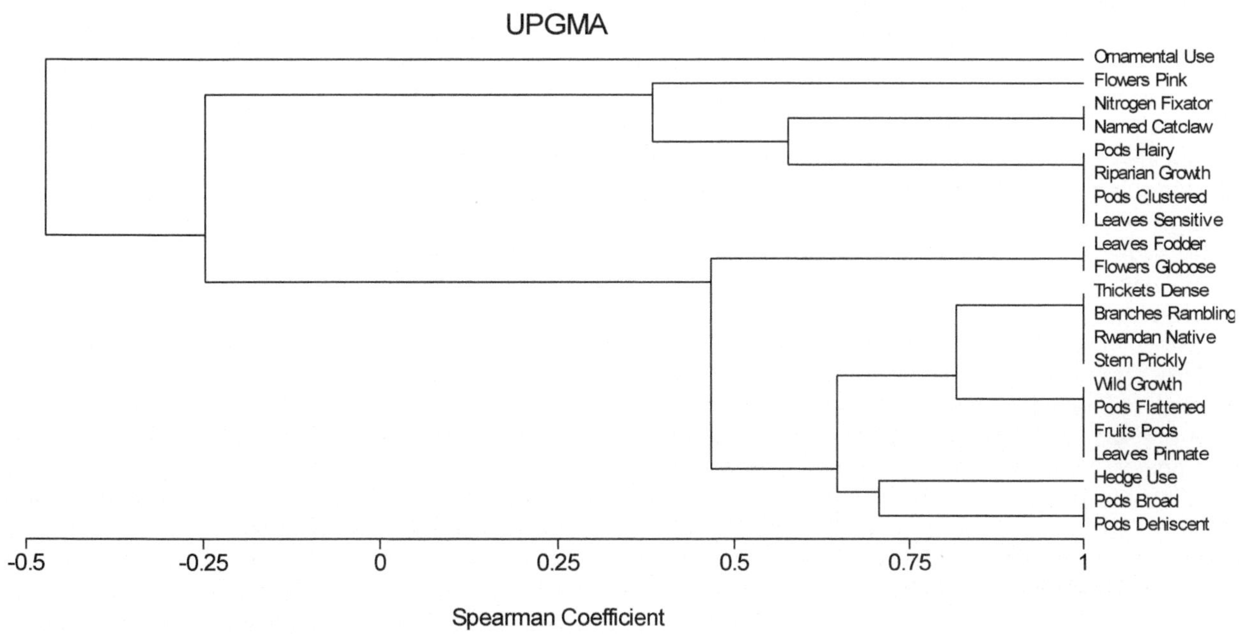

Figure 7. Clustering of *M. pigra* and its three name sake plant species' scores on key characters.

floodplains, tributaries and associated lakes, which, we believe, is in support of the first scenario. For instance, *M. pigra* was not recorded in Kenya's Nakuru National Park (Ng'weno et al., 2009) nor was reported in Uganda's Kibale forest (Chapman et al., 2001), two regions of East Africa naturally disconnected from the Akagera-Victoria-Nile system. Similarly, in their study, Hall et al. (2011) made a list of invasive non-native plant species occurring in agroforests and protected forests of East Usambara Mountains in neighboring Tanzania, but no reference to *M. pigra* was made. If the first scenario, is retained, and if taken into account the time needed for this species to move and migrate from Kagitumba (the probable port of entrance) to Icyanya locality in Kigali (the place where it was first recorded in Rwanda in 1932), we suggest that it should have been in the country not later than the second half of 19th century, well before the arrival of Europeans.

Furthermore, that many samples were taken in the Akagera Basin may indicate that, at the time of arrival of European botanists in Rwanda, *M. pigra* was more frequent there than in any other place in Rwanda. If this was the case, other factors hold constant, Akagera region could have been infested before upstream sites of the Nyabarongo Basin. This could also, in part, explain why colonial botanists did not report *M. pigra* as an invasive species in Rwanda. If they found it relatively more abundant in the park and surrounding natural forests than in the human-dominated landscapes, they would hardly imagine they were faced with a plant invasion case, unless they had prior knowledge of its status in other parts of the world. In addition, contrary to Simon et al. (2011) who suggested that African mimosas are most likely to have originated from South America through late Miocene trans-Atlantic dispersal, *M. pigra* was seen by many scientists obviously including Rwanda's folk ecologists as an African native species, and the conventional knowledge at that time was that an invasive species should be an exotic taxon having escaped from cultivation and expanding its range from human dominated landscapes to natural ecosystems.

Adoption of a name

That *M. pigra* was found to share the same Kinyarwanda name with *A. brevispica* Harms and *C. decapetala* (Roth) Alston [Leguminosae] which is understandable. They belong to the same family of Leguminosae, and highly resemble each other. For instance, they all have pinnate leaves, prickly stems and fruits in the form of pods. Instead, intriguing was to realize that *Rosa* sp., a species that belongs to such a distant family as that of Rosaceae, shares the same name.

The analysis of similarity between the four species showed that the possession of prickly stems, the rambling growth of branches, and the ability to form dense thickets are among the unifying characters of the four species- well in line with Barneby (1991) and Simon et al. (2011) who used morphological and ecological traits to explain affinities within taxonomic groups. This finding is supported by the fact that even in English, two of them, *M. pigra* and *C. decapetala* (Roth) Alston [Leguminosae], are referred to as 'catclaws', a name given after their 'prickles' that resemble the claws of a cat. In Burundi, Rwanda's neighboring country; both species have a similar denomination 'u-mu-bamb-a-n-gu-e'. Baerts and Lehmann (1989) translated this name into French 'l'épineux qui cloue le léopard au sol', which in English would be referred to as 'the spiny plant that nails the leopard to the ground'. Well in line with this, the third species, *A. brevispica* Harms [Leguminosae], has an alternative Kinyarwanda name 'Bwara', which if translated into English would mean 'small nails', of course not of a cat, because cats themselves are a recent introduction in Rwanda. It is important to note that, unfortunately, the Kinyarwanda name applied to the four species 'u-mu-gey-o' has no clear meaning in Kinyarwanda Language. However, it was also interesting to find that it could have been derived from 'l-gey-e', the equivalent denomination of this in the Tanzanian Kisukuma dialect (herbaria.plants.ox.ac.uk, 29.08.2012). These taxonomic relationships and identity confusions have a deeper implication for prevention and control of *M. pigra* invasion. In a developing country, due to high levels of taxonomic illiteracy, people often confuse newly introduced taxa with similar ones, be they native or exotic. This is problematic because it gives the species the opportunity to expand its range before the local communities come to realize that it is actually an alien invader and report it. For instance, *Lantana camara* L. [Verbenaceae], a notorious terrestrial invasive species in Rwanda, is called 'umu-henger-i' as its counterpart *Lantana trifolia* L. [Verbenaceae], a relatively less threatening taxon. Interestingly, as soon as its threat was noticed, local people named it 'a-ka-tey-e', denoting its 'alien' and 'invasive' attributes. Another example is that of *Eichhornia crassipes* (Mart.) Solms [Pontederiaceae], a globally known aquatic weed that shares its local name 'a-ma-reb-e' with species that belong to other families such as *Nymphaea nouchalii* Burm. [Nymphaeace], *Caldesia reniformis* (D. Don) Makino [Alismataceae], and, in some places, the North American-native *Pistia stratiotes* L. [Araceae], and even with the terrestrial *Begonia meyeri-johannis* Engl. [Begoniaceae]. In the case of *M. pigra*, we presume that folk taxonomists in pre-colonial Rwanda were unable to draw a clear demarcation line between *M. pigra* and its cousin species, probably due to the high interrelatedness between taxa within the legume family (Lewis et al., 2003). As a result, under the cover of a non-specific name, it spread almost unnoticeably to become a notorious invader shrub in Rwanda's eastern floodplains'.

Conclusion

In this study, we discussed the status of *M. pigra* in Rwanda, and pleaded for its inclusion in the invasive flora of Rwanda. Our findings suggest that its invasion success in the last decades can be better explained with reference to its 'grain' expansion than to the change in the 'extent' of its distribution patterns. We suggested that Egypt through Nile floodplains and riparian communities of Uganda's Lake Victoria may have been the primary source of *M. pigra* of propagules that infested Rwanda's Akagera-Nyabarongo river system. The altitude, distance to the source of infestation and land use systems in upstream areas best predict its establishment. We presumed that under the shadow of its Kinyarwanda name 'u-mu-gey-o', *M. pigra* was able to spread and become an invader before scientists could take notice of its true identity: an invasive [not environmentally harmless] alien [not Rwandan-native] Mimosa [not Acacia, Ptelorobium, Caesalpinia nor Rosa] species.

REFERENCES

Baerts M, Lehmann J (1989). Les guérisseurs et plantes médicinales de la région des Crêtes Zaïre-Nil au Burundi. Musée Royale de l'Afrique Centrale, Tervuren, Belgique.

Barneby RC (1991). Sensitivae censitae: A description of the genus Mimosa Linnaeus (Mimosaceae) in the New World. Memoirs of the New York Botanical Garden, 65:1-835.

Bernez I, Aguiar F, Violle C, Ferreira T (2006). Invasive river plants from Portuguese floodplains: What can species attributes tell us? Hydrobiologia. 570:3-9.

Bloesch U, Troupin G, Derungs N (2009). Les plantes ligneuses du Rwanda: Flore, ecologie et usages. Shaker Verlag, Aachen, Netherlands.

Byenkya S, Ebong C, Eilu G, Ogwang J, Kiwuso P, Molo R, Mutumba G, Kabanyoro R (2004). Implementation of invasive plant prevention and control Programmes in Uganda.Technical Report, CAB International, Uganda.

Chapman LJ, Balirwa J, Bugenyi FWB, Chapman C, Crisman TL (2001). Wetlands of East Africa: Biodiversity, exploitation, and policy perspectives. In: Gopal B, Junk WJ and Davis, JA (Eds.). Biodiversity in wetlands: assessment, function and conservation. Backhuys Publishers, Leiden, The Netherlands, 2:101-131.

Colautti RI, MacIsaac HJ (2004). A neutral terminology to define 'invasive' species. Diversity and Distributions 10:135-141.

Corlett RT (2009). Seed dispersal distances and plant migration potential in tropical East Asia. Biotropica 41(5):592-598.

Davis MA, Grime JP, Thompson K (2000). Fluctuating resources in plant communities: a general theory of invisibility. J. Ecol. 88:528-534.

Douthwaite RJ, van Lavieren LP (1977). A description of the vegetation of Lochinvar National Park, Zambia. Technical Report 34, National Council for Scientific Research, Lusaka, Zambia.

EIAR (2010). Weeds and invader plants in Ethiopia. Ethiopian Institute Of Agricultural Research, retrieved on 24th August 2012 from http://www.eiar.gov.et/IAS-Publications/WEEDS%20AND%20INVADER%20PLANTS%20Revised.pdf

Eilu G, Hafashimana DLN, Kasenene JM (2004). Density and species diversity of trees in four tropical forests of the Albertine rift, western Uganda. Diversity and Distributions, 10:303-312.

Eilu G, Oriekot J, Tushabe H (2007). Conservation of indigenous plants outside protected areas in Tororo District, Eastern Uganda. Afr. J. Ecol. 45(Suppl. 3):73-78.

Ellis EC, Wang HQ, Xiao HS, Peng K, Liu XP, Li SC, Ouyang H, Cheng X, Yang LZ (2006). Measuring long-term ecological changes in densely populated landscapes using current and historical high resolution imagery. Remote Sensing of Environ. 100:457-473.

Fowler SV, Syrett P, Hill RL (2000). Success and safety in the biological control of environmental weeds in New Zealand. Austral Ecol. 25:553-562.

GISP (2007). Development of case studies on the economic impact of invasive species in Africa: Draft Report – Mimosa pigra. Draft prepared by Psi-Delta, Australia.

Gulezian PZ, Nyberg DW (2010). Distribution of invasive plants in a spatially structured urban landscape. Landscape and Urban Planning 95:161-168.

Hall JM, Gillespie TW, Mwangoka M (2011). Comparison of Agroforests and Protected Forests in the East Usambara Mountains, Tanzania. Environ. Manage. 48:237-247.

Huebner CD (2007). Detection and monitoring of invasive exotic plants: A comparison of four sampling methods. Northeastern Naturalist 14(2):183-206.

Indira TJ (2007). Mapping and modeling of Mimosa pigra expansion in Lochinvar National Park, Zambia. MSc Thesis, International Institute for Geo-Information Science and Earth Observation, Enschede, The Netherlands.

Ipou J, Toure A, Adou LM, Kouame KF, Gue A (2011). A new invasive species of the agrosystems in the south of Côte d'Ivoire: Tithonia diversifolia (Hemsl.) A. Gray (Asteraceae). Afr. J. Food Sci. Technol. 1(6):146-150.

Kalema J, Ssegawa P (2007). The flora of highly degraded and vulnerable wetland ecosystems of Nyamuriro and Doho, Uganda. Afr. J. Ecol. 45(1):28-33.

Kapustka L (2007). Invasive species: A real, but largely ignored threat to environmental security. In: Linkov *et al.* (Eds.). Managing critical infrastructure risks, Springer, pp.175-188.

Lewis G, Schrire B, Mackinder B, Lock M (eds.) (2005). Legumes of the World. Royal Botanic Gardens, Kew.

Lonsdale WM (1993). Rates of spread of an invading species - Mimosa pigra in northern Australia. J. Ecol. 81:513-521.

Masharabu T (2011). Climate change risks to ecosystems and biodiversity in Ruvubu National park, Burundi. UDSM/IRA, Dar es Salaam, Tanzania retrieved on 25th August 2012 from http://start.org/download/2012/biodiv/masharabu.pdf.

Moorhouse TM, Agaba P, McNabb TJ (2001). Recent efforts in biological control of water hyacinth in the Kagera River headwaters of Rwanda. In: Julien MH, Hill T D, Ding J-Q (Eds.). Biological and integrated control of water hyacinth, *Eichhornia crassipes*. ACIAR Proceedings p.102.

Ng'weno CC, Mwasi SM, Kairu JK (2009). Distribution, density and impact of invasive plants in Lake Nakuru National Park, Kenya. Afr. J. Ecol. 48:905-913.

Pimentel D, Lach L, Zuniga R, Morrison D (2005). Update on the environmental and economic costs associated alien-invasive species in the United States. Ecol. Econ. 52:273-288.

Plumptre AJ, Davenport TRB, Behangana M, Kityo R, Eilu G, Ssegawa P, Ewango C, Meirte D, Kahindo C, Herremans M, Peterhanse JK, Pilgrim JD, Wilson M, Languy M, Moyer D (2007). The biodiversity of the Albertine Rift. Biol. Conser. 134:178 -194.

Plumptre AJ, Williamson EA (2001). Conservation-oriented research in the Virunga Region. In: Robbins, M.M., Sicotte, P., Stewart, K.J. (Eds.), Mountain Gorillas: Three Decades of Research at Karisoke. Cambridge University Press, Cambridge, pp.361-389.

Pysek P, Richardson DM, Rejmánek M, Webster GL, Williamson M, Kirschner J (2004). Alien plants in checklists and floras: towards better communication between taxonomists and ecologists. Taxon. 53:131-143.

Richardson DR, Pysek P, Rejmánek M, Barbour MG, Panetta FD, West CJ (2000). Naturalization and invasion of alien plants: concepts and definitions. Diversity and Distributions 6:93-107.

Sheded MG, Hassan L (1999). Reproductive ecology of Mimosa pigra L. in Egypt. Feddes Repertorium 110(34):219-224.

Sheppe WA (1985). Effects of human activities on Zambia's Kafue Flats Ecosystems. Environ. Conser. 12(1):49-57.

Simberloff D (2010). Invasions of plant communities – More of the same, something very different, or both? The American Midland Naturalist. 163:220-233.

Simon MF, Grether R, de Queiroz LP, Särkinen TE, Dutra VF, Hughes CE (2011). The evolutionary history of Mimosa (Leguminosae): toward a phylogeny of the sensitive plants. Am. J. Bot. 98(7):1201-1221.

Sobrino E, Sanz-Elorza M, Dana ED, González-Moreno A (2002). Invasibility of a coastal strip in NE Spain by alien plants. J. Vegetation Sci. 13:585-594.

Theoharides K, Dukes JS (2007). Plant invasion across space and time: factors affecting nonindigenous species success during four stages of invasion. New Phytologist, 176:256-273.

Troupin G (1966). Etude phytocénologique du Parc National de l'Akagera et du Rwanda oriental. Liège, Belgique.

Troupin G (1971). Syllabus de la flore du Rwanda : spermatophytes. Musée Royal de l'Afrique Centrale, Annales IN-8o, Sciences Economiques no 7, Tervuren, Belgique.

Troupin G (1978). Flore du Rwanda, spermatophytes, volume I. Musée Royal de l'Afrique Centrale, Tervuren, Belgique.

Troupin G (1982). Flore des plantes ligneuses du Rwanda. Musée Royal de l'Afrique Cent5ale, Tervuren, Belgique.

Troupin G (1983). Flore du Rwanda, spermatophytes, volume II. Musée Royal de l'Afrique Centrale, Tervuren, Belgique.

Troupin G (1985). Flore du Rwanda, spermatophytes, volume III. Musée Royal de l'Afrique Centrale, Tervuren, Belgique.

Troupin G (1985). Flore du Rwanda, spermatophytes, volume IV. Musée Royal de l'Afrique Centrale, Tervuren, Belgique.

Tutin TG, Heywood VH, Burges NA, Valentine DH, Walters SM, Webb DA (1964). Flora Europaea, Vol. 1. Lycopodiaceae to Platanaceae. Cambridge University Press, Cambridge.

Vilà M, Meggaro Y, Weber E (1999). Preliminary analysis of the naturalized flora of northern Africa. Orsis 14:9-20.

Westhoff V, van der Maarel E (1978). The Braun-Blanquet approach,

2nd Ed. In: Whittaker RH (ed.). Classification of plant communities. The Hague: Junk, pp.287–399.
Williams JA, West CJ (2000). Environmental weeds in Australia and New Zealand: issues and approaches to management. Austral Ecol. 25:425-444.
Yates ED, Levia Jr DF, Williams CL (2004). Recruitment of three non native invasive plants into a fragmented forest in southern Illinois. Forest Ecol. Manage. 190(2-3): 119-130.

Genetic diversity and disease response of rust in bread wheat collected from Waziristan Agency, Pakistan

Mohammad Nisar[1]*, Naeem Khan [1], Nausheen[1,2], Zakia Ahmad[1] and Abdul Ghafoor[2]

[1]Department of Botany[1] and Biotechnology[2], University of Malakand, Chakdara Dir (Lower), Khyber PakhtunKhwa, Pakistan.
[2]Plant Genetic Resources Program (PGRP), National Agricultural Research Centre (NARC), Islamabad, Pakistan.

In the present study, thirty indigenous landraces exclusively collected from Waziristan Agency, Khyber Pakhtunkwa, Pakistan along with 14 cultivars were investigated for genetic diversity and resistance to fungal diseases. The germplasm was evaluated for morphological traits, response to rust and smut, and seed protein markers. A low level of allelic variation was observed for morphological trait, while the agronomic traits exhibited higher level of coefficient of variation. In the present study, six lines (*MIRALI (ECDOCK) NWA*, *MIRALI NWA-1*, *MIRALI NWA-2*, *BANNU PROPER*-1 and *FR-BANNU DOMEL*) were highly resistant to rust disease. The SDS-PAGE was carried out to assess the genetic diversity and selection of elite genotypes based on proteomic homology of high molecular weight (HMW) glutenin (*Glu-A*1) using Jaccard's similarity coefficient. Almost 16% genetic diversity (low level) was observed in HMW glutenin protein and the germplasm was grouped into five clusters. The cluster 5 sorted resistant lines, while the others were inter-spread, although few of these were grouped on the basis of collection sites. The genotype from Chakdara-2 of Cluster 1 was unique due to the presence of all the polypeptides of HMW glutenin protein. Based on HMW glutenin, the resistant line from *Bannu Proper*-1 and *FR-Bannu Domel* were similar to the improved line *RAJ* and *BAKHTAWAR-92*. On the basis of this initial investigation it is suggested to investigate this germplasm for their adaptability and for molecular markers to employ this unique germplasm in modern wheat cultivars.

Key words: Diversity in *Glu-A*1 loci, novel immune lines, Pakistani wheat, *Puccinia triticina*, SDS-PAGE,

INTRODUCTION

Wheat is a staple food for 35% of the world's population and is grown on 17% of the cultivated area in the world (Kronstad 1998). It occupies 70% of Rabbi (winter season) and 37% of total cropped area in Pakistan (Shah et al., 2003). Wheat is affected by leaf rust and smut, serious fungal diseases that is a continuous threat to its cultivation. Infection can cause up to 20% yield losses and resulting grain to shrivel. Pakistan is rich with unique germplasm variation in many crop species, especially wheat. Efficient genetic evaluation is the first step to any crop breeding program. Once the genetic nature of the germplasm is investigated, then one can develop a possible outcome for the future generation for food security that is experiencing a lot of risks at present.

The knowledge of genetic diversity is a useful tool in gene-bank management and breeding experiments like tagging of germplasm, identification and/or elimination of duplicates in the gene stock and establishment of core collections. Due to vitality characterization, Dotlacil et al. (2002) studied 123 wheat lines; Gupta et al. (2002) and Pawar et al. (2003) reported genetic divergence among different wheat advanced cultivars to document the morphological and agro-economical data.

During recent years, biochemical and molecular genetic techniques have emerged as a complementary strategy in conjunction with traditional approaches in the management of plant genetic resources. Lot of work had been published for estimation of genetic diversity using SDS-PAGE for Low as well as high molecular weight glutenin subunits in wheat (Benmoussa et al., 2000; Popa et al., 2006; Sultana et al., 2007; Caballero et al., 2009). Wiser (2007) indicated that gluten proteins play a key role

*Corresponding author. E-mail: mnshaalpk@yahoo.com, nisarhu@gmail.com.

in determining the unique baking quality of wheat by conferring water absorption capacity, cohesively, viscosity and elasticity on dough. Tahir et al. (1996) identified novel high molecular weight glutenin subunit in a hexaploid wheat landrace collected from Baluchistan that is known for wider genetic diversity for most of the cereal. The wheat endosperm storage proteins are widely associated with the bread making quality (Wrigley et al., 2006). There is no report on indigenous wheat genetic resources from Waziristan Agency, Pakistan that is highly vulnerable to biological as well as geo-political instability.

The work was devoted to explore the untouched agro-ecological zones of Pakistan for wheat collection and to document the genetic resources of wheat using morphological and seed protein markers for future utilization for wheat improvement program.

MATERIALS AND METHODS

Plant materials

In the present study, thirty indigenous landraces collected from Waziristan Agency, Khyber Pakhtunkhwa, Pakistan, along with 14 cultivars (Pakistani origin) was also included (Table 1). Different experiments were conducted to investigate the genetic diversity and resistance to fungal diseases. The genetic diversity was investigated through morphological characterization and biochemical markers.

Morphometrics study

The morphological characterization was split into qualitative and quantitative traits. The qualitative traits were scored through general visualization (phenotypic observation) on the genotype basis and 3 qualitative traits (plant color, growth habit and logging activity) were studied at the relevant stage of development in the available germplasm. Data on quantitative traits were recorded for plant height (cm), spike length (cm), Awn length (cm), 1-leaf length, 2-leaf length, 3-leaf length, 4-leaf length, biomass (g), 100-grain weight, grain yield and harvest index.

Disease status

The disease status was examined visually and the severity of disease was presented in percentage for rust disease. The genotypes showing resistance under natural infestation were artificially inoculated with *Puccinia triticina* spores by tapping heavily infected plant parts over the leaves. Severity of the disease was categorized into 0, 2.32, 2.85, 5.26, 5.88, 6.06, 6.66, 6.97, 10, 10.6, 14.28, 15.15, 17.39, 21.7, 28.99, 43.9, 48, 53.19, 54.83, 59.25, 61.1, 61.76, 70, 90.9, 91.3 and 100%, respectively. Effect on different phases as well as on different parameters was observed.

Biochemical assay

To explore the genetic diversity on the basis of protein, sodium dodecyl sulfate polyacrylamide gel electrophoresis (SDS-PAGE) was carried out. For SDS-PAGE analysis, grains of each genotype were crushed into a fine powder with mortar and pestle. Protein extraction buffer (400 µl) was added to 0.01 g of grain flour and mixed well with Automatic Lab-Mixer DH-10. The extraction buffer contained the following final concentrations: 0.5 M Tris-HCl (pH 6.8), 2.5% SDS, 10% glycerol and 5% 2-mercaptoethanol. Bromophenol blue (BPB) was added to the protein extraction buffer as tracking dye to watch the movement of protein in the gel. The molecular weight of the dissociation polypeptides was determined by using molecular weight protein marker MW-SDS-70 Kit (Sigma).

The SDS-PAGE of total grain protein was carried out in the discontinuous buffer system according to the method of Laemmli (1970). Acrylamid gel concentration 12.5% and 6 µl of sample were used for analyzing germplasm. After staining and de-staining, the gel was dried using (Atto, Rapidery-Mini Japan) gel drier. In order to check the reproducibility of the method, two separate gels were run under similar electrophoretic conditions. The data was scored on the presence (1) or absence (0) of protein bands and high intensity of glow was considered as major bands and low intensity as minor bands.

Data analysis

Similarity indices of 43 genotypes were calculated for all possible pairs of protein types and used to construct dendrogram by the UPGMA (Sneath and Sokal, 1973) by computer software NTYSIS Version -2.0. To find out genetic similarity between the collected genotypes from North Waziristan Agency and obtained cultivars, Jaccard's similarity index (S) was calculated for possible pairs of protein types by $S = W/(A+B-W)$ (Sneath and Sokal, 1973). Whereas W = the number of bands of common mobility. A = the number of bands in protein type 'A' and B = The number of bands in protein type 'B'. Cluster analysis and principal components analysis was calculated using the un-weighed pair-group method arithmetic average (UPGMA) using NTYSIS vir-2.0.

RESULTS

Morphometrics study

The results regarding qualitative and quantitative traits are presented in Table 1 and 2, respectively and all the characters showed significant level of genetic divergence. Thirteen genotypes that is FR (Frontier Region)-Bannu Marghali Pirbakhel; FR- Bannu Domel-1; Mirali NWA-3; Miranshah NWA-1; Miranshah NWA-2; Bannu Proper-2; Mirali NWA-4; Mirali NWA-5; North Waziristan Agency-3; Waziristan NWA; Bannu-2, Mardan, Chakdara showed best performance in the present collection. The lodging ability of all the collected genotypes was also investigated. In the collected germplasm, the genotype of Mirali (ECDOCK) NWA, Bannu Plains Areas, FR-Bannu, FR-Bannu (Dry Area), FR-Bannu Domel, Miranshah NWA lodged after > 100 days (Table 1).

Plant height was positively and significantly correlated with 1st leaf length ($r = 0.64$), 2nd leaf length ($r = 0.58$), 3rd leaf length ($r = 0.61$), biomass ($r = 0.71$) and grains yield ($r = 0.79$). However it was negatively correlated with 100-grain weight. Spike length showed positive correlation with awn length, while it was negatively correlated with 100-grain weight. First leaf length showed highly positive correlation with 2nd leaf length, 3rd leaf length, and

Table 1. Genetic diversity based on qualitative traits in wheat germplasm.

R/No.	Genotypes	Plant color	Growth habit	Logging
1	Mirali (ECDOCK) North Waziristan Agency	Green	Weak/normal	Susceptible
2	Mirali North Waziristan Agency-1	Green	Normal	Resistant
3	North Waziristan Agency-1	Green	Weak/normal	Resistant
4	FR-Bannu Marghali Pirbakhel	Green	Vigorous	Resistant
5	FR-Bannu-1	Green	Weak	Resistant
6	Mirali North Waziristan Agency-2	Green	Weak/normal	Resistant
7	FR-Bannu-2	Dark green	Normal	Resistant
8	Bannu proper-1	Green	Weak/normal	Resistant
9	Bannu plain areas-1	Green	Weak	Resistant
10	Bannu plain areas-2	Pale Green	Weak	Susceptible
11	FR-Bannu-1	Green	Weak	Susceptible
12	FR-Bannu (dry area)	Green	Weak	Susceptible
13	FR-Bannu Domel-1	Green	Weak/normal	Resistant
14	FR-Bannu Domel-2	Green	Weak	Susceptible
15	FR-Bannu Marghali Pirbakhe-l	Green	Normal	Resistant
16	Mirali North Waziristan Agency-3	Dark green	Vigorous	Resistant
17	FR-Bannu-2	Dark green	Weak/normal	Resistant
18	FR Bannu Domel-3	Dark green	Normal	Resistant
19	Miranshah North Waziristan Agency-1	Dark green	Vigorous	Susceptible
20	Miranshah North Waziristan Agency-2	Dark green	Vigorous	Resistant
21	Bannu Proper-2	Dark green	Vigorous	Resistant
22	North Waziristan Agency-2	Dark green	Vigorous	Resistant
23	Mirali North Waziristan Agency-4	Dark green	Vigorous	Resistant
24	North Waziristan Agency-3	Dark green	Vigorous	Resistant
25	Mirali North Waziristan Agency-5	Dark green	Vigorous	Resistant
26	Bannu-1	Dark green	Normal	Resistant
27	Waziristan North Waziristan Agency	Dark green	Vigorous	Resistant
28	Bannu-2	Dark green	Vigorous	Resistant
29	Mardan-1	Dark green	Vigorous	Resistant
30	Mardan-2	Dark green	Vigorous	Resistant
31	Chakdara-1	Dark green	Vigorous	Resistant
32	Chakdara-2	Dark green	Vigorous	Resistant
33	Check 29	Dark green	Vigorous	Resistant
34	Check 30	Dark green	Weak	Resistant
35	Check 31	Dark green	Normal	Resistant
36	Check 32	Dark green	Weak	Resistant
37	Check 33	Dark green	Vigorous	Resistant
38	Check 34	Dark green	Vigorous	Resistant
39	Check 35	Dark green	Vigorous	Resistant
40	Check 36	Dark green	Vigorous	Resistant
41	Check 37	Dark green	Vigorous	Resistant
42	Check 38	Dark green	Weak	Resistant
43	Check 39	Dark green	Weak/normal	Resistant
44	Check 40	Dark green	Weak	Resistant

biomass and with grain yield, while it was negatively correlated with 100-grain weight and harvest index. Second leaf length had positive relationship with 3rd leaf length, 4th leaf length, biomass and grain yield, while it was inversely associated with 100-grain weight and harvest index. The 3rd leaf length showed high positive correlation with 4th leaf length, biomass and grain yield, while it was negatively correlated with 100-grain weight and harvest index. The fourth leaf length had positive correlation with biomass and grain yield, while it was

Table 2. Comparative descriptive statistics of eleven quantitative traits of Pakistan wheat genotypes and cultivars.

Traits	Mean ± St-dev		St-EE		Range		Min		Max		CV %	
	Genotype	Check	Genotype	Check	Genotype	Check	Genotype	Check	Genotype	Check	Genotype	Check
Plant height (cm)	58.7 ± 19.33	48.12 ± 6.90	3.47	2.08	89.42	24.48	27.6	36.66	117.02	61.14	*32.92***	14.35
Spike length (cm)	8.1 ± 4.47	8.19 ± 1.40	0.8	0.42	26.6	4.98	2	5.74	28.6	10.72	*54.62***	17.12
Awn length (cm)	5.9 ± 1.15	6.23 ± 0.80	0.21	0.24	5.9	2.42	1.2	5.06	7.1	7.48	19.42	12.84
1-leaf length	19.05 ± 3.6	22.82 ± 1.23	0.65	0.37	16.2	3.52	13.6	20.98	29.8	24.50	*18.91*	5.39
2-leaf length	23.28 ± 4.47	25.72 ± 1.63	0.8	0.49	21.18	6.20	15.54	23.32	36.72	29.52	*19.19*	6.32
3-leaf length	22.15 ± 4.27	22.77 ± 1.53	0.77	0.46	19.54	4.42	13.56	20.52	33.1	24.94	*19.27*	6.70
4-leaf length	20.5 ± 3.89	19.36 ± 2.61	0.7	0.79	16.4	7.54	11.7	15.32	28.1	22.86	*18.96*	13.47
Biomass (g)	18.63 ± 6.91	19.96 ± 5.40	1.24	1.63	28.84	17.12	9.2	14.40	38.04	31.52	*37.08***	27.07
100-grain weight	4.64 ± 0.35	5.10 ± 0.43	0.06	0.13	1.57	1.50	3.79	4.19	5.36	5.69	7.53	8.49
Grain yield	5.42 ± 1.9	6.33 ± 1.01	0.34	0.30	7.79	3.50	2.6	4.77	10.39	8.27	*35.09***	15.90
Harvest index	30.29 ± 8.04	32.79 ± 5.89	1.44	1.78	41.2	18.53	18.75	25.06	59.96	43.60	*26.53***	17.97

Standard error-St-EE; Standard deviation-St-dev; CV %- Coefficient of variance in percent ** Genotypes represent significant variation with respect to cultivars; highlighted box showed high level of %CV in comparison to checks/cultivars/improved lines

negatively correlated with 100-grain weight and harvest index. Biomass had significant correlation with grain yield, while it was negatively correlated with 100-grain weight and harvest index. The 100-grain weight had negative correlation with grain yield, whereas grain yield had highly significant correlation with harvest index (Table 3).

Disease response

The genotypes collected from Mirali (ECDOCK) NWA, Mirali North Waziristan Agency-1, FR-Bannu-1, Mirali North Waziristan Agency-2, Bannu Proper1 and FR-Bannu Domel-1, were resistance to rust caused by *Puccinia triticina*, while the remaining were affected at the range of 9.2% and considered as escaping lines. All the check varieties (from serial # 31-44) were highly susceptible to rust disease and they were averagely affected at 78.5% (Table 4). After artificial inoculation, the lines showed immunity to pathogen was genetically resistant and hence genotypes (Mirali (ECDOCK) North Waziristan Agency, Mirali North Waziristan Agency-1, FR-Bannu-1, Mirali North Waziristan Agency-2, Bannu Proper-1 and FR Bannu Domel -1) were selected as immune lines.

Molecular characterization

The SDS-PAGE was carried out on 15% polyacrylamide gel concentration. To check reproducibility of the polypeptide bands, the experiment was repeated twice and only consistent markers were reported. The binary data was analyzed for similarity indices to investigate proteomic homology using Jacquard's similarity index. The genotypes at 83% coefficient of similarity were divided into two main lineages, while the genotypes at 90% coefficient of similarity were divided into five clusters. In cluster analysis, cluster 1 and 2 explained 89%, cluster 1 with 2 and 3 indicated 84%, and cluster 3 and 4 resolved 86%; whereas cluster 4 with 3 and 5 explained 86% genetic similarity in the glutenin protein profile (Figure 1). Based on HMW glutenin, Jaccard's similarity coefficient was calculated, which showed that the genotypes of Bannu Proper-1, FR Bannu Marghali Pirbakhel and Mirali North Waziristan Agency-4 were similar to Raj. While the genotypes of Mirali (ECDOCK) North Waziristan Agency, Mirali North Waziristan Agency-1, Bannu Plain Areas-2, FR Bannu Domel-1, FR Bannu Domel-3, Miranshah North Waziristan Agency-1, Miranshah North Waziristan Agency-2, Bannu Proper-2 Waziristan Agency-2 and Waziristan North Waziristan Agency were less (80%) similar to Raj. Among the germplasm, the genotype of Bannu Proper-1, Mirali (ECDOCK) North Waziristan Agency, Mirali North Waziristan Agency-1, and FR Bannu Domel-1

Table 3. Correlation coefficient of 44 wheat genotypes/cultivars among quantitative traits cultivated during 2008 to 2009.

S/No.	Correlation	Plant height	Spike length	Awn length	1 leaf length	2 leaf length	3 leaf length	4 leaf length	Biomas/5	100-grain wt.	Grain yield/5
1	Spike length	0.25*									
2	Awn length	0.32*	0.28*								
3	1 leaf length	0.64**	0.23*	0.31*							
4	2 leaf length	0.58**	0.22*	0.15*	0.76**						
5	3 leaf length	0.61**	0.13*	0.09*	0.66**	0.94**					
6	4 leaf length	0.31*	0.13*	0.05	0.32*	0.65**	0.77**				
7	Biomas/5	0.71**	-0.01	0.15*	0.70**	0.79**	0.77**	0.54**			
8	100-grain wt.	-0.32*	-0.23*	0.08	-0.35*	-0.59**	-0.56**	-0.36*	-0.29*		
9	Grain yield/5	0.79**	0.07	0.14	0.58**	0.61*	0.64**	0.53**	0.83**	-0.19	
10	Harvest index	-0.02	0.08	-0.06*	-0.31*	-0.35*	-0.30*	-0.11*	-0.40**	0.1	0.13

*significant at the 0.05 probability level; ** significant at the 0.01 probability level.

were resistant to disease. Therefore, attention is needed to recommend these varieties for cultivation. The genotypes ofFR Bannu-1, FR Bannu-2, FR Bannu (Dry Area), FR Bannu Domel-2, North Waziristan Agency-1, Mirali North Waziristan Agency-2, FR Bannu Marghali Pirbakhel (Inqilab), Bannu Plain Areas-1, FR Bannu-3, Mirali North Waziristan Agency-3, North Waziristan Agency-3, Mirali North Waziristan Agency-5, Bannu-1 and Bannu-2 were 60% similar with Raj. Among these germplasm, the genotypes of Mirali North Waziristan Agency-2 and FR Bannu-1 were resistant to disease. Therefore, genes should be transferred from these resistance varieties to high yielding susceptible varieties through conventional breeding (Table 5).

The Jaccard's similarity coefficient for Bakhtawar-92 was also calculated for proteomic homology. It was calculated that the genotypes of FR Bannu (Dry Area), FR Bannu Marghali Pirbakhel, FR Bannu Domel-1, FR Bannu Domel-2, Mirali North Waziristan Agency-3, North Waziristan Agency-3, Bannu-1 and Bannu-2 displayed 100% similarity with Bakhtawar-92. Similarly, the genotypes of FR Bannu Domel-3, Bannu Proper-2, North Waziristan Agency-2 and Waziristan North Waziristan Agency showed 80% homology. Among them, the genotypes of Mirali North Waziristan Agency-2 and FR Bannu Domel-1 were resistant. Genotypes of the North Waziristan Agency-1, Mirali North Waziristan Agency-2, Bannu Proper-1, Bannu Plains Areas-1, FR Bannu-3, FR Bannu Marghali Pirbakhel (Inqilab), FR Bannu Marghali Pirbakhel, Mirali North Waziristan Agency-4, Mirali North Waziristan Agency-5 , FR Bannu-1 and FR Bannu-2, showed 60% and that of Mirali (ECDOCK) North Waziristan Agency-1, Mirali North Waziristan Agency-1 and Bannu Plain Areas-2 showed 40% proteomic homology with Bakhtwar-92. Among these varieties Mirali (ECDOCK) North Waziristan Agency-1, Mirali North Waziristan Agency-1, Mirali North Waziristan Agency-2, FR Bannu-1 and Bannu Proper-1 were resistant (Table 5).

DISCUSSION

A low level of allelic variation was observed in qualitative traits, while the quantitative traits showed higher magnitude of variation among landraces as compared to improved lines. Green plant colour existed predominantly in landraces, while the improved lines were comparatively dark green. The qualitative traits are frequently used for varietals identification and germplasm cataloging. Due to plant vigour, all the improved lines were vigorous and similarly, five genotypes were identified. Wheat plants with better vigour have been reported as promising and high yielding. Sixty eight percent of the total population had a plant height ranging from 42 to 60 cm, while 23% were 61.1 to 68.9 cm height. Plant height exhibited a positive correlation with leaf length, biomass and grain yield. Thirty percent of the total population had 8.06 to 8.6cm spike length and 50% ranged from 6.1 to 7.94 cm. Spike length showed a positive correlation with awn length (Spagnoletti and Qualset, 1987).

Table 4. Immune response against rust disease reported in Pakistani wheat germplasm collected from North Waziristan Agency.

Row No.	DS after 87 days	DS after 120 days	Plants resistance /row	Susceptible plants/row	% Susceptibility during harvesting
1	Absent	Absent	34	0	0
2	Absent	Absent	23	0	0
3	Absent	Present	36	2	5.26
4	Present	Present	34	1	2.85
5	Absent	Absent	16	0	0
6	Absent	Absent	37	0	0
7	Absent	Present	32	2	5.88
8	Absent	Absent	21	0	0
9	Absent	Present	28	2	6.66
10	Present	Present	42	5	10.6
11	Absent	Present	36	10	21.7
12	Absent	Present	42	3	6.66
13	Absent	Absent	38	0	0
14	Absent	Present	56	10	15.15
15	Absent	Present	44	4	10
16	Absent	Present	38	8	17.39
17	Absent	Present	40	3	6.97
18	Absent	Present	42	7	14.28
19	Absent	Present	23	18	43.9
20	Absent	Present	42	1	2.32
21	Absent	Present	27	11	28.94
22	Present	Present	31	2	6.06
23	Absent	Present	33	7	17.5
24	Absent	Present	23	9	28.12
25	Present	Present	19	8	29.62
26	Present	Present	13	9	34.61
27	Present	Present	19	8	29.62
28	Present	Present	13	12	48
29	Present	Present	NA	25	53.19
30	Present	Present	14	17	54.83
31	Present	Present	6	14	70
32	Present	Present	11	16	59.25
33	Present	Present	NA	22	61.11
34	Present	Present	13	21	61.76
35	Present	Present	2	20	90.9
36	Present	Present	2	21	91.3
37	Present	Present	0	17	100
38	Present	Present	NA	25	100
39	Present	Present	0	23	100
40	Present	Present	0	24	100

DS-disease status; absent - rust disease absent; present - rust present after specific days.

The leaf length showed positive correlation with biomass and grain yield that indicated influence of photosynthesis and thus providing the greatest proportion of photosynthates to the developing grain that had also been reported by Dammania et al. (1996). High level of genetic diversity was observed in leaf length, 100 grain mass and harvest index that indicated the scope of improvement for these traits and the selected genotypes from this germplasm could be used for wheat improvement program. Due to positive linear relationship of biomass with grain yield, five percent germplasm ranging from 35.42 to 38.04 gm for biomass is suggested to exploit

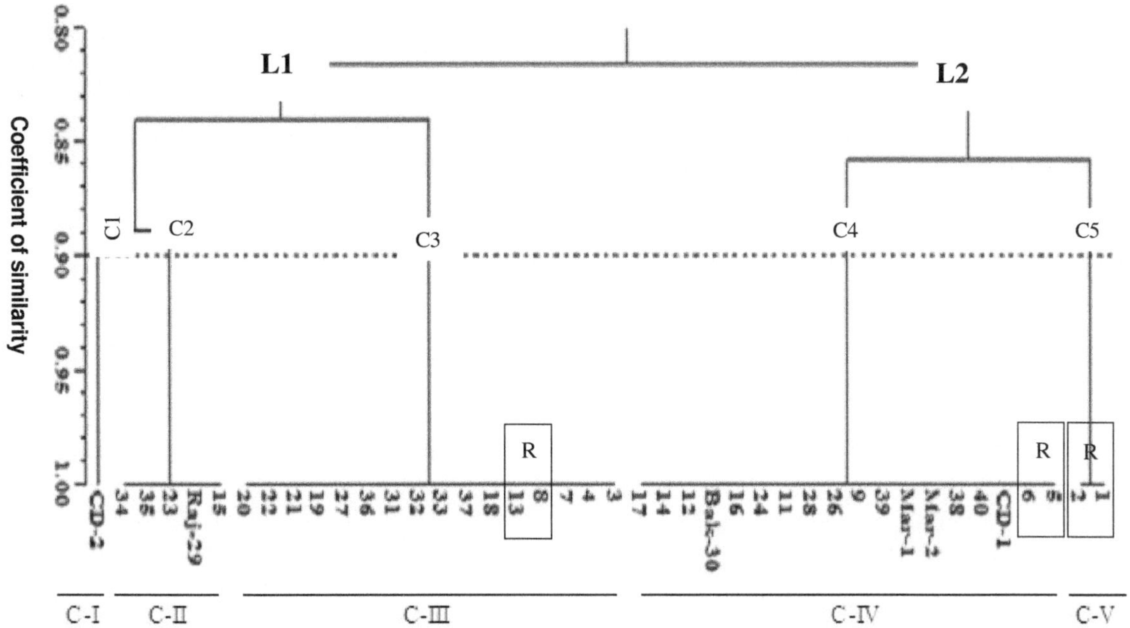

Figure 1. Cluster analysis of 43 lines collected from different parts of Pakistan based on high molecular weight HMW glutenin protein.

Table 5. Jaccard's similarity coefficient (genetic similarity-GS) between genotypes Raj and Bakhtawar-92 on the basis of HMW glutenin protein.

S/No.	Genotypes	GS-Raj (%)	GS- Bakh-92 (%)
1	Mirali (ECDOCK) North Waziristan Agency	80	40
2	Mirali North Waziristan Agency-1	80	40
3	North Waziristan Agency-1	60	60
4	FR Bannu Marghali Pirbakhel (Inqilab)	60	60
5	FR Bannu-1	60	60
6	Mirali North Waziristan Agency-2	60	60
7	FR Bannu-2	60	60
8	Bannu Proper-1	100	60
9	Bannu plain areas-1	60	60
10	Bannu plain areas-2	80	40
11	FR Bannu-3	60	60
12	FR Bannu (dry area)	60	100
13	FR Bannu Domel -1	60	10
14	FR Bannu Domel -2	60	10
15	FR Bannu Marghali Pirbakhel	100	60
16	Mirali North Waziristan Agency-3	60	100
17	FR Bannu-4	60	100
18	FR Bannu Domel -3	80	80
19	Miranshah North Waziristan Agency-1	80	80
20	Miranshah NW North Waziristan Agency-2	80	80
21	Bannu Proper-2	80	80
22	North Waziristan Agency-2	80	80
23	Mirali North Waziristan Agency-4	100	60
24	North Waziristan Agency-3	60	100
25	Mirali North Waziristan Agency-5	60	60

Table 5. Continued.

26	Bannu-1	60	100
27	Waziristan North Waziristan Agency	80	80
28	Bannu-2	60	100
31	Check 31 (Bakkar)	80	80
32	Check 32 (Watan)	80	80
33	Check 33 (Saleem-2000)	80	80
34	Check 34 (Yecura)	100	60
35	Check 35	100	60
36	Check 36	80	80
37	Check 37	80	80
38	Check 38	60	100
39	Check 39	60	100
40	Check 40	60	100
41	Mardan-1 (cultivars)	60	100
42	Mardan-2 (cultivars)	60	100
43	Chakdara-1 (cultivars)	60	100
44	Chakdara-2 (cultivars)	80	80

for breeding wheat cultivars with higher biomass that could effectively partition in to economic yield. In the present study, new source of rust resistance genotypes namely *Mirali* (ECDOCK) North Waziristan Agency, Mirali North Waziristan Agency-1, *FR Bannu-1*, *Mirali North Waziristan Agency-2*, *Bannu Proper-1and FR Bannu Domel-1*, were identified. These lines are suggested to investigate against various races of rust for their wider utilization that will ultimately help to combat serious threats due to rust for wheat production. It was also suggested that molecular characterization of this germplasm may be accorded to select new markers against rust disease.

The analysis of HMW glutenin subunits in wheat is useful tool not only for diversity studies but also to optimize the variation in germplasm collections and, to breed cultivars with improved bread-making quality (Sultana et al., 2007). Our results revealed a preliminary investigation on HMW glutenin genes in the landraces collected from a remote area and several novel alleles detected at all the *Glu-1* loci could help in improving wheat quality. Generally, the "null" allele at the *Glu-A1* locus is predominantly observed in hexaploid wheat cultivars and the germplasm accessions (Cross and Guo, 1993). We also observed a similar pattern in wheat landraces possessing "null" allele followed by the allele 2* and 1. This contrasting pattern of *Glu-A1* allele distribution between wheat cultivars and the landraces is an example of genetic erosion caused by man (Martin et al., 2008; Caballero et al., 2009). Detection of novel *Glu-1* alleles in the germplasm in both the hexaploid and tetraploid wheat had been reported by Lagudah et al. (1987) and Ciaffi et al. (1993)

Another interesting allelic variant encoded at the *Glu-D1* locus was the subunit pair 2**+12' with slow mobility as compared to the standard subunits (2+12) at this locus and, was found frequently distributed in the accessions collected from the Baluchistan province of Pakistan (Tahir et al., 1996). The *Glu-D1xy* (D1x+ D1y) variants with slower mobility as compared to 2+12 were also detected in some hexaploid landraces from Afghanistan and this protein was quite expected in the present landraces. The allelic variants (*Glu-D1xy*) detected in the germplasm from Afghanistan seem similar to the subunits 2**+12' detected in our study on the basis of description in the Cross and Gou (1993). The material being cultivated in Afghanistan might share common parentage/landraces.

In conclusion, wheat germplasm reported in this manuscript was collected from a very remote area of Pakistan that has never been explored. These genotypes showed high range of genetic divergence and significant response to rust diseases. New wheat genotypes, resistant to rust disease, were identified, which posses' new allele for rust response and further investigation of the sources through molecular markers is needed. Polymorphism for HMW glutenin loci was observed that could be associated with collection site of landraces. Since the HMW glutenin subunits indicated higher bread-making quality, hence the variation at these loci is essential for the plant breeders to develop cultivars with improved bread-making quality. The preliminary information was quite encouraging and genetic markers must employ to broaden the horizon for germplasm utilization.

REFERENCES

Benmoussa ML, Vezina P, Page M, Yelle S, Laberge S (2000). Genetic polymorphism in low-molecular-weight glutenin genes from *Triticum aestivum*, variet Chin. Spring Theory Appl. Genet., 100(5): 789-793.

Caballero L, Peña RJ, Martín LM, Alvarez JB (2009). Characterization of Mexican Creole wheat landraces in relation to morphological characteristics and HMW glutenin subunit composition. Genet. Resour. Crop Evol., 10.1007/s10722-009-9501-8.

Caballero LM, Martın A , Alvarez JB (2009). Genetic diversity for seed storage proteins in Lebanon and Turkey populations of wild diploid wheat (*Triticum urartu* Thum. ex Gandil.) Genet. Resour. Crop Evol., DOI 10.1007/s10722-009-9434-2.

Ciaffi M, Lafiandra D, Porceddu E, Benedettelli S (1993). Storage-protein variation in wild emmer (*Triticum turgidum* ssp.dicoccoides) from Jordan and Turkey 11 Patterns of allele distribution. Theory Appl. Genet., 8: 518-525.

Cross RJ, Guo B (1993). Glutenin variation in adverse pre-1935 world wheat germplasm collection. In: Biodiversity and Wheat Improvement (eds.) Damania AB, ICARDA Syria.

Dammania AB, Pecetti L, Qualset CO, Humeid BO (1996). Diversity and geogrtaphic distribution of adaptive traits in *Triticum turgidum* L. (durum group) wheat landraces from Turkey. Genet. Resour. Crop Evol., 43: 409-422.

Dotlacil L, Gregova E, Hermuth J, Stehno Z, Kraic J (2002). Diversity of HMW Glu alleles and evaluation of their effects on some characters in winter wheat land races and old cultivars. Czech J. Genet. Plant Breed., 38(3-4): 109-116.

Gupta RS, Tiwari DK, Doel SS, Singh RP (2002). Genetic divergence in bread wheat (*Triticum aestivum* L. em Thell). New Bot., 60: 237-243.

Kronstad WE (1998). Agricultural development and wheat breeding in 20th century. Braun HJ et al. (eds) Wheat prospects for global improvement, pp. 1-10.

Laemmli UK (1970). Cleavage of Structural Proteins during the Assembly of the Head of Bacteriophage T4 Nat., 227: 680–685.

Lagudah ES, Floor RG, Halloran GM (1987). Variation in high molecular weight glutenin subunits in landraces of hexaploid wheat from Afghanistan. Euphytica, 36: 3-9.

Martin MA, Martin LM, Alcarez JB (2008). Polymorphism at the Gli-A^u 1 and Gli- A^u 2 in wild diploid wheat (*Triticum urartu*). Euphytica, 163: 303-307.

Pawar SV, Patil SC, Naik RM, Jambhale VM (2003). Genetic variability and heritability in wheat. J. Maharashtra Agric. Uni., 27(3): 324-325.

Popa M, Gregova E, Kraic J (2006). Romanian Wheat (*T. aestivum* L.) landraces characterized by seed storage-protein electrophoresis. PGR newsletter, 135: 53-58.

Shah MI, Jabeen M, Ilahi I (2003). *In vitro* callus induction, its proliferation and regeneration in seeds explants of wheat (*T. aestivum* L.) Var lu-26s. Pak. J. Bot, 35(2): 209-217.

Sneath PHA, Sokal RR (1973). Numerical Taxonomy: The Principals and Practical of Numerical Classification. San Francisco: Freeman.

Spagnoletti APL, Qualset CO (1987). Geographical diversity for quantitative spike characters in world collection of durum wheat. Crop Sci., 34: 774-783.

Sultana T, Ghafoor A, Ashraf M (2007). Genetic variability in bread wheat (*Triticum aestivum* L.) of Pakistan based on polymorphism for high molecular weight glutenin subunits. Genet. Resour. Crop Evol., 54: 1159-1165.

Tahir M, Turchetta T, Anwar R, Lafiandra D (1996). Assessment of genetic variability in hexaploid wheat land races of Pakistan based on polymorphism for HMW glutenin subunits. Genet. Resour. Crop Evol., 43: 211-220.

Wiser H (2007). Chemistry of Gluten Proteins. Food Microb., 24(2): 115-119.

Wrigley C, Bekes F, Bushuk W (2006). Gliadin and glutenin: the unique balance of wheat quality. AACC International Press, St Paul, MN.

Evaluation of pesticide tolerant strain of *Cotesia flavipes* Cameron (Hymenoptera: Braconidae) on maize stem borer, *Chilo partellus* (Swinhoe)

K. Srinivasa Murthy, R. Rajeshwari, S. K. Jalali and T. Venkatesan

Division of Insect Molecular Entomology, National Bureau of Agriculturally Important Insects, P.B.No. 2491, H. A. Farm Post, Bellary Road, Bangalore - 560 024, Karnataka, India.

Field evaluation of pesticide tolerant strain of *Cotesia flavipes* Cameron, a gregarious larval endoparasitoid of the maize stem borer, *Chilo partellus* (Swinhoe) was carried out and the economics of such evaluation was worked out. Releases of the pesticide tolerant parasitoid strain, prior to application of pesticide or after application resulted in significantly higher net returns and yield, as compared to the releases of susceptible parasitoid strain and untreated control. Four releases of the tolerant parasitoid at 2000/ha at 500 per release registered maximum number of parasitized cocoons (52.6 per larva), higher yields (4019 kg/ha) and net profits (Rs. 7807 kg/ha), followed by two releases of the tolerant strain with two sprays of insecticide (40.8, 3809 kg/ha and Rs. 6847/ha, respectively). The insecticidal treatment gave higher net profits and yields than all other treatments. However, the additional costs, ecological hazards and pesticide residues (likely to occur) do not outweigh the benefits of biological control.

Key words: *Cotesia flavipes, Chilo partellus*, population, pesticide tolerance.

INTRODUCTION

Lepidopteran stem borers are among the most damaging pests of maize in the semi-arid countries and are a major constraint in realizing the yield potential of the crop. In India, the stem borer, *Chilo partellus* (Swinhoe) (Crambidae: Lepidoptera) is one of the most serious insect pests causing 24.3 to 36.3% loss in different agroclimatic regions of the country (Ganguli et al., 1997). A large number of insecticides belonging to different groups viz., the organochlorines, organophosphates, carbamates and synthetic pyrethroids have been widely used to curtail the pest (Dharmasena, 1993; Teli et al., 2007) with limited success. Chemical control is often effective on the neonate instars, before the larvae enter the stem (Reyes, 1987), therefore the timing of application is crucial for the successful management of the pest (Nwanze and Mueller, 1987). Indiscriminate use of pesticides had resulted in several ecological hazards; therefore bio-intensive pest management strategies have been advocated. Among the biocontrol agents, *Cotesia flavipes* (Cameron) (Braconidae: Hymenoptera) a gregarious endo larval parasitoid has been widely used in maize, sorghum and sugarcane ecosystems (Kalra and Srivatsava, 1996; Tanwar and Varma, 2002). The parasitoid is reported to cause 32 to 55% decrease in stem borer densities in various crops (Kfir et al., 2002). Parasitism up to 40% on sugarcane borer in Texas (Setamou et al., 2003), 11 to 21% in Brazil and 17 to 21% in Africa on Maize stalk borer *Busseola fusca* (Ngi-Song et al., 2012) was reported. In India, the extent of parasitism was 36.67%in Pantnagar (Choudhary and Sharma, 1987), 35 to 50% in Himachal Pradesh (Nirmala Devi and Desh raj, 1996), 6.6 to 21% in Assam (Borah and Arya, 1995) and 2 to 33% in Haryana (Mohan et al., 1999). The extent of parasitism is often low, due to their susceptibility to insecticidal applications on the target pest (Songa et al., 1999). Therefore, the use of tolerant strain of the parasitoid would result in effective suppression of the pest. A pesticide tolerant population of the parasitoid (*C. flavipes*) was developed and evaluated under the field conditions on the maize stem borer, *C. partellus*. The laboratory experiments were conducted at the Division of insect Molecular Entomology, National Bureau of Agriculturally Important Insects and the field experiments were carried out at the Biocontrol Research Farm of the institute at Attur in Bangalore, India during the year 2009-2010.

MATERIALS AND METHODS

Rearing and maintenance of culture of the stem borer and the parasitoid

Field populations of *C. partellus* were collected from different geographic locations (Aurangabad, Bangalore, Chindwara, Devaganahalli, Dindigul, Hoishiarpur, Malur and Shimla) in India. The populations were maintained in the laboratory on semi-synthetic diets as prescribed by Ballal et al. (1995). The larvae were reared individually in glass vials (10 x 2.5 cm) at ambient temperature of 26 ± 1°C and 65% RH. The populations maintained in the laboratory were utilized for toxicity bioassay, rearing of the parasitoid and field experiments. Fourth-fifth instar larvae of *C. partellus* were offered to two day old adults of the parasitoids kept in plastic jars (10 x 15 cm), covered with muslin cloth. Cotton swab soaked with honey (30%) was placed inside the jar as food for the parasitoids. The larvae were exposed to the parasitoid for two hours and the parasitized larvae were individually transferred into glass vials (10 x 2.5 cm) containing the semi-synthetic diet and were reared till the formation of cocoons. The cocoons were then transferred into another vial for adult emergence.

Laboratory evaluation of pesticide tolerant strain of *C. flavipes*

In vitro toxicity bioassays were carried out with the widely used insecticides in maize ecosystem (fenvalerate and endosulfan), on field populations of *C. flavipes* obtained from different geographic populations of the *C. partellus* (as mentioned above), Laboratory bioassays were carried out by dry film method (Kao and Tzeng, 1992; Keimeesuke et al., 1994). The insecticidal spray was prepared by taking dosages higher than field recommended dosages and corresponding serial dilutions were made to test the level of toxicity to the insecticides. Thus, the concentrations for endosulfan were: 21.88, 43.75, 87.50, 175.0, 300.0 and 750 ppm and for fenvalerate the dosages were 1.26, 2.5, 5.0, 10.0, 25 and 40 ppm. The assays were carried out following the IOBC protocols (Hassan et al., 1985).

The insecticides were prepared as aqueous solutions based on the dosages and sprayed with the help of atomizer. Modified glass tube with both sides open was used as testing unit. The sprayed glass tubes were dried in the shade. After drying, one end of tube was covered with double-layered black cloth and was fastened with rubber bands. From the other side, about 50 adults were released in each vial in each concentration. There were 10 replications for each treatment.

The mortality of the parasitoid was recorded after 15 and 30 min and after 1, 2, 4, 6 and 24 h of constant exposure. The data obtained on mortality were subjected to probit analysis by statistical program SPSS version 8.0. The data were transformed to log base 10 before probit analysis and antilog of calculated values gave actual LC_{50} and LC_{90}. The fiducial limits slope and χ^2 values were also calculated.

Field evaluation of pesticide tolerant strain of *C. flavipes*

A field experiment was conducted during 2009-10, rabi season (October to February) to evaluate the pesticide tolerant strain of *C. flavipes* (Shimla population) at the Biocontrol Research Farm, Attur (Bangalore). A composite maize (variety: Varalakshmi), was sown in randomized block design following the recommended package of agronomic practices. There were eight treatments and three replications with a plot size of 5 x 4 m each. The following treatments were imposed; T1, Four sprays of fenvalerate (0.1 ml/L); T2, Four releases of PTS of *C. flavipes* (Shimla population); T3, four releases of susceptible *C. flavipes* population; T4, two sprays of fenvalerate + 2 releases of tolerant *C. flavipes;* T5, two sprays of fenvalerate + 2 releases of susceptible *C. flavipes*; T6, two releases of tolerant *C. flavipes* + 2 sprays of fenvalerate; T7, two releases of susceptible *C. flavipes* + 2 sprays of fenvalerate and T8, untreated check (control).

Release of the stem borer and parasitoids

The population of *C. partellus* maintained in the laboratory was utilized for the field experiments. The different treatments were created by releasing 10 first instar larvae into the leaf whorl of maize plants with the help of "0" sized camel hair brush. Five plants (tagged) were chosen at random from three central rows of each plot. Application of the insecticide and releases of parasitoid as per the treatments scheduled were done after 3 to 4 days of larval implantation as advocated by Teli et al. (2007). The parasitoids were released in the field as adults in release stations as prescribed by Songa et al. (2001). The treatment of release of parasitoid following the insecticidal sprays was done immediately after the scheduled spray. Darkened cocoon masses (likely for adult emergence) obtained from the laboratory culture (Shimla population) were placed in plastic Petri plates (9 x 1.5 cm) lined with moist filter paper and then placed in the centre of each plot. The parasitoids were released at 2000 cocoons/ha which works out to 500 cocoons per release, for an individual plot size of 5 x 4 m replicated thrice. There were two release stations per plot. The various treatments were far apart by a distance of 3 m on each side. Releases were not a onetime application since the insect pest, *C. partellus* has over lapping generations and the parasitism is more effective on the 2-3rd instar larvae of the host. The releases commenced at about three weeks after emergence of the maize plants to synchronise the host larval stages (2nd to 3rd instar larva). The study included the releases of the parasitoid to be made either prior to the application of the insecticide and vice versa, to generate information on the pesticide tolerant strain of the parasitoid and its efficiency in terms of parasitism as compared to the susceptible strain.

The parasitism was noted by destructive sampling. Observations were recorded on percent dead hearts, number of cocoons formed per parasitized larvae and yield (kg/ha) in each of the treatments and compared with the untreated control. Parasitized larvae obtained from the different treatment plots were collected, taken to the laboratory and reared individually in glass vials containing artificial diets. The results after angular transformation were subjected to statistical scrutiny by ANOVA. The cost economics of the different treatments was worked out.

Table 1. Field evaluation of tolerant strain of *C. flavipes* on *C. partellus*.

Treatment	Per cent dead hearts	(No. of parasitized cocoons/larva)	Yield (kg/ha)
Four sprays of Fenvalerate	5.72[a]	---	4280[f]
Four releases of tolerant *C.flavipes*	27.5[c]	52.6[g]	4019[e]
Four releases of susceptible *C.flavipes*	22.2[bc]	21.8[c]	3719[c]
Two sprays of Fenvalerate + two releases of tolerant *C. flavipes*	18.6[b]	28.3[d]	3768[c]
Two sprays of Fenvalerate + two releases of susceptible *C. flavipes*	26.2[c]	17.2[b]	3702[c]
Two releases of tolerant *C. flavipes* + two sprays of Fenvalerate	32.6[d]	40.8[f]	3809[d]
Two releases of susceptible *C. flavipes* two sprays of Fenvalerate	38.4[e]	33.2[e]	3692[b]
Untreated check (Control)	79.98[f]	10.6[a]	2448[a]
CD (5%)	5.22	3.57	107.89
SEM	2.283	1.17	67.51
CV (%)	12.9	7.98	4.73

Mean of three replications.

RESULTS

The laboratory bioassays indicated that population of the parasitoid from Shimla was more tolerant (with LC_{50} 2.77 ppm) as compared to others (Coimbatore LC_{50} 2.24 ppm, Chindwara LC_{50} 1.98 ppm). The population from Shimla cons obtained from different locations, the population tolerant was evaluated under field conditions

Efficacy of pesticide tolerant strain (PTS) of *C. flavipes* on maize stem borer

The efficiency of PTS of the parasitoid on the stem borer *C. partellus* in the various treatments revealed that higher percentage of dead hearts (dead central shoot of the plant was formed as a result of feeding by the stem borer) was observed in the releases of the susceptible parasitoid strains than releases of tolerant strains of the parasitoid, either alone or in combination with pesticides. Releases of the susceptible strains of the parasitoid followed by insecticidal sprays recorded 38.4% dead hearts, while releases made after the insecticidal sprays registered 26.2% dead hearts. Interventions with insecticides alone recorded the lowest percent of dead hearts (5.72%).

Effective parasitism of the stem borer was observed in the treatments, where releases of the parasitoid were followed by insecticidal sprays in both the tolerant and susceptible populations of the parasitoid. Exclusive releases of the tolerant parasitoid population (T2), however recorded the maximum number of parasitized cocoons (52.6 per larva). Two releases of the tolerant parasitoids followed by insecticidal sprays recorded 40.8 cocoons/larva, while a low parasitism (17.2 cocoons/ larva) was recorded in the treatment of insecticidal spray followed by releases of the susceptible parasitoid populations (Table 1).

The treatment with two sprays of fenvalerate followed by two releases of the tolerant parasitoid population recorded lower proportion of dead hearts (18.6), while releases of the susceptible parasitoid followed by insecticidal sprays registered the maximum percentage of dead hearts (38.4) (Table 1). However, spray of insecticide alone recorded the least percent dead hearts.

Cost Economics of the field evaluation of PTS

Significant differences in the yields obtained were observed in the different treatments. Releases of the susceptible parasitoid population followed by insecticidal sprays recorded low yield (3692 kg/ha). Four releases of the tolerant population of the parasitoid alone gave higher yield (4019 kg/ha) followed by two releases of the tolerant parasitoid populations with two sprays of insecticides (3809 kg/ha), while the control plots registered the minimum yield (2448 kg/ha).

The cost economics of the field evaluation of the tolerant strain of *C. flavipes* was worked out considering the market value of the produce, cost of the inputs and labour to arrive at the incremental cost-benefit ratio (ICBR) for the various treatments. Among the different treatments, four releases of *C. flavipes* recorded higher net profits (Rs. 7807 or 4,21,578 USD), additional returns (Rs. 10,997 or 5,92,758 USD) and yields (4019 kg/ha), followed by the treatment with two releases of tolerant *C. flavipes* along with two sprays of fenvalerate where the

Table 2. Economics of field evaluation of tolerant strain of *C .flavipes* on *C. partellus*.

Treatment	Yield (kg/ha)	Market value (Rs.700/Q)	Addl. returns Over control	Total cost/ha (Rs.)	Net profit/ha (Rs)	ICB ratio
Four sprays of Fenvalerate	4280	29960 (16,17,840)	12824 (6,92,696)	3254 (1,75,716)	10650 (5,75,100)	1:3.27
Four releases of tolerant *C. flavipes*	4019	28133 (15,19,182)	10,977 (5,92,758)	3170 (1,71,180)	7807 (4,21,578)	1:2.46
Four releases of susceptible *C. flavipes*	3819	26773 (14,45,742)	9597 (5,18,238)	3170 (1,71,180)	6427 (3,47,058)	1:2.02
Two sprays of Fenvalerate + two releases of tolerant *C. flavipes*	3768	26376 (14,24,304)	9240 (4,98,860)	2680 (1,44,720)	6560 (3,54,240)	1:2.44
Two sprays of Fenvalerate + two releases of susceptible *C. flavipes*	3702	25914 (13,99,356)	8778 (4,74, 012)	2680 (1,44,720)	6098 (3,29,292)	1:2.27
Two releases of tolerant *C. flavipes* + two sprays of Fenvalerate	3809	26663 (14,39,802)	9527 (5,14,458)	2680 (1,44,720)	6847 (3,69,738)	1:2.55
Two releases of susceptible *C. flavipes* two sprays of Fenvalerate	3692	25884 (13,97,736)	8748 (4,72,392)	2680 (1,44,720)	6068 (3,27,672)	1:2.26
Untreated check (Control)	2448	17136 (9,25,344)	-	1598 (86,892)	-	-

Figures in parenthesis correspond to value in US $).

net profits, additional returns and yields were Rs. 6847, Rs. 9527 and 3809 kg/ha, respectively. However, ICBR (1:2.55) was higher in the latter. Exclusive insecticidal applications with four sprays of fenvalerate resulted in higher yields (4280 kg/ha), and additional returns (Rs. 12,824 or 6, 92,496 USD) with highest ICBR (1:3.27) over the rest of the treatments (Table 2).

DISCUSSION

Our studies have indicated that pesticide tolerant strain of the braconid parasitoid, *C. flavipes* proved more effective in the suppression of the maize stem borer, *C. partellus*. Utilisation of the parasitoid for the management of the pest was carried out with releases made either prior to or after the application of insecticidal sprays. The parasitism was more effective when releases of the parasitoid were made preceeding spray application of insecticides.

Parasitization by *Cotesia* spp. induces host immune suppression in the sugarcane borer (*Diatraea saccharalis*) and the Diamond back moth *Plutella xylostella*. The immune suppression could impair insect defence capacity against pathogens and insecticides (Mahmoud et al., 2012). The immune suppression factors include the polyadnavirus, venoms, proteins and teratocytes. The parasitism may cause changes in insecticide-susceptibility of the parasitized host larvae, and that insecticide tolerance may result from changes in metabolic activities of enzymes involved. The enhanced activity of detoxifying enzymes such as glutathione-*S*-transferase (GST), cytochrome P450 monooxygenase (CYP) and carboxyesterase (CE) were proven to be responsible for many cases of alteration in insecticide-susceptibility (Tomoko et al., 2006).

In addition, the effective host searching ability, establishment and development of the parasitoid in a pesticide free habitat coupled with tolerance to the pesticide applied later, had played role in the reduction in the percentage dead hearts and increased number of parasitoid larvae than in other treatments. Further, changes in the host selection behavior (either due to repelling odors or avoidance) by the parasitoid due to application of the insecticide prior to their releases (treatments T4 and 5), would have resulted in reduced parasitism. Successful parasitisation includes host finding based on the frass excreted by the tunnelling larvae and the probe and biting strategy by the parasitoid (White et al., 2004). The chemical cues present in stemborer larvae are also involved in host recognition and acceptance (Meshack et al., 2010). The inability of the parasitoid to properly discriminate the cues due to the altered receptivity, after the application of the insecticide had contributed to the decreased parasitism.

Higher parasitism (parasitized cocoons/larva) due to *C. flavipes* contributed to higher yields in the various treatments. Low percent dead hearts was observed with the use of synthetic pyrethroids (Deltamethrin) in maize (Deepti and Sharma, 2004; Teli et al., 2007), in sweet sorghum (Deepti and Patil, 2008) and sugarcane (Tanwar and Varma, 2002),

In all the treatments, releases of the susceptible strain

of the parasitoid either alone or in combination with insecticide registered lower net profits, returns and yield as compared to those with releases of the tolerant strain. All the treatments were superior to the control. The additional costs and net returns obtained in the insecticidal treatment, coupled with environmental hazards and residues, do not outweigh the benefits of ecofriendly biocontrol agents in our attempts to manage the insect pest in a sustainable manner. Although, the releases of the parasitoid either alone or in combination with pesticides resulted in a relatively lower accrued benefits, the increased parasitism in the absence of the insecticide or its reduced quantum, could eventually contribute to higher net returns and profits over a period of time.

REFERENCES

Ballal CR, Kumar P, Ramani S (1995). Laboratory evaluation, storability and economics of an artificial diet for rearing *Chilo partellus* (Swinhoe) (Lepidoptera: Pyralidae). J. Ent. Res. 19:135-141.

Borah BK, Arya MPS (1995). Natural parasitisation of the Sugarcane Plassey borer (*Chilo tumidicostalis* Hmpsn) by braconid larval parasitoid in Assam. Ann. Agril. Res. 16:362-363.

Choudhary RN, Sharma VK (1987). Parasitisation in diapausing larvae of *Chilo partellus* (Swinhoe) by *Apanteles flavipes* (Cameron). Ind. J. Ecol. 14:155-157.

Deepti P, Sharma VK (2004). Efficacy of some insecticides against major insect pests of kharif maize. Ind J. Appl Ent. 18:59-62.

Deepti J, Patil SRK (2008). Evaluation of Biorational Pesticides for the management of Stemborer, *Chilo partellus* Swinhoe in Sweet Sorghum. Kar. J. Agri. Sci. 21:293-294.

Dharmasena CMD (1993). Chemical control of maize stem borer, *Chilo partellus* (Swinhoe), Pestology, 17:25-27.

Ganguli RN, Chaudhary RN, Jayalakshmi Ganguli (1997). Effect of time of application of chemicals on management of maize stem borer, *Chilo partellus* (Swinhoe). Int. J. Pest Mng. 43:253-259.

Hassan SA, Bigler F, Blaisinger D, Bogensehutz H, Brun J, Chiverton P, Dicker E, Easterbrook MA, Edwards PJ, Englert WD, Firth SI, Hung P, Inglesfield C, Klingauf F, Kuhner C, Ledieu MS, Naton E, Oomen PA, Overmeer WPJ, Pleots P, Robonlet JN, Rieckmann W, Samsoe-Peterson L, Shives SW, Sttaubli A, Steenson J, Tusset JJ, Vanwetsinkel G, Van Zon AQ (1985). Standard methods to test the side-effects of pesticide on natural enemies of insects and mites developed by the IOBC/WPRS working group "Pesticides and beneficial organisms" Bull. OEPP/EPPO 15:214-255.

Kfir R, Overholt WA, Khan ZR, Polaszek A (2002). Biology and management of economically important lepidopteran cereal stem borers in Africa. Annu. Rev. Ent. 47:701-731.

Kalra AN, Srivastava RC (1996). Biology of *Apanteles flavipes* Cameron (Braconidae: Hymenoptera) a parasite of moth borers of sugarcane, under micro-ecological conditions. Proc. V All India Conf. on Sugarcane Research Workshop, Coimbatore. pp. 606-609.

Mahmoud AMA, De Luna-Santillana EJ, Guo X, Mario A Rodríguez-Pérez (2012). Parasitism by *Cotesia flavipes* alters the haemocyte population and phenoloxidase activity of the sugarcane borer, *Diatraea saccharalis*. Can Entomol. 44:599-608.

Meshack Obonyo AB, Fritz Schulthess C, Bruno Le Ru BDE, Johnnie van den Berg A, Jean-François Silvain DE, Paul-André Calatayud (2010). Importance of contact chemical cues in host recognition and acceptance by the braconid larval endoparasitoids *Cotesia sesamiae* and *Cotesia flavipes*. Biol. Contl. 54:270-275.

Mohan BR, Verma AN, Singh SP (1999). Periodic parasitisation of *Chilo partellus* (Swinhoe) larvae on forage sorghum in Haryana. J. Insect. Sci. 47:167-169.

Muirheed KA, Murphy NP, Sallam MN, Donnellan SC, Austin AD (2006). Mitochondrial DNA phylogeography of the *Cotesia flavipes* complex of parasitic wasps (Hymenoptera:Braconidae). Ann. Ent. Soc. Amer. 42:309-318.

Ngi-Song AJ, Kimani-Njogu S (2012). Multiple parasitism by *Cotesia flavipes* (Hymenoptera: Braconidae) on *Busseola fusca* (Lepidoptera:Noctuidae). Biocont. Sci. Tech.11:381-390.

Nirmala Devi, Desh Raj (1996). Extent of parasitisation of *Chilo partellus* (Swinhoe) on maize by *Apanteles* sp., in mid hill zone of Himachal Pradesh (India). J. Ent. Res. 20:171-172.

Nwanze KF, Mueller RAE (1987). Management options for sorghum stem borer for farmers of the semi arid tropics. Proc. Int. Workshop on sorghum stem borer, 17-20 November, ICRISAT Centre, Patancheru, A.P, India, pp.105-116.

Reyes R (1987). Sorghum stem borer in central and South Africa. Proc. Int. Workshop on sorghum stem borer, 17-20 November, ICRISAT Centre, Patancheru, A.P, India. pp.49-58.

Sarup P, Marwaha KK, Panwar VPS, Siddiqui KH (1973). Evaluation of some exotic and indigenous maize germplasm for resistance to *Chilo partellus* under artificial infestation. J. Ent. Res. 2:98-105.

Setamou M, Bernal JS, Lagaspi JC, Mirkov TE (2003). Parasitism and location of sugarcane borer (Lepidoptera: Pyralidae) by *Cotesia flavipes* (Hymenoptera: Braconidae) on Trnasgenic and conventional sugarcane. Environ. Ent. 31:1219-1225.

Songa J, Songa W, Kavoi J, Kimotho L, Nzioki S (1999). Integrated management of stem borers in maize in semi arid eatern Kenya. In Towards Increased Use of Demand Driven Technology (Ed) Sutherland J A. pp.939-395.

Tanwar RK, Varma A (2002). Field Trials with *Cotesia flavipes* Cameron against Sugarcane Borers in sub Tropical India. Sugar Tech. 4:153-156.

Teli VS, Chavan BP, Ankalkoppe MN, Khot, RB, Harers PN (2007). Evaluation of some insecticides for the control of maize stem borer, *Chilo partellus* (Swinhoe). J. Ent. Res. 31:323-326.

Tomoko Takeda, Yutaka Nakamatsu, Toshiharu Tanaka (2006). Parasitization by *Cotesia plutellae* enhances detoxifying enzyme activity in *Plutella xylostella*. Pestic. Biochem. Phy. 86:15-22.

White WH, Regan JW, Smith Jr, Salazar JA (2004). Releases of *Cotesia flavipes* (Hymenoptera: Braconidae) in to the Louisiana sugarcane ecosystem. Environ. Ent. 33:627-632.

Proximate and mineral composition of four edible mushroom species from South India

J. Manjunathan[1]*, N. Subbulakshmi[2], R. Shanmugapriya[2] and V. Kaviyarasan[1]

[1]Centre for Advanced Studies in Botany, University of Madras, Guindy Campus, Chennai-600 025, Tamil Nadu, India.
[2]Dhanalakshmi Srinivasan College of Arts and Science for Women, Perambalur-12, Tamil Nadu, India.

Results of proximate analysis of four edible species of mushroom collected from different parts of IIT Chennai, Tamil Nadu, India., indicate that *Auricularia polytricha* was a very good source of crude protein (37%), crude fiber (21.97%), ash (6.87%), calcium (607 mg/g dry weight) and manganese (136 mg/kg dry weight (dw)). *Macrolepiota rhodocus* was the richest in carbohydrate (48%) and magnesium (250 mg/g), while *Clitocybe sp* was richest in moisture content (21%), iron (1230 mg/g) and copper (9.0 mg/g). It is also a good source of carbohydrate (42%), calcium (208 mg/g) and manganese (120 mg/g). *Lentinus tigrinus* was, however, the richest in dry matter (94%), and is also rich in carbohydrate (60%), magnesium (14 mg/g) and copper (1.2 mg/g). It was observed that lipids, sodium and phosphorus contents of the four species were generally very low.

Key words: Edible mushrooms; food composition.

INTRODUCTION

Mushrooms are saprophytes. They include members of the Basidiomycota and some members of the Ascomycota. Mushrooms have been a food supplement in various cultures and they are cultivated and eaten for their edibility and delicacy. They fall between the best vegetables and animal protein source. Mushrooms are considered as source of proteins, vitamins, fats, carbohydrates, amino acids and minerals (Jiskani, 2001). All essential amino acids are present as well as water soluble vitamins and all the essential minerals (Buigut, 2002). Mushroom are good sources of vitamins like riboflavin, biotin and thiamine (Chang and Buswell, 1996). Ogundana and Fagade (1981) indicated that mushroom is about 16.5% dry matter out of which 7.4% is crude fiber, 14.6% is crude protein and 4.48% is fat and oil.

Protein contents vary between 4 to 9% in *Auricularia* sp. and between 24 to 44% in *Agaricus* species. The protein value of mushrooms is twice as that of asparagus and potatoes, four times as that of tomatoes and carrots, and six times as that of oranges (Jiskani, 2001). Their energy value also varies according to species, which is about equal to that of an apple. *Pleurotus tuber-regium* is a common species and it is useful in some combinations to cure headache, stomach ailments, colds and fever (Oso, 1977) asthma, smallpox and high blood pressure (Fasidi and Olorunmaiye, 1994; Oso, 1977), while *Lentinus tuber-regium* and *Lentinus tigrinus* are used for treating dysentery and blood cleansing respectively. *Auricularia* species have been traditionally used for treating hemorrhoids and various stomach ailments (Chang and Buswell, 1996). Chanterelles, *Boletus edulis* and *Lactarius* spp. are used for killing flies, while the puffballs are used for healing wounds (Harkonen, 1998; Delena, 1999).

They are also recommended to diabetic and anemic persons, owing to their low carbohydrate and high folic acid content. Some mushrooms are reputed to possess anti-allergic, anticholesterol, anti-tumor and anti-cancer (Jiskani, 2001). Mushrooms are now marketed along major highways and urban centers. They are also relatively much cheaper than beef, pork and chicken that contain similar nutrients. A comparative evaluation of the nutritional potentials of four edible mushrooms is reported in this paper.

*Corresponding author. E-mail: jmanjunathan@gmail.com, manikavi53@gmail.com.

Table 1. Proximate compositions of four wild mushrooms (per 100 g of sample).

Parameters	*Auricularia polytricha*	*Macrolepiota rhacodes*	*Clitocybe Sp.*	*Lentinus tigrinus*
Moisture (%)	6.48	11.10	21.0	2.51
Ash %	6.87	11.80	15.73	5.14
Crude fibre (%)	21.97	4.78	13.04	14.69
Fat (%)	0.74	2.25	1.24	2.25
Protein (%)	37	34.31	24.8	18.07
Carbohydrate (%)	38.48	48	42	60
Sodium (mg/100g)	858.4	274.4	858.4	37.3
Potassium (mg/100g)	588.4	294.3	1369.1	90.8
Calcium (mg/100g)	607	195	208	248
Magnesium (mg/100g)	136	250	120	14
Copper (mg/100g)	0.3	9.0	9.0	1.2
Manganese (mg/100g)	1.3	3.4	2.7	0.6
Zinc (mg/100g)	1.0	3.8	6.2	4.9
Iron (mg/100g)	16.3	85.6	61.4	36.2
Energy (kcl)	274	260	265	318

MATERIALS AND METHODS

Sampling of mushrooms

The fully matured mushroom species were collected from different parts of IIT Chennai, Tamil Nadu, India. Collections were made at different times of the day: Morning, afternoon and sometimes mid-day by uprooting its substratum with the aid of a scalpel.

Proximate analysis

Four edible mushroom species were analyzed for food composition according to the Association of Official Analytical Chemists (AOAC, 1995). These include the determination of crude protein, crude fat, moisture content, dry matter, ash, crude fiber, carbohydrate and minerals. The minerals include sodium, potassium, calcium, phosphorus, magnesium, iron, copper and manganese. Values for, Fe, Cu and Mn were read on Atomic Absorption Spectrophotometer after standardizing with respective elements. The percentage of all the fractions (crude protein, crude fat, minerals and ash) were added together and subtracted from 100 to obtain the total carbohydrate percentage, while the nitrogen free extract (dry weight) was calculated as the percentage of the total carbohydrate and crude fiber.

RESULTS AND DISCUSSION

Proximate analysis was carried out on four edible mushroom species: *Auricularia polytricha, Macrolepiota rhodocus*, *Clitocybe sp.* and *L. tigrinus.* They were selected based on their availability at the time of analysis. Results of proximate composition are presented in Table 1. *A. polytricha* had the highest concentration of protein (37%) followed by *M. rhodocus* and *L. tigrinus*, while *Clitocybe sp.* had the least (24.8%). With respect to moisture content, *Clitocybe* had the highest value (21.0%) and *A. polytricha* the least value (6.48%). *M. rhodocus* had the highest carbohydrate (48%) and crude fiber was highest in *A. polytricha* (21.97%). Results in Table 1 show values of the macro mineral compositions. Calcium content was 208 mg/g dry weight for *Clitocybe sp.*, and 195 mg/g dw for *M. rhodocus*. *Clitocybe sp* and *A. polytricha* had the highest sodium content of 858.4 and 296 mg/g dry weights, respectively. *Clitocybe sp.* had the higher potassium concentration (1369.1 mg/g) and *M. rhodocus* had the least content (294.3 mg/g). *L. tigrinus* had the highest fat content (2.25%), followed by *Clitocybe sp.* and *A. polytricha. M. rhodocus* had the highest magnesium content (250 mg/g). Table 1 shows micromineral composition. Iron content varied from *A. polytricha* with 16.3 mg/g dw to *M. rhodocus* with 85.6 mg/g. Copper content ranged from *A. polytricha* (0.3 mg/g) to *M. rhodocus* 9.0 mg/g. Manganese content in *M. rhodocus*, *Clitocybe sp*, *A. polytricha*, and *L. tigrinus* were 3.4, 2.7, 1.3 and 0.6 mg/g, respectively. The results of mineral values of the four edible species of mushrooms clearly indicate the potential for their use as sources of good quality food. The crude protein, ash and crude fiber values of most mushrooms compared favourably with and in some instances surpassed those reported for most legumes except groundnut and soybeans grown in West Africa (FAO, 1970; Aletor and Aladetimi, 1995). The mineral levels, mainly potassium, phosphorous, sodium and iron in these mushrooms were higher than those reported for several cowpea varieties (Aletor and Aladetimi, 1989), but lower than those reported for fish, snails and broiler meat (Imevbore, 1992). Using this proximate analysis, the mineral and analytical food value as approximate indices of nutritional quality, it would appear that some of these mushrooms fall between most legumes and meat. In earlier studies, Gruen and Wong (1982) indicated that edible mushrooms were highly nutritional and compared favorably with meat, egg and milk. Some of the mushrooms are known to possess antitumorigenic and hypocholesterolaemic agents, which implies that mushrooms could hold special attraction for and may be recommended for people with cholesterol-related ailments (Chihara, 1993). The moisture contents of some of the mushrooms analyzed are high, indicating that mushrooms are highly perishable. High moisture contents promote susceptibility to microbial growth and enzyme activity. However, among those mushrooms analyzed *A. polytricha* had the highest protein content, followed by *M. rhodocus* and *L. tigrinus*, and then *Clitocybe* sp. The author reported 21.97% fiber and 6.87% ash contents for *A. polytricha*. The results of nutritionally valuable minerals show that the four mushroom species were rich in potassium, calcium, magnesium, iron and manganese. This is in agreement with the report of analysis of some cultivated mushrooms like *Agaricus bisporus*, *Lentinus edodes*, and *Pleurotus ostreatus* (Mattila et al., 2001). They were generally low in sodium, phosphorus and copper. Minerals in the diet are required for metabolic reactions, transmission of nerve impulses, rigid bone formation and regulation of water and salt balance among others. From the study, it was observed that these four edible mushrooms hold tremendous promise in complementing the protein and mineral supply deficits prevalent in developing countries. Edible mushrooms are grown with little efforts in their husbandry: They are grown on strawbased compost, and sawdust supplemented with other nutrients. For their full nutritional potentials to be realized, intensive efforts must be geared towards their husbandry and popularization of the more nutritious species like *Agaricus* spp., *P. tuber-regium* and *Termitomyces* spp.

REFERENCES

Aletor VA, Aladetimi O (1989). Compositional evaluation of some cowpea varieties and some under utilized edible legumes in Nigeria. Die Nahrun, 33: 99-1007.

Aletor VA, Aladetimi OO (1995). Compositional studies on edible tropical species of mushrooms. Food Chem. 54(3): 265-268.

AOAC (1995). Official Methods of Analysis. Association of Official Analytical Chemists. 16th Ed., Arlington. VA.

Buigut SK (2002). Mushroom production in sustainable small-scale farming system-opportunities and constraints: a survey of Uasin Gishu District. In: Proceedings of the Holticulture seminar on Sustainable Horticultural Production in the Tropics at Jomo Kenyatta

Chang ST, Buswell JA (1996). Mushroom Nutriceuticals. World Rapid Science publishers. J. Microbiol. Biotechnol., 12: 473-476.

Chihara G (1993). Medicinal aspects of Lentian Isolated from *Lentinus edodes* (Berk). Hong Kong, Chinese University Press, pp. 261-266.

Delena T (1999). Edible and Useful Plants of Texas and the SouthWest - A Practical Guide. 2003-2005 University of Texas Press, p. 542.

FAO (1970). Food and agricultural Organization (No 12) FAO. Rome.Italy.

Fasidi IA, Olorunmaiye KS (1994). Studies on the requirements for vegetative growth of *Pleurotus tuber-regium* (Fr.) Singer, a Nigerian mushroom. Food Chem., 50: 397-401.

Gruen VEC, Wong HX (1982). Immunodulatory and Antitumour activities of a polysaccharide-peptide complex from a mycelial culture of *Trichoderma* sp. Sci., 57: 269-281.

Harkonen M (1998). Uses of mushrooms by Finns and Karelians. Int. J. Circumpolar Health. 57(1): 40-55.

Imevbore EA (1992). Perspectives of snail farming in tropical Africa: the Nigerian situation. In: Proc. Invertebrates (Microlivestock) Farming Seminar. La Union. Philippines.

Jiskani MM (2001). Energy potential of mushrooms. DAWN Econ. Bus. Rev., Oct. 15-21, p. 4.

Mattila P, Konko K, Eurola M, Pihlava JM, Astola J, Vahteristo L, Hietaniemi V, Kumpulainen J, Valtonen M, Piironen V (2001). Contents of vitamins, mineral elements, and some phenolic compounds in cultivated mushrooms. J. Agric. Food Chem., 49(5): 2343-2348.

Ogundana SK, Fagade O (1981). The nutritive value of some Nigerian edible mushrooms. In: Mushroom Science XI, Proceedings of the Eleventh International Scientific Congress on the Cultivation of Edible Fungi. Australia, pp. 123-131.

Oso BA (1977). *Pleurotus tuber-regium* from Nigeria. Mycologia. 69: 271- 279, University of Agriculture & Technology, Juja, Kenya 3rd - 6th October, 2001. Eds. Wesonga JM, Losenge T, Ndung'u CK, Fricke A, Hau B, Stützel H (2002), pp. 1-5.

Migratory status of the goral, *Naemorhedus goral* (Hardwicke) (Artiodactyla: Bovidae) in Kohistan, Pakistan

Farzana Perveen and Anzela Khan

Department of Zoology, Shaheed Benazir Bhutto University (SBBU), Main Campus, Sheringal, Khyber Pakhtunkhwa (KP), Pakistan.

The Himalayan goral, *Naemorhedus goral* (Hardwicke) (Artiodactyla: Bovidae) is classified as endangered internationally and vulnerable in Pakistan. Its migratory status was determined from 24th May to 10th July, 2010 in the Pattan and Keyal Valleys, Kohistan, Pakistan. The 165 questionnaires were filled in 15 study sites (Pattan, eight study sites; Keyal, seven study sites). It is very common in the two valleys and almost everyone has seen it more often. In both Pattan and Keyal valleys, *N. goral* migrates during winter season to lower areas (populated areas), in search of food and water, due to cold temperature and threats from predators and hunters. If it migrates due to predators and hunters, then it will never return to its native habitat. Its main predators are wolf, jackal and others; its hunter is man. Its life-span is 10 to 20 years. It gives birth throughout the year. Some others wildlife, such as sparrow, *Passer domesticus* Linnaeus, chukor, *Alectoris chukar* (Gray); koklass pheasant, *Pucrasia macrolopha* (Lesson), jackal, *Canis aureus* Linnaeus, monkey, *Macaca mulatta* Zimmermann and wolf, *Canis lupus* Linnaeus etc. were also found during the present survey. Its native habitat and migratory areas should be protected.

Key words: Keyal valley, Kohistan, migration, *Naemorhedus goral,* native habitat, Pattan valleys, vantage point.

INTRODUCTION

Kohistan is a district in Pakistan's Khyber Pakhtunkhwa (KP) with a total area of 7,492 km^2 and a population of 472,570. It forms border with Azad Kashmir that extends from the eastern Afghanistan province of Nuristan in the west (Biddulph, 2004). Pattan, Keyal and Dubair are the valleys of Kohistan. Pattan lies on the right side of River Indus and links with Karakorum highway through a 20 km long road. Altitude of the valley varies from 3000'-11000' above the sea level. The area consists of rocky terrain with steep mountains. People experience seasonal migration with their livestock. The Keyal Valley is situated in the center of Pattan sub-division of Kohistan District. Both valleys are selected due to their floral diversity that supports the goral, *Naemorhedus goral* (Hardwicke) (Artiodactyla: Bovidae) population. It has the characteristics of both goats and sheep; therefore, it is a true antelope. At the present, it is found at middle slops of Himalayas and is endemic to Asia. It migrates during winter season to lower areas to escape from hunters, predators, low temperature and in search of food (Figure 1) (Zhiwotsehenko, 1990).

It is classified as endangered with decreasing population

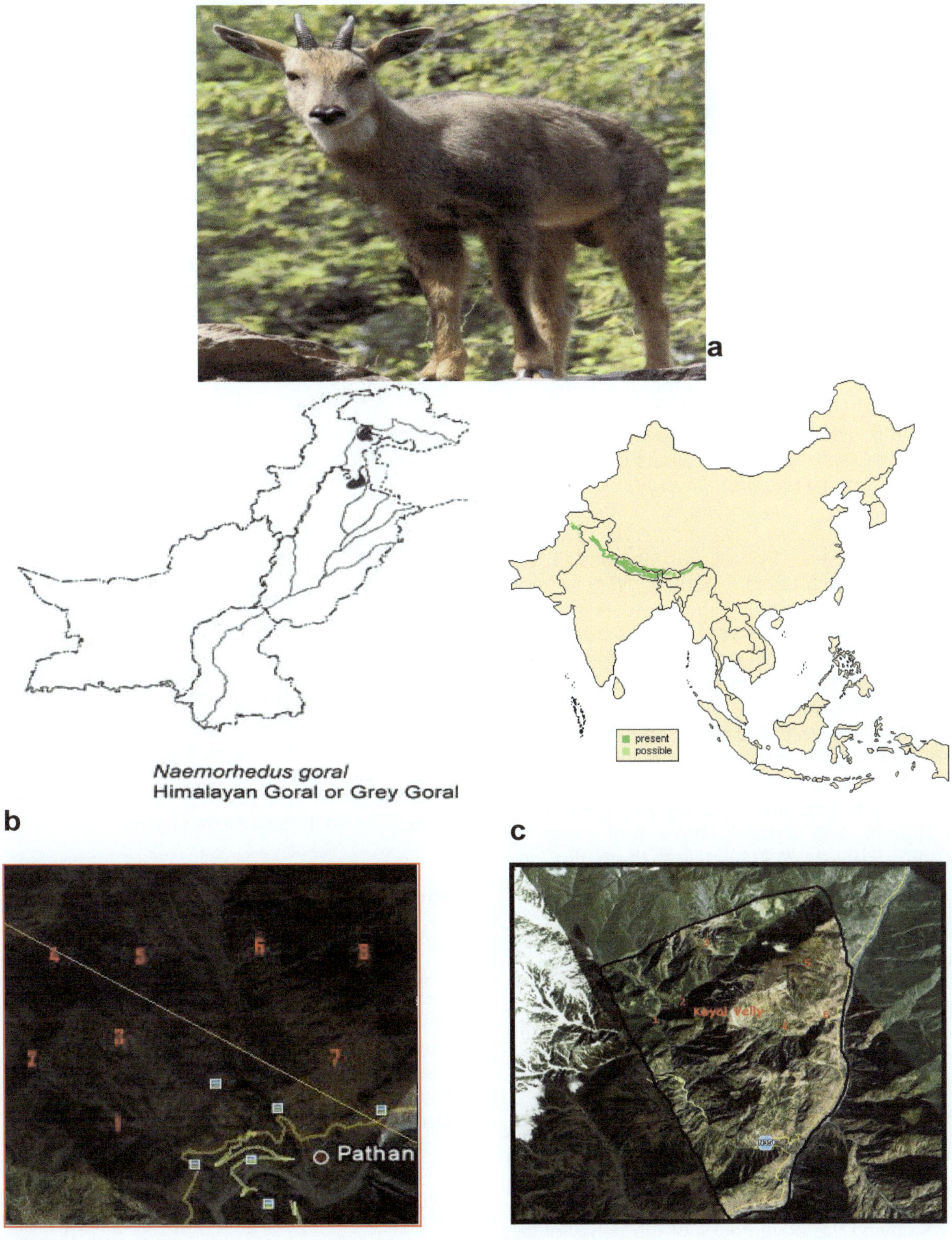

Figure 1. Distribution map of grey goral, *Naemorhedus goral* (a; Online, 2012) in 2 study areas of Kohistan district within Khyber Pakhtunkhwa, Pakistan; b: distribution map, black dots shows the distribution of *N. goral* in Kohistan and Margalla Hills, Islamabad, Pakistan; c: distribution of global population of *N. goral* (Shackleton, 1997); d: Pattan Valley, 1: Tankor Janchil; 2: Rasta Dong Janchil; 3: Barho Kandogay; 4: Landai Sar Bohil; 5: Hawery Kamar Bohil; 6: Barho Bohil; 7: Barho Gulkand; 8: Nabaz; e: Keyal Valley, 1: Baroon Nala Fagaiel; 2: Galto Fagaiel; 3: Shaig Bhapobanda; 4: Keero Keyal; 5: Balkhun; 6: Rodair (highlighted in red) (Perveen, 2013).

(Mead, 1989) and threatened species by the International Union for Conservation of Nature (IUCN) (Shackleton, 1997). It is known as vulnerable in Pakistan based on information gathered from different parts disseminated (Sheikh and Molur, 2005). In KP, its range extends from Abbottabad, Mansehra, Mardan, Kohistan, Swat, Dir, Malakand and Nowshera which form the western border of its distribution range. The main surviving population in Pakistan is probably in the area among the Indus Kohistan, Swat valley and watershed Kunhar. It belongs to the order, Artiodactyla or cloven-hoofed mammals (Robert, 1977) and shares the sub-order, Ruminantia with deer, antelopes, sheep, goats; it belongs to the family, Bovidae (Walker, 1964; Robert, 1977). Its sub-family, Caprinae (sheep and goats) is characterized by stickily built species, adapted for climbing on mountains (Robert, 1977; Halt and North, 1963).

Presently, the genus *Naemorhedus* Smith is represented by six species: Japanese serow, *Capricornis crispus* or *Naemorhedus crispus* (Temminck); Sumatran serows, *Naemorhedus sumatraensis* (Bechstein); Taiwan serow, *Naemorhedus swinhoei* (Gray); red goral, *N. bailey* Pocock (Groves and Grubb, 1985); Chinese long-tailed goral, *Naemorhedus caudarus* (Edwards) (Zhang, 1987) and Himalayan goral, *N. goral* (Hardwicke). Its population is divided into 3 sub-species: the goral, *N. goral goral* (Hardwicke); brown goral, *N. goral hodgsoni* (Pocock) and grey goral, *N. goral bedfordi* (Roberts).

It has broad and bell shaped ears; both sexes have slender and cylindrical horns, which are not divergent but curving backward and bearing inconspicuous annulations. The males have no beard, have small sub-orbital glands and females have four mammae (Pocock, 1908). Roberts (1977) called them the grey goral, *N. goral bedfordi* based on accounts in the literature. The small sized ones are like goat and adults are 65 to 71 cm. Its shoulders' height, head and body length average are 105 cm and weigh 25 to 28 kg (Primrose, 1911; Stockley, 1928).

The breast and belly are lighter gray with a white spot in the throat and one or two white spots on the lower muzzle and cheeks. The horns are 12.5 to 15.5 cm with slender feet (Stockley, 1928). During summer, *N. goral* goes upward the hills and in winter returns downward the hills. Sometime due to specific circumstances, it migrates to far places (Perveen, 2013). Therefore, the objective of the present research is to determine the migratory status of *N. goral* in Kohistan, Pakistan by studying the advantages of the behavior and ecology of the species.

MATERIALS AND METHODS

Study areas

In Kohistan, two potentially bio-diversified valleys, Pattan with eight study sites: Barho Bohil, Barho Gulkand, Barho Kandogay, Hawery Kamar Bohil, Landai Sar Bohil, Nabaz, Rasta Dong Janchil and Tankor Janchil and Keyal with 7 study sites: Balkhun, Baroon Nala Fagaiel, Galto Fagaiel, Lotos, Keero Keyal, Rodair and Shaig Bhapobanda were selected for the present survey (Figure 1).

Collection of data

Data were collected during 24th May to 10th July, 2010. Field survey, direct sighting, informal discussion, interviews with community and questionnaire were the major tools of data collection. The local people and hunters were asked 32 questions in questionnaire, which explained everything about the *N. goral* hunting, migration, predators etc. The 165 questionnaires were filled in the study sites of Pattan and Keyal valleys, respectively (Perveen and Husain, 2012a).

Data analysis

For data analysis, % and Computer Program Microsoft Excel (CPME) were used (Perveen and Husain, 2012b). For the life span of the species, Boxplots analysis was employed.

RESULTS

The present survey was conducted during 24th May-10th July, 2010 to investigate the migratory status of the grey goral, *Naemorhedus goral* (Hardwicke) (Artiodactyla: Bovidae) in two valleys of Kohistan District, Khayber Pukhtunkhwa, Pakistan: 1) Pattan valley with eight study sites; 2) Keyal valley with 7 study sites.

In both valleys, 165 questionnaires were filled by the local people and hunters, respectively. All of them have seen *N. goral*. This was a great evidence of the presence of *N. goral* in these valleys (Table 1).

Migratory status

All the local people and hunters (100%) said that *N. goral* migrated from Pattan Valley. However, in Keyal Valley, most of them have the same answer but very few (n = 2) said that it was resident of the valley (Figure 2a). In both Pattan and Keyal valleys, *N. goral* migrates during winter season towards lower area (populated area). But few of the local people and hunters did not agree with this statement (Figure 2b and c). The reasons for its migration include: search for food and water, running from cold temperature and threats from predators. Whenever it migrates due to threat from predators, then it will never return to its original habitat (Figure 2d). According to the local people and hunters, *N. gorals are* threatened by predators in Pattan Valley; the predators are man, wolf, jackal e tc. In Keyal valley, they are man, jackal, wolf etc. (Figure 2e).

N. goral has different life-spans. According to the local people and hunters, the most probable life-span in Pattan Valley is: 15 > 10 > 20 years; however, in Keyal valley: 15 >20 > 10 years (Figure 2f). When the local people

Table 1. Sighting of the grey goral, *Naemorhedus goral* (Hardwicke) (Artiodactyla: Bovidae) by the local people and hunters in Pattan and Keyal valleys of district Kohistan during the survey, 24th May-10th July 2010

S/N	Valley	Sits[1]	n[2]	Watching of *Naemorhedus goral*	
				Yes (%)	No (%)
1.	Pattan	8	90	100	00
2.	Keyal	7	75	100	00

[1]Pattan valley with eight study sites: Barho Bohil, Barho Gulkand, Barho Kandogay, Hawery Kamar Bohil, Landai Sar Bohil, Nabaz, Rasta Dong Janchil and Tankor Janchil; Keyal valley with 7 study sites: Balkhun, Baroon Nala Fagaiel, Galto Fagaiel, Lotos, Keero Keyal, Rodair and Shaig Bhapobanda where questionnaire were distributed. [2]n: number of peoples and hunters in Pattan and Keyal valleys from which data were collected; %: percentage

and hunters were asked about the birth time of fawns , most of them replied that it might be January, February or December in both valleys. It also reported that these can give birth to their fawns throughout the year (Table 2).

Some others wildlife animals like mammals: cat, *Felis catus* Linnaeus; fox, *Vulpes vulpes* (Linnaeus); jackal, *Canis aureus* Linnaeus; monkey, *Macaca mulatta* Zimmermann; squirrel, *Sciurus carolinensis* Gmelin and wolf, *Canis lupus* Linnaeus; and birds like chukor, *Alectoris chukar* (Gray); Himalayan monal pheasant, *Lophophorus impejanus* Latham; koklass pheasant, *Pucrasia macrolopha* (Lesson); parrot, *Psittacula eupatria* (Linnaeus); ram chukor, *Tetraogallus himalayensis* Gray and sparrow, *Passer domesticus* Linnaeus were also observed in the survey sites of the two valleys (Table 3).

DISCUSSION

The present survey was conducted in Kohistan, Pakistan during May-July, 2010. The suitable season for goral survey is winter, but due to academic schedule, it was conducted in summer.

This survey could not be compared with previous data because such type of survey has not been conducted before in this area. Abbas (2006) suggested that *N. goral* still persists at favorable altitudes of Mardan, Buner, Central Kohistan, Abbotabad, western Mansehra, Margalla hills and the central and southern parts of the Azad Kashmir. The present findings are in line with the general remarks of Roberts (1977) who suggests that the main *N. goral* population is present in Indus Kohistan, between Swat and Kunhar valleys and a limited population in Margalla hills, Islamabad, Pakistan. Although, *N. goral* is also found in most places, but Pattan and Keyal were selected for the present survey in Kohistan.

The local people and hunters perceive *N. goral* by direct sighting, fecal materials, sounds and foot prints. They hunt them for food, recreation and earning. In both valleys, the methods for hunting were mostly applying of coloring cloth, whistling by mouth, using special little yellow dogs, gun shots and pellet bow. In Pattan Valley, the local people and hunters have hunting permit but in Keyal, they do not have. They have been hunting for 1 to 5 years in Pallas and 6 to 10 years in Keyal (Perveen, 2013). Therefore, *N. goral* migrates to escape from cold temperature, predators, hunting and in search of food during winter season.

For the survey, firstly, Pattan valley was visited, local hunters , watchers and villagers were contacted to obtain their guidance and formal knowledge. Eight vantage points were selected from Pattan Valley. The highest vantage point was Landi Sar Bohil while the lowest was Tankor Janchil in Pattan Valley. In these vantage points, 21 *N. gorals* were found, in which 13 adults, six sub-adults and 2 fawns were identified (Perveen and Hussain, 2012a). In Pattan Valley, the numbers of adult were more in vantage points with lower height than in Keyal Valley's vantage points.

The Keyal Valley with seven vantage points supports *N. goral* population due to its floral diversity. The highest vantage point was Shaig Bhapobanda while the lowest was Baroon Nala Fagaiel. In this valley, a total of 40 *N. goral* were found, among them 22 adults, 14 sub-adults and eight fawns were identified (Perveen and Hussain, 2012b). They were greater in number in Keyal than in Pattan. Therefore, it is clear that *N. goral* likes to live in higher than lower places because people can kill them easily in lower places. Thus, they are protected in higher places. Secondly, biodiversity is found more in higher places than lower places. Finally, higher places are more natural than lower places, that is why more natural food sources are available in higher places.

There are three methods of hunting *N. goral*, by the hunters of these valleys. First one is the use of special little yellow dogs to hunt *N. goral*. When they see them, they do not flee. Second, white cloth of one meter is encircled around them; then the hunters easily capture them. The color of cloth is changed to yellow in summer, but to dark brown in winter. When they show these cloths to *N. goral*, it stops to run and then the hunters hunt it. Third, whistling by mouth is also a technique for hunting *N. goral*. Another method which is prohibited and rarely

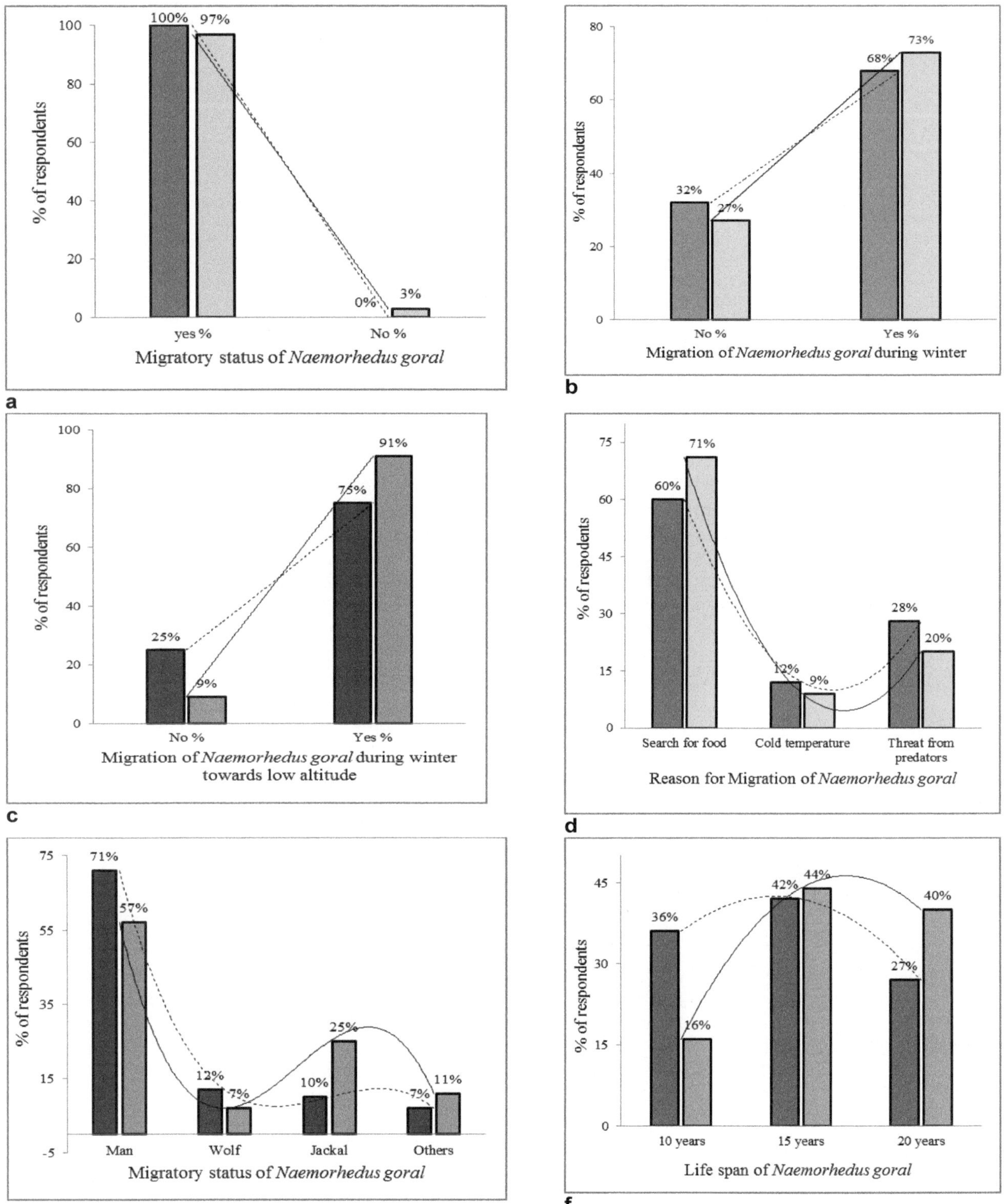

Figure 2. The survey for determining migratory status about the grey goral, *Naemorhedus goral* (Hardwicke) (Artiodactyla: Bovidae) in 2 valleys of Kohistan, Pakistan during 24th May-10th July 2010; migratory status of *N. goral*: a; migration during winter: b; migration during winter to low altitude: c; reason for migration: d; ■: Pattan valley with 8 study sites, that is, Tankor Janchil, Rasta Dong Janchil, Barho Kandogay, Landai Sar Bohil, Hawery Kamar Bohil, Barho Bohil, Barho Gulkand and Nabaz; ▒: Keyal valley with 7 study sites, that is, Baroon Nala Fagaiel, Galto Fagaiel, Shaig Bhapobanda, Keero Keyal, Balkhun and Rodair); n: number of questionnaire filled by local people and hunters; n=90 in Pattan; n=75 in Keyal, Kohistan, Pakistan; polynomial trend line for Pattan valley: ---; for Keyal valley.

Table 2. Birth time of fawns of the grey goral, *Naemorhedus goral* (Hardwicke) (Artiodactyla: Bovidae) during the survey, 24th May-10th July 2010.

Valleys	Sites[1]	n[2]	January (%)	February (%)	December (%)	Throughout the year (%)	M±SD[3]
Pattan	05	90	7	11	4	78	23±27
Keyal	07	75	9	15	8	68	19±18

[1]Pattan valley with 8 study sites: Barho Bohil, Barho Gulkand, Barho Kandogay, Hawery Kamar Bohil, Landai Sar Bohil, Nabaz, Rasta Dong Janchil and Tankor Janchil; Keyal valley with 7 study sites: Balkhun, Baroon Nala Fagaiel, Galto Fagaiel, Lotos, Keero Keyal, Rodair and Shaig Bhapobanda where questionnaire were distributed. [2]n: number of peoples and hunters in Pattan and Keyal valleys from whom data were collected; %: percentage [3]M±SD: Mean± standard deviation; the data was analyzed by t-test and $P < 0.05$; 2 populations were not significantly different.

Table 3. Other wildlife animals and birds were observed in study area during the survey, 24th May-10th July 2010.

S/N	Common name	Scientific name	Pattan[1](%)^P	Keyal[2](%)^K
	Animals			
1.	Cat	*Felis catus* Linnaeus	3	-
2.	Fox	*Vulpes vulpes* (Linnaeus)	3	3
3.	Jackal	*Canis aureus* Linnaeus	8	3
4.	Monkey	*Macaca mulatta* Zimmermann	10	6
5.	Squirrel	*Sciurus carolinensis* Gmelin	3	3
6.	Wolf	*Canis lupus* Linnaeus	3	6
	Birds			
7.	Chukor	*Alectoris chukar* (Gray)	13	20
8.	Himalayan monal pheasant	*Lophophorus impejanus* Latham	5	3
9.	Koklass pheasant	*Pucrasia macrolopha* (Lesson)	7	6
10.	Parrot	*Psittacula eupatria* (Linnaeus)	8	-
11.	Ram chukor (Himalayan snow cock)	*Tetraogallus himalayensis* Gray	5	3
12.	Sparrow	*Passer domesticus* Linnaeus	32	47

[1]Pattan valley with 8 study sites: Barho Bohil, Barho Gulkand, Barho Kandogay, Hawery Kamar Bohil, Landai Sar Bohil, Nabaz, Rasta Dong Janchil and Tankor Janchil. [2]Keyal valley with 7 study sites: Balkhun, Baroon Nala Fagaiel, Galto Fagaiel, Lotos, Keero Keyal, Rodair and Shaig Bhapobanda where questionnaire were distributed. (%)^P and (%)^K: number of animals and birds found in Pattan and Keyal valleys

used by the hunters is gunshot which does not easily damage their coat because the coat of *N. goral* is very hard. Mostly little children and very few adults of both valleys use pellet bow as game, which they hang in their necks. They throw stones at them with power by pellet bow, which injured *N. goral* found during the present survey (Verbal Communication with Community, 2010). The people of the study area were very religious and so said everything very clearly and firmly.

They said that they have seen 12 to 18 *N. goral*. Every hunter hunts at least 4 to 5 *N. goral* per day in winter; therefore, the population status of *N. goral* has dropped down. Most of the people do not care about the wildlife of the areas (interviews and informal discussion). However, if any hunter has hunted *N. goral*, then he cannot escape from wildlife department (interviews and discussion with community). Children like to hunt birds as well as *N. goral* in their childhood, because they like to use guns for shooting targets. Their natural attachment to guns also makes them to spoil wildlife.

In winter season, a number of *N. goral* are attacked by a disease which kills many of them. In this disease, the eyes of *N. goral* are affected seriously. They cannot open their eyes. Thus, it is impossible for them to get food and water, which leads to their death. If they are alive, it is very easy to capture or kill them.

Some people get them in ill- condition. The people said that this disease does not transmit to the meat eaters of *N. goral* (informal discussion with community). However, they are the natural food resource in backward area of Kohistan, Pakistan.

Conclusion

N. goral migrates during winter season towards lower areas, that is, populated areas, to escape from cold temperature, predators, hunters and in search of food. If it migrates due to predators, then it will never return to its native habitat. Its main predators are man, wolf, jackal and others. Human beings are it chief predators; they hunt it for food, recreation and earning.

Recommendations

For the conservation of *N. goral*, its migrated sites, i.e., lower and populated areas, should be protected. The protected area should be established within the Kohistan. Local committees should monitor the population and hunting of *N. goral*. A comprehensive conservation action plan is required to develop the species conservation in its historic habitat. Awareness and education should be increased towards the importance of *N. goral*; therefore, attitude of the local people should be changed towards them. Deforestation should be controlled in the area.

ACKNOWLEDGEMENTS

The authors are grateful to Mr. Iftikhar-uz-zaman, DFO and Mr. Masood Ali, Range Officer, Wildlife Kohistan, Pakistan for providing all possible information and cooperation during the survey. The experiments are in line with the current laws of the country in which they were performed.

REFERENCES

Abbas F (2006). A study on ecobiology of grey goral *(Naemorhedus goral)* with reference to Pakistan. PhD Thesis, Department of Zoology University of Punjab, Lahore, Pakistan, pp. 1-179.

Biddulph J (2004). Tribes of the Hindoo Koosh. Sang-e-Meel Publications, Lahore, Pakistan. 2:1-8.

Groves CP, Grubb P (1985). Reclassification of the serows and goral (Naemorhedus: Bovidae). In: Lovari, S. (ed) The biology and management of mountain ungulates. Croom Helm, United State of America pp. 71-76.

Halt E, North T (1963). Die Klassifikation der Saugetiere: Artiodactyla. Handbook of Zoology, Berlin, Germany 32(8):1-167.

Mead JI (1989). *Naemorhedus goral*. Mammalian specie 335:1-5.

Perveen F (2013). Population status of goral, *Naemorhedus Goral* (Hardwick) (Artiodactyla: Bovidae) in Pattan and Keyal valleys of Kohistan, Pakistan. Inter. J. Farming and Allied Sci. 2(6):133-139.

Perveen F, Husain A (2012a). Vantage points of goral, *Naemorhedus goral* (Hardwick) (Artiodactyla: Bovidae) in Pattan and Keyal valleys of district Kohistan, Pakistan. Europ. J. Zool. Res. 1(1):16-22.

Perveen F, Hussain Z (2012b). Use of statistical techniques in analysis of biological data. Basic Res. J. Agricul. Sci. Rev. 1(1):1-10.

Pocock RI (1908). On the generic names of the Rupicaprine Ruminant known as Serows and gorals, Ann. Taki India and the Straits settlements. J. Bot. Nat. History Soc. 19:807-821.

Primrose AM (1911). Weight and measurements of a goral. J. Bombay Nat. History Soc. India 21(1):11-246.

Robert TJ (1977). The Mammals of Pakistan. Ernest Benn Ltd. London and Tonbridge, England. pp. 200-542.

Shackleton DM (1997). Wild sheep and goats and their relatives: Status survey and conservation action plan for caprinae. Gland, Inter. Union for Conser. Nat. pp. 1-390.

Sheikh MK, Molur S (2005). (eds) Status and red list of Pakistan mammals based on Pakistan`s conservation assessment and management plan for mammals. Inter. Union for Conser. Nat. Pak. pp.1-344.

Stockley G (1928). Big game shooting in the Indian empire. London: Constable. Symposium. Kalispell, Montana, United State pp.1-179.

Walker EP, Warrick F, Kenneth I, Lange HE, Patricia FW (1964). Mammals of the world 3:50-140.

Zhang C (1987). Nemorhaedus cranbrooki Hayman. In: Soma, H. H. (ed) The biology and management of Capricorns and related mountain antelopes, Croom Helm, London pp. 213-223.

Zhiwotschenko V (1990). Gorals (genus Naemorhedus) Grzinek`s encyclopedia of mammals. McGraw Hill Publishing Company, New York. 5:506-507.

23

Survey of medicinal plants used to treat human ailments in Hawzen district, Northern Ethiopia

Gidey Yirga*, Mekonen Teferi and Mezgebe Kasaye

Department of Biology, Mekelle University P. O. Box 3072, Mekelle, Ethiopia.

Traditional medicine is an integral part of the culture of the Ethiopian people due to its long period of practice and existence. The present study was conducted to identify the medicinal plants used to treat human ailments in Hawzen district, Northern Ethiopia. Twenty informants including six females were selected purposefully for semi-structured interview with the help of local elderly people. Ethnobotanical data was collected between January and February 2011 on seven field trips made to the site. Thirty-three species of medicinal plants were collected and identified for treating 25 human ailments. Most (17, 51.5%) of the traditional medicinal plants were collected from the wild stands and leaves (65%) and roots (17%) were the most commonly used plant parts for herbal preparations. Most (85.7%) of the traditional medicinal plant preparations were used in fresh form. Oral, dermal and nasal were the routes of application of remedies. Squeezing, grinding, boiling, chewing, crushing and tying were the methods of remedy preparation. Leaves were mainly harvested part for traditional medicine preparation in the area; the practice does not affect the sustainable utilization of the medicinal plants.

Key words: Ethnobotany, Ethiopia, Hawzen, medicinal plants, ailments.

INTRODUCTION

The plant kingdom is the most essential to human well being in providing basic human needs. Human beings used plants for the purpose of disease control and prevention since time immemorial. Early humans acquired the knowledge on the utilization of plants for disease prevention and curative purposes through many years experience, careful observations and trial and error experiments (Sofowora, 1982; Martin, 1995). Such ethnomedicinal knowledge involves traditional diagnosis, collection of raw materials, preparation of remedies and prescription to the patients.

More than 35,000 plant species are reported as being used across the globe for medicinal purposes (Lewington, 1993) and in Ethiopia more than 800 plant species have been employed as medicinal plants (Tesema et al., 2002). There is a considerable global interest in tapping the accumulated knowledge of traditional medicine; researches are being carried out in many countries with the aim of increasing the use of traditional medicine to the welfare of the human population (Demissew and Dagne, 2001). The documentation of traditional knowledge of medicinal uses of plants has provided many important drugs of modern day (Balick and Cox, 1996) as a source of direct therapeutic agents, as raw materials for the manufacture of complex semi-synthetic compounds and as taxonomic markers in the search for new compounds (WHO, 1998).

It is estimated that about 75 to 90% of the rural population in the world excluding Western countries depend on traditional medicines as their only healthcare system (Fassil, 2001). About 80% of the population in Africa primarily depends on traditional medicinal plants for their healthcare (WHO, 2002). Similarly, plants have been used as a source of traditional medicine since ancient times in Ethiopia to combat different ailments and human sufferings (Asfaw et al., 1999). Traditional medicine was the only option available for healthcare prior to the introduction of modern medicine for prevention, diagnosis and treatment of social, mental and physical illness (Dawit, 1986). Traditional medicine is an integral part of the culture of the Ethiopian people due to its long history of practice and existence (Mirgissa, 1998).

*Corresponding author. E-mail: gidey1998ec@yahoo.com.

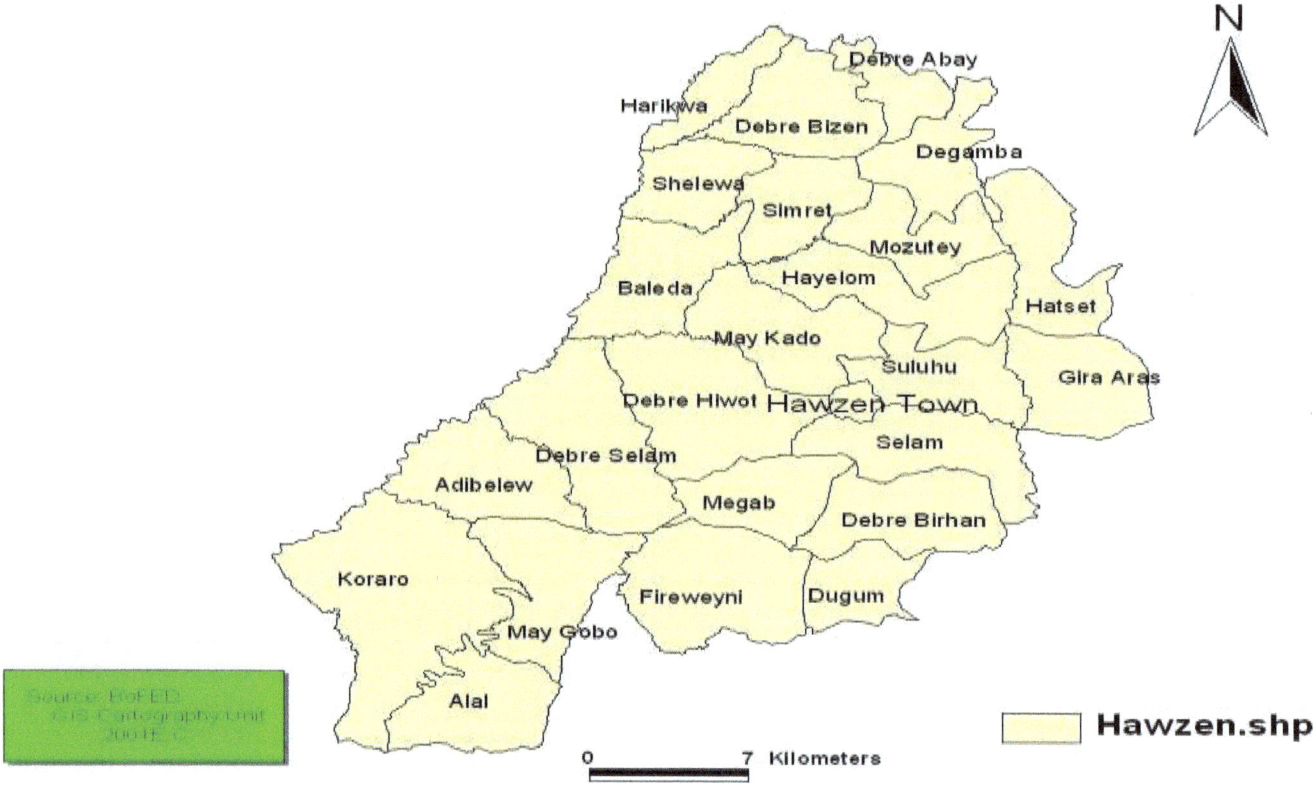

Figure 1. Map of Hawzen district showing the study site, Hayolom (Source: BoFED GIS-category unit, 2009.

There is a large magnitude of use and interest of medicinal plants in the country mainly due to acceptability, accessibility and biomedical benefits (Dawit, 2001). The majority of Ethiopians still depend on traditional medicine mainly due to shortage of pharmaceuticals, inadequate coverage of modern medical system and unaffordable prices of modern drugs (Zerabruk and Yirga, 2011).

The present study was conducted to identify the medicinal plants used to treat human ailments in Hawzen district, northern Ethiopia.

Study area

The study was conducted in Hayolom sub-district, in the Northern Ethiopian highlands (Figure 1), in Hawzen district located at about 75km from Mekelle, capital city of Tigray regional state of Ethiopia. The mean, minimum and maximum annual rain fall of the district is 580 and 750 mm respectively and the temperature ranges between 14 and 27°C. The total human population is 117,338 (56,616 males and 60722 females). Agriculture is the main economic source of the community. About 95% of the population of the district depends on agriculture practicing mixed farming, livestock husbandry and crop production simultaneously.

MATERIALS AND METHODS

A total of 20 informants (14 males and 6 females) were selected purposefully with the help of local area administrator and elderly people from Hayolom sub-district. Recommended traditional medicine practitioners were identified as potential informants and participated in personal interviews which were based on a checklist of questions prepared before hand in English and translated to the local language (Tigrigna). The majority (12, 60%) of the respondents were illiterate, with age range between 28 to 65 years. Ethnobotanical data was collected using semi-structured interviews between January and February 2011 on seven field trips made to the site. Information regarding local names of medicinal plants, methods of preparation, part(s) used, diseases treated, route of application, uses other than medicinal uses was recorded. Identification of the medicinal plants was done in Mekelle University using Flora of Ethiopia and Eritrea, by comparison with authentic specimens, illustrations and taxonomic keys.

RESULTS

Thirty-three plant species were collected and identified for treating 25 human ailments (Table 1). Most (17, 51.5%) of the traditional medicinal plants were wild, five species (12.1%) were cultivated and the remaining 12 species (36.4%) were obtained both from cultivation and the wild. Most of the traditional medicinal plant preparations were used in fresh form (85.7%). It was found that 57.1% of medicinal plants have values other

Table 1. List of medicinal plants, disease treated, ingredients added, and condition of plant used in the study area.

Vernacular name	Scientific name	Family name	Disease treated	Ingredients added	Condition of the plant used
Seraw	*Acacia etbaica* Schweinf.	Fabaceae	Wart	None	Dry
Muchelo	*Achyranthes aspera L.*	Amaranthaceae	Foot strain	None	Fresh
Tsaeda shgurti	*Allium sativum L.*	Alliaceae	Gland TB	None	Fresh
Ere	*Aloe barbadensis* Miller.	Aloeaceae	Foot strain	None	Fresh
Endirur	*Balanites aegyptiaca L.*	Zygophyllaceae	Snake bite	None	Fresh
Hemligbo	*Chenopodium oplifolium* Auct.	Chenopodiaceae	Wound	None	Fresh
Lemon	*Citrus aurantifolia L.*	Rutaceae	Mouth smelling	None	Fresh
Awuhi	*Cordia Africana Lam.*	Boraginaceae	Abdominal pain	Coffee	Fresh
Shinfa	*Coronopus didymus L.*	Brassicaceae	Common Cold	None	Dry
Tanbuk	*Croton macrostachyus* Hochst.	Euphorbiaceae	Abdominal pain	Chili	Fresh
Tanbuk	*Croton macrostachyus* Hochst.	Euphorbiaceae	Hepatitis	None	Fresh
Haffaflo	*Cucumis pustulatus* Hook.	Cucurbitaceae	Eye infection	None	Fresh
Duba	*Cucurbita pepo L.*	Cucurbitaceae	Tapeworm	None	Dry
Tenege	*Cynoglossum coeruleum* Hochst.	Boraginaceae	Skin cramp	None	Fresh
Acerkuka	*Cyphostemma adenocaule* Baker.	Vitaceae	Anthrax	Honey	Fresh
Acerkuka	*Cyphostemma adenocaule* Baker.	Vitaceae	Snake Bite	None	Fresh
Mestenagir	*Datura innoxia Mill.*	Solanaceae	Leishmaniasis	None	Fresh
Mezerbai	*Datura stramonium L.*	Solanaceae	Hair fungal infection	None	Fresh
Tsaida Baharzaf	*Eucalyptus globulus* Labill.	Myrtaceae	Common cold	None	Fresh
Kuliaw	*Euclea natalensis L.*	Ebenaceae	Rabies	Cheese	Fresh
Shlanata	*Foeniculum vulgare L.*	Apiaceae	*Entamoeba histolytica*	None	Fresh
Metere	*Glinus lotoides L.*	Molluginaceae	Oestrus ovis	None	Fresh
Entati	*Linum usitatissimum L.*	Linaceae	Sewage around neck	None	Fresh
Andel	*Maytenus senegalensis Lam.*	Celastraceae	Evil eye	None	Fresh
Awli	*Olea europaea L.*	Oleaceae	*Entamoeba histolytica*	None	Fresh
Shibti	*Phytolacca dodecandra L.*	Fabaceae	Abortion	Holy water	Fresh
Tetaelo	*Rhus natalensis* Bernh.	Anacardiaceae	Depression	None	Fresh
Hahot	*Rumex nervosus Vahl.*	Polygonaceae	Vomiting	Honey	Fresh
Chena-Adam	*Ruta chalepensis L.*	Rutaceae	Cholera	None	Fresh
Tickure berebere	*Schinus molle L.*	Anacardiaceae	Abdominal pain	Garlic	Fresh
Tfreria	*Sida schimperiana* Hochst	Malvaceae	Foot strain	None	Dry
Engule	*Solanum incanum L.*	Solanaceae	Lumbar strain	None	Fresh
Abaeke	*Trigonella foenum-graecum L.*	Fabaceae	Abdominal pain	None	Fresh
Sebere	*Vicia sativa L.*	Fabaceae	Scabies	Cheese	Dry
Agol	*Withania somnifera L.*	Solanaceae	Common cold	None	Fresh

Table 2. Methods of Preparation of remedies and other uses of the traditional medicinal plants in the study area.

Methods of preparation	Frequency	Percentage
Squeezing	12	34.3
Grinding	8	22.9
Chewing	4	11.4
Boiling	5	14.2
Crushing	3	8.6
Tying	3	8.6
Total	**35**	**100**

Other uses of the medicinal plants	Frequency	Percentage
Fire wood	10	28.6
Food	5	14.2
None	15	42.9
Washing	2	5.7
Splice	3	8.6
Total	**35**	**100**

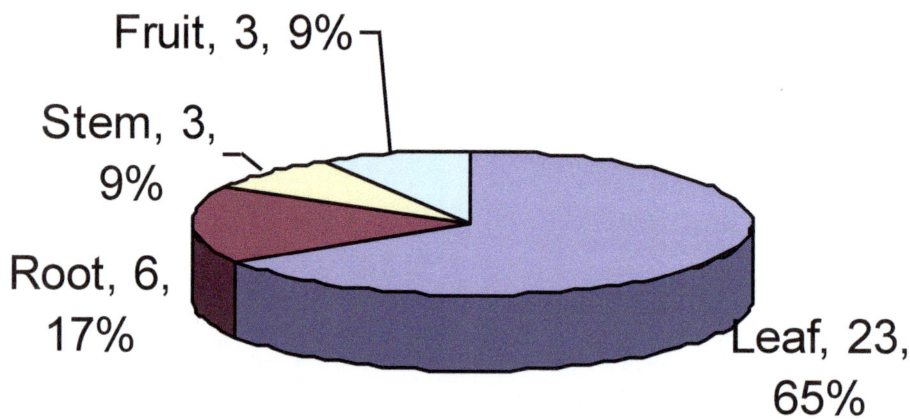

Figure 2. Parts of the medicinal plants used in the preparation of remedies.

than their medicinal role (Table 2). The most commonly used plant parts for herbal preparations were leaves (65%) and roots (17%) (Figure 2). Oral, dermal, nasal, oral or dermal were the routes of application of remedies in decreasing order (Figure 3). Squeezing, grinding, boiling, chewing, crushing and tying were methods of preparation/application used in the study area (Table 2).

DISCUSSION

In Ethiopia, more than 800 plant species have been used as medicinal plants (Tesema et al., 2002). However, only 33 species of medicinal plants were recorded and identified for treating 25 human ailments in Hayolom sub-district. Large magnitude of use and interest in medicinal plants in Ethiopia is due to acceptability, accessibility and biomedical benefits (Dawit, 2001). The long history of use of medicinal plants in the country is believed to have originated several centuries ago (Kibebew, 2001).

In various parts of the world, medicinal plants are mostly harvested from the wild sources (Lange, 1998). Various researchers in Ethiopia have also reported that most of the medicinal plants are obtained from the wild vegetation (Awas and Asfaw, 1999; Mesfin, 2007; Yirga, 2010b). Similarly, in the study area, most of the medicinal plants were collected from the wild stands. It is therefore important that the healers in consultation with government officials should take care not to deplete medicinal plant species altogether. Moreover, awareness

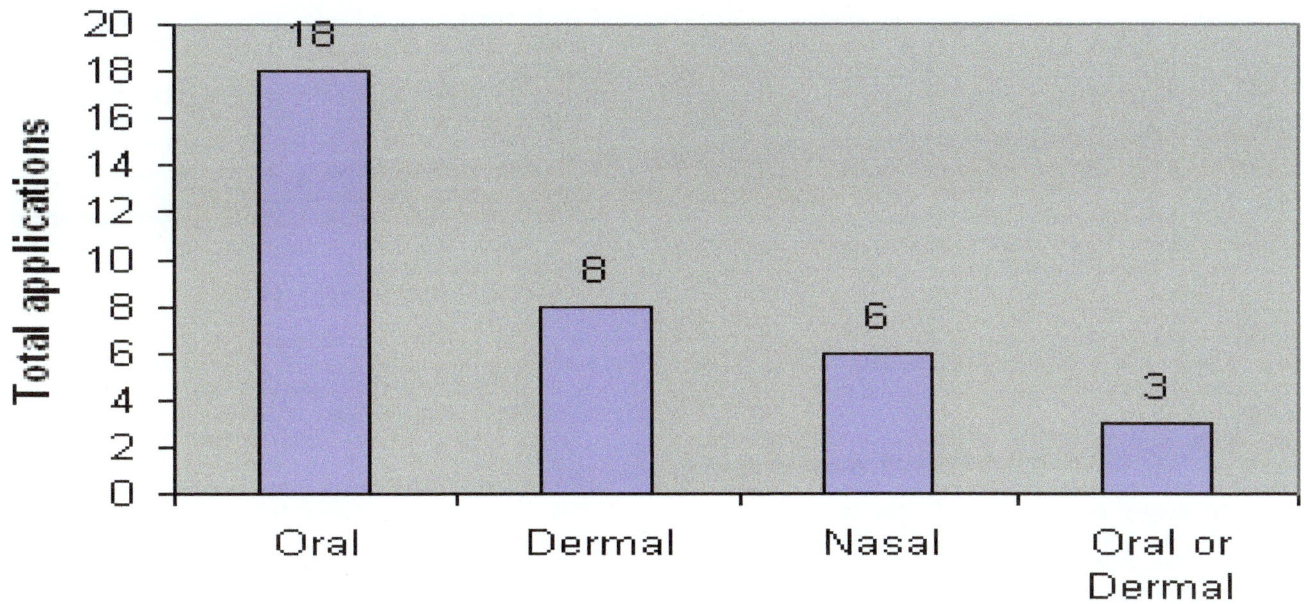

Figure 3. Route of application of traditional medicinal plants.

creation among the traditional healers and community at large is important in order to preserve the indigenous medicinal plant species (Yirga, 2010a).

Leaves were the most harvested plant part of remedy preparation in the area which does not affect the sustainable utilization of the medicinal plants. Collecting leaves does not cause great danger to the continuity of an individual plant compared to the collection of roots, bark, stem or whole plant. However, the dependence on roots of plants results in consequences from both ecological point of view and from the survival of the medicinal species (Abebe and Ayehu, 1993). Further research in the use, management and conservation of medicinal plants should target conservation in the area.

Oral application was the leading route of application of traditional medicine in northern Ethiopia followed by dermal. Various ethnobotanical researchers elsewhere in Ethiopia have also indicated oral as the predominant route of application (Abebe and Ayehu, 1993; Gidey, 1999; Hunde, 2001; Addis et al., 2001; Balemie et al., 2004; Lulekal, 2005; Mesfin, 2007; Yirga, 2010a).

Squeezing, grinding, boiling, chewing, crushing and tying were also other methods of preparation used. Grinding or crushing and soaking or boiling different parts of plants are common methods for drug extraction (Deeba, 2009). Application routes and the medical preparation methods differ depending on the active ingredient to be extracted (McCorkle and Mathias-Mundy, 1992).

Traditional medicine plays a central role in health care needs of the people of developing countries. This condition would remain for long, because modern medicine continues to be unable to meet the health care of the people effectively (Jansen, 1981). The dependence of traditional medicinal plants and their role in health care system will increase in the future as they are culturally viable and expected to remain affordable (WHO, 1998) and the modern health care service is limited and expensive compared to the traditional medicine.

ACKNOWLEDGMENTS

The authors greatly acknowledge people of the study area especially, the traditional healers for their hospitality and kind response.

REFERENCES

Abebe D, Ayehu A (1993). Medicinal plants and enigmatic health practice of north Ethiopia. Berhanina Selam Printing Enterprise, Addis Ababa. p. 341.

Addis G, Abebe D, Urga K (2001). A Survey of Traditional Medicinal Plants in Shirka District, Arsi Zone, Ethiopia. Ethiop. Pharm. J., 19: 30-47.

Asfaw D, Dawit A, Kelbessa U (1999). An overview of traditional medicine in Ethiopia: Prospective and Development Efforts. In: Tamirat Ejigu, (ed.). *EthiopianPharmaceutical Association*. Ethiopian Pharmaceutical Association. Addis Ababa, Ethiopia. Silver Jubilee Anniversary, Special Issue. pp. 45-53.

Awas T, Asfaw Z (1999). Report on Ethnobotanical Study Nations, of Nationalities and People in Gambella and Benishangul Gumuz Regional States. Progress Report to Research and Publication Office, Addis Ababa University, Addis Ababa.

Balemie K, Ensermu K, Asfaw Z (2004). Indigenous Medicinal Utilization, Management and Threats in Fentale Area, EasternShewa, Ethiopia. Ethiop. J. Biol. Sci., 3: 1-7.

Balick MJ, Cox PA (1996).Plants, people and Culture: Science of Ethnobotany. NewYork, USA.

Dawit A (2001). The role of medicinal plants in Health care Coverage of

Ethiopia, the possible benefits of integration. In: Medhin Zewdu and Abebe Demissie (eds.). *Coservation and Sustainable Use of Medicinal plants in Ethiopia*. Proceeding of the National workshop on Biodiversity Conservation and Sustainable use of medicinal plants in Ethiopia, 28 April- 01 May 1998, IBCR, Addis Ababa. pp. 107-118.
Dawit A. (1986). Traditional medicine in Ethiopia. The attempt being made to promote it for effective and better utilization. SINET: Ethiop. J. Sci., 9: 61-69.
Deeba F (2009). Documentation of ethnoveterinary practices in urban and peri-urban areas of Faisalabad, Pakistan, PhD Thesis University of Agriculture, Faisalabad Pakistan
Demissew S, Dagne E (2001). Basic and Applied Research in Medicinal Plants. In: Zewdu M, Demissie A (Eds.), Conservation and Sustainable Use of Medicinal Plants in Ethiopia Proceeding of The National Workshop on Biodiversity Conservation and Sustainable Use of Medicinal Plants in Ethiopia. IBCR, AA, 28 April-01 May 1998. pp. 29–33.
Fassil K (2001). The status and availability of oral and written knowledge on traditional health care in Ethiopia. In: Medhin Z, Abebe D (eds.). Conservation and Sustainable Use of Medicinal plants in Ethiopia. Proceeding of the National workshop on Biodiversity Conservation and Sustainable use of medicinal plants in Ethiopia, 28 April- 01 May 1998, IBCR, Addis Ababa pp. 107-119.
Gidey M (1999). An Ethno-botanical study of Medicinal Plants Used by the People in Ethiopia. M. Sc. Thesis, Uppsala, Sweden.
Hunde D (2001). Use and Management of Traditional Medicinal Plants by Indigenous People of Boosat Woreda, Wolenchiti Area: An Ethnobotanical Approach. M.Sc. Thesis, Addis Ababa University.
Jansen PCM (1981). Spices, Condiments and Medicinal plants in Ethiopia, their Taxonomy and Agricultural Significance. Center for Agricultural Publishing and Documentation,Wageningen, Netherlands.P. 327.
Kibebew F (2001). The status and availability of oral and written knowledge on traditional health care on traditional health care in Ethiopia. In: Zewdu M, (editor): Conservation and sustainable use of medicinal plants in Ethiopia, Proceedings of the National workshop, 28 April – 01 May 1998. Addis Ababa Demissie A: Institute of Biodiversity Conservation and Research. pp. 107-119.
Lange D (1998). Europe's medicinal and aromatic plants: their use, trade and conservation. Cambridge: TRAFFIC Europe.
Lewington A (1993). Medicinal plants and plant extracts: A Review of their Importation into Europe. Cambridge, UK. P. 92.
Lulekal E (2005). Ethnobotanical Study of Medicinal Plants and Floristic Composition of Mana Angatu Moist Montane Forest, Bale, EthiopiaM.Sc. Thesis. Addis Ababa University.
Martin GJ (1995). Ethnobotany: A method Manual. Chapman and Hall, London. Pp. 265-270.
McCorkle CM, Mathias-Mundy E (1992). Ethno veterinary medicine in Africa, Afr. J. Int. African Institute (London), 62(1): 59-93.
Mesfin F (2007). An ethnobotanical study of medicinal plants in Wonago oreda, *SNNPR*, ETHIOPIA. Msc thesis Addis Ababa University, Ethiopia.
Mirgissa K (1998). Utilization of plant medicine for the treatment of health problems. The case of Oromo of Chora District Illubabor Zone, Western Ethiopia. Ethiop. J. Health Dev., 10: 161-166.
Sofowora A (1982). Medicinal plants and Traditional medicine in Africa. John Wiley and Sons. New York. Pp. 255-256.
Tesema T, Mirutse G, Nigusu A (2002). National Biodiversity Strategy and Action Plan Project. Resource base of medicinal plants of Ethiopia, first phase report.
WHO (2002). Traditional Medicines Strategy *2002-2005*. World Health Organization, Geneva: WHO. http://www.who.int/medicines/organization/trm/orgtrmmain.shtml.
WHO (1998). Regulatory situation of herbal medicines: A World wide Review. WHO/TRM/98.1,Geneva. Pp. 1-9.
Yirga G (2010a). Assessment of indigenous knowledge of medicinal plants in Central Zone of Tigray, Northern Ethiopia, Afr. J. Plant Sci. 4: 006-011.
Yirga G (2010b). Assessment of traditional medicinal plants in Endrta District, South-eastern Tigray, Northern Ethiopia,Afr. J. Plant Sci., 4: 255-260.
Zerabruk S, Yirga G (2011). Traditional knowledge of medicinal plants in Gindeberet district, Western Ethiopia, South Afr. J. Bot., doi:10.1016/j.sajb.2011.06.006.

24

Using odonates as markers of the environmental health of water and its land related ecotone

D. Acquah - Lamptey , R. Kyerematen and E. O. Owusu

Department of Animal Biology and Conservation Science, P. O. Box LG67, University Of Ghana, Legon, Ghana.

The study of Odonata communities along wetlands requires the basic understanding of the abundance, distribution and number of species present. As habitat conditions change, they also exhibit changes in their diversity and distribution. Odonata assemblages were surveyed along the Densu River at Atewa Range Forest Reserve (ARFR) and Nsawam in the Eastern Region of Ghana and Weija in the Greater Accra Region of Ghana. Of the 177 species recorded for Ghana, 66 species (43 dragonfly and 23 damselfly species) were sampled along the Densu River. These belonged to eight families of which the Libellulidae dominated. The distribution of species was significantly different between the sites with the most diverse area being ARFR with 47 species. The various environmental variables along the river were recorded and their effects discussed.

Key words: Atewa range forest reserve, environmental quality, diversity indices, Densu River, dragonfly, damselfly, conservation.

INTRODUCTION

Dragonflies and Damselflies are insects belonging to the order Odonata. There are a total of 177 Odonata species that have been recorded so far in Ghana (Dijkstra, 2007a) with 72 species recorded from the streams and rivers that have their headwaters within the Atewa Range Forest Reserve (and associated standing water habitats), although only 31 (43%) were found strictly within the reserve's boundaries (Dijkstra, 2007a).

The presence of dragonflies is an important indicator for ecological balance. By way of reproduction, these insects lay their eggs in or near only freshwater (Corbet, 1999) and thus, their high abundance in an area is a good indication of the quality of freshwater. Odonata (dragonflies and damselflies) are used as bioindicators for wetland quality in Europe, Japan, the USA, Australia (Clausnitzer and Jodicke, 2004) and in South Africa (Clark and Samways, 1996; Stewart and Samways, 1998). The greatest numbers of species are found at sites that offer a wide variety of microhabitats, though dragonflies tend to be much more sensitive to pollution than damselflies (Ameilia et al., 2006). Many ecological factors affect the distribution of the nymphs. The acidity of water, the amount and type of aquatic vegetation, the temperature, and whether the water is stationary or flowing all affect the distribution of odonate nymphs. Some species can tolerate a wide range of conditions whiles others are very sensitive to their environment (Chovanec and Waringer, 2001; Schindler et al., 2003; Chovanec et al., 2004; Ameilia et al., 2006; Smith et al., 2006). Clark and Samways (1996), and Stewart and Samways (1998) who studied adult dragonfly site selection, habitat preferences, disturbance response, and bioindicator potential across a variety of wetland and aquatic systems in South Africa concluded that, adult dragonflies prefer habitats with heterogeneous vegetation and are reliable indicators of human perturbation. They based pond conservation on species-habitat associations because of predictable species turnover along gradients of vegetation structure, light regime, and hydrology (Steytler and Samways, 1995; Osborn and Samways.

1996).

The use of odonates as indicators offers several advantages: they are widespread and represent one of the historically most studied insect groups, and so there is a good knowledge of the ecological requirement of a large number of species and their distribution and seasonality; they are relatively easy to observe and indentify, and finally they are well dependent on the ecological conditions of the environment (Corbet, 2004).

Although odonatological studies in Ghana dates back to 1871 (Dijkstra, 2007a), no studies have been conducted on their suitability as indicators or on their relationship with the environment. Hence this study thus sought to: 1) identify the Odonata species occurring along the Densu River; 2) estimate and compare Odonata species diversities along the river; 3) investigate the suitability of Odonata as indicators of water and environmental quality.

MATERIALS AND METHODS

Study areas

The Atewa range located in the eastern region of Ghana, lies between Latitude 4°30' N and 11°00' N and straddles the Greenwich Meridian from Latitude 1°10'E to 3°15'W and consist of a range of hills aligned approximately north-south with steep-sided slopes and flat summits (McCullough et al., 2007). The Atewa Range Forest Reserve covers an area of 232 km² and the surrounding Atewa Range Extension. Geologically, the area is underlain by Birimian formations rich in minerals such as gold, diamond, bauxite and kaolin. The Atewa Range forest is an Upland Evergreen Forest and one of the highest forest covered hills in Ghana and also serves as headwaters for the Ayensu, Sumatua, Suhen, Kuia, Birim (Brim) and the Densu river basins.

The Weija Reservoir (0° 20" W 0° 25' W and 5° 30' N 5° 45' N) was created in 1977 as a replacement for an earlier one which was washed away in 1968 by Ghana Water Company Limited (GWCL) through damming River Densu mainly to satisfy the demand for potable water supply (Asante et al., 2008). The Weija Reservoir, located about 17 km west of Accra, is almost at the mouth of the 116 km long River Densu which takes its source from the Atewa-Atwiredu mountain range in the Eastern Region of Ghana (Asante et al., 2008). The current reservoir provides water to western parts of Accra, supports irrigation projects, as well as fisheries. The Weija Reservoir is 14 km long, 2.2 km wide and has a total surface area of 38 km^2 with mean depth of 5 m (Vanden Bossche and Bernacsek, 1990). The normal surface elevation is estimated at 14.37 km with maximum of 15.24 km (Nukunya and Boateng, 1979). The main economic activities in the catchment are fishing and crop farming with fertilizers (nitrate and phosphate based) being used on most of these farms (Ansa-Asare and Asante, 2005). Major crops include maize, cassava, sugarcane and vegetables. Untreated domestic effluents are discharged into the Reservoir (Asante et al., 2008).

The sampling stations were at the Aqua Vitens Rand Ltd (AVRL) dam site, Oblogo, Tetegu and water fetching site. Nsawam is located in the Akuapem South District of the Eastern Region of Ghana. The township is about 40 km from the capital Accra, on coordinates 05° 48' 15" N 00° 21' 49" W. It is highly populated with industries. The river Densu flows through the township with fishing and farming being the main activities along it. Some of the industries as well as homes have their drainage systems terminating into the river. Sewage discharge points from industries and homes were observed along the stream.

Data collection

Permit was sought from the Forestry Commission of Ghana and the management of the Atewa Range Forest Reserve (ARFR) and Aqua Vitens Rand Ltd (AVRL) at Weija Dam site before sampling commenced. The other sampling stations were in Nsawam and Weija. Sampling stations were chosen within 100 m of the main access points and were at least 400 m apart. GPS points were recorded using a Garmin GPS at a central location for each site, and sampling took place within a 50 m radius of this point.

Data collection was done 5 days per month starting from September, 2010 to April, 2011 with a sampling effort of 2-man per hour. Capturing of Odonates was done along transects with width of 50 m perpendicular from the water systems to maximize the area along the water systems (Bried, 2005).

All the odonates were identified visually with the aid of a pair of close-focus binoculars or caught with an aerial net when necessary and identified using the keys from Dijkstra and Clausnitzer (2013), and by Olsvik (1993) as well as by using comparative material in the Biodiversity and Entomology Museum of the Department of Animal Biology and Conservation Science, University of Ghana, Legon. When it was necessary to transport captured specimen to the laboratory for further identification, pressure was applied on the thorax to paralyze it, the wings folded to one side and put in plastic envelops which were labelled.

Subsequent captures were identified, photographed with a digital camera for further identification and confirmation and released from the aerial nets. The geographical position and altitude were taken with a Garmin GPS. The type of land use (bank side vegetation) was also recorded.

Environmental parameters were recorded for the sample points which included the colour of the stream, type of immediate vegetation use/cover and shade cover. Type specimens of the dragonflies and damselflies are being kept at the Biodiversity and Entomology Museum of the Department of Animal Biology and Conservation Science, University of Ghana, Legon, Ghana.

Data analyses

Species accumulation curve, species richness estimators (Colwell and Coddington, 1994; Colwell, 2000; Magurran, 2010) as well as actual species diversity indices (Shannon-Wiener, Margalef, Pielou and Simpson) were computed and treated as surrogates for biodiversity (Marshall et al., 2006).

The program PRIMER (Clarke and Warwick, 2001; Clarke and Gorley, 2006) was employed to conduct all multivariate analysis. Overall fourth root transformation was applied to the species abundance data and Bray -Curtis similarity resemblance matrix was then generated and then the data was analyzed through ordination by Non-metric Multidimensional Scaling (NMDS) (Clarke and Warwick, 2001). Analysis of Similarity (ANOSIM) was applied to these factors to test significance (Chao et al., 2000).

RESULTS

A total of 5143 individuals belonging to 66 species from eight families comprising of 43 Anisopterans and 23 Zygop terans species were recorded at all 22 sites (4 Weija, 4 Nsawam and 14 within the Atewa Range Forest)

Table 1. Species diversity indices for the study sites.

Site	S	N	d	J'	H'(loge)	1- λ	Rank
Weija	22	3190	2.603	0.4052	1.252	0.5310	3
Atewa	47	1215	6.477	0.8463	3.258	0.9478	1
AOR	39	791	5.694	0.8649	3.169	0.9467	-
AR	27	424	4.298	0.8757	2.886	0.9283	-
Nsawam	34	738	4.997	0.8456	2.982	0.9321	2
All sites	66	5143	7.606	0.6221	2.606	0.8067	-

S = Species Richness, N =species Abundance, d = Margalef index, J' = Pielou's evenness, H' = Shannon's index, 1-λ=Simpson's index
Rank, 1- Highest diversity, 2 - Medium diversity, 3- Lowest diversity.

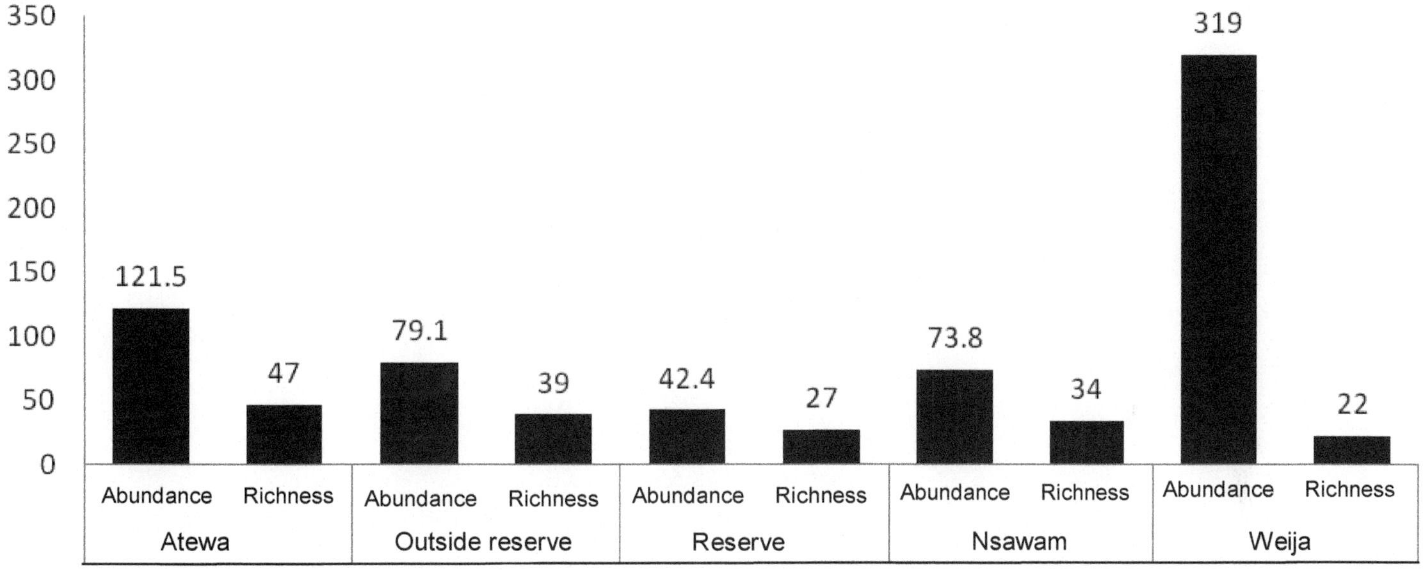

Figure 1. Distribution of Odonata over the entire study area. Abundance frequency is (values are in multiples of 10).

along the Densu River system. The Atewa Range Forest exhibited the highest number of species (47) with 1215 individuals, compared to 34 species (738 individuals) occurring at Nsawam, and 22 species (3190 individuals) occurring at Weija respectively (Table 1, Figure 1). Within the Atewa Range Forest there was a difference in species composition within the protected area (27 species, 424 individuals) and outside the protected area (39 species, 791 individuals) (Figure 1). These differences in species numbers correspond to diversity indices (Shannon-Wiener Index) calculated for these three areas: Atewa (3.258), Nsawam (2.982) and Weija (2.603) (Table 1).

An alternative diversity descriptor, the Simpson index, classified the sites in the same order: Atewa Range Forest as the most species rich habitat (0.9478) followed by Nsawam (0.9321) and then Weija (0.531) (Table 1). Of all the diversity indices calculated, Atewa (outside reserve and within reserve) was the most diverse followed by Nsawam and then Weija (Table 1). The overall Shannon's diversity index for the whole study region was 2.606 (Table 1).

The most abundant Anisopteran species was *Brachythemis leucosticta* (2113 individuals) present at all sites especially along the Densu River at Weija (2092 individuals), whereas the most dominant Zygopteran was *Sapho ciliata* (139) which was only encountered at the Atewa Range Forest Reserve. The Gomphidae, Aeshnidae and Corduliidae families recorded five (3), three (3) and one (1) species respectively.

A summary of estimated species per site as well as the observed species richness is given (Table 2) and species accumulation curve (Figure 2) for the entire sample area and the three major areas. The latter complement the estimates of observed species richness in the three habitats (Table 2), with the Chao 1 and 2, Jackknife 1, Jackknife 2, Bootstrap, Abundance-based Coverage Estimate (ACE) and the Incidence-based Coverage Estimate (ICE) of expected species richness. These estimates emphasize the representativeness of sampling effort in this study, based on the occurrence of rare species (especially singletons and doubletons), as

Table 2. Species richness estimators for Atewa Range Forest, Nsawam, Weija and all sites combined.

Site	N	Sobs Mau Tao	Chao 1	Chao 2	ICE	ACE	Jack knife 1	Jack knife 2	Bootstrap
All sites	22	66	66	68.15	70.72	66	74.59	69.79	71.54
Atewa	14	47	47	50.65	56.41	47	57.21	56.36	52.67
Weija	4	22	21	21.64	23.2	21.44	24	24	22.66
Nsawam	4	34	34	35.05	38.28	34.46	39.25	38.08	37.11

N - Number of sites.

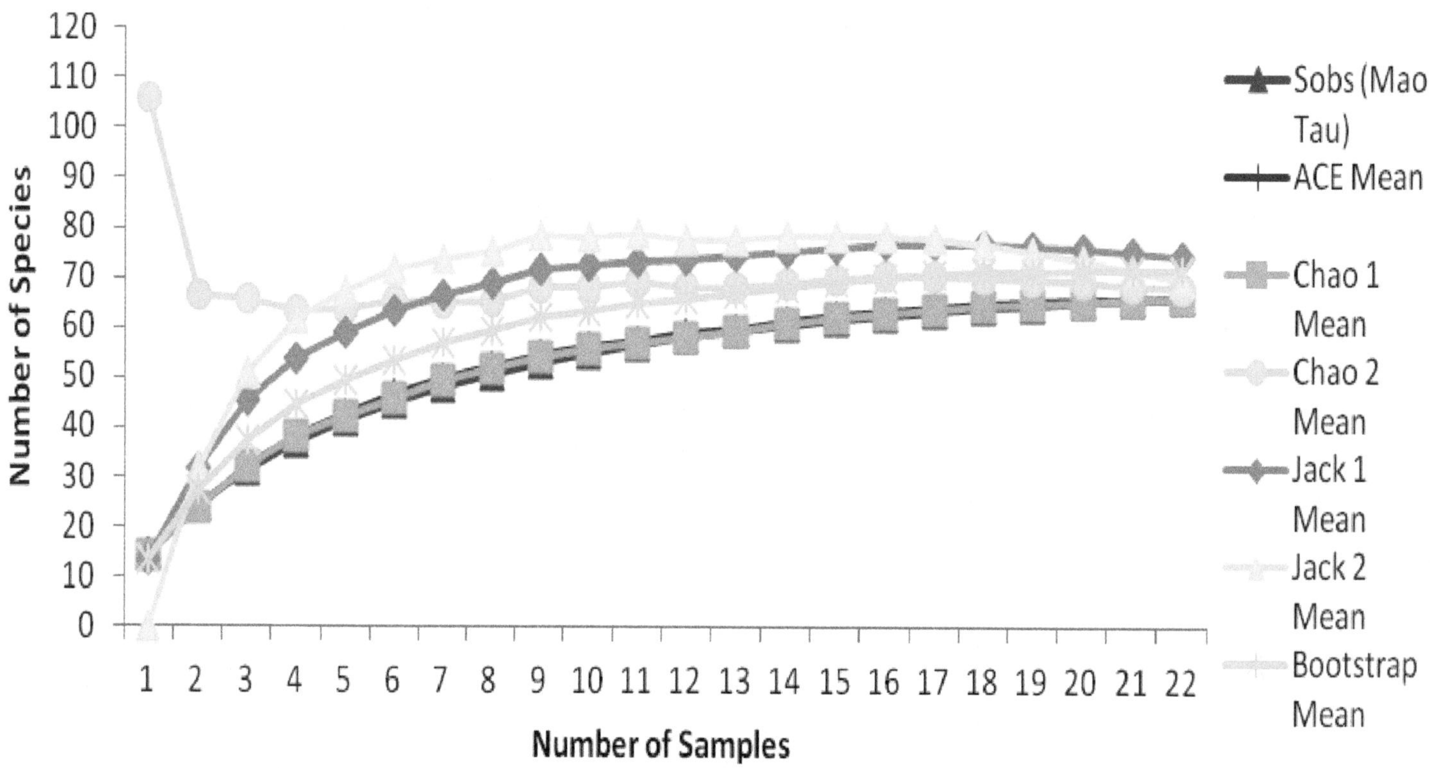

Figure 2. Species accumulation curve for Odonata assemblages along the Densu river system with its associated species richness estimators. Average species richness is based on 50 randomizations (Colwell, 2000).

explained by Colwell et al. (2004) and Colwell and Coddington (1994).

The spatial agglomerative dendrogram of pooled sites species abundance revealed that the study sites can be grouped into six significant groups distinguished at a Bray-Curtis similarity of 63.68% (Figure 3). The SIMPROF test based on the spatial variation in Odonata assemblage composition showed no significant structure at 47.03, 47.7, 51.19, 58.68 and 63.68% similarity respectively.

However, an evidence of structure was found at 42.96% (P=1.6%), 39.36% (P=0.1%), 34.77% (P=0.3%), 34.69% (P=0.1%), 17.07% (P=0.1%), and at 12.9% (P=0.1%) similarity (Figure 4). The following groupings can thus be obtained: Group A: SITES (5, 8, 9), B: SITES (7, 17, 18), C: SITES (10, 11, 12, 13, 14, 15, 16), D: SITES (19, 20, 21, 22), E: SITES (1, 2, 3, 4) and group F: SITE (6).

In comparison, Analysis of similarities (ANOSIM) (Chao et al., 2000) demonstrated a significant difference in species composition between the main geographic areas (Atewa, Nsawam and Weija) (R=0.809, p= 0.001) pair wisely: Atewa-Weija (p=0.002), Atewa-Nsawam (p=0.029) and Weija-Nsawam (p=0.001) (Table 3). In terms of the colour of the stream, the sites with their various abundance were significantly different from each other (R=0.619, p=0.001) (Table 4).

The colour of a water body is an indication of its transparency and transparency is directly affected by the level of suspended particles and dissolved materials in the water Similarly there were significant differences between species composition of the type of immediate

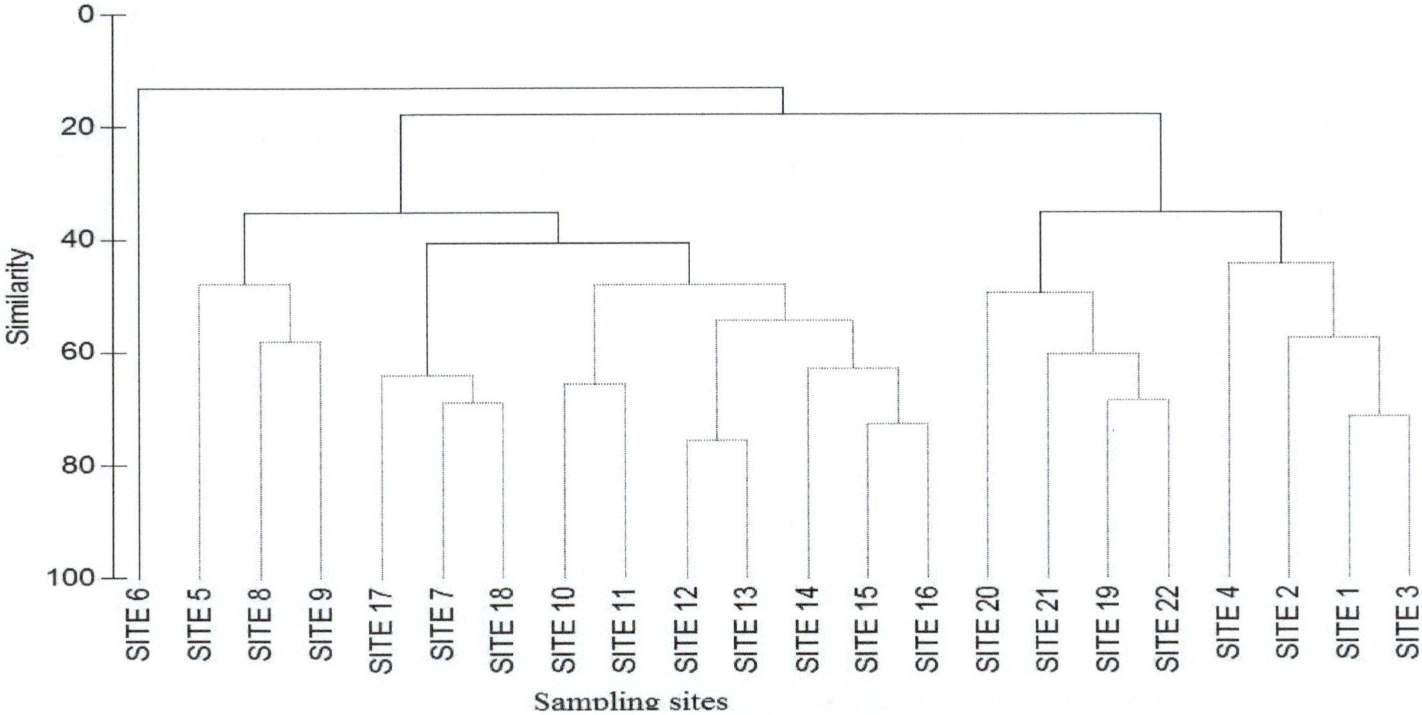

Figure 3. Group average agglomerative dendrogram of Bray-Curtis Similarity of Odonata Species abundance data for 22 different sites along the Densu River system. Thinner lines show evidence of structure (SIMPROF test, $p<0.05$) whereas the darker lines show no evidence of structure.

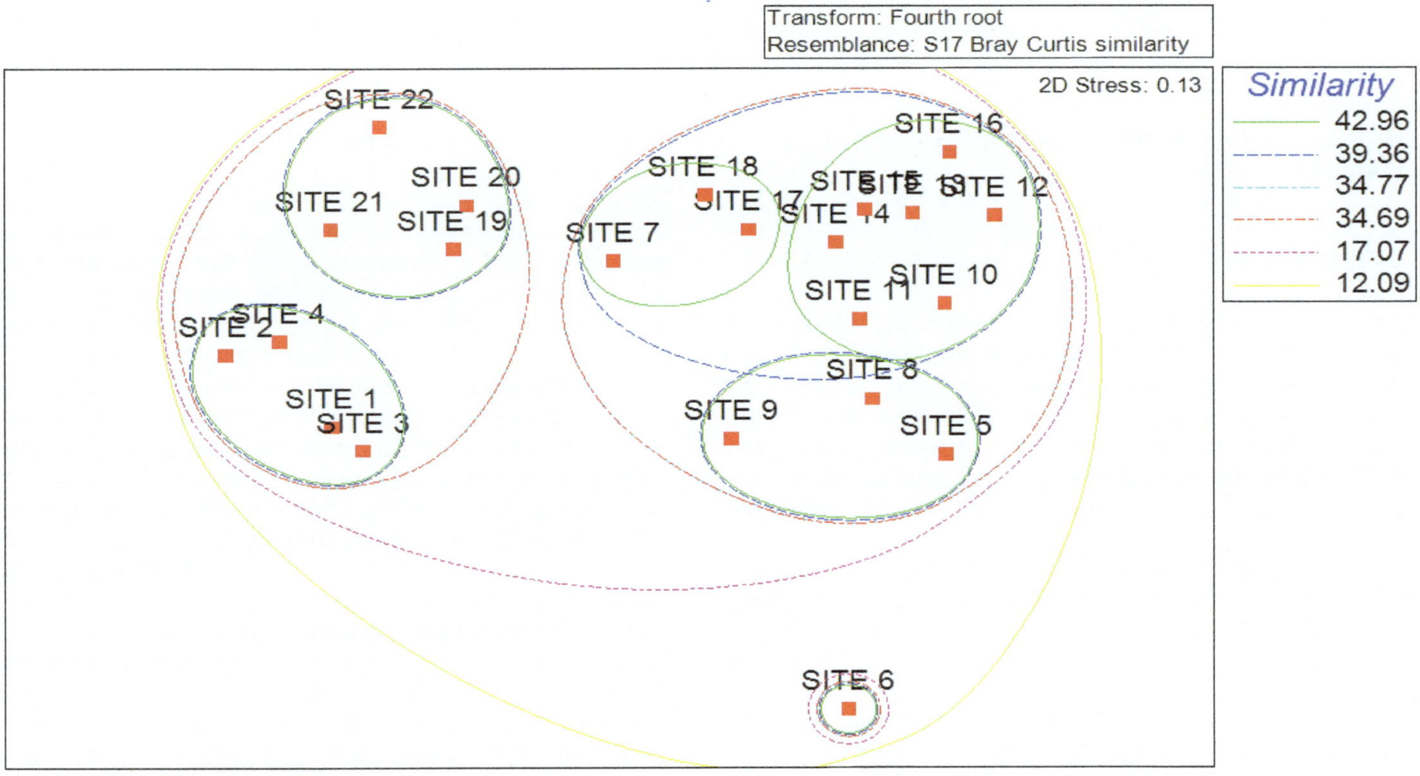

Figure 4. Multidimensional scaling (MDS) ordination plot of the Odonata community structure, based on Bray-Curtis similarity of fourth root transformed species abundances. The significantly separated sample groups by SIMPROF are indicated in blue-green circles.

Table 3. Pair wise analysis of similarity (ANOSIM) comparison of the three stations along the Densu River with Global R=0.809, significance level =0.0001 and 999 permutations.

Group	R Statistic	Significance level %	Possible permutations	Actual permutations	Number >= observed
Weija, Atewa forest	0.893	0.2	3060	999	1
Weija, Nsawam	0.823	2.9	35	35	1
Atewa forest, Nsawam	0,823	0.1	3060	999	0

Table 4. Pair wise analysis of similarity (ANOSIM) comparison of species resemblance matrix with respect to colour of stream with Global R=0.619, significance level =0.0001 and 999 permutations.

Group	R Statistic	Significance level %	Possible permutations	Actual permutations	Number >= observed
Green, brown	0.397	15.6	45	45	7
Green, clear	0.956	1.1	91	91	1
Brown, clear	0.552	0.1	125970	999	0

Table 5. Pair wise analysis of similarity (ANOSIM) comparison of species resemblance of sampled sites with respect to the type of bank side vegetation type with Global R=0.479, significance level =0.0001 and 999 permutations.

Group	R Statistic	Significance level %	Possible permutations	Actual permutations	Number>= observed
Grassland, farmland	0.453	0.3	3003	999	2
Grassland, secondary forest	0.976	0.1	792	792	1
Farmland, secondary forest	0.215	2.2	19448	999	21

vegetation cover along the stream (R=0.479, p=0.001) (Table 5).

DISCUSSION

Of the 5143 individuals of Odonata sampled, a total of 66 species belonging to 8 families were identified in this inventory from 22 sampled sites along the Densu River system, which takes up its source from the Atewa Range Forest Reserve, flowing through several towns and settlements and discharging into the sea at the Densu Delta at Weija. Dijkstra and Clausnitzer (2006) suggest that, the odonatofauna of Ghana is estimated to be about 226 species even though till date only 177 species have been discovered (Neville, 1960; Pinhey, 1962; Frempong and Nijjhar, 1973; Marshall and Gambles, 1977; Olsvik, 1993; D'Andrea and Carfi, 1994; O'Neill and Paulson, 2001; Dijkstra, 2007a).

This study recorded 37% of the species discovered so far in Ghana. This can be pooled from the species richness estimators provided (Table 1) where for instance, for the entire study, jack knife, a non-biased estimator looking at rare/unique species, estimates a total richness of 72 as opposed to the 66 recorded which is the lower boundary of the true number (Smith and Pontius, 2006). Another non-biased estimator of species richness accounted for is the bootstrap which also compensates for underestimating. The other estimators of species richness include the Abundance - based Coverage Estimator (ACE), Incidence-based Coverage Estimator (ICE), Chao 1, Chao 2 and they all give a fair idea of the actual number of species (Chao, 1984; 1987; Chao and Lee, 1992) which is necessary if conservation measures are to be put in place. The species richness recorded for the study however falls within the range of species richness estimators indicating a fair sampling effort (or uniformity of sampling)

The highest species diversity (S = 47) as well as high diversity indices recorded at Atewa (Table 1) can be attributed to the high biodiversity at Atewa as well as its spatial heterogeneity. Streams within the Atewa Forest Reserve have thicker vegetation along their banks which serve as perching sites for less mobile species and the fact that more sites were sampled here due to its heterogeneity compared to the other sites may also account for the highest species diversity at Atewa. Forests are known and expected to have the greatest diversity of Odonata (Dijkstra and Clausnitzer, 2006; Kalkman et al., 2008). In terms of species occurrence, Atewa had 5 singletons where as Nsawam had 1 singleton and 2 doubletons with Weija having 1 singleton and 1 doubleton. This could be as a result of poor 'detectability' of species within the forest as it is expected to harbour rare and low abundance species (Dijkstra and Clausnitzer, 2006). The diversity of Odonata is higher in the forest with some generalist species occurring in both forest and non forested regions.

Sites similar in community composition and or environmental variables were grouped together which is an indication of the environmental variables that accounted for their grouping. This demonstrates the suitability of using the insect order Odonata to indicate the environmental conditions (variables) of the Densu River and its land related ecotones.

In addition, non metric multidimensional plots showed the similarity between sites based on a Euclidean distance matrix on encountered species and their abundance at each site. The relative distance between the sites demonstrates their level of dissimilarity, however, this enhances straightforward, quantitative and objective interpretation, whereby inductive inferences on environmental structure can be readily formulated as proposed by Tong (2002a, 2002b).

Upon analysis of sites based on species abundance matrix using Bray - Curtis dissimilarity, the geographic areas were significantly different. Species abundance also differed between the sites. These could be due to the difference in environmental structures such as shade cover (Dijkstra and Lempert, 2003; Dijkstra and Clausnitzer, 2006; Remsburg et al., 2008), land use type (Kalkman et al., 2008) and micro-climate (Oertli, 2010). Also, Dijkstra and Lempert, (2003) point out that the degree of shadiness seems to be a principal cue for odonates to select a forest habitat, but the forest type (species composition) as well as the age of the forest bordering running water sites is of little influence for the Odonata assemblages as long as the required cover is present. This is also mentioned by Hofmann and Mason (2005), when they found shade to strongly affect adult dragonflies near a river in Eastern England. According to them, vegetation type is an important influence on controlling shading and temperature of streams and thus there was a relationship between shade, temperature and vegetation type with species diversity.

A similar pattern was observed in this study, which reinforces the argument that shade, as well as vegetation cover play an important role in determining Odonate assemblages in the study area as the highest species richness is observed in areas with shade cover and forest as the principal land use type. When species compositions were grouped by similar environmental conditions, it was observed that differences between them were mostly significant. This is probably an indication that, these environmental conditions (land use type and water colour) serve as boundaries distinguishing Odonate assemblages. These factors again are considered by odonates in selecting for ovipositing sites (Corbet, 1999; Thompson et al., 2003), mating sites (Remsburg et al., 2008), roosting and resting sites (Askew, 1982; Pritchard and Kortello, 1997), feeding sites (Corbet, 1999) as well as requirement for nymphs (Samways and Steytler, 1996; Corbet, 1999; Thompson et al., 2003).

Out of the eight families (4 Anisoptera, 4 Zygoptera) of Odonata recorded, the most represented families were Libellulidae and Coenagrionidae with 33 and 12 species respectively (Figure 1), confirming Dijkstra (2007b) results. The dominance of the Anisopterans by the Libellulidae family is also mentioned in (Dijkstra, 2007b; Dijkstra and Clausnitzer, 2013; Lawler, 2001; Silsby, 2001; Kadoya et al., 2004 Suhling et al., 2004; Suhling et al., 2005).

The Gomphidae and Aeshnidae families are known to be very sensitive to ecological, habitat and environmental changes eventhough they do not occur in large numbers (Ameilia et al., 2006; Dijkstra and Clausnitzer, 2013; IUCN, 2010). They are relatively larger in size, are very fast flyers and were mostly observed flying at dusk. According to Clausnitzer (2003) they hunt above streams and along forest paths at dusk. Hence could have resulted in their low encounter rates.

The dominance of the Zygopterans by the Coenagrionidae family could be as a result of the following reasons: Coenagrionidae is the largest of the Zygoteran odonates, they have a shorter life cycle and a wide distribution (Norma-Rashid et al., 2001) and can tolerate a wide range of habitats (Samways, 1989). The low numbers of the Zygopterans is generally due to their low dispersal ability. *Pseudagrion nubicum, Ceriagrion glabrum*, *Ischnura senegalensis* and *Elattoneura nigra* were the only species found only in open areas in high abundance, whereas all the other Zygopterans such as *Pseudagrion torridum*, the Chlorocypha and Phaon species had higher numbers in shaded areas. This confirms work done by Subramanian (2005) and Arulprakash and Gunathilagaraj (2010) who proposed that shade could favour damselflies more than dragon-flies.

Conclusion

This study has shown that, Odonata diversities along the Densu River vary hence these populations can be monitored, related and used as indicators of the physical structure of the river and its surrounding water related ecotones.

The Odonata is an insect order suitable for measuring the environmental quality of the Densu River and its land related ecotones. Though sections within the Atewa Forest showed signs of high disturbance, with illegal felling of trees on daily basis, it still harboured diverse fauna, including typical forest Odonates like *Umma cincta*, *Sapho ciliata*, *Chlorocypha luminosa*, *C. radix* and *Cyanothemis simpsoni*. This suggests that the water quality of the Densu River is still good enough to support these species despite extensive damage to the surrounding landscape. The detection of migratory and open habitat species like *Crocothemis erythraea* serves as an indicator of an anthropogenic change in habitats in the Atewa Range Forest Reserve.

The study has also shown that at least 66 species of Odonate can be found along the Densu River with the Atewa Range Forest Reserve recording the highest diversity, a finding that is significant in terms of ecotourism potential for Dragonfly enthusiasts.

Recommendations

This inventory has served as the baseline for Odonata communities along the Densu River hence can be a measure of monitoring in the near future.

This inventory can also serve as a point of contact to tourists interested in odonatofauna in Ghana, therefore it is recommended that Managers of such reserves and protected areas must strengthen and increase measures for deterring illegal human activities within these areas.

District authorities should embark on a massive re-education campaign to address the low levels of environmental awareness that pertains in the basin.

Finally, there must be an increase in education on the importance of using local insect species as first level indicators of environmental health which when improved upon can save the nation a lot of money otherwise used in the chemical evaluation and monitoring of water quality.

ACKNOWLEDGEMENTS

We are grateful to IdeaWild (www.ideawild.org) for supporting this study with equipment and to the Tropical Biology Association (www.tropical-biology.org) for their advice and mentoring. We are also grateful to Klaas-Douwe-Dijkstra for his mentoring, advice and guidance with species identifications.

REFERENCES

Ameilia ZS, Che Salmah MR, Abu Hassan A (2006). Distribution of Dragonfly (Odonata: Insecta) in the Kerian River Basin, Kedah-Perek, Malaysia. USU Repository. 14 p.

Ansa-Asare OD, Asante KA (2005). Changes in the Chemistry of the Weija Dam Reservoir in Ghana, Twenty Years after impoundment. West Afr. J. Appl. Eco. 8, 35-47.

Arulprakash R, Gunathilagaraj K (2010). Abundance and diversity of Odonata in temporary water bodies of Coimbatore and Salem districts in Tamil Nadu. *J. Threatened Taxa* 2(8): 1099-1102.

Asante KA, Quarcoopome T, Amevenku FYK (2008). Water quality of the Weija Reservoir after 28 years of impoundment. West Afr. J. Applied Ecol. 13:1-8.

Askew RR (1982). Roosting and resting site selection by coenagrionid damselflies. Adv. Odonatol. 1:1-8.

Bried JT (2005). Community and Conservation Ecology of Dragonfly and Damselfly Adults in Mississippi Wetlands. An MSc. Thesis Submitted to the Faculty of Mississippi State University.

Chao A (1984). Non-parametric estimation of the number of classes in a population. Scand. J. Stat. 11:265-270.

Chao A (1987). Estimating the population size for capture - recapture data with unequal catchability. *Biometrics.* 43:783-791.

Chao A, Hwang WH, Chen YC, Kuo CY (2000). Estimating the number of shared species in two communities. *Statistica Sinica* 10, 227 - 246.

Chao A, Lee SM (1992). Estimating the number of classes via sample coverage. J. Am. Stat Assoc. 87:210-217.

Chovanec A, Waringer J (2001). Ecological integrity of river-floodplain systems- assessment by dragonfly surveys (Insecta: Odonata). Regul. River 17 (4-5):493-507.

Chovanec A, Waringer J, Rajj R, Laister G (2004). Lateral connectivity of a fragmented large river system: assessment on a macroscale by dragonfly surveys (Insecta: Odonata). Aquat. Conser. - Marine and Freshw. Ecosyst. 14:163-178.

Clark TE, Samways MJ (1996). Dragonflies (Odonata) as indicators of biotope quality in the Kruger National Park. South Afr. J. Appl. Ecol. 33:1001-1012.

Clarke KR, Gorley RN (2006). PRIMER v6: User manual/tutorial, PRIMER-E, Plymouth UK, 192pp.

Clarke KR, Warwick RM (2001). Change in Marine Communities: An Approach to Statistical Analysis and Interpretation. (2nd ed): PRIMER-E, Plymouth UK. 172 p.

Clausnitzer V (2003). Odonata of African humid forests - a review. *Cimbebasia*, 18:173-190.

Clausnitzer V, Jödicke R (2004). Guardians of the watershed. Int. J. Odonatol. 7 (2):111-112.

Colwell RK (2000) EstimateS- Statistical estimation of species richness and shared species from samples. Version 6. User's guide and application published at http://viceroy.eeb.uconn.edu/EstimateS.

Colwell RK, Chang X, Chang J (2004). Interpolation, extrapolating and comparing incidence-based species accumulation curves. Ecology 98:2717-2727.

Colwell RK, Coddington JA (1994). Estimating terrestrial biodiversity through extrapolation. *Philos. T. R. Soc. Lond.* B 345, 101-118.

Corbet PS (1999) *Dragonflies:* Behav. Ecol. Odonata. Cornell University Press, New York.

Corbet PS (2004) *Dragonflies: Behav. Ecol. Odonata.* Comstock Publishing Associates, Cornell University Press, Ithaca, NY. 303pp.

D'Andrea M, Carfì F (1994). Annotations on a small dragonfly collection from Ghana, West Africa, with six new species for the national fauna (Odonata). Opuscula Zool. Fluminensia 125:1-7.

Dijkstra KDB (2007a). Dragonflies and Damselflies (Odonata) of the Atewa Range, Ghana. *In* McCullough, Alonso J, Naskrecki LE, Wright P, Osei-Owusu HE, (eds.) A Rapid Biological Assessment of the Atewa Range Forest Reserve, Eastern Ghana. RAP Bull. Biol Assess. 47. Conservation International, Arlington, VA. 50-54.

Dijkstra KDB (2007b). Demise and rise: the biogeography and taxonomy of Odonata of tropical Africa. PhD Thesis, 2007, Leiden University. pp. 143-187.

Dijkstra KDB, Clausnitzer V (2006). Thoughts from Africa: how can forest influence species composition, diversity and speciation in tropical Africa? In Cordero-Rivera, A. (ed.), Forests *and Dragonflies*, 127-151. Pensoft Publishers, Sofia.

Dijkstra KDB, Lempert J (2003). Odonate assemblages of running waters in the Upper Guinean forest. *Archiv für Hydrobiologie* 157:397-412.

Dijkstra KDB, Clausnitzer V (2013). The dragonflies and damselflies of eastern Africa: handbook for all Odonata from Sudan to Zimbabwe. *Studies in Afrotropical Zool.* Belgian Royal Museum for Central Africa (in press).

Frempong E, Nijjhar B (1973). Some preliminary observations on the fauna and flora of Barekese Lake, Ghana. Buleltin de l'Institut Fondamental de l'Afrique Noire 35 (1):67-78.

Hofmann TA, Mason CF (2005). Habitat characteristics and the distribution of Odonata in a lowland river catchment in eastern England. Hydrobiologia 539:137-147.

IUCN (2010). IUCN Red List of Threatened Species. Version 2010. 3. Retrieved from http://www.iucnredlist.org. Accessed on 05 October 2010.

Kadoya T, Suda S, Washitani I (2004). Dragonfly species richness on man-made ponds: effects of pond size and pond age on newly established assemblages. *Ecol. Res.* 19:461-467.

Kalkman VJ, Clausnitzer V, Dijkstra KDB, Orr AG, Paulson DR, van Tol J (2008). Global diversity of dragonflies (Odonata) in freshwater. *Hydrobiologia* 595, 351-363.

Lawler SP (2001). Rice fields as temporary wetlands: a review. Israel J. Zool. 47:513-528.

Magurran AE (2010). Measuring Biological Diversity. Blackwell Publishing, Malden, MA. 256pp

Marshall AG, Gambles RM (1977). Odonata from the Guinea Savanna Zone in Ghana. J. Zool. Lond. 183, 177-187.

Marshall JC, Steward AL, Harch BD (2006). Taxonomic resolution and quantification of freshwater macroinvertebrate samples from an Australian dryland river: the benefits and costs of using species abundance data. Hydrobiologia, 572:171-194.

McCullough J, Alonso LE Naskrecki P, Wright HE, Osei-Owusu Y (eds.) (2007). A Rapid Biological Assessment of the Atewa Range Forest Reserve, Eastern Ghana. RAP Bull. Biolog. Assess. 47. Conservation International, Arlington, VA.

Neville AC (1960). A list of Odonata from Ghana, with notes on their mating, flight and resting sites. *Proc. Roy. Entomolog. Soc. Lond.* 35: 124-128

Norma-Rashid Y, Mohd-Sofian A, Zakaria-Ismail M (2001). Diversity and distribution of odonata (dragonflies and damselflies) in the fresh water swamp Lake Tasek Bera, Malaysia. Hydrobiologia, 459 p.

Nukunya GK, Boateng EO (1979). Weija Dam Project: An Evaluation/Study of Water Utilisation Behaviour and its Related Socio-economic Impact. Report submitted to Canadian International Development Agency (CIDA).

O'Neill G, Paulson DR (2001). An annotated list of Odonata collected in Ghana in 1997, a checklist of Ghana Odonata, and comments on West African Odonate biodiversity and biogeography. Odonatologica 30:67- 86.

Oertli B (2010). The local species richness of Dragonflies in mountain water bodies: an indicator of climate warming? In: Ott, J. (Ed) (2010) Monitoring Climatic Change With dragonflies. *BioRisk* 5:243-251.

Olsvik H (1993). Adult Odonata. A preliminary Key to Afrotropical Families and Genera with a checklist and Bibliography of West Africa Species. Unpublished. 29-38.

Osborn R, Samways MJ (1996). Determinants of adult dragonfly assemblage patterns at new ponds in South Africa. Odonatologica 25:49-58.

Pinhey E (1962). Some records of Odonata collected in Tropical Africa. J. Entomol. Soc. South Afr. 25:20-50.

Pritchard GD, Kortello A (1997). Roosting, perching and habitat selection in *Argia vivida* and *Amphiagrion abbreviatum* (Odonata: Coenagrionidae), two damselflies inhabiting geothermal springs. Canadian Entomol. 129:733-743

Remsburg AJ, Olson AC, Samways MJ (2008). Shade Alone Reduces Adult Dragonfly (Odonata: Libellulidae) Abundance. *J. Insect Behav.* 21(6):460-468.

Samways MJ (1989). Taxon turnover in Odonata across a 3000m altitudinal gradient in Southern Africa. *Odonatologica* 18:263-274.

Samways MJ, Steytler NS (1996). Dragonfly (Odonata) distribution patterns in urban and forest landscapes, and recommendations for riparian management. Biol. Conserv. 78:279-288.

Schindler M, Fesl C, Chovanec A (2003). Dragonfly associations (Insecta: Odonata) in relation to habitat variables: a multivariate approach. Hydrobiologia 497(1):169-180.

Silsby J (2001). Dragonflies of the World. Smithsonian Institution Press, Washington, DC.

Smith CD, Pontius JS (2006). Jackknife estimator of species richness with S-PLUS. J. Stat. Software, 15 (3):12pp.

Smith J, Samways MJ, Taylor S (2006). Assessing Riparian Quality Using Two Complementary Sets of Bioindicators. *Biodiversity and. Conservation,* 16 (9):2695-2713.

Stewart DAB, Samways MJ (1998). Conserving dragonfly (Odonata) assemblages relative to river dynamics in an African savannah game reserve. Conserv. Biol. 12, 683-692.

Steytler NS, Samways MJ (1995) Biotope selection by adult male dragonflies (Odonata) at an artificial lake created for insect conservation in South Africa. Biol. Conserv. 72: 381-386.

Subramanian KA (2005). Dragonflies and Damselflies of Peninsular India: A Field Guide. Project Lifescape. Indian Academy of Science, Bangalore, India. 118 p.

Suhling FF, Sahlen G, Kasperski J, Gaedecke D (2005). Behavioural and life history traits in temporary and perennial waters: comparisons among three pairs of sibling dragonfly species. Oikos 108:609-617.

Suhling, FF, Schenk K, Padeffke T, Martens A (2004). A field study of larval development in a dragonfly assemblage in African desert ponds (Odonata). Hydrobiologia 528:75-85

Thompson DJ, Rouquette JR, Purse BV (2003). Ecol. Southern Damselfly. Conserving Natura 2000 Rivers Ecology Series. English Nature. Peterboroug. 24 p.

Tong STY (2002a). The use of non-metric multidimensional scaling as an ordination technique in resource survey and evaluation: a case study from southeast Spain. Appl. Geogr. 12(3):243-260.

Tong STY (2002b). On non-metric multidimensional scaling ordination and interpretation of the matorral vegetation in lowland Murcía. Plant Ecol. 79 (1-2) 65-74.

Vanden Bossche JP, Bernacsek GM (1990). Source Book for the Inland Fishery Resources of Africa, 1. CIFA Technical Paper No 18/1. 240 p.

Ethnomedicinal studies on Bondo tribe of Malkangiri District, Odisha, India

Suchitra Rout[1], Sitaram Prasad Panda[2] and Hemant Kumar Patra[1]

[1]Post Graduate Department of Botany, Utkal University, Vanivihar, Bhubaneswar-751004, Odisha, India.
[2]Mangrove Forest Division (WL), Rajnagar, Kendrapara, Odisha, India.

Ethnomedicinal studies revealed the use of 34 angiospermic species by the Bondo tribe of Malkangiri district of Odisha. The traditional medicinal uses were observed to be wide spread and prevalent over modern medicine in the study area. In the present paper, plants for various diseases have been discussed togrther with dosage and mode of administration. Concurrently, correct name, family, local name are also provided for correct identification of the species.

Key words: Bondo, ethnomedicine, Malkangiri district, Odisha.

INTRODUCTION

In the past decade, there was renewed attention and interest in the use of traditional medicine globally. The World Health Organization (WHO) has pointed out that traditional medicine is an important contributor to its health goals. Today, according to the WHO, as many as 80% of the world's people depend on traditional medicine and in India, 65% of the population in the rural areas especially those residing in the remote forests mainly rely on traditional health practices as it is cost effective. Nowadays, the modern people are also using traditional medicinal practices as it has no side effects. Ethnobotany is a very broad field and it is an age old process starting from human civilization and medicine is one of its facet. In Odisha, 62 tribes are living in harmony with the nature in the mountainous forests of different districts. Out of these, 13 tribes are notified as Primitive Tribal Groups (PTG) while two tribes namely Bondo of Malkangiri and Dongaria Kandha of Kalahandi and Raygada districts are very rare and remote tribes under Particularly Vulnerable Tribal Group (PVTG). The present study focused on Bondo tribes of Khairaput block of Malkangiri sub-division. The Bondo with a population of 5895 is one of the primitive tribes found only in the Malkangiri district of Odisha. They are confined to the hill ranges of Khairaput block of this district. Because of the long inhabitation of the tribe in this region, it is known as the Bondo country and the hills present in this region are known as the Bondo hills. The Bondo belongs to south Munda group of austraic family. They are sometimes referred to as the Bondo paraja. There are two important villages namely Mudulipada and Andrahal flanked by a distance of 14 km from each other where the Bondo tribes are living in harmony with nature.

In earlier times, Malkangiri was endowed with rich floristic diversity due to less population pressure coupled with dense forest coverage. In spite of rich and diverse floristic composition, Haines (1921-1925), the pioneer plant explorer for the state of Bihar and Orissa and Mooney

(1950), the subsequent worker could not botanize this region thoroughly and as such various uses of plants by the natives were not documented in their respective treatise. There is considerable amount of genetic erosion in the plant wealth with the march of urbanization, establishment of factories coupled with the interferences of various categories such as shifting cultivation, illegal felling of tree species for timber and fuel, etc. Paradoxically, the magnitude of devastation is in increasing order during recent times, quite a good number of medicinal and food plants have already been wiped out. On the other hand, establishment of hospitals in remote areas has fascinated the aboriginal people to adopt the modern medical practices. Hence, traditional therapy is almost ignored.

Similarly, their food habit and utilization of plants and plant products for cultural activities have also been changed due to influence of modern civilization. Hence, critical identification as well as conservation of these fast disappearing elements is highly essential. Realizing this, Jain (1970-1971), Pal and Banerjee (1971), Sexena and Dutta (1975), Mudgal and Pal (1980), Subudhi and Choudhury (1985), Sahoo (1986) and Anonymous (1989, 1996) have made some sporadic reports on the ethnobotanical aspects.

Study area

Malkangiri is the southernmost district of Odisha. The district is named after its headquarters town Malkangiri. During the formation of Orissa province in 1936, Malkangiri was a 'Taluk' of Nabarangpur sub-division of Koraput district of Odisha. In 1962, it was upgraded to a sub-division of Koraput district and later on designated as a separate district. Malkangiri lies in between 17°45'N and 18°40' N latitudes and 81°10'E to 82.00' E longitude and situated at an elevation of 641'(feet) from the mean sea level (Map 1). The district spreads over an area of 5791 sq. km (48. 43 km radius) as per the census of 2011. Malkangiri has a population of 613,192, out of which 5895 are Bondo.

The area is of broken mountains intersected by large riverbeds and water courses. The altitude varies from 400 m near western side to 1200 m on the eastern side with mountain peaks and ridges. Sandy and clay type of soil is found to be predominant in the district. The climate of the major portions of the district is influenced by its varied elevation ranges. The minimum and maximum temperatures are 13 and 47°C in the month of December and May, respectively (Indian Meteorological Department, Bhubaneswar). Humidity is generally high especially in the monsoon and post monsoon months. The average annual rainfall is 1,700 mm (67 inches).

METHODOLOGY

Exhaustive seasonal field visits were undertaken (2010-2011) in the aegis of Post Graduate Department of Botany, Utkal University, Bhubaneswar for collection of plants and their relevant information. The plants collected, have been identified in consultation with the regional floras (Haines 1921-1925; Saxena and Brahmam, 1994-1996) and monographs and preserved in the herbarium of the Post Graduate Department of Botany, Utkal University, Bhubaneswar. The information was gathered through oral interviews of the local tribal men, especially older persons, local medicine men and herbalists. During discussions with the tribal peoples, several data like local names, usable parts, medicine preparation methods, mode of application and dosage were recorded. The plants were arranged under respective diseases on which these are used along with family, local name and mode of preparation as well as administration. The medicinal uses of the species have also been confirmed with standard literatures viz. Agarwal and Ghosh (1985), Anonymous (1948-1972), Chopra et al. (1956), Kirtikar and Basu (1935), Satyavati et al. (1987), Saxena and Dutta (1975) and Warrier et al. (1994-1996).

RESULTS AND DISCUSSION

A total of 34 plant species belonging to 33 genera and 25 families were identified as being used for treatment of approximately 17 ailments or therapeutic indications including headache, toothache etc. (Table 1). Fabaceae, Caesalpiniaceae and Acanthaceae families had the largest number of plants used; and the largest numbers of preparations used were for dysentery and skin diseases. Mostly, leaves are used for preparation of medicines for different ailments. Water is used exclusively in the preparation of medicines. The use of *Diospyros melanoxylon* Roxb., *Nyctanthes arbor-tristis* L., *Cassia fistula* L., *Holarrhena pubescens* Wall. ex G.Don and *Hemidesmus indicus* R.Br. by the Bondo tribes for healing different ailments are recorded for the first time and marked with asterisk in Table 1.

Conclusion

Malkangiri, one of the richest botanical treasures in Odisha is under intense biotic pressure owing to forest fire, over exploitation of medicinal plants for drugs and conversion of forest to agricultural lands. These factors of degradation results not only in the depletion of certain valuable interesting medicinal taxa but also leads to substantial reduction in the density and genetic diversity of forest cover. This rapidly damages the natural habitats threatening the very survival of several indigenous species.

Presently, developing nations such as India have an imperative need of a systematic document on the traditional knowledge of various medicinal uses of plants in all autonomous areas or communities, many of which are still largely unexplored. Such documentation is necessary because older people are usually the only custodians of such information and the fast disappearance of traditional cultures and natural resources arising from urbanization and industrialization of such areas suggest that unrecorded information may be lost for ever.

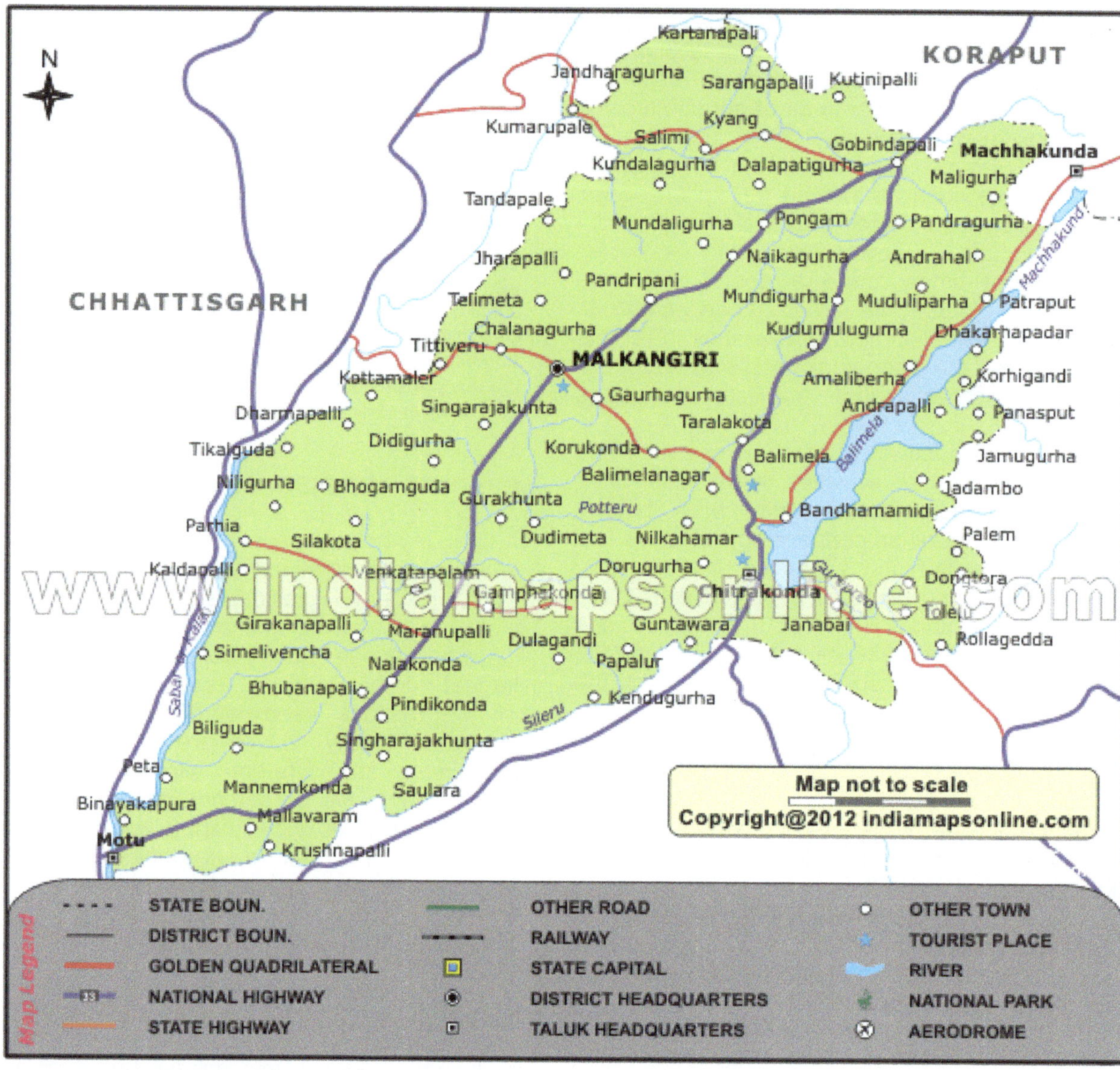

Map 1. Map of Malkangiri District of Odisha.

Table 1. Traditional uses of plants against various diseases.

Therapeutic indication **Botanical and Family name**	**Local name**	**Plants parts used**	**Medicinal preparation and dosage**
Abortifacient			
Abrus precatorius L. (Fabaceae) (Figure 1)	*Kaincha*	Seeds	Three seeds taken for 4 consecutive days to induce abortion
Plumbago zeylanica L. (Plumbaginaceae)	*Sitaparu*	Root	Two grams of root power with water taken orally for 2 times.
Cough			
Adhatoda zeylanica Medik. (Acanthaceae)	*Basongo*	Leaves	Decoction of seven leaves taken orally for one week.
Barleria prionitis L. (Acanthaceae)	*Tamresa*	Leaves	15 ml of leaf juice mixed with honey given orally for one week.

Table 1. Contd.

Cuts and wounds			
Semecarpus anacardium L. f. (Anacardiaceae)	*Kalabhalia*	Fruits	Oil extracted from the fruits applied externally.
Sida acuta Burm. f. (Malvaceae)	*Bajarmuli*	Leaves	Juice of the leaves applied locally.
Tridax procumbens L. (Asteraceae) (Figure 7)	*Bisal gacha*	Leaves	7-8 leaves rubbed in hand and the juice applied externally.
Diabetes			
Pterocarpus marsupium Roxb. (Fabaceae)	*Bija*	Stem bark	Decoction of the stem bark (2 g) given orally every day.
Syzygium cuminii (L.) Skeels (Myrtaceae)	*Jamukoli*	Leaves and fruits	Juice of leaves and powdered fruit with water taken orally every day.
Dysentery			
Aegle marmelos (L.) Correa ex Roxb. (Rutaceae)	*Bel*	Fruits	10-15 ml of infusion of fruit taken orally twice daily.
Bauhinia vahlii Wight & Arn. (Caesalpiniaceae)	*Sialpatar*	Pods	Decoction of seven pods taken orally for 2 days.
**Diospyros melanoxylon* Roxb. (Ebenaceae)	*Duri*	Stem bark	Crushed stem bark with water given orally for 3 days.
Phyllanthus emblica L. (Euphorbiaceae)	*Anlakoli*	Fruits	Decoction of fruits (seven) taken orally for 3 days.
Fever			
Andrographis paniculata (Burm.f) Wall. ex Nees (Acanthaceae)	*Bhuingkara*	Leaves	50 ml of infusion of leaves taken orally for 3 days.
Argemone mexicana L. (Papaveraceae) (Figure 2)	*Kantakusum*	Root	Fresh root grounded with black pepper and the extract given orally one spoonful twice in a day for 5 days.
**Nyctanthes arbor-tristis* L. (Oleaceae)	*Kokra*	Leaves	Leaf juice together with black pepper is given internally till fever cures.
Gastro–intestinal disorders			
Achyranthes aspera L. (Amaranthaceae)	*Chir-Chir*	Aerial parts	Half glass of decoction of aerial parts taken orally for 2 days.
Asparagus racemosus Willd. (Liliaceae) (Figure 5)	*Pinaspari*	Root	Dried root powder crushed with turmeric and the filtrate taken orally, 2 spoonfuls twice a day for 3 days.
Terminalia chebula Retz. (Combretaceae)	*Harida*	Bark	5 ml decoction of bark taken orally for one week.
Headache			
Ricinus communis L. (Euphorbiaceae)	*Jada*	Seeds	Oil from the seeds applied on head for cooling effect.
Jaundice			
Vitex negundo L. (Verbenaceae)	*Begunia*	Leaves	20 ml of infusion of leaves taken orally for 3 times with a gap of 4 days.
Leprosy			
**Cassia fistula* L. (Caesalpiniaceae)	*Sonari*	Leaves	Decoction of leaves applied on the affected area.

Table 1. Contd.

Costus speciosus (Koenig.) Sm. (Zingiberaceae)	*Keokanda*	Rhizome	Paste of the rhizome with water applied locally.
Liver disorder			
**Butea monosperema* (Lam.) Taub. (Fabaceae) (Figure 3)	*Palaso*	Leaves	A glass of leaf extract taken orally for 2 days.
Rheumatic pain			
Holarrhena pubescens Wall. ex G.Don (Apocynaceae)	*Khucchi*	Leaves	Two to three leaves are attached with the latex of the same plant over back bone and fomented externally.
Lantana camara L. *var. aculeata* (L.) Mold. (Verbenaceae) (Figure 6)	*Nagiri*	Leaves and flowers	Decoction of both parts given orally for 21 days.
Skin diseases			
Azadirachta indica A. Juss. (Meliaceae)	*Nima*	Leaves	Crushed dried leaves with water applied locally till cure.
Calotropis gigantea (L.) R. Br. (Asclepiadaceae) (Figure 4)	*Arakha*	Root	Dried root powder sprayed locally.
Cassia tora L. (Caesalpiniaceae)	*Chhota Chakunda*	Leaves	Leaf paste applied locally for 3 days.
Mimosa pudica L. (Mimosaceae)	*Lajuli*	Root	5 ml of root juice applied locally for one week.
Snakebite			
Strychnos nux-vomica L. (Loganiaceae)	*Kochila*	Root	Root paste applied locally.
Stomach pain			
Ocimum tenuiflorum L. (Lamiaceae)	*Tulsi*	Leaves	A handful of leaves boiled in water and the infusion taken orally till pain relief.
Toothache			
**Hemidesmus indicus* (L.) R. Br. (Asclepiadaceae)	*Chirmar*	Root	Root paste is applied on teeth or a piece of root is crushed by teeth.

Figure 1. *Abrus precatorius.*

Figure 2. *Argemone mexicana.*

Figure 3. *Butea monosperma.*

Figure 4. *Calotropis gigantea*

Figure 5. *Asparagus racemosus.*

Figure 6. *Lantana camara var. aculeatae.*

Documentation of plant materials used in traditional medicine could benefit general health care and promote forest conservation and ecological research. Such plants could also be incorporated into primary health care, as people generally feel safer with indigenous cures and also cost effective as compared to modern drugs. So this will pave way for conservation and to restore the genetic diversity.

Conflict of Interests

The author(s) have not declared any conflict of interests.

Figure 7. *Tridax procumbens.*

ACKNOWLEDGEMENTS

The authors are thankful to Prof. and Head, Post Graduate Department of Botany, Utkal University for providing necessary facilities. Thanks are also due to the local medicine men, vaidyas of Bondo tribes for their valuable information.

REFERENCES

Agarwal YS, Ghosh B (1985). Drug plants of India (Root Drugs) Kalyani Publishers, New Delhi.

Anonymous (1948-1972). Wealth of India (Raw materials), 9 vols, CSIR, New Delhi.

Anonymous (1989). Medico-Botanical Exploration of Puri District, Odisha, CCRAS, New Delhi.

Anonymous (1996). Medico-Botanical Exploration of Phulbani and Koraput District of Odisha.

Chopra RN, Nayar SL, Chopra IC (1956). Glossary of Indian Medicinal Plants, CSIR, New Delhi.

Haines HH (1921-1925). The Botany of Bihar and Orissa. 6 parts, London.

Jain SK (1970-1971). Some magico-religious beliefs about plants among Adibasis of Orissa 12(1-4):38-43.

Kirtikar KR, Basu BD (1935). Medicinal plants of India. Vol (1-2), ICMR, New Delhi.

Mooney HF (1950). Supplement to the Botany of Bihar and Orissa, Catholic Press, Ranchi.

Mudgal V, Pal DC (1980). Medicinal plants used by tribals of Mayurbhanja (Orissa), 22(1-4):38-43.

Pal DC, Banerjee DK (1971). Some less known plant foods among the tribals of Andhra Pradesh and Orissa State. Bull. Bot. Surv. India. 13(3&4):221-223.

Sahoo AK (1986). Plant Resources of Kandhamal district (Orissa) some suggestion to develop cottage industries in tribal localities, Orissa Review 43 (4):39-44.

Satyavati GV, Gupta AK, Tandon N (1987). Medicinal plants of India Vol.(1-2), ICMR, New Delhi.

Saxena HO, Brahman M (1994-1996). The flora of Orissa. Vol. (1-4). Orissa Forest Development Corporation, Bhubaneswar.

Saxena HO, Dutta PK (1975). Studies on the ethnobotany of Orissa. Bull. Bot. Surv. India 17 (1-4):124-131.

Subudhi HN, Choudhury BP (1985). Ethnobotanical studies in the district of Kandhamal (Orissa-1)-1 Bio-Sci. Res. Bull. 1(1-2):26-32.

Warrier PK, Nambiar VPK, Ramankutty C (1994-1996). Indian Medicinal Plants (Vol.1-5), Orient Longman Ltd., Madras.

26

Effect of pre-sowing treatments on seed germination in *Quercus serrata* Thunb. and *Quercus semecarpifolia* Sm.

Aseesh Pandey and Sushma Tamta*

Plant Tissue Culture and Molecular Biology Laboratory, Department of Biotechnology, Bhimtal Campus, Kumaun University, Nainital-263136, Uttarakhand, India.

***Quercus* species are major forest forming species and helps in sustaining an ecosystem. Natural regeneration of these species is very poor due to various biological and physical factors. The present study was conducted to evaluate the effect of different pre-sowing treatments (scarification and cold-stratification) on seed germination of two *Quercus* species, that is, *Quercus serrata* (deciduous) and *Quercus semecarpifolia* (evergreen). On the basis of the present study, it was observed that pretreated seeds germinated better in comparison with control, and both species responded differently to the treatments. The deciduous, lower altitude *Q. serrata* gave better results in comparison with the evergreen, high altitude *Q. semecarpifolia* in terms of percent germination and time taken for germination. Being a simple and cost effective method, the results of the present study could be used for developing nurseries for commercial purposes.**

Key words: Pre sowing treatments, seed germination, stratification, scarification, *Quercus*.

INTRODUCTION

Seeds provide the most natural means of plant reproduction, preservation of genetic variability, transportation and propagation in angiospermic plants (Vazquez and Rojas, 1996). Propagation through seeds is considered to be one of the most reliable, efficient and universally applied methods (Hartmann and Kester, 1990).

Oaks are naturally propagated through seeds, but regeneration through seeds in nature is poor (Troup, 1921). In many cases, viable seeds do not germinate even under favorable environmental conditions; this phenomenon is termed seed dormancy (Taiz and Zeieger, 2002). Several internal factors cause dormancy which include seed coat, embryo or inhibitors which influence the seed germination rate (Agrawal and Dadlani, 1995). Different methods such as heating (Herranz et al., 1999), stratification, scarification (Narbona et al., 2003) and gibberellin application are well known to overcome these problems depending on the type of plant species and dormancy. Moist chilling or cold stratification is widely used for breaking seed dormancy and increasing the rate of germination percentage of dormant seeds of many species (Wang and Berjak, 2000). Influence of various concentrations of growth substances in seed germination and seedling growth was reported by Dhoran and Gudadhe (2012). However, the responses are species specific and no one treatment has been reported to be effective universally (Clemens et al., 1977).

Since the scarification and cold-stratification treatment are less expensive methods for seed germination with promising outcomes, these were tested to improve seed germination rate and establish a cost-effective method for enhancing the germination rate of two *Quercus* species

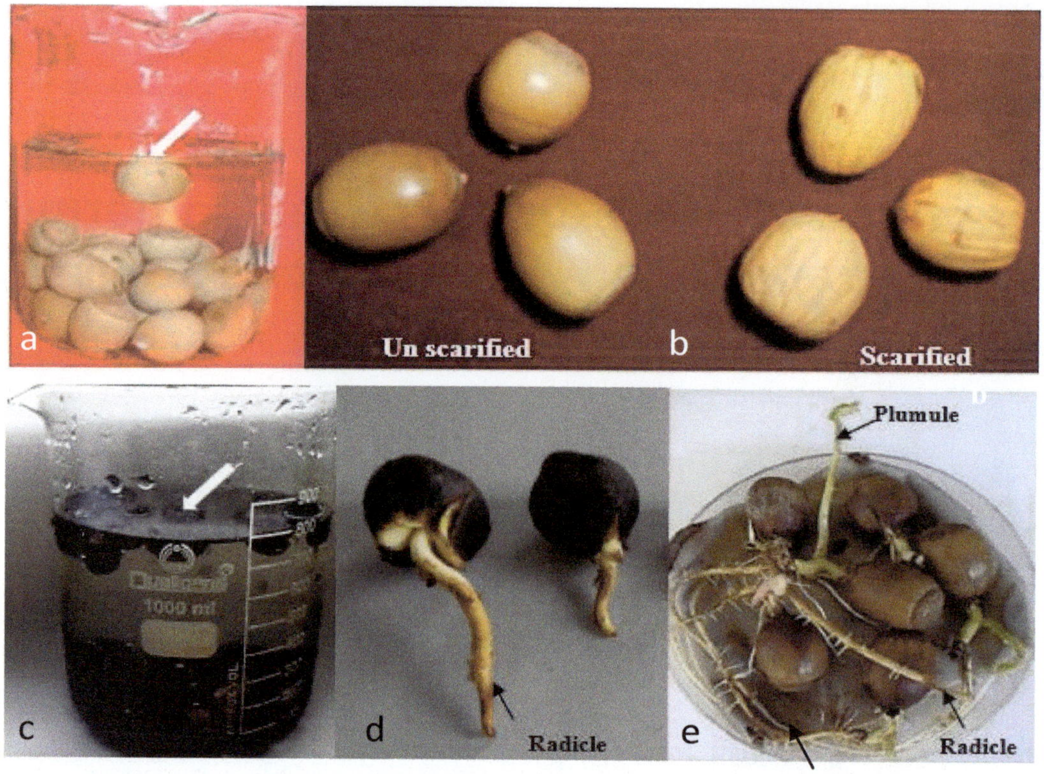

Plate 1. Effect of pre-sowing treatments on seed germination in *Q. serrata* and *Q. semecarpifolia*. a and c: Viability test of seeds by soaking in water, arrow indicate viability difference in both species b: Unscarified and scarified seeds; d-e: comparison of seed germination rate between *Q. semecarpifolia* and *Q. serrata* after 30 days of inoculation.

viz. *Quercus serrata* and *Quercus semecarpifolia.*

MATERIALS AND METHODS

Seed collection

Mature seeds (acorn) of *Q. serrata* were collected (handpicked) from Bhimtal (1,370m asl) and Pithoragarh (1,635m asl) area of Kumaun region, while seeds of *Q. semecarpifolia* were collected from Kunja-khark and Chinapeak area of Nainital (2,619m asl), Kumaun region and Srinagar (2,680m asl) area of Garhwal region, Uttarakhand, India. Acorns of *Q. serrata* and *Q. semecarpifolia* were collected during November-December and July-August 2009, respectively.

Seeds were removed from inflorescence, seeds of *Q. serrata* were dried at room temperature (24±5°C) and stored in airtight polybags at room temperature until use while, seeds of *Q. semecarpifolia* were used as soon as possible because of low viability.

Test for viability

After seed collection, the seed viability was checked by dipping seeds in a container containing tap water (frequently used method for acorns). Seeds having embryo settled down and embryoless seeds floated in the water. Settled down seeds were considered as viable and selected for the present study (Plate 1a and c).

Pre sowing treatments

Scarification

Seed coat was removed with the help of forceps, needle and scalpel. The seed without seed coat served as scarified seed and the seed with seed coat called un-scarified seed served as control (Plate 1b).

Cold-stratification

Both scarified and un-scarified seeds were kept at 4 ± 2°C for 5, 10, 20 and 30 days for cold- stratification.

Laboratory conditions and maintenance

Before giving cold-stratification treatment, viable seeds (scarified as well as un-scarified) were treated with 1% bavistin for 20 min, rinsed with distilled water. After 5, 10, 20 and 30 days interval seeds were taken from the refrigerator and placed in sterile petri dishes (17 x 95 mm) containing moistened filter paper and incubited at room temperature (24 ± 2°C). Humidity was maintained by adding distilled water over filter paper in one day interval. Seeds having emerged radicle were considered germinated. The whole experiment was monitored upto two months.

Mean germination time was calculated according to the method described by Hartman and Kester (1990) by using the following formula:

$$\text{Mean germination time} = \Sigma(n \times d) \div N$$

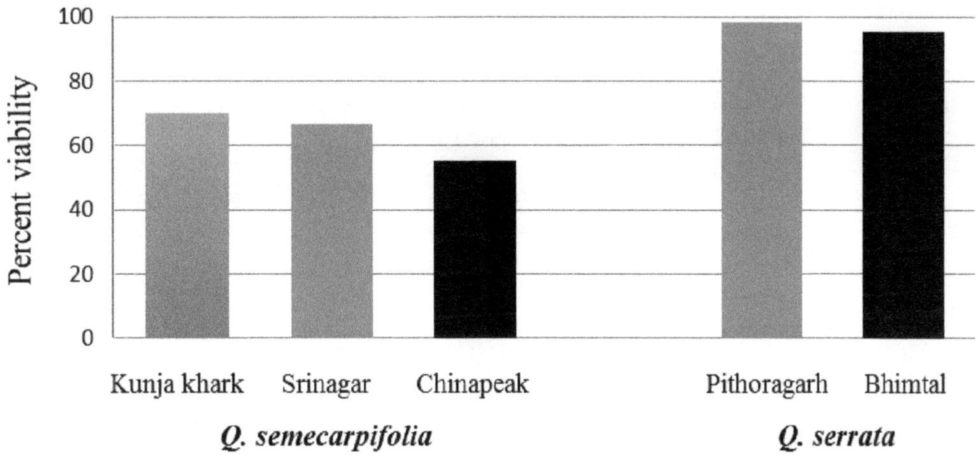

Figure 1. Percent viability of seeds of target species collected from different sites.

Where, n = number of seeds that germinated at time d; N = total number of seeds that germinated; d = days from the beginning of the germination.

Statistical analysis

All experiments were performed in replicates. Three replicates of 6 seeds were analyzed in each treatment. Mean values of treatments were subjected to one way analysis of variance (ANOVA) with the help of SYSTAT (Wilkinson, 1986).

RESULTS

Viability

Viability results showed difference in viability among the target species (Figure 1); seeds of *Q. serrata* collected from Pithoragarh showed higher viability (98.33%) than those collected from Bhimtal (95% viability). This may be due to the seed-size, because seeds which were collected from Pithoragarh were large in comparison with seeds collected from Bhimtal (data not shown). *Q. semecarpifolia* seeds collected from Kunja-khark showed maximum viability (70%) whereas seeds collected from China peak possessed least viability (only 55%) (Figure 1). It was observed that percent viability was influenced by seed size, as the larger seeds showed greater percentage of viability in comparison with smaller seeds (detail data not presented).

Pre sowing treatments

Quercus serrata

The results of present study revealed that all the pretreatments significantly increased the average percentage of seed germination when compared with un-scarified and un-stratified seeds (control) (Table 1). In the control no significant effect on germination was observed. Among the treatments, scarification with 20 days stratification was found to be the most effective treatment for breaking the dormancy of *Q. serrata* having 88.67 ± 11.33 percent germination in average time of 11.33 ± 0.67 days as compared to 14.67 ± 3.67 percent germination in average time of 27.67 ± 1.45 days in untreated seeds (control) (Table 1).

Scarification also improved the percent germination rate up to 55.00% and stratification improved germination up to 66.67% when applied for 20 days. Further increase in stratification time decreased the germination rate (44.00%) (Table 1). The results showed that scarification and stratification treatments when applied separately had low effect on the germination rate but scarification with stratification for 20 days at 4 ± 2°C had significantly increased the rate of seed germination and was considered the best treatment. Plate 1e shows germination after 30 days of incubation.

Quercus semecarpifolia

The results of present study showed that pretreatments significantly enhanced the average seed germination rate in comparison to control. Among the treatments, scarification with 10 days stratification was as the found most effective treatment for breaking the dormancy in *Q. semecarpifolia* having 69.67% germination in average time of 12.67 ± 3.18 days as compared with the untreated seeds (control) (33.33% germination in 25 ± 2.89 days) (Table 1). Stratification alone improved germination up to 51.33% when applied for 10 days. Further increase in stratification time decreased the germination rate up to 25.67%. The results showed that scarification had comparatively low effect in the germination rate than stratification and scarification with stratification treatments. Due to high moisture content, after certain point of stratification the germination decreased drastically, was less than control and lost viability. Plate 1d shows germination after 30 days of incubation.

Table 1a. Effect of pre-sowing treatments on seed germination in *Q. serrata and Q.semecarpifolia*.

Treatment (time in days)	*Q. serrata*		*Q. semecarpifolia*	
	Germination (%)	Average germination time (days)	Germination (%)	Average germination time (days)
Scarified	55.00 ±6.35	16.67±2.40	44.44±6.35	21.67±4.41
Stratified (5)	25.67±7.33	21.00±3.79	47.67±3.67	25.00±2.89
Stratified (10)	44.44±11.11	19.00±2.08	51.33±9.70	19.00±3.06
Stratified (20)	66.67±12.70	18.00±2.00	36.67±3.67	20.83±2.85
Stratified (30)	44.00±12.70	18.67±0.88	25.67±3.67	19.67±1.76
Scarified+ Stratified (5)	44.44±6.35	16.67±2.40	62.33±3.67	14.67±2.40
Scarified+ Stratified (10)	69.67±9.70	15.00±0.58	69.67±3.67	12.67±3.18
Scarified+ Stratified (20)	88.67±11.33	11.33±0.67	40.33±3.67	14.00±2.31
Scarified+ Stratified (30)	77.33±11.33	14.00±1.15	29.33±3.67	17.67±0.88
Control	14.67±3.67	27.67±1.45	33.33±12.70	25.00±2.89
LSD (p=0.05)	0.0002*	0.0013*	0.0022*	0.0528^{ns}

Table 1b. Anova.

Source of variation	df	*Q. serrata*				*Q. semecarpifolia*			
		Germination (%)		Average germination time (days)		Germination (%)		Average germination time (days)	
		Mean Square	F-ratio	Mean Square	F-ratio	Mean Square	F-ratio	Mean Square	F-ratio
Between Groups	9	1743.041	6.638	58.82963	4.999	772.61	9.12	55.67	2.36
Within Groups	20	262.6		11.76667		84.7		23.6	
Total	29								

Values represent means ± standard error *significant at 0.5 level, ^{ns}not significant at 0.05 level, Str.- Stratification, twelve explants were used per treatment, the experiments were repeated thrice, and data scored after 4-week.

DISCUSSION

The seed coat is very much hard and thick in *Q. serrata* and may inhibit the rate of seed germination. During this period, it comes in contact with various insects and wild animals which also inhibits the germination, but in the case of internal dormancy stratification is necessary to overcome it. The results of present study showed that scarification (total removal of seed coat) along with stratification significantly improved the germination rate of target species. In *Q. serrata*, scarification contribute to improving the regeneration like other Himalayan oaks viz. *Q. glauca* and *Q. leucotrichophora* (Rawat et al. 1998) where pericarp resisted cotyledon expansion and water uptake of seed during radicle emergence.

Abscisic acid has also been reported as control agent of dormancy (Tillberg, 1983), as stratification proceeds, abscisic acid decreases and giberellic acid is synthesized or released from bound forms (Diaz and Martin, 1972). Giberellic acid accelerates germination by inducing the production of hydrolytic enzymes (Mozer, 1980) and cell plasticity (Taylor and Cosgrove, 1989). It is probably because the scarified seeds when subjected to stratification, the level of giberellic acid increases with the increase of stratification and leads to germination enhancements because gibberellic acid activates the reserve food mobilizing system (Hartman and Kester 1990) while in the control hard seed coat prevents this process and causes germination hindrance of the embryo caused by inability to absorb either water or oxygen.

Some earlier workers viz. Vetaas (2000), Shrestha (2003), Singh and Rawat (2012) and Bisht et al. (2012) also studied the regeneration status of *Q. semecarpifolia* and found very low natural regeneration rate under different microclimatic conditions. Seed germination process in *Q. semecarpifolia* completes in several steps viz. emergence of radicle followed by thickening of radicle and swelling of approximately 2 cm part of radicle takes place, from this swollen portion emergence of root and shoot occur. Bisht (2001) presumed that swollen portion of radicle requires chilling treatment for seedling emergence while Bisht et al. (2012) observed that light has significant effect on seed germination. In *Q. ilex,* Ghasemi and Khosh-Khui (2007) observed that stratification of scarified seeds positively affected germination rate. In *Q. semecarpifolia,* seeds germinated best in 10 days stratification but as duration of stratification increases the germination rate deteriorated. Bisht (2001) reported that at/below 0°C seeds of *Q. semecarpifolia* lose their viability due to the formation of ice crystal in the cells but seed germination of *Jasminus fruticans* was enhanced by the application of three month cold stratification (Pipinis et al., 2009). Scarification and stratification treatments were able to promote germination and growth of seedling in many of the angiosperms and gymnosperms (Esen et al., 2007), not only in herbaceous but also in tree species like *Solanum nigrum* (Suthar et al., 2009), *Prunus scoparia* and *Prunus webbii* (Heidari et al., 2008) and *Aeluropus lagopoides* (Zaman et al., 2011).

Conclusion

Q. serrata responded considerably better than the *Q. semecarpifolia* in applied germination enhancement experiment. Although, some seeds of *Q. semecarpifolia* in control failed to develop shoot and root even after emergence of radicle and died but it was found that stratification along with scarification improved germination rate significantly in both the target species in comparison with individual stratification and scarification treatments. The present study was able to enhance the seed germination rate and improve germination time. Being a simple and cost effective method, the results of the present study could be used for developing nurseries for commercial purposes.

ACKNOWLEDGEMENTS

Authors are thankful to GBPIHED Kosi-Katarmal Almora for financial support and head, Department of Botany, D.S.B. Campus Kumaun University Nainital for providing necessary facilities.

REFERENCES

Agrawal PK, Dadlani M (1995). Techniques in Seed Science and Technology, 2nd Edn., South Asian Publishers, New Delhi, pp.109-113.

Bisht H (2001). Physiobiochemical aspects of seed viability in *Quercus semecarpifolia* smith: A possible recalcitrant seed. Ph.D. Thesis, H.N.B. Garhwal University, Srinagar (Garhwal), 246 174, Uttarakhand, India, pp.45-48.

Bisht H, Prakash V, Nautiyal AR (2012). Factors Affecting Regeneration Potential of Quercus semecarpifolia, Smith: A Poor Regenerated Oak of Himalayan Timberline. Res. J. Seed. Sci. 5:63-70. DOI: 10.3923/rjss.2012.63.70.

Clemens J, Jones PG, Gilbert NH (1977). Effect of seed treatments on germination in *Acacia*. Aust. J. Bot. 25:269–276.

Dhoran VS, Gudadhe SP (2012). Effect of Plant Growth Regulators on Seed Germination and Seedling Vigour in *Asparagus sprengeri* Regelin. I. Res. J. Bio. Sci. 1(7):6-10.

Diaz DH, Martin GC (1972). Peach seed dormancy in relation to endogenous inhibitors and applied growth substances. J. Am. Soc. Hortic. Sci. 97:651-654.

Esen D, Yildiz O, Sarginci M, Isik K (2007). Effects of different pretreatments on germination of *Prunus serotina* seed sources. J. Environ. Biol. 28:99-104.

Ghasemi M, Khosh-Khui M (2007). Effect of stratification and growth regulators on seed germination and seedling growth of *Quercus ilex* L. J. Plant Sci. 2:341-346.

Hartmann HT, Kester DE (1990). *Plant Propagation*, Englewood Cliffs, NJ: Prentice-Hall.

Heidari M, Rahemi M, Daneshvar MH (2008). Effect of mechanical, chemical scarification and stratification on seed germination of *Prunus scoparia* (Spach.) and *Prunus webbi* (Spach.) Vierh. Am.-Eurashian J. Agric. Environ. Sci. 3:114-117.

Herranz JM, Ferrandis P, Martínez Sánchez JJ (1999). Influence of Heat on Seed Germination of Nine Woody *Cistaceae* Species. Int. J. Wildl. Fire, 9:173-182.

Mozer TJ (1980). Control of protein synthesis in barley aleurone layers by the plant hormones gibberellic acid and abscisic acid. Cell, 20:479-485.

Narbona E, Arista M, Ortiz PL (2003). Seed germination of *Arbutus unedo* L. (*Ericaceae*). Acta Bot. Mala, 28:73–78.

Pipinis E, Milios E, Aslanidou M, Mavrokodopoulou O, Smiris P (2009). The effect of stratification on seed germination of *Jasminu fruticans* L.(Oleaceae): A contribution to a better insight on the species germination ecology. Int. J. Bot. 5:181-185.

Rawat DCS, Thapliyal P, Nautiyal AR (1998). Pericarp delays germination in *Quercus glauca* Thunb seeds. J. Trop. For. Sci. 10:472-477.

Shrestha BB (2003). *Quercus semecarpifolia* Sm. in the Himalayan Region: ecology, exploitation and threats. Him. J. Sci. 1:126-128.

Singh G, Rawat GS (2012). Depletion of Oak (*Quercus* spp.) Forests in the Western Himalaya: Grazing, Fuelwood and Fodder Collection, Global Perspectives on Sustainable Forest Management, Dr. Clement A. Okia (Ed.), ISBN: 978-953-51-0569-5,

Suthar AC, Naik VR, Mulani RM (2009). Seed and seed germination in *Solanum nigrum* Linn. Am-Eurasian J. Agric. Environ. Sci. 5:179-183.

Taiz L, Zeiger E (2002). Plant Physiology, Third Ed. Sinauer Associates, 10:33-46.

Taylor A, Cosgrove DJ (1989). Gibberellic acid stimulation of cucumber hypocotyl elongation. Plant Physiol, 90:1335-1340.

Tillberg E (1983). Levels of endogenous abscisic acid in achenes of *Rosa rugosa* during dormancy release and germination. Physiol. Plant. 58:243-248.

Vazquez C, Rojas M (1996). *Ex situ* conservation of tropical rain forest seed: problems and perspectives. Interciencia, 21(5):293-298.

Vetaas OR (2000). The effect of environmental factors on the regeneration of *Quercus semecarpifolia* Sm. in central Himalayan, Nepal. Plant Ecol. 146:137-144.

Wang BSP, Berjak P (2000). Beneficial Effects of Moist Chilling on the Seeds of Black Spruce (*Picea mariana* [Mill.] B.S.P.). Ann. Bot., 86(1):29-36. doi: 10.1006/ anbo. 2000.1150

Wilkinson LL (1986). SYSTAT: The system for statistics, Systat, Evanston.

Zaman S, Padmesh S, Tawfiq H (2011). Selected seed pretreatments on germination of Kuwait`s native perennial plant species. Int. J. Bot. 7:108-112.

Generic relationship among *Cassia* L., *Senna* Mill. and *Chamaecrista* Moench using RAPD markers

Vivek Tripathi[1]* and Sandhya Goswami[2]

[1]Delhi Technological University,Shahbad, Daulatpur Bawana Road,Delhi-110042, India.
[2]Betel Vine Biotechnology Laboratory,National Botanical Research Institute,Lucknow-226 001, India.

Generic relationships were examined among twenty-four species belonging to genus *Cassia* L., *Senna* Mill. and *Chamaecrista* Moench using RAPD marker. Total 80 primers were initially screened, 514 amplification products obtained with 38 informative primers, of which 514 were polymorphic. A vary high degree of polymorphism (100%) was observed among them. UPGMA cluster analysis of genetic similarity indices grouped all the species into three major clusters. Cluster I included four species of *Cassia* L., Cluster II included eighteen species of *Senna* Mill. and Cluster III included two species of *C.* Moench. Highest similarity (0.9%) was observed between *Cassia fistula* L. and *Cassia fistula* with nodded filaments and least (0.001%) between *Cassia fistula* L. and *Senna splendida*. The Polymorphic information contents (PIC) of the twenty-four species with RAPD marker varied from 0.08 to 0.49 with an average of 0.005. The result confirms the statement of Irwin and Barneby, they divided the genus *Cassia* L. into three subgenera; *Cassia* L., *Senna* Mill. and *C.* Moench on the basis of morphological characters. The results obtained from the present study support the previous taxonomic classification of the genus *Cassia* L. and showed large diversity among the species of three newly created genera. Our results suggested that RAPD marker is a sensitive, precise and efficient tool for genomic analysis of Cassia L. that may be useful in future studies by assigning new unclassified germplasm to specific taxonomic groups and reclassify previously classified species and genera.

Key words: *Cassia, Senna, Chamaecrista*, genetic relationship, RAPD.

INTRODUCTION

Cassia L. is the largest genus in the subfamily Caesalpiniodeae of the Caesalpinioceae. It contains about 600 flowering species which are distributed in most continents (Singh, 2001). *Cassia* L. species show large diversity related to habit, ranging from delicate, annual herbs to tall trees. Bentham (1871) divided the genus *Cassia* into three genera and nine sections. Britton and Rose (1930) splited the genus *Cassia* into twenty-eight genera. Recently Irwin and Barneby (1982) splited the genus *Cassia* L. into three subgenera; *Cassia* L., *Senna* Mill. and *Chamaecrista* Moench. Irwin and Baneby (1981 to 1882) realizing diversity and complexity and proposed an improved classification proposing new delimitation based on persistent suit of characters. They have raised the genus *Cassia* L. to the levels of subtribe (Cassiinae) and raised the subgenera *Senna* Mill. and *C.* Moench to genetic level.

The plants of *Cassia* L. are used as fodder, purgatives, timber and medicine (Tomlinson, 1981; Tiwari, 1983). The taxonomy and nomenclature of *Cassia* L. species are quite complex and intriguing. They are not easily differentiated from closely related species due to the large variation in similarities range causing mis-identification and misinterpretation of the components. The cultivars identification and assessing of diversity using phenotypic markers have several limitations especially in perennial crops (reference). In plants as general, due to the overlapping of morphological characters, a great amount of confusion persists for the selection of ideal plant. Generally morphological characters are influenced by environmental changes, and the changes are not constant in the species to species and differ from place to place.

Structural changes in DNA, that is, translocation, deletion, inversion mutation have been able to

*Corresponding author. E-mail: genevivek@rediffmail.com.

Table 1. Collected *Cassia, Senna* and *Chamaecrista* species, accessions no. and their location.

S/N	Accessions no.	Accessions code.	Accessions name	Location
1	200382	Cf1	*C. fistula*	Utterkashi (Uttranchal)
2	200390	Cf2	*C. fistula* (white)	NBRI, Lucknow India
3	200435	Cj	*C. javanica*	NBRI, Lucknow India
4	200436	Cr	*C. roxburghii*	BSI Jodhpur (Rajasthan) India
5	200437	St	*S. tora*	Kanpur (U.P) India
6	200438	So	*S. obtusifolia*	Unnao (U.P) India
7	200439	Su	*S. uniflora*	Karnataka university Dharwad India
8	200440	Sb	*S. biflora*	Lucknow (U.P) India
9	200441	Ss	*S. sophera*	Karnataka university Dharwad India
10	200442	Sp	*S. purpurea*	Metha (U.P) India
11	200443	Ss1	*S. spetabilis*	Karnataka collage Dharwad India
12	200444	Se	*S. excelsa*	NBRI, Lucknow India
13	200415	Ss2	*S. sulfurea*	Tehari garhwal (Uttranchal) India
14	200432	Ss3	*S. surattensis*	Aligarh (U.P) India
15	200447	Ss4	*S. splendida*	Madgavo Karnataka India
16	200448	So1	*S. occidantalis*	Bhopal (M.P) India
17	200449	Sa	*S. auriculata*	BSI Jodhpur (Rajasthan) India
18	200450	Ss5	*S. siamea*	Faizabad (U.P) India
19	200451	Si	*S. italica*	BSI Jodhpur (Rajasthan) India
20	200452	Sa1	*S. alexandrina*	CIMAP Lucknow
21	200453	Sh	*S. hirsuta*	Karnataka collage Dharwad India
22	200454	Sa2	*S. alata*	Lucknow (U.P)
23	200455	Chm	*C. mimosoides*	K.U Dharwad India
24	200456	Cha	*C. absus*	K.U Dharwad India

change the genomic constitution of species. In decades, these changes in the genome of species, separate the particular species from their ancestors. To assess these changes, parameters based on morphological characters are not sufficient. Now it is possible to single out differences on the basis of molecular markers, which is authentic and less affected by environmental factors. Hence characterization of species at the genetic level supplemented efficient conservation, maintenance and utilization of the existing genetic diversity.

In the present study, RAPD markers were used to assess genetic relationship among the species of *Cassia*, *Senna* and *Chamaecrista* by using a single arbitrary primer (10-mer) and amplifying DNA by polymerase chain reaction (PCR) as the resulting DNA marker can be easily separated on an agarose gel by electrophoresis (Williams et al., 1990). The advantage of RAPD is its simplicity, rapidity and requirement for only a small quantity of DNA, and the ability to generate numerous polymorphism (Stebbins, 1957; Cheng et al., 1997; Khanuja et al., 1998).

MATERIALS AND METHODS

Plant materials

Twenty-four species belongs to three genera; *Cassia*, *Senna* and *Chamaecrista* were collected from different ecological locations of Uttrakhand, Uttar Pradesh, Madhya Pradesh, Karnataka, Rajasthan and Himachal Pradesh, India (Table 1). Out of which some species are cultivated due to the medicinal and ornamental properties and some are wild.

Preparation of genomic DNA

Total genomic DNA from the leaves of *Cassia, Senna* and *Chamaecrista* species was extracted using the protocol proposed by Doyle and Doyle (1987) with slight modification. 1 g of leaves around to powder with liquid nitrogen in a morter and pestle, then transferred to a 30 ml centrifuge tube containing 12 ml of CTAB buffer (2% CTAB, 100 mM Tris-Cl, 20 mM EDTA and 1.4 M NaCl) and 120 µl β-mercaptoethanol mixed thoroughly and incubated in water bath at 68°C for 3 h. The tube was cool at room temperature, mix with equal volume of chloroform and tube was gently inverted repeatedly. The tube was centrifuged at 10,000 rpm for 10 min. The supernatant was transferred to a corex tube and 0.7% volume of 2-propanol was added. Tube was put at -20°C for 2 h. The tube was centrifuged at 12000 rpm 10 min at 4°C to collect precipitated DNA. The pellet was resuspended with 700 µl TE (10 mM, Tris-Cl pH-8.0, 1 mM EDTA) and incubated with 2 µl (10 mM DNase free RNase A) for 40 min at 37°C. The RNase A and the remaining protein were extracted with equal volume of phenol: chloroform 1:1 and centrifuged at 12000 rpm for 15 min at room temperature. The supernatant was transferred to a new tube and the DNA was precipitated by the addition of 0.7 volume of 2-propanol. Precipitated DNA was collected by centrifugation at 12000 rpm for 15 min at 4°C washed with 70% ethanol twice and dried before redissolving in 100 µl of TE. DNA yield were calculated by fluorometer (DyNa Quant [200]) (Amersham pharmacia Biotech, USA).

RAPD reactions

Sixty-eight decamer Oligonucleotide primers (Operon, USA) were screened by polymerase chain reaction (PCR). PCR reactions were performed by using a 20 µl mixture, conataing 50 ng template (genomic DNA), dNTPs (0.2 mM each), 0.5 mM primer, 1.5 U Taq DNA polymerase, 25 mM $MgCl_2$ and remain Mili Q water. The DNA thermocycler (Applied Biosystem, 9700) was programmed as follow: Incubation at 94°C for 2 min, 44 cycles at 94°C for 1 min. 36°C for 1.30 min and 72°C for 1.30 min followed by electrophoresis in 1.4 % (w/v) agarose gel with 0.5 × TBE stained with diluted ethidium bromide (10 mg/ml) and photographed in gel documentation unit (AlphaImager™ 3400).

Data analysis

Amplification of DNA was repeated at least three to six only reproducible and unambiguous fragment were scored as (1) for its presence or (0) for its absence. A fragment was considered polymorphic if both the presence and absence of that fragment were observed in the same species and monomorphic if it was present among all individual within a species. To reduce the possibility of comparing non-homologous bands, a positive control (an individual possessing the band to be scored) was included on each agarose. Analysis of RAPD markers was based on the following three assumptions: (1) Each RAPD marker represented a single locus comprising two alleles, a marker allele (amplified product present) and a non-marker alleles (amplified product absent). (2) RAPD marker is inherited in a dominant fashion with the marker allele dominant to the non-marker allele. (3) Co-migrating bands from different populations present homologous amplified products (Allan et al., 1997; Hadrys et al., 1992).

The genetic associations among species were evaluated by calculating the Jaccard similarity coefficient for pair-wise comparisons based on the proportion of shared bands (alleles) produced by primer. Similarity matrices were generated using 'Simqual subprogram, similarity coefficients were used for cluster analysis of accessions performed using the 'SHAN' sub program, dendrogram was built by the un-weighted pair group method with arithmetic average (UPGMA). The computer program used was NTSYS-pc Version 2.02 (Rohlf, 1998).

The polymorphic information content (PIC) was calculated by applying the formula given by Powell et al. (1996) and Smith et al. (1997).

$$PIC = 1-\sum_{i=1}^{n} f i^2$$

where f i is the frequency of the ith alleles and the summation extends over n alleles.

RESULTS

A total 80 primers were initially screened among twenty-four species belonging to *Cassia*, *Senna* and *Chamaecrista*. Out of these 80 primers, 38 primers exhibited amplification pattern with all the species. Examples of amplification pattern of these species with primer OPAP-17, OPAP-16 and C-19 are shown in Figure 1. A total 514 bands were scored from PCR amplification of genomic DNA with all the species. The RAPD markers showed 100% polymorphism (Table 2). Homology of the RAPD bands with the same molecular weight was confirmed using EcoRI and Hind III digested marker. Average number of 14 bands was obtained per primer and amplification product ranged in size from 100 bp to 4.2 kb. Maximum numbers of 21 amplification products were obtained with primer OPC-19 followed by 20 products with primer OPAP-1 and OPAP-20. Minimum numbers of RAPD products were generated with primers OPAP-1, OPAP-17 and OPAP-16. The polymorphic information contents (PIC) ranged from 0.08 to 0.49 with an average of 0.005.

Genetic relationship measured through analysis of RAPD data of twenty-four species of *Cassia*, *Senna* and *Chamaecrista*. Highest similarity (0.9%) was observed between *Cassia fistula* L. and *Cassia fistula* with nodded filaments and least (0.001%) between *C. fistula* L. and *Senna splendida* (Table 3). Dendrogram showed that RAPD markers distinguished all species in three major groups (Figure 2). Cluster I included four species of the genus *Cassia* (*C. fistula* L., with nodded filament, *Cassia javanica* L, *Cassia roxburghii*). Cluster II included eighteen species belong to genus *Senna* (*Senna tora* L., *Senna obtusifolia* L., *Senna uniflora* mill., *Senna biflora*, *Senna sophera* L., *Senna purpurea*, *Senna spetabilis*, *Senna spetabilis* var. *excelsa*, *Senna sulfurea* DC, *Senna surattensis* Burm.f., *Senna splendida*, *Senna occidantalis* L., *Senna auriculata*, *Senna siamea* L., *Senna hirsuta* L., *Senna italica*, *Senna alata* and *Senna alexandrina*, and cluster III included two species of the genus *Chamaecrista* (*Chamaecrista absus* and *Chamaecrista mimosoides*).

DISCUSSION

When dealing with morphologically similar taxa, study of molecular characters has exemplified a more definitive approach than morphological observations. Among the methods used in such studies, DNA markers have proved to be an excellent parameter to resolve the problems of identification of critical taxa and to understand their relationships and taxonomic status (Esen and Hilu, 1991; Khan, 1992). The groups of natural populations produced distinct group than other groups (Mayer, 1970). The reorganization and appreciation of the dynamic variations in the genus inter and intra-specific levels necessitate thecharacterization of genetic variation in order to determine the genetic base and phylogenetic status of its species.

Genus *Cassia* was considered to consist of three subgenera (Linnaeus, 1754) but recently, each of these has been raised to the generic level. Classification into these groups had often been confused, because of the absence of a clearly defined set of taxonomic characters.

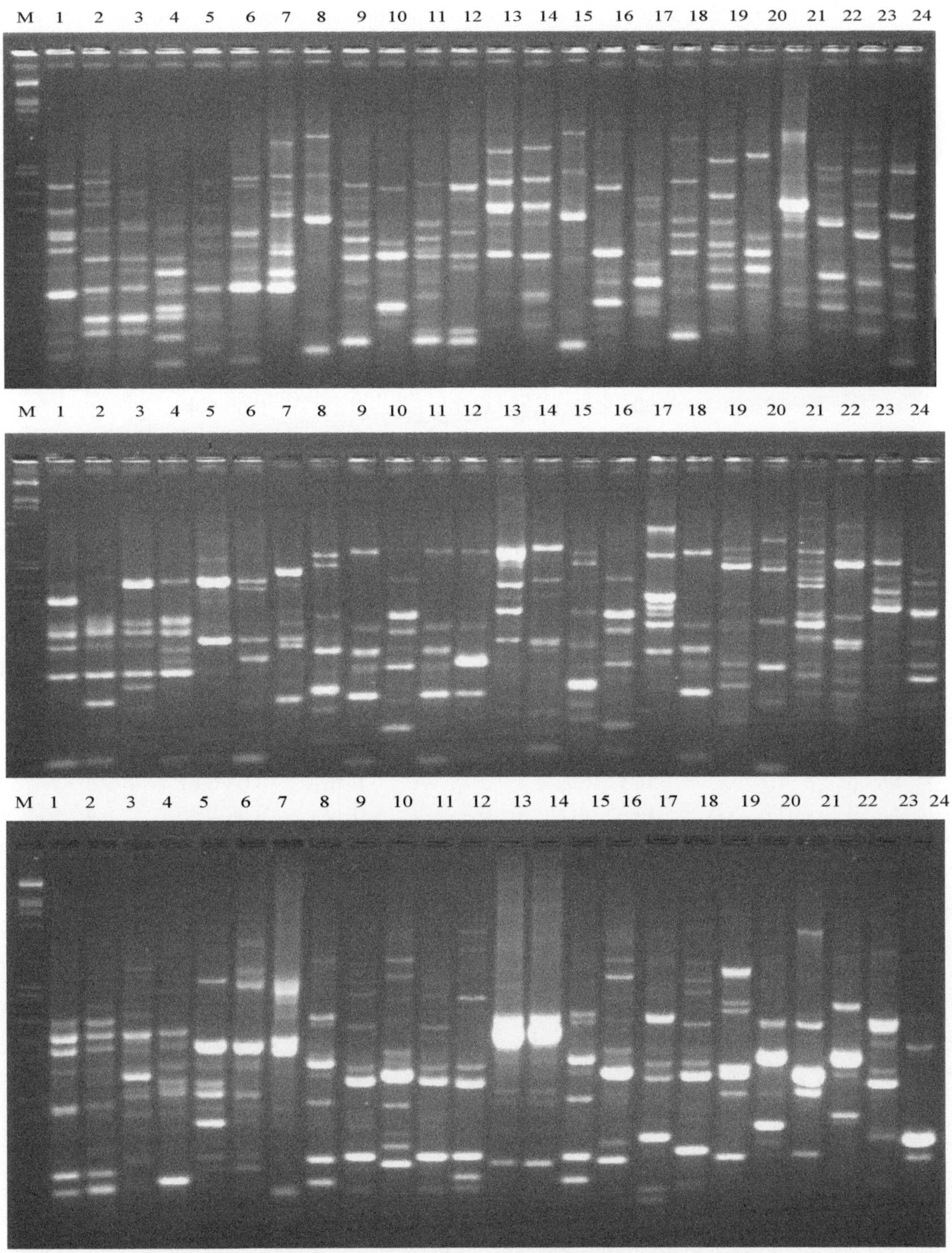

Figure 1. Lane-M Eco RI and Hind III digested λ DNA. Lane-1-24: RAPD profile of *Cassia, Senna* and *Chamaecrista* with primers OP AP-17 (Upper), OP AP-16 (Middle) and OPC-19 (Lower).

Irwin and Barneby (1982) described a suite of characteristics which members of the Cassiinae, could be consistently classified into one of the three genera; *Cassia* L., *Senna* sensu stricto and *C.* Moench. Initially, *Chamaecrista* separated from *Senna* and *Cassia* using characters relating to the androcium. Subsequently separation of *Cassia* and *Senna* involves consideration of a number of floral characters.

Todaria et al. (1983) had investigated the electrophoresis protein profile of nodulated and non-nodulated *Cassia* species (*C. fistula*, *C. occidentalis*, *C. tora*, *C. laevigata, C. glauca, C. absus* and C. *dimidiala*)

Table 2. Analysis of polymorphism among species of *Cassia, Senna* and *Chamaecrista* obtained with random primers

Primer no.	Total no. of amplicon	Total no. of bands	Polymorphic bands	Monomorphic bands	PIC value	Average	Average no of bands	Size range of amplified product (bp)
AP-1	146	20	20	0	0.30	0.16 -0.49	6	200-2500
AP-2	53	11	11	0	0.13	0.16-0.48	2.2	100-990
AP-8	80	17	17	0	0.20	0.08-0.47	3.3	100-2700
AP-9	79	14	14	0	0.21	0.16-0.48	3.2	100-3000
AP-12	56	12	12	0	0.14	0.08-0.4	2.3	200-1900
AP-14	95	19	19	0	0.26	0.16-0.49	3.9	250-3530
AP-15	40	13	13	0	0.11	0.08-0.37	1.6	200-1500
AP-16	118	17	17	0	0.23	0.08-0.49	4.9	300-1900
AP-17	126	16	16	0	0.49	0-0.26	5.2	250-4268
AP-20	119	20	20	0	0.26	0.08-0.48	4.9	100-3500
C-6	56	13	13	0	0.13	0.08-0.47	2.3	100-800
C-8	81	17	17	0	0.21	.08-0.45	3.3	300-1000
C-9	62	13	13	0	0.13	0.08-0.48	2.5	200-1500
C-11	47	15	15	0	0.13	.08-0.47	1.9	250-1500
C-12	62	14	14	0	0.16	0.08-0.48	2.5	100-3100
C-13	57	11	11	0	0.09	.08-0.33	2.3	100-1900
C-14	93	16	16	0	0.22	0.08-0.48	3.8	250-3100
C-16	53	16	16	0	0.13	0.08-0.49	2.2	100-1700
C-18	78	18	18	0	0.20	0.16-0.48	3.2	100-2400
C-19	145	21	21	0	0.3	0.08-0.49	6	150-2100
B-5	69	14	14	0	0.13	0.08-0.48	2.8	210-2000
B-8	88	17	17	0	0.21	0.08-0.48	3.6	210-2000
B-15	68	15	15	0	0.16	0.08-0.47	2.8	180-1300
B-18	59	14	14	0	0.13	0.08-0.45	2.4	250-1500
B-19	86	17	17	0	0.21	0.4-0.5	3.5	100-3530
B-20	96	15	15	0	0.23	0.22-0.48	4	150-3530
U-1	52	15	15	0	0.24	0.08-0.48	2.1	100-1900
U-2	52	15	15	0	0.19	0.08-0.49	3.1	200-2000
U-3	80	16	16	0	0.19	0.08-0.47	3.3	100-3500
U-4	43	14	14	0	0.11	0.08-0.45	1.7	250-1500
U-7	50	14	14	0	0.12	0.08-0.47	2	150-1200
U-15	84	17	17	0	0.18	0.08-0.47	3.5	200-2500
U-16	54	12	12	0	0.14	0.16-0.49	2.2	250-2500
U-17	98	16	16	0	0.08	0.03-0.47	4	150-1700
35	2648	514	514	0	0.188		Min. 1.6 Max.6	

and agreed their placement according to the Bentham and Hooker system. Whitty et al. (1994) used RAPD as molecular marker for examining four *Cassia* species, 11 *Chamaecrista* species and 14 *Senna* species including *C. glauca* for separation of the nodulated nitrogen fixing genus *Chamaecrista* from the previously congeneric group *Cassia* and *Senna*. Gareeb et al. (1999) discussed seed protein profile, chromosome number and morphological characters between ten species of genus *Cassia*. According to numerical cluster analysis, the studied taxa were splited into two groups. Group I (belonging to subgenus *fistula*) includes three *Cassia* spp. (*C. fistula, C. javanica,* and *C. nodosa*), while *C. occidentalis, C. sophera, C. siamea, C. didymobotrya, C. italica, C. senna* and *C. surattensis* are included in group II (belonging to subgenus *Senna*). Mondal et al. (2000) evaluated interspecific variation among eight species of *Cassia* L. on the basis of the free amino acid composition, SDS-polyacrylamide gel electrophoresis of total seed protein and mitochondrial DNA restriction fragment length polymorphism to understand their phylogenetic relationships and grouped the eight species into two clusters; clusters 1 consist *C. occidentalis, C. sophera*, *C. mimosoides* and *C. tora,* and clusters 2

Table 3. similarity indices (Jaccards coefficient of 24 species of genus *Cassia, Senna* and *Chamaecrista* obtained with 38 primers)

	1	2	3	4	5	6	7	8	9	10	11	12	13	14	15	16	17	18	19	20	21	22	23	24
1	1																							
2	0.9	1																						
3	0.262	0.042	1																					
4	0.315	0.019	0.163	1																				
5	0.082	0.121	0.156	0.141	1																			
6	0.131	0.111	0.146	0.165	0.065	1																		
7	0.361	0.013	0.161	0.198	0.192	0.214	1																	
8	0.011	0.091	0.057	0.152	0.097	0.235	0.15	1																
9	0.082	0.022	0.131	0.112	0.521	0.174	0.1	0.097	1															
10	0.041	0.091	0.048	0.005	0.061	0.077	0.14	0.114	0.052	1														
11	0.011	0.086	0.125	0.225	0.028	0.036	0.09	0.124	0.981	0.088	1													
12	0.096	0.023	0.005	0.068	0.022	0.029	0.08	0.092	0.023	0.063	0.18	1												
13	0.069	0.059	0.215	0.153	0.091	0.071	0.12	0.034	0.052	0.029	0.17	0.165	1											
14	0.126	0.098	0.031	0.181	0.048	0.142	0.14	0.042	0.109	0.127	0.18	0.178	0.072	1										
15	0.001	0.001	0.051	0.071	0.052	0.097	0.16	0.092	0.071	0.022	0.18	0.147	0.042	0.062	1									
16	0.031	0.047	0.108	0.024	0.048	0.232	0.66	0.035	0.074	0.121	0.08	0.141	0.035	0.016	0.094	1								
17	0.136	0.022	0.125	0.097	0.102	0.209	0.12	0.105	0.019	0.081	0.12	0.051	0.017	0.072	0.134	0.086	1							
18	0.091	0.042	0.162	0.098	0.071	0.128	0.63	0.168	0.061	0.616	0.93	0.155	0.021	0.071	0.101	0.126	0.03	1						
19	0.178	0.037	0.178	0.071	0.972	0.222	0.44	0.156	0.011	0.141	0.12	0.121	0.037	0.055	0.156	0.049	0.017	0.013	1					
20	0.155	0.001	0.149	0.117	0.032	0.292	0.07	0.121	0.009	0.109	0.14	0.151	0.109	0.047	0.074	0.021	0.028	0.062	0.013	0.054	1			
21	0.117	0.108	0.031	0.015	0.036	0.123	0.23	0.063	0.015	0.141	0.11	0.091	0.053	0.116	0.105	0.023	0.091	0.094	0.028	0.052	0.026	1		
22	0.082	0.032	0.188	0.176	0.053	0.194	0.14	0.081	0.137	0.038	0.16	0.104	0.481	0.112	0.085	0.047	0.081	0.034	0.038	0.061	0.291	0.273	1	
23	0.271	0.081	0.062	0.159	0.081	0.112	0.06	0.025	0.109	0.189	0.12	0.041	0.215	0.673	0.053	0.017	0.084	0.045	0.031	0.114	0.262	0.319	0.069	1
24	0.091	0.037	0.163	0.132	0.045	0.237	0.13	0.109	0.074	0.138	0.62	0.011	0.583	0.044	0.071	0.036	0.115	0.188	0.351	0.512	0.254	0.283	0.091	0.183

consist *C. alata, C. siamea, C. fistula* and *C. renigera.* RAPD markers have found wide application due to the easiness of performing the assay and also for the reason of being less time/labor consuming and less expensive. However, some doubts have also been raised

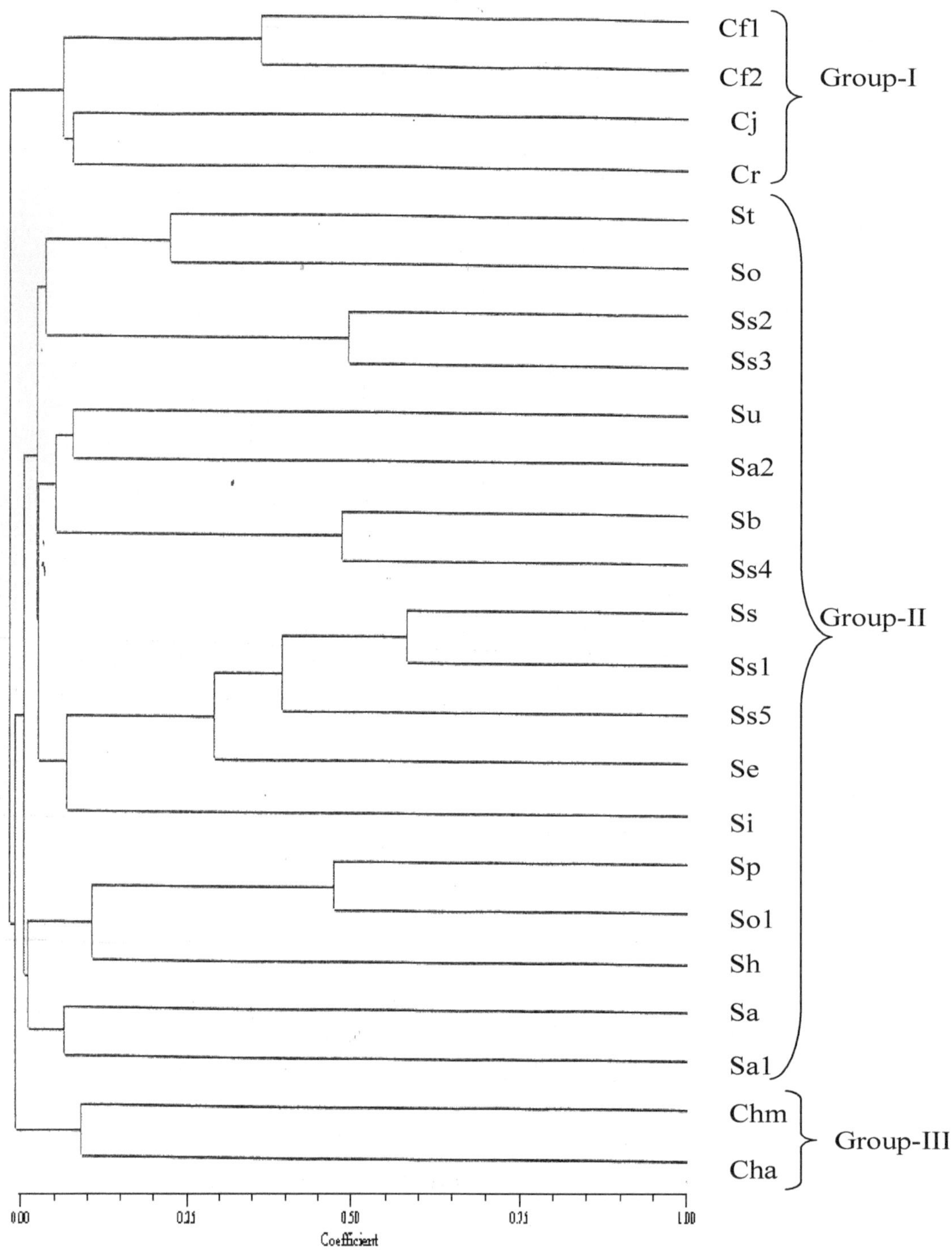

Figure 2. Dendrogram showing the relationship among different species of *Cassica, Senna and Chaemaecrisia* based on UPGMA AND sequential agglomerative hierarchical nested (SHAN) clustering.

regarding the suitability of RAPD for diversity analysis. It is debated mostly that co-migrating RAPD bands may not be allelic or composed of similar sequences (Bowditch et al., 2003). On the other hand, studies in some species of *Glycine* and *Allium* have demonstrated the homology of co-migrating RAPD bands (Williams et al., 1993; Wilkie et al., 1993). Also the use of large number of polymorphic markers minimizes the skewing of results due to non-allellism (Pujar et al., 1999). Another problem often encountered and questioned regarding RAPD analysis is the reproducibility of band patterns. This problem can be solved by thoroughly optimizing PCR reaction conditions and following the same protocol each time. For more accurate analysis, reaction should be performed thrice, scoring only those bands that are reproducible in each reaction. The intra-population genetic variations observed in heterogeneous populations of out breeding plant species complicate the analysis of genetic diversity among populations.

In the present study, RAPD patterns were generated by using genomic DNA from twenty-four species of *Cassia*, *Senna*, and *Chamaecrista*. PCR was done with random primer followed according to Williams et al. (1990). RAPD marker is able to differentiate closely related species of *Cassia*, *Senna*, and *Chamaecrista*. The study has shown RAPD analysis to be a robust and reliable method to detect intra and inter specific genetic relationship among *Cassia*, *Senna*, and *Chamaecrista*. In the case of twenty-four species of *Cassia*, *Senna* and *Chamaecrista* 80 primers were screened, out of which 38 RAPD primers were considered for the data analysis. These primers resulted reproducible profiles in the twenty-four species of *Cassia*, *Senna* and *Chamaecrista*. The high interspecific genetic divergence in species demonstrates that the level of genetic variation with in species is also substantial and suggested that genetic base is quite broad.

Dendrogram showed grouping of species into three major clusters. Cluster I grouped four species (*C. fistula*, *C. fistula* with nodulated filament, *C. javanica* and *C. roxburghii*). Cluster III grouped two species (*C. mimosoides* and *C. absus*) and remaining eighteen species (*S. tora*, *S. obtusifolia*, *S. uniflora, S. biflora*, *S. sophera, S. purpurea*, *S. spectabilis*, *S. excelsa*, *S. sulfurea*, *S. surattensis, S. splendida*, *S. occidantalis*, *S. auriculata, S. siamea*, *S. italica*, *S. alexandrina*, *S. hirsuta* and *S. alata*) were grouped in cluster II. Cluster II has divided into seven sub groups and show high closeness among the *Senna* species, *S. tora* L. grouped with *S. obtisifolia* L. *S. sulfurea* DC. ex Collad. with *S. surattensis* Burm f., *S. uniflora* with *S. alata*, *S. biflora* with *S. splendida, S. sophera* with *S. spectabilis, S. siamea* with *S. excelsa* and *S. italica*, *S. purpurea* and *S. occidentalis* with *S. hirsuta* and *S. alexandrina*. Hook (1878) considered *Senna purpurea* (Roxb. ex Lindl) is a variety of *S. sophera* L. but the present study clearly indicated that there are no grouping between *S. purpurea* and *S. sophera*. Irwin and Barneby (1982) separated *S. excelsa* and considered as a variety of *S. spectabilis* but they are not grouped together, while *S. sophera* is very close to *S. spectabilis*. Dendrogram showed clear separation of the three genera *Cassia*, *Senna* and *Chamaecrista*. The polymorphic information content (PIC) of *Cassia*, *Senna* and *Chamaecrista* with RAPD showed high level of genetic diversity. The study has revealed clear groups of three genera *Cassia*, *Senna* and *Chamaecrista* but do not show any clear pattern according to the location in which they were collected. This results is in conforimty with Irwin and Barneby (1982); they divided the genus *Cassia* into three major groups; *Cassia*, *Senna* and *Chamaecrista* on the basis of morphological criteria. This result supported the previous taxonomic classification of the genus *Cassia* L. and show great diversity among the species of *Cassia*, *Senna* and *Chamaecrista*.

From the previous report, on basis of morphological and cytological studies, twenty-four species divided into three groups (Irwin and Barneby, 1982). RAPD maker revealed high degree of polymorphism (100%) among the twenty-four species belonging to genus *Cassia*, *Senna* and *Chamaecrista*. RAPD markers have been used earlier for taxonomic and phylogenetic relationships (Demcke et al., 1992; Millan et al., 1996). The use of molecular marker can help in establishing the limits among the defined groups on a more objective basis. In the case of genus *Scaevola*, which was initially misclassified by Linnaneus in 1753 and further rearranged several times by other scientist (Bentham, 1868; Krauze, 1912; Carolin, 1992) has now been reclassified, resolving the previous confusions through RAPD analysis (Swoboda and Bhalla, 1997).

The use of RAPD analysis has also been successful for resolving phylogenetic relationship in other plant groups. For example, phylogenetic relationship investigated using RAPD analysis among the *Rosa* species accessions, proved useful in assigning unclassified accessions to specific taxonomic groups. In another case Virk et al. (1995) have analysed the germplasm collection of rice accessions by RAPD markers and classified the unclassified rice accessions as *indica* or *japonica* types. Similarily, Pipe et al. (1995) supported the separation of two groups of *Opiostoma piceae* into two species based on the clear-cut divergence revealed by RAPD maker. In a typical defining a species, that is, whether the plant *Eucalptus granticola* is a relict species or a hybrid of existing species, the RAPD data show that *E. granticola* is 40% similar to *Eucalptus rudis* and *Eucalptus drummondii* are 25% similar among themselves. This is combination with morphological data revealed that the plant *E. granticola* is a hybrid and not a relict species (Rosseto et al., 1997). RAPD analysis of Tibetan wheat, common wheat and European spelt wheat supported the previous classification of the Tibetan wheat as a subspecies of common wheat (Sun et al., 1998).

AKNOWLEDGEMENT

We thank Dr. S. Hiremath and Dr. Kotresh, Karnatak University Dharwad for their help during plant material collection and species identification. We also thanks Dr. Rakesh tuli, Director NBRI Lucknow India for providing necessary facilities related to this work and Department of Biotechnology New Delhi India for financial assistance.

REFERENCES

Allan GJ, Clark C, Rieseberg LH (1997). Distribution of parental DNA markers in*Encelia virginenesis* (Asteraceae-Heliantheae) a diploid species of putative hybrid origin. Plant Sys. Evol. 205: 205-221.

Hook F (1878). Flora British India. London and Beccle. 2.

Bentham G (1871). A revision of the genus *Cassia*. Trans. Linn. Soc. London, 27: 503-593.

Bentham G (1868). *Scaevola* in: Flora Australiensis,. Reeve and co. London, 3: 83-104.

Bowditch BM, Albright GD, Williams KGJ, Braun JM (2003). Use of RAPD and amplified polymorphic DNA marker in comparative genomic studies. Methods Enzym., 224: 294-309.

Britton NL, Rose JN (1930). Caesalpiniacae. North Am. Flora, 23: 201-349.

Carolin (1992). *Scaevola*. In : Flora of Australia, Australian Govt. Pub. Service, Canberra, Australia. 35: 84-146.

Cheung WY, Champagne G, Hubert N, Landry SB (1997). Comparison of the genetic maps of *Brassica napus* and *Brassica oleracea*. Theor. Appl. Genet., 94: 569-582.

De-meke T, Adam PR, Chibbar NR (1992). Potential taxonomic use of random amplified polymorphic DNA (RAPD): A case study in *Brassic*. Theor. Appl. Genet., 84: 990-994.

Doyle JJ, Doyle JL (1987). Method of isolation of plant DNA from fresh tissue. *Phytoche. Bull.*, 19: 11.

Esen A, Hilu KW (1991). Electrophoretic and immunological studies of prolamines in Poaceae-II phylogenetic affinities of the Aristideae. Taxon., 40: 5-17.

Gareeb A, Khalifa PS, Fawzi N (1999). Molecular systamatics of some *Cassia* species, cytologia, 64: 11-16.

Hadrys H, Balick M, Scierwater B (1992). Application of random amplified polymorphic DNA (RAPD) in molecular ecology. Mol. Ecol., 1: 55-63.

Irwin HS, Barneby RC (1981). Tribe *Cassieae* Brronn. In Polhil, R. M. and Raven, P. H. (eds.) Recent advantages in legume systematics, Royal Botanic Garden, Kew. Pt, 1: 97-106.

Irwin HS, Barneby RC (1982). The American Cassiinae-A synoptical revision of Leguminosae tribe Cassieae subtribe Cassiinae in the New world. Mem. New York Bot. Gar., 35: 1-918.

Khan MA (1992). Seed protein electrophoretic pattern in brachypodium, P. Beauv. species. Ann. Bot., 70: 61-68.

Khanuja SPS, Shasany KA, Darokar PM (1998). Molecular taxonomy: the tools and relevance in plant Research. J. Med. Arom. Pl. Sci., 20: 996-999.

Krauze K (1912). In: Goodiniaceae and Branoniaceae Das pflanzenreich Regni vegetable concepts, Engelmann, Berun. pp. 117-168.4.

Linnaeus C (1754). Genera plantarum, 5th edition Stockholm.

Linnaeus C (1753). Species plantarum vol.1. Laurentii. Holmiae.

Mayer E (1970). Population, species and Evolution. Harvard University Press, Cambridge.

Millan TF, Osuna S, Cobos AM, Torres CIJ (1996). Using RAPD to study phylogenetic relationship in *Rosa*. Theor. Appl. Genet., 92: 273-277.

Mondal AK, Mondal S, Mandal S (2000). Molecular taxonomy of the genus *Cassia* L based on seed protein and mitochondria DNA RFLP. Phytomorphol., 50(1): 15-25.

Pipe ND, Buck WK, Braster MC (1995). Genomic fingerprinting supports the separation of *Ophiostoma piceae* into two species. Mycol. Res., 99: 1182-1186.

Powell W, Morgante M, Andree C, Hanagfey M, Vogel J, Tinglay S, Rafalski A (1996). A comparison of RFLP, RAPD, AFLP and SSR marker for germplasm. Mol. Breed., 2: 225-238.

Pujar S, Tauihauker AS, Rao SV, Gupta SV, Naik S, Ranjekar KP (1999). Arbitrarily primed-PCR based diversity assessment reflects in hierarchical grouping of Indian tetraploids wheat genotypes. Theor. Appl. Genet., 99: 868-876.

Rohlf FJ (1998). Numerical taxonomy and multivariate analysis system Version 2.0 New York.

Rosseto M, Lucarotti F, Hooper DS, Dixon WK (1997). DNA fingerprinting of Eucalyptus granticola: A critically endangered relict species or a rare hybrid. Hered., 79: 310-318.

Smith JSC, Chin LC E, Shu H, Smith SO, Wall JS, Senior LM, Michell ES, Kresovick S, Ziegle J (1997). An evaluation of the utility of SSR loci as molecular markers in maize (*Zea mays*) comparison with data from RFLPs and pedigrees. Theor. Appl. Genet., 95: 163-173.

Singh V(2001), Critical taxonomic notes on some species of *Cassia* L. found in india. J. Bombay Nat. Hist. Soc., 75(2): 434-444.

Stebbins GL (Jr.) (1957). Genetics, Evolution and plant breeding. Proc. Symp on genet and Pl. breed. in southest Asia. Jan 1957 New Dehli. Ind. J. Genet., 17: 129-141.

Sun Q, Ni Z, Liu Z, Gao J, Huan T (1998). Genetic relationship and diversity among Tibetan wheat comman wheat and European spelt wheat revealed by RAPD markers. Euphytica, 99: 205-211.

Swaboda I, Bhalla LP (1997). RAPD analysis of genetic variation in the Australion fern flower *Scaevola*. Genome, 40: 600-606.

Tiwari KM (1983). Fuelwood-presence and future with special reference to conditions in developing countries, In: strib, A., charter, p. and Schleser, G (eds.), Energy from Biomass, 2nd E.C. conference. Applied Science Publishers, London, New York, pp. 682-698.

Todaria NP, Nautiyal AR, Semval JK (1983). Electrophoresis protein profile of nodulated and non-nodulated *Cassia* species in relation to taxonomy. Biochem. Syst. Ecol., 11(3): 217-219.

Tomlinson PB (1981). The biology of trees native to tropical florida Allston. Massachusetts, USA., p. 209.

Virk PS, Ford-Llyod VB, Jackson TM, Newburg JH (1995). Use of RAPD for the study of diversity within plant germplasm collections. Hered., 74: 170-179.

Whitty WP, Powell W, Sprent I J (1994). Molecular separation of genus in Cassiinae (Leguminosae) and analysis of variation in the nodulating species of *Chamaecrista*. Mol. Ecol., 3: 507-515.

Wilkie SE, Isaac GP, Slattar JR (1993). Random amplified polymorphic DNA (RAPD) markers for genetic analysis in Allium. Theor. Appl. Genet., 99: 868-876.

Williams JGK, Kubelik AR, Livak KJ, Rafalski JA, Tingey SV (1990). DNAs polymorphisms amplified by arbitrary primers are useful as genetic markers. Nucl. Acids Res., 18: 6531-6535.

Williams JGK, Hanafey KM, Rafalski JA, Tingey SV (1993). Genetic analysis using random amplification polymorphic DNA markers. Methods Enzymol., 218: 704-740.

Propagation studies in *Farsetia aegyptia* Turra

Majda Khalil Suleiman*, Narayana Ramachandra Bhat, Mehdi Saleh Abdal, Sameeha Zaman, Sheena Jacob and Rini Rachel Thomas

Arid land Agriculture and Greenery Department, Kuwait Institute for Scientific Research, P. O. Box 24885, 13109 Safat, Kuwait.

Native plants are key components of the global biological diversity, and are highly adapted to the local environmental and climatic conditions. As Kuwait's native plants are threatened in their natural habitat and have begun to disappear at an alarming rate, their use in landscape projects will help in conserving biodiversity and heritage. *Farsetia aegyptia* Turra. is a grey green, exceedingly rare woody perennial with potential to adapt to urban landscape conditions. In this study, efforts were made to standardize the techniques for mass propagation of *F. aegyptia*. Results indicated that pre-treating *F. aegyptia* seeds with 1000 ppm (78%), 750 ppm (77%), and 500 ppm (76%) GA_3 was effective in enhancing the germination, compared to the control (49%).

Key words: Urban landscape, greenery development, bio-diversity, conservation of native plants, *Farsetia aegyptia*.

INTRODUCTION

Semi arid regions of the world are most susceptible to land degradation and desertification with serious implications for sustainable use of the natural environment (Kassas, 1995; Brown, 2003). The term 'desertification' is somewhat controversial, but in this study is used in accordance with Dregne (1986), referring particularly to a reduction in plant productivity (due to the loss of perennial shrub cover), a decline in species diversity and the loss of soil resources (Brown and Mazrooei, 2003). The arid eco-systems of the world can support plants with minimum water requirements that are drought and salt tolerant, for use in afforestation, landscaping and gardening projects, as drought has always been a normal recurrent event in arid and semi-arid lands (Le Houérou, 1996). Perennial vegetation cover in Kuwait is sparse, and is usually less than 10% (Brown, 2001). As in other arid and semiarid countries, annuals are the most dominant species (256 species) followed by herbaceous perennials (83 species), shrubs and under shrubs (34 species) and tree (one) species (Omar and Bhat, 2008). Kuwait's native vegetation is of enormous scientific value as it represents a transition between semi-desert and desert vegetation and it is highly tolerant to harsh environmental conditions such as extreme temperatures, drought and salinity. Several native desert plants are being threatened and are facing danger of extinction due to anthropogenic causes (Sudhersan et al., 2003). In addition to their contribution to the integrity of the environment, native plants are invaluable sources of useful genes for genetic improvement of crop plants (AboEl-Nil, 1997). Shrubs in specific are significant for desert rangeland vegetation and have the potential for urban landscape utilization (Abo El-Nil et al, 1993).

The climate of Kuwait is characterized by extremely hot summer, with daytime temperature exceeding 50°C and winter, cooler and at times wet, with temperature sometimes falling below 4°C (Annual Statistical Report, 2006). The mean annual rainfall is 113 mm (Omar et al., 2007). The total conventional fresh water resources available in Kuwait is six million m^3/ year while the total water demand has exceeded 350 million m^3/ year in the year 2000 (Hamoda, 2001). Native soil in Kuwait is mostly sandy in texture with high infiltration rate and is calcareous in nature (Omar et al., 2007).

Al Lbanah (*Farsetia aegyptia* Turra.) is one of the native perennial species that belong to the Cruciferae family. It is of fodder value and has potential to adapt to urban landscape conditions where several exotic species are now being used. It is a grey green, woody perennial about 30 cm high. It has slender, smooth and multi-branched stems. The flowers are creamy brown with four petals. Two rows of seeds are formed in an oval-shaped

*Corresponding author. E-mail: mkhalil@safat.kisr.edu.kw.

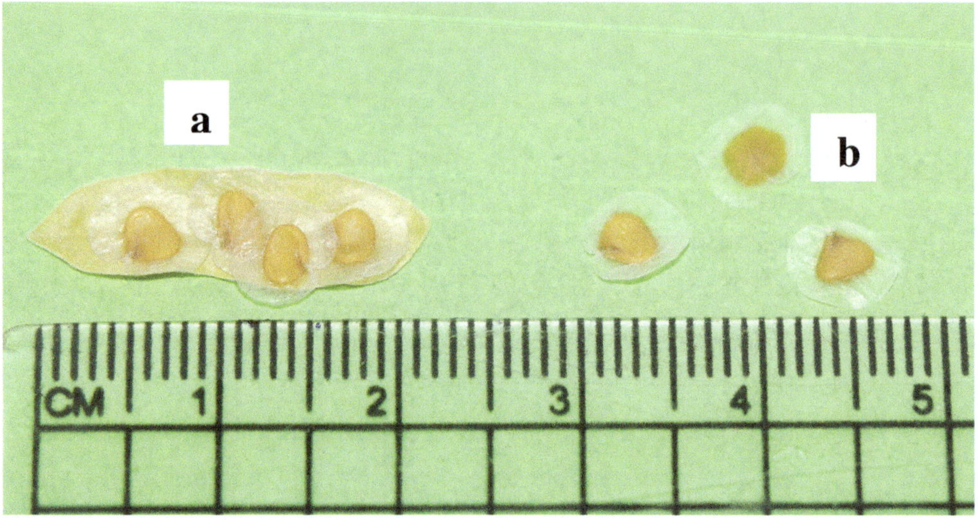

Figure 1. *Farsetia aegyptia* Turra. (a) pod; (b) seeds.

seed pod (Figure 1). It flowers during April and the propagation is through seeds (Omar et al., 2007). *F. aegyptia* is now an exceedingly rare perennial. Literature study indicated that there are no previously published reports on mass propagation of this species (Brown and Mazrooei, 2003).

Unfortunately, efficient propagation and establishment techniques that are crucial for both conservation of native plants and large-scale use in landscape programs are currently unavailable. Hence, this study investigates the possibility of mass propagation via seeds.

MATERIALS AND METHODS

Seed source

Seeds used in this study were obtained from the seed bank of Kuwait Institute for Scientific Research (KISR) in November 2007. These seeds were collected from Sulaibiya (Kuwait) in April, 2007. Prior to their use in the experiment, seed viability was determined using Triphenyl Tetrazolium Chloride (TTC) Test (ISTA, 1999). In this test, viable embryos are stained pink due to the reduction of 2, 3, 5- TTC by respirative activity in the cell. For this, 100 seeds (four replicates of 25 seeds each) were soaked in distilled water overnight, placed in a Petri dish, soaked in 0.1% TTC solution, covered the Petri dish with aluminum foil and left for 24 h at room temperature (25°C). Following the treatment, seeds were washed thoroughly with distilled water to remove excess stain and examined under the microscope.

Germination studies

In this experiment the combined effects of dry heat (50°C) exposure of seeds for 10 or 20 days and treatment with 0, 500, 750 or 1000 ppm GA_3 solution for 24 h prior to sowing on germination of *F. aegyptia* were studied. For this the seeds from room temperature storage (25°C) were kept in an oven at 50°C for 10 or 20 days, after which they were removed from the oven and treated with GA_3 for 24 h. The individual effect of various concentrations of GA_3 (500, 750 or 1,000 ppm) in enhancing the germination of *Capparis spinosa* seeds was also assessed. The control seeds were not subjected to either heat or GA_3 treatment. Seeds with radicle that are at least 2 mm in length considered germinated and were recorded weekly in each treatment. The total germination was calculated when no more seeds germinated. In all there were 12 treatments which were replicated five times in a randomized design. Seeds were sown in a soil substratum comprising of three parts agricultural soil and one part peat moss (v/v basis) in jiffy pots and maintained in laboratory conditions at 25°C. The data were analyzed using R analysis procedure of Analysis of Variance (ANOVA) and Dunkan's Multiple Range Test to ascertain the significant differences among treatments (Little and Hill, 1978; Crawley, 2005).

RESULTS AND DISCUSSION

Viability of seeds used in these studies was found to be 90%. The highest germination (78%) occurred when the seeds were subjected to 1000 ppm GA_3 for 24 h (Table 1). Pre-treatment with 750 and 500 ppm of GA_3 also resulted in 77 and 76% respectively. Forty nine percent of the untreated seeds germinated. While heat treatment alone for ten days resulted in slight reduction in germination (38%), exposure to extended duration of heat (20 days) increased the germination marginally. It is interesting to note that germination decreased with the increase in GA_3 concentrations in seeds that were exposed to dry heat for 10 days. The improvement in germination by dry heat and GA_3 was significant at $P \leq 0.001$ level compared to control.

The findings of this study are essential as this is the first attempt to standardize the propagation techniques of *F. aegyptia*, and to promote its usage in urban landscape projects. It is of ultimate necessity to develop and enhance seed and vegetative propagation techniques to ensure their adoption by the agricultural sector in Kuwait, including local agricultural nurseries. Results of this study are imperative as desert rehabilitation and biodiversity conservation is crucial to prevent the extinction of valuable

Table 1. Effect of dry heat (50°C) exposure and GA3 on germination of *Farsetia aegyptia* seeds

Heat treatment	Germination (%) [i] GA$_3$ Treatment (ppm) 0	500	750	1000
	24 h			
RT	49 ab [ii] ± 4.29	76 de± 7.31	77 e ± 3.39	78 e± 2.55
50°C 10 Days	38 a ± 4.67	71 cde± 7.18	55 abc ± 10.20	41 a ± 12.22
50°C 20 Days	56 abcd ± 6.18	61 bcde ± 3.74	64 bcde ± 3.40	52 abc ± 4.29

[i] Seeds with 2 mm or longer radicle or shoot are considered as germinated. [ii] The means followed by the same letter are not statistically different at p 0.001.

native plants in Kuwait (Suleiman et al., 2004: Omar et al., 2007).

Conclusions

For achieving re-establishment of the natural vegetation which is on the verge of extinction, restoration practices like artificial establishment of the same type of vegetation is essential. Utilizing native plants in the urban landscape promotes their sustainability in that region, facilitates conservation of natural plant diversity and imparts a fully natural appeal to the landscape. This is the initial step for the commercial production of these plants in Kuwait's nurseries. Mass propagation of native plants like *F. aegyptia,* a rare perennial in Kuwait, can be standardized for further commercial use. Pre treating the seeds of *F. aegyptia* with 500, 750 or 1000 ppm GA$_3$ was capable of increasing the germination percentage effectively.

ACKNOWLEDGEMENTS

The authors would like to thank the Kuwait Foundation for the Advancement of Sciences (KFAS) and Kuwait Institute for Scientific Research for providing funds and encouragement during the investigation.

REFERENCES

AboEl-Nil M (1997). Tissue Culture of Native Plants in the Developing Countries. III International Symposium on in Vitro Culture and Horticultural Breeding, ISBN 9066059095. ISHS Acta Horticulturae 447: 507-514.

Abo El-Nil M, Al-Sabah L, Al-Menaie H, Bastaki S, Zaman S (1993). Mass propagation of stress tolerant desert plants by tissue culture. Phase I/Biotechnology Department, Aridland Agriculture Department, Food Resources Division. KISR 4300. pp: 83-86.

Annual Statistical Report (2006). Ministry of Planning. (Kuwait: The State of Kuwait).

Brown G (2003). Factors maintaining plant diversity in degraded areas of northern Kuwait. J. Arid Environ., 54:183-194.

Brown G, Al-Mazrooei S (2003). Rapid Vegetation Regeneration in a Seriously Degraded *Rhanterium epapposum* Community in Northern Kuwait after 4 Years of Protection. J. Environ. Manage. 68(4): 387-395.

Brown G (2001). Vegetation ecology and biodiversity of degraded desert areas in north-eastern Arabia.-Habilitation thesis. Rostock University. pp. 9-22.

Crawley MJ (2005). Statistics- An Introduction using R. (England: John Wiley & Sons Ltd.),pp: 155-185.

Dregne HE (1986). Desertification of Arid Lands. In: El-Baz F, Hassam MHA, Editors, Physics of Desertification, Martinus Nijhoff Publishers, Dordrecht.

Hamoda MF (2001). Desalination and water resource management in Kuwait. Desalination, pp. 138-165.

International Seed Testing Association (ISTA) (1999). International Rules for Seed testing. Seed Sci. Technol., 27: 333.

Kassas M (1995). Desertification: a general review. J. Arid Environ. 16: 125-145.

Le Houerou HN (1996). Climate change, drought and desertification. J. Arid Environ. 34: 249-258.

Little TM, Hills FJ (1978). Agricultural Experimentation Design and Analysis. (U.S.A: John Wiley & Sons Inc), pp. 63-65.

Omar SAS, Bhat NR (2008). Alteration of the *Rhanterium epapposum* plant community in Kuwait and restoration measures. Int. J. Environ. Stud., 65(1): 139-155.

Omar SAS, Al-Mutawa Y, Zaman S (2007). Vegetation of Kuwait, Kuwait: Kuwait Institute for Scientific Research. pp: 47-89.

Sudhersan C, AboEl-Nil M, Hussain J (2003). Tissue culture technology for the conservation and propagation of certain native plants. J. Arid Environ., 54: 133-147.

Suleiman MK, Bhat NR, Bellen R (2004). Various attempts in greening the State of Kuwait. Arch. Agron. Soil Sci. 50: 353-360.

Physico-chemical factors affecting the distribution of wetland birds of Barna Reservoir in Narmada Basin, Central India

Satish Balapure, Sumana Dutta and Vipin Vyas

Department of Environmental Sciences and Limnology, Barkatullah University, Bhopal, 462026, India.

The ecology of wetland birds of the Barna Reservoir was studied from March 2009 to February 2011. Species composition, diversity and abundance of birds were assessed. Sixty four (64) species of wetland birds were enumerated. Fourteen (14) environmental variables were correlated with the wetland birds species richness among which most of the variables were positively correlated with bird species richness except water depth, dissolved oxygen, total hardness and chloride. A strong correlation of bird species was noted with benthos density ($r = 0.98$). Bird species richness was also positively correlated with macrophytic biomass ($r = 0.8$), orthophosphate ($r = 0.75$), and conductivity ($r = 0.64$). It was observed that bird species richness had a strong negative correlation with water depth and dissolved oxygen.

Key words: Barna reservoir, wetland birds, environmental variables, macroinvertebrates, macrophytes.

INTRODUCTION

Birds are important components of our ecosystem and play a major role in maintaining the natural balance in the food chain in nature. There are several importance of birds as they act as browsers, pollinators and seed dispersers (Clout and Hay, 1989).

To study any ecosystem the birds serve as important component as they have the ability to fly away and avoid any obnoxious condition. Hence, they are considered as important health indicators of the ecological conditions and productivity of an ecosystem (Newton, 1995; Desai and Shanbhag, 2007). Birds also play an important role in wetland ecosystem. Wetlands are important especially for bird habitats. Birds use wetlands for breeding, nesting and teaching young, as a source of drinking water, for feeding, resting, shelter and for social interaction. Wetlands provide food for birds in the form of plants, vertebrates and invertebrates.

Some feeders forage for food in wetland soils. Some feed on water column, some feed on the vertebrates and invertebrates that live on submerged and emergent plants. Birds have daily and seasonal dependence on wetlands for food and other life supporting systems (Stewart, 2001).

Distribution and abundance of water birds was affected by several factors. Little change in physical, chemical or biological properties put forth intense effects on bird's habitats (Murphy et al., 1984). Any change in the physical, chemical and biological factors in the catchment exerts severe impact on the wetland as habitat for aquatic communities. These in turn affect the wetland dependent communities as well as the ecosystem attributes such as species richness, its distribution and density (Burkert et al., 2004). Ultimately, these changes alter the corresponding food web structures at the primary and secondary production levels (Wrona et al., 2006). Madhya Pradesh is having a number of small, medium and large water bodies (Sugunan, 1995) but very little is known about the avifaunal aspect of these water bodies. Vyas (1992) has initiated a survey on wetland birds of Upper Lake in relation to habitats available to migratory birds.

Vyas et al. (2010) worked on avian diversity of Bhoj wetland of Bhopal with seasonal variation. Recently Balapure et al. (2012) studied the wetland birds of Barna reservoir of Central India, but all these works are done based on spatial and seasonal census of the bird. No work has been done in Madhya Pradesh on the relationship between wetland avifauna and trophic status of the wetland.

No systematic and comprehensive report is available on the avifaunal diversity in any water body of Madhya Pradesh. To fulfill this research gap this study was chosen to conduct the surveys of avifauna and their relationship with wetland characteristics including limnological, macrophytes and macroinvertebrate during 2009-2011.

MATERIALS AND METHODS

Description of the study area

Barna is one of the major irrigation projects of Madhya Pradesh constructed by damming river Barna in Raisen district near Bari village. Barna sub-basin is a part of Narmada river system. The total length of Barna River is 105 km with a catchment area of 1789 km^2. Barna reservoir is located at latitude 22° 50' 23.5" N and longitude 77° 50' 78.20" E and falls in the toposheet number 55 I/4 of Survey of India (Map 1). This is located about 100 km from Bhopal. This reservoir is an important source of fish production in the area. Regular fish stoking is done in this reservoir every year. The water of the reservoir is mainly used for the fisheries and irrigation purpose. The reservoir supports a rich biodiversity and provide habitat for wildlife including migratory birds. A major part of the basin falls under Singhori Wild Life Sanctuary.

Description of the site

For the purpose of the bird study, the wetland was divided into three zones Viz- Zone-I, zone-II and Zone III. This categorization was based on the habitat available for birds. Zone-I comes under Jamner river basin situated at left bank of the reservoir. Maximum runoff of this basin is from Singhori Wildlife sanctuary. The majority area of this zone is with steep slope which creates minimum spread area. Zone II is the central zone of the reservoir which receives runoff from Barna River. This zone is with gentle slope and formed a scattered marshy land when water receded from this part. This area is heavy infested with macrophyte. Plants of seasonal occurrence seen during November to March are *Hydrilla verticillata*, *Vallisneria spiralis*, *Sagittaria sangitofolia*, *Ottelia alismoides*, *Najas graminae*, *Najas minor* and *Jussiaea repens*.

During post monsoon maximum area of this zone is used for paddy cultivation which serves as suitable habitat for wintering birds. Zone III which is on the right side comes under flat area. Some part of this area is used for paddy cultivation. Maximum portion of this area is mudflat. *Hydrilla* sp. and *Najas* sp. are the dominant macrophytes of this zone. Land digging and other human disturbances were noticed in this zone. Each zone is again divided into three subzones based on depth measured during initial survey. In zone-I, there were sites-1, 2 and 3, in zone II, sites-5, 6 and 7 and in zone III, there were sites-7, 8 and 9. Sites-1, 4 and 7 were the shallower part of the reservoir having a depth of 0-2 m, site-2, 5 and 8 with medium depth of 2-4 m and sites-3, 6 and 9 come under deeper part with 4-6 m.

Data collection and analysis

Study of avifaunal diversity of Barna wetland was conducted between March, 2009 and February, 2011. Seasonal observations were made during the study. The study period was divided into four seasons namely summer (March to May), monsoon (June to August), post monsoon (September to November) and winter (December to February). Birds were observed within the transect of 300 m in shoreline when watching from boat. Nikon Binoculars of 10 × 50 were used for observations. The field book of Ali and Ripley (1986), Ali (2002) were used to identify bird species. The checklist was prepared using the standardized common and scientific names of the birds of the Indian subcontinent by Manakkadan and Pittie (2001).

Measurement of hydrological variables

Water samples were collected from the surface of the reservoir for water chemistry. Thirteen physicochemical variables were investigated which directly or indirectly involved macrophytic and benthos growth which ultimately affect the bird population. Analytical techniques as described by standard methods for examination of water and waste water (APHA, 1998; Adoni et al., 1985).

Macroinvertebrate and macrophyte estimation

The benthic samples were collected with the help of Peterson grab, mud sampler. The collected samples were sieved through 2 and 0.5 mm mesh size sieve one after the other (Adoni 1985). The material which retained on sieve were sorted out with the help of forceps and brush and collected in narrow mouthed plastic bottle containing 5% formalin as preservative. Attached fauna from stones and macrophytes were also collected. Identification of all macrofaunal organisms was done with the help of Metzer binocular light microscope by using standard keys of Subba Rao and Dey (1989) and Needham and Needham (1962). Density of macro-invertebrates was calculated as per Adoni et al. (1985). For collection of macrophytes, a wooden quadrate (50 x 50 cm) was placed at different sites in different habitats and the entire content of the specified area was uprooted. The collected material was washed thoroughly to get rid of adhering material. The extra water of the plants was soaked with the help of filter paper. Plants of each quadrate were separated species wise and fresh weight was noted after weighing them. The samples were transported to the laboratory in polythene bags and sorted out for identification and biomass estimation. Identification of macrophytes was done by using following keys and manuals by Cook (1996) and Biswas and Calder (1984). Biomass estimation was carried out according to Adoni et al. (1985). Correlation matrix and cluster analysis were carried out using software Paleontological Statistics (PAST) version 2.04 (Hammer et al., 2001) and biodiversity professional (McAleece et at., 1999) version.

RESULTS

A total of 64 water birds belonging to 7 orders and 12 families were recorded between March 2009 to February

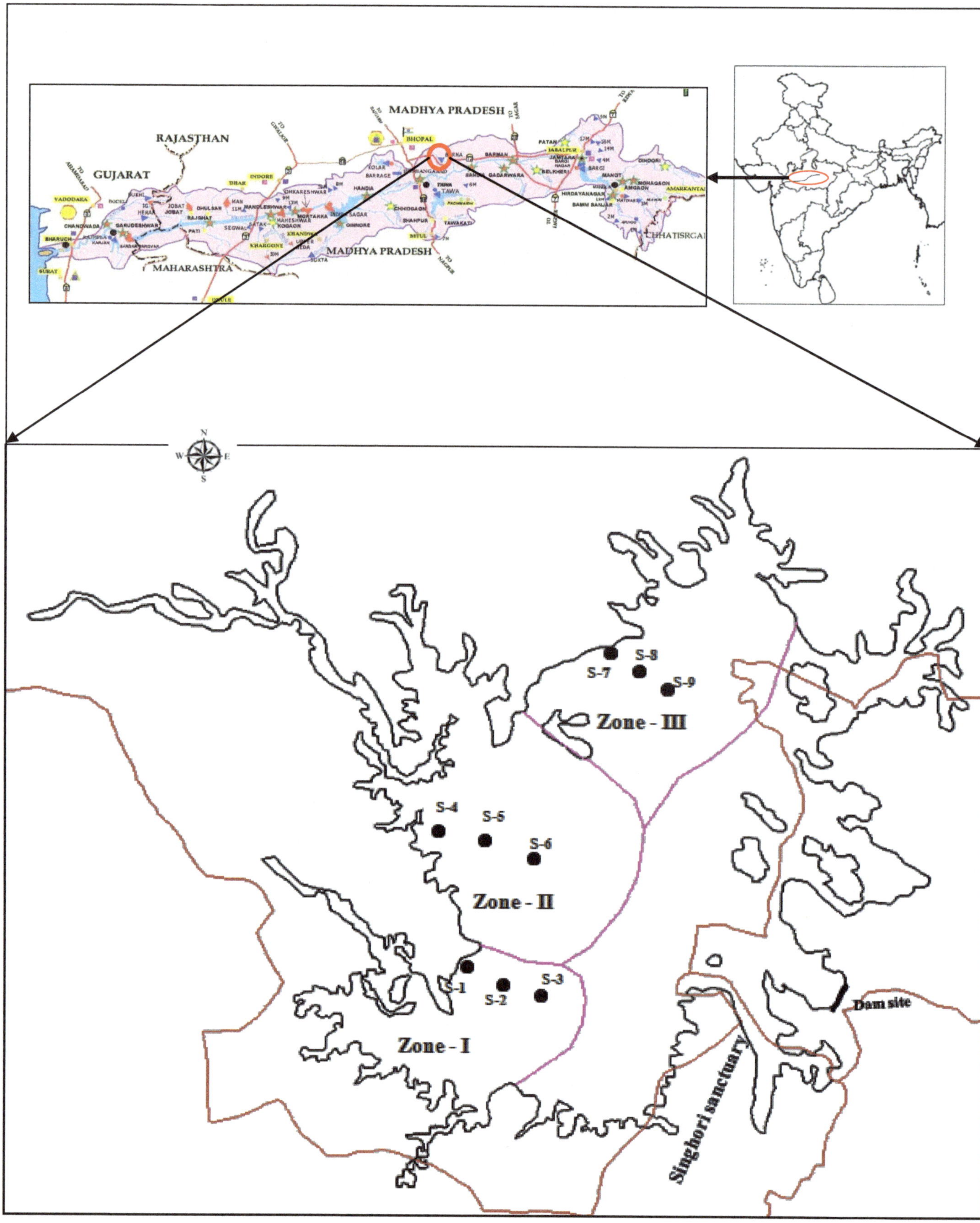

Map 1. Map of Barna reservoir in Narmada basin of central India where study sites are located zone wise.

2011 (Table 1). Waterbirds counting was done by the Transect methods (Eberhardt, 1968) and point count methods (Ralph et al., 1995). The most dominating family was Anatidae contributing 16 species. Charadriidae formed the second dominant family with 10 species followed by Ardeidae, Rallidae, Laridae, Phalacrocoracedae, Ciconiidae, Threskiornithidae, Alcedinidae, Podicipedidae, Gruidae and Recurvirostridae. The most abundant family was Anatidae, having 53456 individuals which contributed 77.58% of the total bird population. *Anas acuta* (Northern Pintail) and *Anas crecca* (Common Teal) were the most abundant species of this family which formed 36.69% of the total Anatidae percentage.

During the present investigation it was noted that shallower sites (sites-1, 4 and 7) showed maximum bird species richness as compared to the other sites. A maximum of 47 water bird species were recorded from sites-4 followed by 45 wetland birds from site-7 Figure 2. These sites were characterized as shallower zones of the reservoir. Minimum water bird assemblage was noted at sites-8, 9 and 2.

Cluster analysis confirms the findings by making different clusters with similar characters. Den-drogram (Figure 3) is showing the similarity of sites based on the water birds among nine sites of Barna reservoir during two years study period. The arrangement of stations produced by sum-of-squares agglomeration depicts the two main groups, that is, group 1 formed by station 1, 4 and 7. A second group was formed by station 2, 3, 5, 6, 8 and 9 which forms sub groups, that is, 2, 5 and 8 and 3, 6 and 9 (Figure 1). Higher similarity was observed between Site 8 and 5 (84%), 3 and 2 (82%) and 7 and 1 (80%).

During the present study, 14 parameters were taken for correlation analysis between each other including depth, water temperature, conductivity, TDS, pH, D.O., total alkalinity, total hardness, chloride, ortho-phosphates, ni-trate, benthos density, macrophytic biomass and bird spe-cies richness (Table 2). Water temperature, conductivity, TDS, total alkalinity, ortho-phosphates and nitrate nitro-gen were positively and significantly correlated with bird species richness where water depth, dissolved oxygen, total hardness and chloride were negatively correlated. The strong correlation of bird species was noted with benthos density

DISCUSSION

Among avian communities, the components of diversity are known to differ between locations and seasons (Kricher, 1972; Bethke and Nudds, 1993). Species richness showed higher ranges in shallower sites (Site-4 and 7) with macrophytic vegetation and lower in deeper zones. Maximum richness was noted at Site-4 having gentle slope and dense macrophytic vegetation. Nelson and Kadlec (1984) described the interactions occurring among macrophytes, macroinvertebrates and water birds in freshwater wetlands as a complex interdependency in which dynamic changes in the abundance and distributional pattern of macrophytes resulting from processes in litter decomposition and macroinvertebrate communities that, in turn, affect avifaunal abundance in water bodies. Species richness, bird abundance and diversity reach higher values in larger and structurally more heterogeneous wetlands (Gonzalez-Gajardo et al., 2009).

Dendrogram show similarity in the number of waterbird species among nine sites of Barna reservoir. Site 4, 7 and 1 which accounted for greater richness of waterbird species belong to one cluster being the marginal part of the reservoir, whereas the rest of the six sites are in the open water with less species richness.

Fluctuation in water level might alter the habitat characteristics that could cause prompt changes in fish, amphibians, invertebrates and waterbird communities (Johnson et al., 2007). In addition to limiting access to foraging habitats, water depth affects the net energy intake of waterbirds because foraging efficiency decreases with increasing water depth. Gawlik (2002) indicated that for wading birds that forage on prey in the water column, the locomotion of the birds might be slowed in deep water because of increased water resistance with depth.

At Barna reservoir, it was noted that temperature and bird richness was positively correlated. Maximum bird richness was noted during winter months with low temperature which was positively correlated. Deshkar et al. (2010) also supports our study. The average water pH was in a slightly alkaline range (7.2-9.2) at our study site during the entire study period.

Longcore et al. (2006) reported that a water pH in the alkaline range supported higher macro-invertebrates and thereby attracted more ducks to the water bodies. Minns (1989) considered pH as an indicator of overall productivity that can cause habitat diversity. He established a significant correlation of pH with species richness of phytoplankton, invertebrates, fishes, amphi-bians and the water birds, which depend on these organisms.

According to Tutle et al. (1984) waterfowl abundance was affected by nitrogen and phosphate fluctuations of water bodies. During the present investigation avifaunal richness showed a positive relationship with nitrate and phosphate in Barna reservoir.

Hoyer et al. (1994) found a close correlation between the density of aquatic birds and the ortophosphate-phosphorus concentration in the water. According to Pip (1979) alkalinity, pH, dissolved organic matter, nitrogen, phosphorous, chloride, and sulfate all affect macrophyte species distribution.

A linear relation-ship between species richness of bird communities with habitat condition, chloride, total phosphorus, temperature, total nitrogen, ortho-phosphate and nitrate was found by Getachew et al. (2012) in Cheffa wetland.

Table 1. List of birds recorded in Barna reservoir during March 2009 - February 2011

S/N	Common name	Scientific name
	Grebes	
1	Little Grebes	*Tachybaptus ruficollis*
2	Great Crusted Grebes	*Podiceps cristatus*
	Cormorants	
3	Great Cormorant	*Phalacrocorax carbo*
4	Indian Shag	*Phalacrocorax fuscicollis*
5	Little Cormorant	*Phalacrocorax niger*
6	Darter	*Anhinga melanogaster*
	Egrets and Herons	
7	Large Egret	*Casmerodius albus*
8	Little Egret	*Egretta garzetta*
9	Median Egret	*Mesophoyx intermedia*
10	Cattle Egret	*Bubulcus ibis*
11	Grey Heron	*Ardea cinerea*
12	Purple Heron	*Ardea purpurea*
13	Little Green Heron	*Butorides striatus*
14	Black-Crowned Night-Heron	*Nycticorax nycticorax*
15	Indian Pond Heron	*Ardeola grayii*
	Storks	
16	Painted Stork	*Mycteria leucocephala*
17	Asian Openbilled Stork	*Anastomus oscitans*
18	Black-necked Stork	*Ephippiorhynchus asiaticus*
19	White-Necked Stork	*Ciconia episcopus*
	Ibis and Spoonbill	
20	Black Ibis	*Pseudibis papillosa*
21	Oriental White Ibis	*Threskiornis melanocephalus*
22	Eurasian Spoonbill	*Platalea leucorodia*
	Ducks	
23	Greyleg Goose	*Anser anser*
24	Barheaded Goose	*Anser indicus*
25	Brahminy Shelduck	*Tadorna ferruginea*
26	Common Shelduck	*Tadorna tadorna*
27	Comb Duck	*Sarkidiornis melanotos*
28	Lesser Whistling -Duck	*Dendrocygna javanica*
29	Northern Pintail	*Anas acuta*
30	Common Teal	*Anas crecca*
31	Spottbill Duck	*Anas poecilorhyncha*
31	Gadwall	*Anas strepera*
33	Eurasian Wigeon	*Anas penelope*
34	Northern Shoveler	*Anas clypeata*
35	Red Carested Pochard	*Rhodonessa rufina*
36	Common Pochard	*Aythya ferina*
37	Tufted Pochard	*Aythya fuligula*
38	Cotton Teal	*Nettapus coromandelianus*

Table 1 Contd.

	Cranes	
39	Common Crane	*Grus grus*
40	Sarus Crane	*Grus antigone*
	Rails and Coots	
41	Whitebreasted Waterhen	*Amaurornis phoenicurus*
42	Watercock	*Gallicrex cinerea*
43	Common Moorhen	*Gallinula chloropus*
44	Purple Moorhen	*Porphyrio porphyrio*
45	Common Coot	*Fulica atra*
	Waders	
46	Black-winged Stilt	*Himantopus himantopus*
47	Red-wattled Lapwing	*Vanellus indicus*
48	River Lapwing	*Vanellus duvaucelii*
49	Little Ringed Plover	*Charadrius dubius*
50	Kentish Plover	*Charadrius alexandrinus*
51	Marsh Sandpiper	*Tringa stagnatilis*
52	Wood Sandpiper	*Tringa glareola*
53	Common Sandpiper	*Actitis hypoleucos*
54	Curlew Sandpiper	*Calidris ferruginea*
55	Little Stint	*Calidris minutus*
56	Dunlin	*Calidris alpina*
	Gulls and Terns	
57	Brown-headed Gull	*Larus brunnicephalus*
58	Yellow-Legged Gull	*Larus cachinnans*
59	Common Tern	*sterna hirundo*
60	River Tern	*Sterna aurantia*
61	Little Tearn	*Sterna albifrons*
	Partially dependent waterbirds	
62	Whitebrested Kingfisher	*Halcyon smyrnensis*
63	Small Blue Kingfisher	*Alcedo atthis*
64	Lesser Pied Kingfisher	*Ceryle rudis*

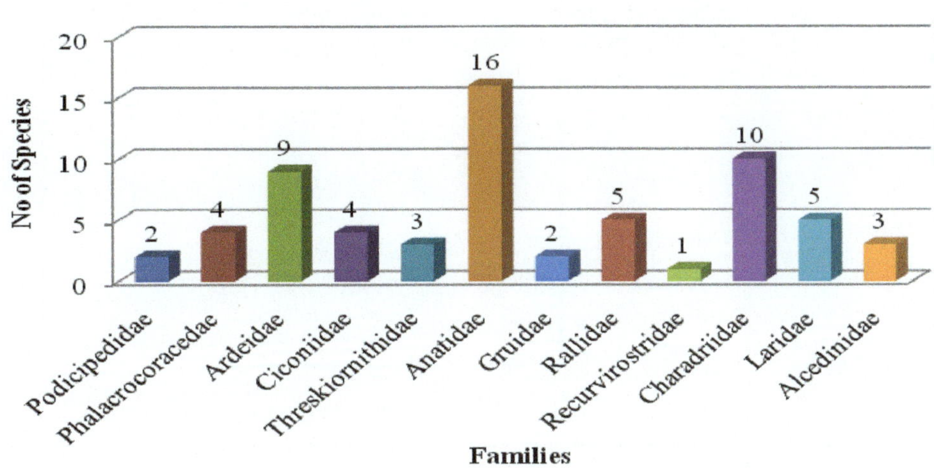

Figure 1. Family wise species composition of water birds in Barna reservoir.

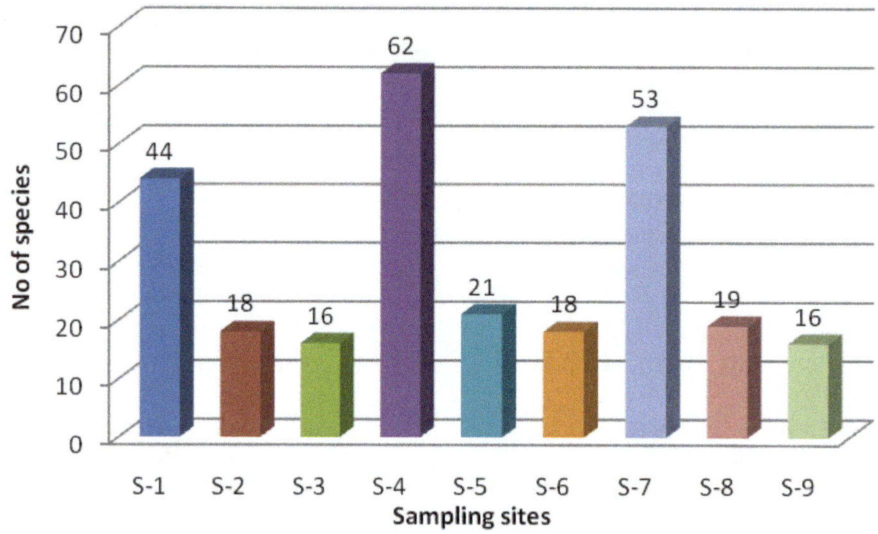

Figure 2. Spatial variation in species richness of wetland birds in Barna reservoir.

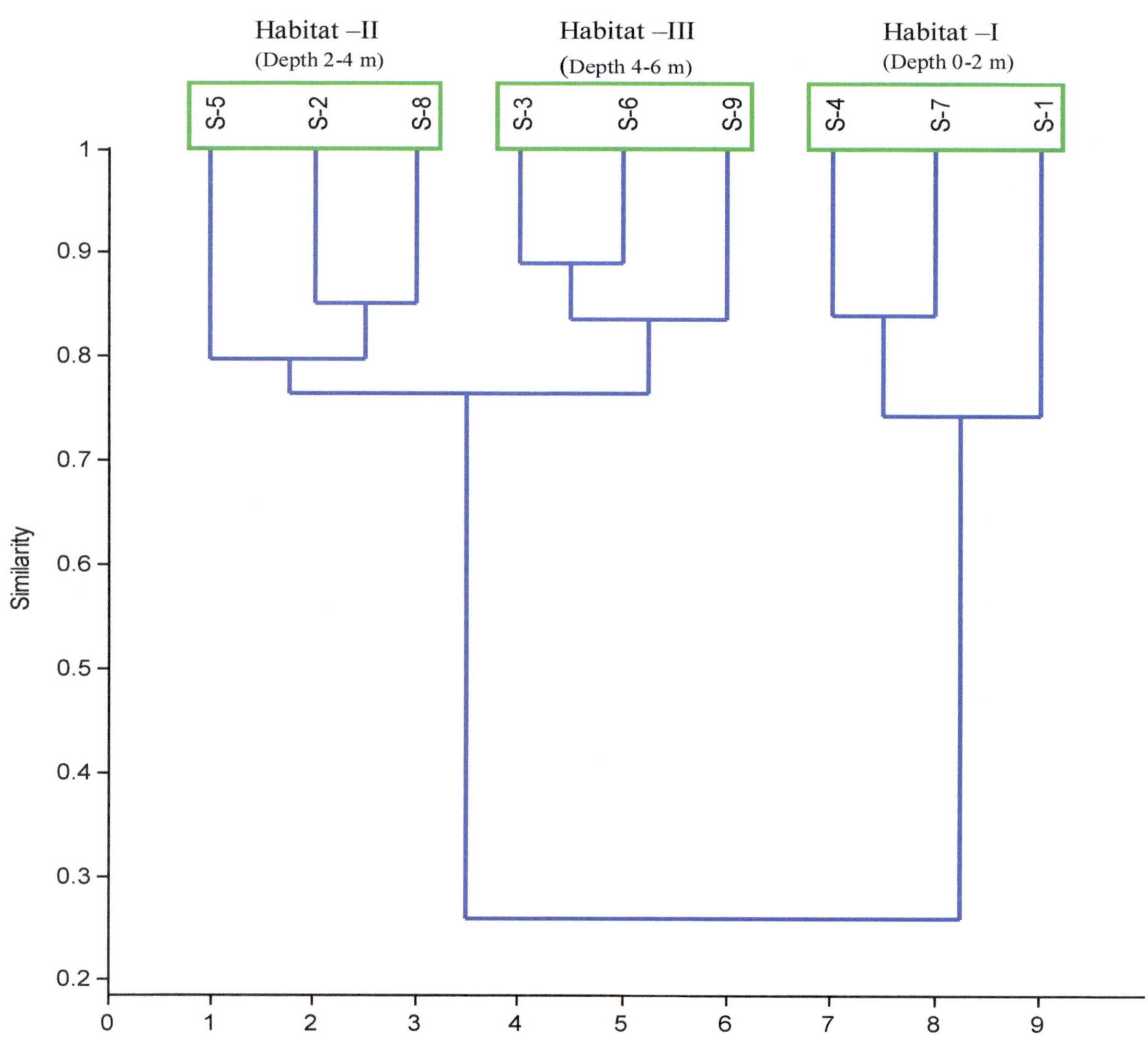

Figure 3. Dendrogram showing Jaccard's similarities between selected sampling sites in Barna reservoir.

Table 2. Correlation matrix for physico-chemical parameters, macrophyte biomass, benthos density and richness of water birds.

	1	2	3	4	5	6	7	8	9	10	11	12	13	14
1	1	*	*	*	*	*	*	*	*	*	*	*	*	*
2	0.15	1.00	*	*	*	*	*	*	*	*	*	*	*	*
3	-0.41	-0.10	1.00	*	*	*	*	*	*	*	*	*	*	*
4	-0.23	0.42	0.75	1.00	*	*	*	*	*	*	*	*	*	*
5	-0.20	0.16	0.64	0.90	1.00	*	*	*	*	*	*	*	*	*
6	0.78	-0.18	-0.04	-0.06	0.10	1.00	*	*	*	*	*	*	*	*
7	-0.21	0.22	0.76	0.74	0.65	0.20	1.00	*	*	*	*	*	*	*
8	0.20	0.47	0.07	0.30	0.15	-0.01	0.01	1.00	*	*	*	*	*	*
9	-0.10	0.47	0.20	0.40	0.32	0.04	0.54	0.41	1.00	*	*	*	*	*
10	-0.01	-0.03	0.14	0.17	0.31	0.28	0.10	-0.12	-0.04	1.00	*	*	*	*
11	-0.18	-0.18	0.73	0.35	0.28	0.29	0.70	0.19	0.27	0.25	1.00	*	*	*
12	-0.86	0.11	0.59	0.59	0.51	-0.67	0.49	-0.21	0.10	-0.02	0.18	1.00	*	*
13	-0.70	0.33	0.75	0.82	0.66	-0.53	0.60	0.23	0.38	0.12	0.37	0.85	1.00	*
14	-0.87	0.01	0.64	0.51	0.45	-0.73	0.32	-0.22	-0.07	0.75	0.08	0.98	0.80	1.00

1, Depth (m); 2, water temperature (°C); 3, conductivity (µs/cm); 4, TDS (mg/l); 5, pH; 6, D.O, (mg/l); 7, total alkalinity (mg/l); 8, total hardness (mg/l); 9, chlorides (mg/l); 10, ortho-phosphates (mg/l); 11, nitrate nitrogen (mg/l); 12, benthos density 13, macrophytic biomass; 14, bird richness.

Birds often have correlation with their habitats (Seymour and Simmons, 2008) and have also been used as surrogates for assessing the impact of habitat changes (Yang et al., 2008). Water bird species richness showed strong positive correlation with macrophyte biomass and benthos density in Barna reservoir. A suite of characteristics reflecting higher trophic status, from water quality to invertebrate and macrophytic biomass, were correlated with the avifaunal density (Staicer et al., 1994).

Total avi-faunal density has been related positively ($P < 0.05$) with total macrophytic biomass in Santragachi Jheel (Patra et al., 2010). Aquatic macrophytes are important to bird populations that use water bodies and the management of aquatic macrophytes has the potential to affect bird populations.

Nelson and Kadlec (1984) described the interactions occurring among macrophytes, macroinverte-brates and water birds in freshwater wetlands as a complex interdependency in which dynamic changes in the abundance and distributional pattern of macrophytes resulting from processes in litter decomposition and macro-invertebrate communities that, in turn, affect avifaunal abundance in water bodies. Well vegetated wetlands seem attractive to wetland bird species (Weins, 1997).

Conclusion

The congregation of large numbers of waterbird species at Barna reservoir for feeding, resting and roosting, is due to the abundance of food (macrophytes, macrobenthic organisms, and free swimming organisms inclusive of fish), accessibility to food resources due to the shallowness in winters, availability of exposed mudflats and shorelines for roosting in an area well protected from human and other disturbances and presence of submerged as well as emergent vegetation patches. Macrophyte diversity is higher in zones II and III than zone I as zones II and III were both having gentle slope which creates maximum mudflat area for the growth of macrophyes.

The different vegetation zones of the Barna reservoir include exclusive zones of individual species such as *Hydrilla verticillata, Cyperus articulatus, Najas minor, Vallisneria spiralis* as well as the mixeture of these species at different proportions. Based on the correlation, it was noted that the strong correlation of bird species was noted with benthos density in all zones.

Barna reservoir supports vulnerable endangered and near threatened species of fishes and birds. It supports one vulnerable bird Sarus crane and four near threatened bird, that is, species Darter, Painted stork, Black necked stork and Oriental white Ibis were also found here. So, attention of the concerned agencies is very much needed for the conservation as well as sustainable use of the reservoir.

ACKNOWLEDGEMENTS

Authors are thankful to Dr. Pradeep Shrivastava, Professor, Department of Environmental Science and Limnology, Barkatullah University, Bhopal for his support during the study period by providing laboratory facility. We are thankful to the local people of Bari village who helped during field visits. We are also grateful to the M. P. Fisheries Federation, Bhopal for the boat facility.

REFERENCES

Adoni AD, Joshi G, Ghosh K, Chourasia, SK, Vaishya AK, Yadhav M, Verma HG (1985). Work book on Limnology, Pratibha Publication Sagar, MP (India). p. 216

Ali S (2002). The book of Indian birds. Bombay Natural History Society. Oxford University Press, Bombay. p. 326.

Ali S, Ripley SD (1986). Handbook of the birds in India und Pakistan Compact Ed., Oxford University Press New Delhi.

American Public Health Association (1998). Standard methods for the examination of water and waste water. 20th Edition.

Balapure S, Dutta S, Vyas V (2012). Avian diversity in Barna Wetland of Narmada basin in Central India. J. Res. Biol. 2(5):460-468.

Bethke RW, Nudds T (1993). Variation in the diversity of ducks along a gradient of environmental variability. *Oecologia* 93:242-250.

Burkert U, Ginzel G, Babenzien HD, Koschel R (2004). The hydrogeology of a catchment area and an artificially divided dystrophic lake - consequences for the limnology of Lake Fuchskuhle. Biogeochemistry 71:225-246.

Clout MN, Hay JR (1989). The importance of birds as browsers, pollinators and seed dispersers in New Zealand forests. New Zealand J. Ecol. 12:27-33.

Cook CDK (1996). Aquatic and wetland plants of India. Oxford, New York.

Desai M, Shanbhag A (2007). Birds breeding in unmanaged Monoculture plantations in Goa, India. Indian Forester 133:1367-1372.

Deshkar S, Rathod J, Padate G (2010). Avifaunal diversity and water quality analysis of an inland wetland. J. Wetlands Ecol. 4:1-32.

Gawlik DE (2002). The effects of prey availability on the numerical response of wading birds. Ecol. Monogr. 72:329-346.

Getachew MA, Ambelub, Tikub S, Legessec W, Adugnad A, Kloose H (2012). Ecological assessment of Cheffa Wetland in the Borkena Valley, northeast Ethiopia: Macroinvertebrate and bird communities Ecol. Indic. 15:63-71.

Gonzalez-Gajardo A, Sepulveda PV and Schlatter R (2009). Waterbird Assemblages and Habitat Characteristics in Wetlands: Influence of Temporal Variability on Species-Habitat Relationships. Waterbirds 32(2):225-233.

Hammer O, Harper DAT, Ryan PD (2001). PAST version 1.39: Paleoontological statistical software package for education and data analysis. Paleontol. Eletronica 4:1-9.

Hoyer MV, Canfield JDE (1994). Bird abundance and species richness on Florida lakes: influence of lake trophic status, morphology, and aquatic macrophytes. Hydrobiologia 297/280:107-119.

Johnson KG, Allen MS, Havens KE (2007). A review of littoral vegetation, fisheries, and wildlife responses to hydrologic variation at Lake Okeechobee. Wetlands 27(1):110-126.

Kricher JC (1972). Bird species diversity: the effect of species richness and equitability on the diversity index. Ecology 53:278-282.

Longcore JR, McAuley DG, Pendelton GW, Bennatti CR, Mingo TM, Stromborg KL (2006). Macroinvertebrate abundance, water chemistry, and wetland characteristics affect use of wetlands by avian species in Maine. Hydrobiologia 567:143-167.

Manakkadan R, Pittie A (2001). Standardized common and scientific names of the birds of the Indian subcontinent. Buceros 6(1):1-37.

McAleece N, Lambshead PJD, Paerson GLG, Gage JD (1999). Biodiversity Pro. A programme research for analyzing ecological data.

Minns CK (1989). Factors affecting fish species richness in Ontario lakes. Trans. Am Fish. Soc. 76:332-334.

Murphy S, Kessel MB, Vining LJ (1984). Waterfowl population and limnological characteristics of Taiga ponds. J. Wildl. Manage. 48(4):1156-1163.

Needham JG, Needham PR (1962). A guide to study the freshwater biology Holden - Rey Inc Francisco. pp. 1-108.

Nelson JW, Kadlec JA (1984). A conceptual approach to relating habitat structure and macroinvertebrate production in freshwater wetlands. Trans. N. Am. Wildl. Conf. 49:262-270.

Newton I (1995). The contribution of some recent research on birds to ecological understanding. J. Anim. Ecol. 64:675-696.

Patra A, Santra KB, Manna CK (2010). Relationship among the abundance of waterbird species diversity, macrophytes, macroinvertebrates and physico-chemical characteristics in Santragachi Jheel, howrah, W. B., India. Acta Zool. Bulg. 62(3):277-300.

Pip E (1979). Survey of the ecology of submerged aquatic macrophytes incentral Canada. Aquat. Bot. 7:339-357.

Seymour CL, Simmons RE (2008). Can severely fragmented patches of riparian vegetation still be important for arid-land bird diversity? J. Arid Environ. 72(12):2275-2281.

Staicer C, Freedman AB, Srivastava D, Dowd N, Kilgar J, Hayden J, Payne F, Pollock T (1994). Use of lakes by black duck broods in relation to biological, chemical and physical features. *Hydrobiologia,* 279/280:185-199.

Stewart RE (2001). Technical Aspects of Wetlands - Wetlands as Bird Habitat. National Water Summary on Wetland Resources. United States Geological Survey. p. 86.

Subba Rao NV, Dey A (1989). Handbook of freshwater molluscs of India. ZSI, Calcutta. p. 289.

Tutle CH, Hanson PR, Owen M (1984). Some ecological factors affecting winter waterfowl distribution on inland waters in England and Wales, and the influence of water based recreation. J. Appl. Ecol. 21:41-62.

Vyas V (1992). Waterfowl community of Bhoj Wetland of Bhopal with reference to its management and conservation. In Environment and Biodegradation (Ed. V.P. Agrawal). pp.155-162.

Vyas V, Vishwakarma M, Dhar N (2010). Avian Diversity of Bhoj Wetland: A Ramsar Site of Central India. Our Nature 8:34-39.

Weins JA (1997). The Ecology of Bird Communities. Foundations and Patterns, Cambridge University Press, Cambridge, 1:539.

Wrona F, Prowse T, Reist J, Hobbie, Levesque L, Warwick F (2006). Climate change effects on aquatic biota, ecosystem structure and function. Ambio 35:359-369.

Yang HY, Chen B, Zhang ZW (2008). Seasonal changes in numbers and species composition of migratory shorebirds in northern Bohai Bay, China. Wader Study Group Bull. 115:133-139.

The hydrochemistry and macrobenthic fauna characteristics of an urban draining creek

Clement A. Edokpayi[1]*, Aveez O. Olowoporoku[2] and Roland E. Uwadiae[1]

[1]Department of Marine Sciences, University of Lagos, Akoka, Yaba, Lagos State, Nigeria.
[2]University of West of England, Bristol, United Kingdom.

Lagos lagoon is the largest of the eight lagoons that make up the lagoon systems of Nigeria and probably the most exposed to anthropogenic influence. The pollution status of the Lagos lagoon is generally attributed to the direct discharge of waste (domestic and industrial) and the contribution from rivers, creeks and drainage canals that empty into the Lagoon at different points. The paucity of information on the pollution status of water bodies that feeds the Lagos lagoon informed this present study. Investigation into the hydrochemistry and benthic macrofauna of Ogbe creek that drains through the main land of Lagos and empty into the Lagos lagoon was carried out. Fortnightly, sample collections between March and August 2002 at three stations along a 2 km stretch of the creek within the University of Lagos were used for the study. The hydrogen ion concentration ranged from 5.4 to 9.4. The electrical conductivity, salinity, alkalinity, nitrate, phosphate, dissolved oxygen and biochemical oxygen demand ranged between 3.65 ± 3.07 to 3.96 ± 3,22 Sμcm^{-1}; 0.51 ± 0.64 and 1.01 ± 0.96 ‰; 33.3 ± 23.2 and 60 ± 32.79 mgl^{-1}; 4.12 ± 0.17 and 7.46 ± 1.02 mgl^{-1}; 1.93 ± 0.53 and 3.65 ± 1.02 mgl^{-1}; 2.83 ± 1.42 and 4.65 ± 0.59 mgl^{-1} and 7.96 ± 1.99 and 8.13 ± 1.61 mgl^{-1}, respectively. The high BOD_5, nitrates and phosphates values are indicative of a perturbed environment. A total of 246 organisms belonging to 16 benthic taxa, 13 genera, 12 families, 8 orders and 4 phyla were collected during the study period. Chironomid larvae and the Naidid worms were the most abundant groups. They accounted for 25.61 and 22.76%, respectively, of the total macrobenthic count. The low number of taxa and numerical abundance of pollution indicator macrobenthos in the study area reflected a perturbed creek.

Key words: Hydrochemistry, benthos, University of Lagos, Ogbe creek, Nigeria.

INTRODUCTION

Many freshwater resources draining urban settlement are usually contaminated through input associated with human activities. Nigerian aquatic systems are subjected to pollution pressures associated with urbanization and population growth (Edokpayi et al., 2000; Nkwoji et al., 2010). Many of the rivers and creeks are used for irrigation, flood control, fisheries development and provision of potable water to towns and the surrounding rural areas. Along the approximately 853 km coastline of Nigeria are estuaries and lagoons receiving effluents from many rivers and creeks most of which are routes for the introduction of pollutants.

Owing to the seasonal distribution of rainfall, the lagoon systems and creeks experience seasonal flooding which introduces a lot of detritus, nutrients as well as other land based pollutants (Ogunwenmo and Osuala, 2004; Edokpayi et al., 2008). Such pollutants arising from land-based activities include domestic and industrial effluents, urban storm and agricultural run-offs. The introduction of these pollutants into aquatic systems flowing through Nigerian cities and villages constitute a major threat to hydrochemical and faunal characteristics of the aquatic ecosystems (Edokpayi et al., 2004; Edokpayi and Nkwoji, 2007). Also of significant impact to the hydrochemistry and hydrobiology of these water bodies are coastal habitat destruction and contamination from garbage and solid waste dump, which are common features along the Nigeria coastal waters (NEST, 1991).

The disappearance of benthic species, notably *Neritina kuramoensis* from the Kuramo water (Edokpayi et al., 2004), and the prolific growth of a cyanobacterium *Micocystis aeruginosa* Kutzing in parts of Ogun River

*Corresponding author. E-mail: klemedokpayi@gmail.com

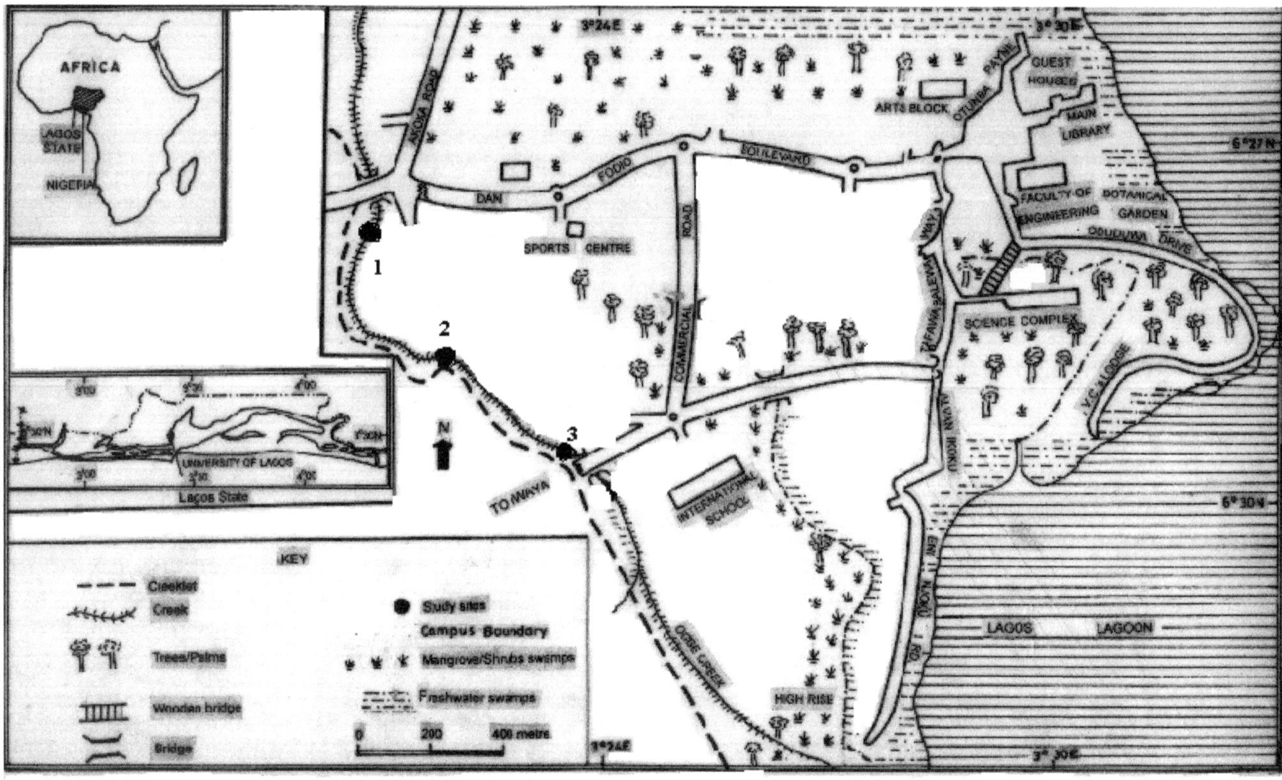

Figure 1. Map of the study area showing the sampling points on the Ogbe creek.

(Nwankwo, 1993), reflects the alteration of the hydrodynamic and hydrobiological characteristics of most aquatic systems in South-western Nigeria due to perturbbation stress.

The increasing levels of anthropogenic perturbation in Ogbe creek (Chukwu and Nwankwo, 2003) underscores the need for this study as there is scanty information on the hydrochemistry and fauna characteristics of this creek. This paper presents the results of the investigation on the hydrochemistry and benthic macrofauna of the creek.

MATERIALS AND METHODS

Description of study area

Ogbe creek is in South-western Nigerian and it empties into the Lagos lagoon (Figure 1). The Lagos lagoon is the largest of the eight lagoons (Yelwa, Badagry, Ologe, Iyagbe, Lagos, Kuramo, Ekpe, Lekki and Mahin) that make up the lagoon system of Nigeria. The creek is non-tidal and experiences flooding in the wet season as a result of surface run-offs (Nwankwo and Akinsoji, 1992). The creek drains the Lagos metropolis and thus carries urban run-off, sewage and other domestic waste from adjoining areas to the Lagos lagoon. Along the bank of the creek are farmlands, waste dumps sites, public toilets and bathroom structures. Other prominent features along the bank of the creek are mechanics and car wash shops. The water had a foul odour and deep dark colour, which suggests a pollution condition during the sampling period. Pockets of aquatic macrophytes at different locations along the creek covered the water surface; notable among them are *Lemna paucicosta, Salvinia molesta, Eichhornia crassipes, Dryopteris* sp., *Vernonia amygdalina, Emilia* sp., *Fucus* sp. and *Pistia stratiotes.* Three sampling stations were purposively selected based on accessibility (Figure 1).

Field studies

Water and benthic macrofauna samples were collected fortnightly between 0800 h and 1200 h on each sampling day. Water samples for dissolved oxygen (DO) and biochemical oxygen demand (BOD_5) determination were collected in two pre-washed 250 ml transparent and amber colored glass bottles respectively. Those for other chemical analyses were collected in one litre plastic bottles.

Three grab hauls for benthic samples were taken from each station; the collected materials were washed through a 0.5 mm mesh sieve in the field. The residue in the sieve was preserved in 10% formalin solution and kept in labeled plastic containers for onward transportation to the laboratory.

Laboratory investigation

The DO, BOD5, alkalinity, pH, nitrate, phosphate, salinity and conductivity of the surface water were determined using the methods recommended by APHA (1980). Preserved benthic samples were washed with tap water to remove the preservative and any remaining sediment in order to facilitate easy sorting.

Invertebrate fauna were sorted on a white tray into taxonomic groups using suitable texts (Michael, 1977; Barnes et al., 1988). The number of species and individuals for each station were counted and recorded. The methods of Slack et al. (1973) and

Table 1. Summary of the physico-chemical characteristics of the Ogbe Creek surface water.

Parameter	Station 1				Station 2				Station 3			
	Mean	SD	Min	Max	Mean	SD	Min	Max	Mean	SD	Min	Max
Conductivity (mg/l)	3.85	3.11	0.45	7.50	3.65	3.07	0.47	7.30	3.96	3.22	0.50	7.60
pH			5.1	8.8			5.6	9.2			7.5	9.4
Salinity (‰)	0.88	0.89	0.08	2.62	1.01	0.96	0.06	2.56	0.51	0.64	0.04	1.97
Alkalinity (mg/l)	33.3	23.2	10	80	60	32.79	10	100	44.6	24.58	8.00	83
Nitrate (mg/l)	7.46	1.02	6.00	9.00	6.05	1.54	4.10	8.00	4.12	0.17	3.80	4.42
Phosphate (mg/l)	3.68	1.02	1.80	5.00	1.93	0.53	0.90	2.80	2.76	0.76	1.90	3.95
DO (mg/l)	4.17	1.03	2.5	5.9	2.83	1.42	1.0	5.2	4.65	0.59	3.8	5.6
BOD_5 (mg/l)	7.96	1.99	4.3	9.8	8.13	1.61	5.0	9.9	7.98	1.14	5.6	9.5

Victor and Ogbeibu (1991) were used to define the dominant and subdominant groups. Diversity (Zar, 1984) of benthic macrofauna was estimated using Margalef's taxa richness (d), Shannon-Weiner's index of general diversity (H) and equitability using evenness index (E).

RESULTS

Physical and chemical conditions

Summary of the values measured for the physico-chemical parameters are summarized in Table 1. The highest value of DO (5.9 mgl^{-1}) was recorded in March at Station 1, while the least value (1.0 mgl^{-1}) was observed in station 2 in August. The values of BOD_5 ranged between 4.3 mg/l at station 1 and 9.9 mgl^{-1} in August at station 2. Station 2 had the highest values (100 mgl^{-1}) for alkalinity, while the least value (6 mgl^{-1}) was recorded at station 3 in August. The pH range for the period of study was 5.1 – 9.4. The water was acidic at station 1 in the dry months and alkaline in the wet months. The highest pH value (9.4) was recorded in March at station 3.

The lowest value of nitrate (3.8 mgl^{-1}) was observed in July at station 3 while the highest value (9.0 mgl^{-1}) was in July at station 1. Generally, nitrate concentrations were higher at station 1 than stations 2 and 3.The highest value (3.95 mgL^{-1}) of phosphate was recorded in July at station 3, while the lowest (0.9 mgL^{-1}) was recorded at station 2 in April. Salinity was high in the dry months and generally low in the rainy months. However, there was an increase in the values of the salinity in late July and August at the study sites except in station 3 that recorded a slight decline in August. The maximum value of 2.62‰ was recorded in March in station 1 while the minimum value of 0.1‰ was recorded also at station 1 in June. Electrical conductivity was low for all stations from March to May. There was an increase in the values of this parameter in all the stations in June after which it followed a steady pattern till the end of the sampling. Station 3 recorded the highest value of 7.6 μScm^{-1} in June while the least value of 0.26 μScm^{-1} was recorded in station 1 in May.

Benthic macrofauna

Community structure

A total of 246 organisms belonging to 16 benthic taxa, 13 genera, 12 families, 8 orders and 4 phyla were collected during the study period. The benthic faunal list is presented in Table 2. Of all the phyla encountered, the Arthropoda represented by insects were the most abundant in the study stretch. They accounted for 60% of the taxa and 58% of benthic animals enumerated. The family Chironomidae was the most abundant group contributing 63 individuals (26%) of total benthic macrofauna count. The Naididae, of the phylum annelida ranked second in abundance with the 56 individuals, which accounted for 23% of the total benthic macrofauna count. The phylum mollusca and platyhelminthes contributed 14 (5.69%) and 4 (1.62%) individuals, respectively (Figures 2, 3 and Table 2).

Composition and distribution

The plathyhelminthes was represented by 4 individuals of *Crenobia alpine* of the family Planaridae that occurred in stations I (3 individuals) and 3 (1 individual). The Annelida was represented by *Nais* sp and *Dero* sp of the Naididae and *Tubifex* sp of the family Tubificidae. They occurred in all the study stations. Also encountered in this study are 7 members of the Arthropoda which include *Chironomous* sp of the family Chironimidae, *Anopheles* sp, *Culex* sp and *Aedes* sp of the family Culicidae, *Arthripsodes aterrimus* (leptoceridae) and *Platambus maculatus* of the family Dysticidae. Arthropoda was represented in all the stations studied except the non occurrence of the family Dysticidae in stations 2 and 3, and leptoceridae in stations 1 and 3. The phylum mollusca were represented by 3 families (hydrobiidae, lymnaeidae and unionidae), and each of the 3 families had a single representative species, these were *Potamopyrgus jenkinsi*, *Limnaea truncatatula* and *Unio pictorum* respectively. Hydrobiidae was recorded in stations 1

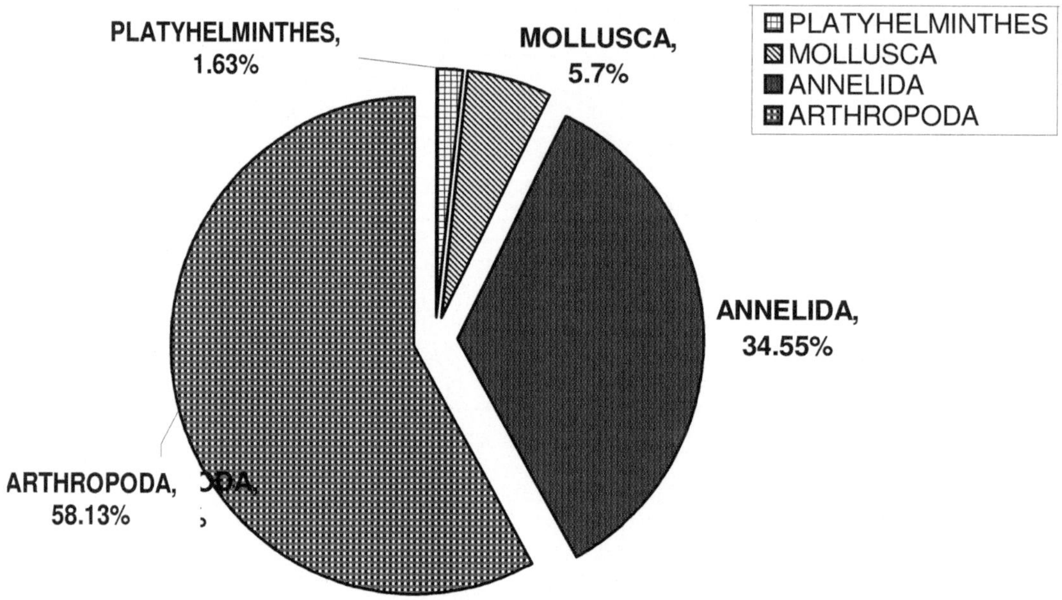

Figure 2. Overall percentage abundance of benthic macrofauna in the study area.

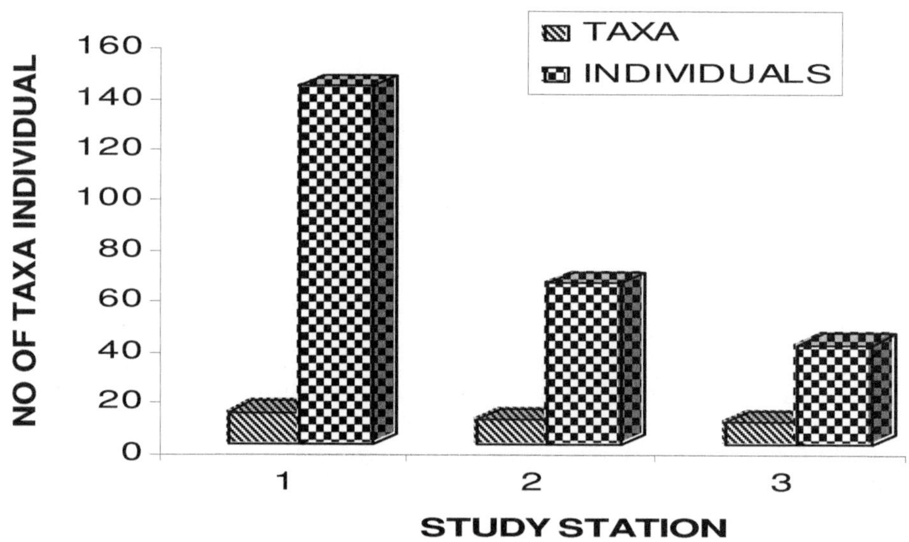

Figure 3. Overall number of individuals and taxa at the study stations in Ogbe creek.

and 2 while, lymnaeidae and unionidae occurred only in stations 1 and 3, respectively.

The distributions of individuals and taxa at the study stations are presented in Figure 2. Station 1 had the highest number (13) of taxa and individuals (142). This station accounted for 58% of the total number of individuals. Station 2 which recorded 64 individuals comprising of 10 taxa accounted for 26%. Station 3 had the least number (9) of taxa and individuals (40), and accounted for 16% of the total number of individuals found during the sampling period.

Dominant and subdominant groups

The percentage contributions of major macrobenthic order at the study stations are presented in Figure 4. Plesiopora (annelida) and Diptera (Arthropoda) constituted the dominant groups in all the stations sampled in the study area (Table 3). Coleoptera was dominant in station 1 but recorded a subdominant status in stations 2 and 3. Order Triclidida of the plathyhelminths was subdominant in stations 1 and 3, while Prosobranchia (mollusca) was subdominant in stations 1 and 2.

Table 2. Abundance of macrobenthic invertebrates in Ogbe creek.

Fauna	Station 1		Station 2		Station 3		Total	
	No. of Taxa	No. of individuals	No. of Taxa	No. of individuals	No. of Taxa	No. of individuals	No. of Taxa	No. of individuals
Platyhelminthes								
Planariidae	1	3			1	1	1	4
Annelida								
Naididae	2	27	2	17	2	12	2	56
Tubificidae	1	12	1	10	1	7	1	29
Arthropoda								
Dysticidae	1	16					1	16
Gyrrinidae	1	27	1	7	1	3	1	37
Chironomidae	1	39	1	17	1	7	1	63
Culicidae	4	9	2	5	2	6	4	20
Leptoceridae			1	2			1	2
Hydroptilidae			1	5			1	5
Mollusca								
Hydrobiidae	1	3	1	1			1	4
Limnaeidae	1	6					1	6
Unionidae					1	4	1	4
Total	13	142	10	64	9	40	16	246

General diversity (H) and equitabilty

The values of taxa richness, diversity and equitability are presented in Table 4. Taxa richness was highest at station 1 and lowest at station 3. Diversity was highest at station 1 and lowest at station 3, while equitability was highest at station 3 and lowest at station 5.

DISCUSSION

Physical and chemical parameters

Conductivity of a water body is an index of the total ionic content, therefore it can be used to indicate the freshness or otherwise of the water. Egborge (1993) reported that for many Nigerian inland water bodies, the conductivities are much less than 500 μScm^{-1} at the peak of dry season. In this study, a range of 5.1 – 9.4 μScm^{-1} was recorded. This indicates low ionic concentration of the water during the study period. Also, the sharp increase in the value of conductivities observed in June can be attributed to the flushing terrain during the rainy season, which increases the allochthonous input into the creek.

The pH range (5.1 – 9.4) at the study stations generally falls within ranges reported for rivers flowing through areas with thick vegetation (Awuchie, 1981; Osimen, 1997; Uwadiae et al., 2009). The increase in acidity in station 2 during the rainy months may be attributed to the allochthonous input consisting mainly of wash back of the spoils and chemical wastes from auto-mechanic and car wash shops at the bank of the creek. Pidgeon and Cains (1987) observed that organic acids resulting from decaying vegetation might be responsible for the low pH in most aquatic ecosystems.

Salinity range (0.1 - 2.62‰) observed in this study was slightly brackish and fell within the range reported by other workers within the vicinity (Edokpayi et al., 2008; Onyema and Nkwoji, 2009). The salinity values of the study stations were relatively higher during the dry months; this may be due to dilution of the water by the increase freshwater input during these raining months. Alkalinity of a water body is a measure of its capacity to neutralize acids to a designated pH (APHA, 1980; Edokpayi, 2005). High alkalinity results in physiological stress on aquatic organisms and may lead to loss of biodiversity.

The levels of nitrate observed here are within the ranges reported for some Nigerian waters (Edokpayi and Nkwoji, 2007; Onyema et al., 2009). The values of phosphate recorded were higher during the rainy season and lower during the dry months. This observation is supportive of the claims of Stewart et al., (1975) and Osimen (1997) for most tropical waters. This can be

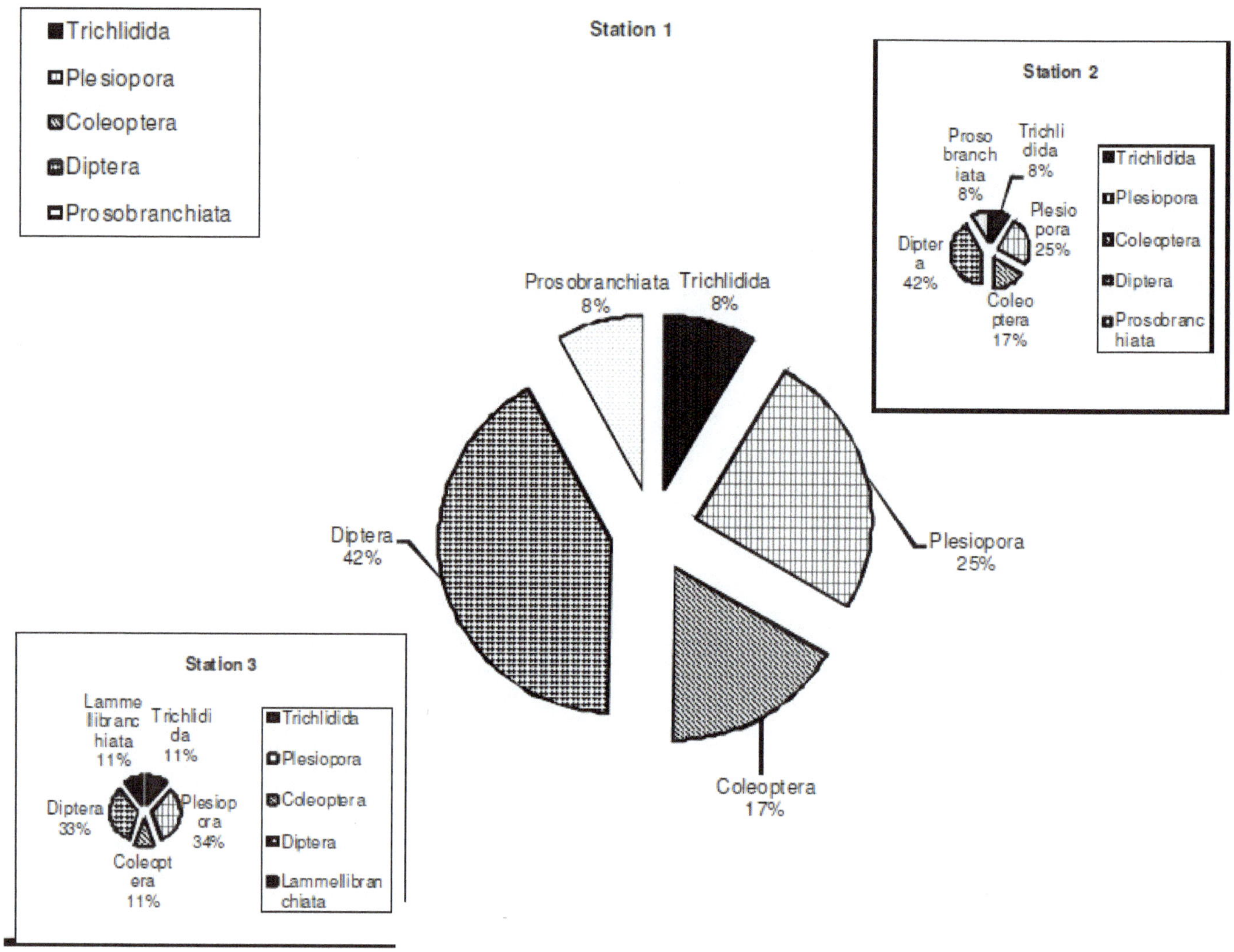

Figure 4. Percentage contribution of major macrobenthic order in the study stations.

Table 3. Summary of the dominant and subdominant (figures in %) macrobenthic invertebrate groups at the study station of Ogbe creek (dormant ≥ 15%; subdominant > 15% ≥ 5%; Rear > 5%, Slack *et al.*, 1973; (** indicates dominance, * indicates subdominance).

Major taxonomic group	Station 1	Station 2	Station 3
Trichlidida	7.69*	-	11.11*
Plesiopora	23.08**	30**	33.33**
Coleoptera	15.39**	10*	11.11*
Diptera	38.46**	30**	33.33**
Trichoptera	-	20**	-
Prosobranchiata	7.69*	10*	-
Pulmonata	7.69*	-	-
Lammellibranchiata	-	-	11.11*

accounted for by high amount of organic wastes, which enters the creek through run-offs. The stream was moderately oxygenated as observed in the DO values ($1.0 - 5.9\ mgl^{-1}$). The range was similar to those reported for many other polluted Nigerian waters including that (6.9 - 8.8 mg/l) of Lagos lagoon (Ogunwenmo and

Table 4. Summary of the diversity and faunal indices of macrobenthic invertebrates at the study stations.

Indices	Station 1	Station 2	Station 3
Number of samples	12	12	12
Number of taxa	13	10	9
Number of individual	142	64	40
Margalef's taxa richness	1.91	1.78	1.64
Shannon-Weiner index of diversity (H)	1.75	1.40	1.24
Equitability	0.62	0.65	0.68

Kusemiju, 2004) and those observed (1.20 - 9.40 mg/l) for some polluted water bodies in Nigeria (Victor and Onmivbori, 1996; Edokpayi and Osimen, 2001). The low DO observed in the study area could be attributed to the high organic content, mainly human faeces, decayed plant and animal materials and domestic effluents flushed into the creek at different points. This requires large quantities of DO for aerobic decomposition.

Biochemical oxygen demand is a measure of the potential of biologically oxidisable matter for deoxygenating water, and thus provides a measure of the effect of pollution on a receiving water body (Mason, 1991). The BOD_5 values (4.3 – 9.9 mgl^{-1}) recorded in this study was higher than those (0.42 - 8.0 mg/l) reported for some Nigerian streams (Osimen, 1997; Edokpayi and Osimen, 2001; Ogbeibu, 2001; Ogunwenmo and Osuala, 2004). This is an indication of the pollution status of Ogbe creek. The high organic load was responsible for the high BOD_5 especially in the rainy months.

Macrobenthic invertebrates

The important factors that affects the abundance of macrobenthic fauna in a given community includes: the physicochemistry of the water, immediate substrate of occupation and food availability (Dance and Hynes, 1980). In this study, it was desirable to identify environmental factors that were responsible for the observed structure of the benthic communities. The high BOD_5 due to human and domestic wastes discharges appear to be responsible for the structure of the benthic macrofauna community of Ogbe creek. Chironomidae and Naididae as recorded in this study dominate the benthic macrofauna community of Ogbe creek. This observation agrees with the report of Edokpayi et al., (2004) on Kuramo Water. Chironomid larvae have been reported to dominate aquatic benthic invertebrates' communities (Hynes, 1998; Victor and Ogbeibu, 1985) as they hardly show any habitat restriction (Victor and Ogbeibu, 1991) and are known to replace other invertebrate taxa in streams perturbed by human activity. Chironomidae were represented in all the stations. The elevated levels of organic matter in Ogbe creek sediment may be responsible for the dominant status of a chironomid taxon, *Chironomus* (Edokpayi et al., 2004) at the study sites. The high abundance of the chironomids in the study area is a confirmed phenomenon in polluted water bodies in both temperate and tropical zones (Edokpayi and Nkwoji, 2007; Cardoso et al., 2010).

The naidid taxon, *Nais* also occurred in significant number at the study stations. Edokpayi et al., (2004) reported that Naidid oligochaetes have been found in polluted tropical waters and prefer soft sediment rich in organic content (Ogunwenmo and Kusemiju, 2004), as observed for the Ogbe creek. Furthermore, naidid worms have been reported to respond to organic pollution by increasing in abundance in polluted water bodies (Bishop, 1973).

The relatively lower taxa observed in station 3 could be attributed to the resultant effect of continuous dredging activities at this station and lack of vegetation cover which is the primary source of allochthonous material that may be used as food by the benthos. Anthropogenic activities such as dredging results in substratum instability and increased siltation. Suspended silt has the ability of reducing light penetration and primary productivity and can clog the gills of aquatic fauna thereby smothering them (Edokpayi and Nkwoji, 2007). The occurrence of relatively higher taxa and individuals' in station 1 may be an indication of lower degree of anthropogenic activities at this station compared to other stations.

Overall diversity has been reported to be the product of all dynamic spatial and temporal changes affecting the community (Victor and Ogbeibu, 1991). It could also be a reflection of the extent, to which the ecosystem has been perturbed by human activity. The very low diversity at the study stations is an indication of environmental perturbation.

REFERENCES

American Public Health Association (APHA) (1980). Water Pollution Method for the Examination of Water and Wastewater. 18th ed., Washington D.C. p. 1437

Barnes RSK, Calow P , Olive PJW (1988). The invertebrates: A new synthesis. Blackwell Scientific publications, Oxford, pp: 582.

Cardoso I, Granadeiro JP, Cabral H (2010). Benthic macroinvertebates vertical distribution in the Tagus estuary (Portugal): The influence of tidal cycle. J. Estuarine Coastal Shelf Sci., 86: 580 – 586.

Chukwu LO , Nwankwo DI (2003). The impact of land-based pollution on the hydrochemistry and macrocommunity of a tropical West African creek. The Ekol., 2:(1-2): 1 - 19
Dance KW , Hynes HBN (1980). Some effects of Agricultural land use on stream insect communities. Environ. Pollut. Ser., 22: 19 – 28.
Edokpayi CA, Okenyi JC, Ogbeibu AE , Osimen EC (2000). The effect of human activities on the macrobenthic invertebrates of Ibiekuma stream, Ekpoma, Nigeria. Bioscience Res. Commun. 12(1): 79 – 87.
Edokpayi CA , Osimen EC (2001). Hydrobiological studies on Ibiekuma River at Ekpoma, Southern Nigeria after Impoundment: The faunal characteristics. Afr. J. Sci. Tech. pp: 72 – 81.
Edokpayi CA, Lawal MO, Okwok NA , Ogunwenmo CA (2004). Physico-chemical and macrobenthic faunal characteristics of Kuramo water, Lagos, Southern Nigeria. Afr. J. Aquat. Sci., 29(2): 235 – 241.
Edokpayi CA (2005). Variation of chemical constituents of a brackish water prawn habitat in southern Nigeria. actaSATECH 2(1): 11 – 18.
Edokpayi CA , Ayorinde AO (2005). Physical, chemical and Macrobenthic invertebrate fauna characteristics of swampy water bodies within University of Lagos, Nigeria. West Afr. J. Appl. Ecol., 8: 129 – 139.
Edokpayi CA, Nkwoji JA (2007). Annual changes in the physico-chemical and macrobenthic invertebrate characteristics of the Lagos lagoon sewage dump site at Iddo, Southern Nigeria. Ecol. Environ. Conserv. 13(1): 13 – 17.
Egborge ABN (1993). Biodiversity for Aquatic Fauna in Nigeria. National Resources Conservation Council, Abuja. p. 173
Hynes KE (1998). Benthic macroinvertebrate diversity and biotic indices for monitoring of 5 urban and urbanizing lakes within the Halifax Regional Municipality (HRM), Nova Scotia, Canada. Soil & Water Conservation Society of Metro Halifax. 14: 114
Mason TT (1991). Biology of Freshwater Invertebrates Animals. Longman, London. p. 111
Michael Q (1977). Invertebrates of streams and Rivers. A key to identification. Edward Arnold Publishers Ltd., London. p. 84
Nigerian Environmental Study /Action Team-NEST (1991). Nigeria's Threatened Environment. A national profile. Intec. Printers Ltd. Ibadan. p. 288
Nkwoji JA, Yakub A, Ajani GE, Balogun KJ, Renner KO, Igbo JK, Ariyo AA , Bello BO (2010). Seasonal variations in the water chemistry and benthic macroinvertebrates of a south western Lagoon, Lagos, Nigeria. J. Am. Sci., 6(3): 85 – 92.
Nwankwo DI , Akinsoji A (1992). Epiphyte community on water hyacinth *Eichornia crassipes* (MART) solutions in coastal water of South Western Nigeria. Arch Hydrobio. 124(4): 501 – 511.
Nwankwo DI (1993). Cyanobacteria bloom species in coastal waters of South Western Nigeria. Arch. Hydrobiol., 4: 533 – 542.
Ogbeibu AE (2001). Distribution, density and diversity of dipterans in a temporary pond in Okomu forest reserve, Southern Nigeria. J. Aquat. Sci., 16: 43 – 52.
Ogunwenmo CA , Kusemiju K (2004). Annelids of a West African estuarine system. J. Environ. Biol., 25(2): 227 – 237.
Ogunwenmo CA , Osuala IA (2004). Physico-chemical parameters and macrobenthos of an estuarine creek and an artificial pond in Lagos southwestern Nigeria. actaSATECH 1(2): 128 – 132.
Onyema IC, Lawal-Are AO, Akinremi TA , Bassey OB (2009). Water quality parameters, chlorophyll a and zooplankton of an estuarine creek in Lagos. J. Am. Sci., 5(6): 76 - 94
Osimen EC (1997). The effects of impoundment on the physico-chemical parameters and macrobenthic invertebrate fauna of Ibiekuma River, Ekpoma, Edo State. Environ. Pollut., 64: 24 – 88.
Pidgeon RWJ , Cains SC (1987). Decomposition and colonization by invertebrates of nature and exotic leaf material in a small stream in New England (Australia). Hydrobiologia, 77: 113 – 124.
Slack KV, Nauman JW , Tilley LJ (1973). Benthic inveretebrates in a North flowing stream and a South flowing stream, Brooks range. Alaska. Water. Res. Bullet, 15: 108 – 135.
Stewart WDP, May E , Tuckwell (1975). Eutrophication and algal growth in Scottish Freshwater Lochs. In: The Ecology of resource degradation and renewal. Chadwick WJ , Goodman CT, Blackwell Scientific Oxford, 7 0-80.
Uwadiae RE, Edokpayi CA, Adegbite O, Abimbola O (2009). Impact of sediment characteristics on the Macrobenthic invertebrates community of a perturbed tropical lagoon. Ecol. Environ. Conserv. 15(3): 441 – 448.
Victor R , Ogbeibu AE (1991). Macroinvertebrate communicates in the erosional biotope of an urban stream in Nigeria. Nig. J. Sci., 4: 1 – 2.
Victor R , Onomivbori O (1996). The effects of urban perturbation on the benthic macroinvertebrates of a southern Nigeria stream. In: Schiomer F , Biland. (Eds)..SPB Academic Publishing Amsterdam, The Netherlands. pp: 223-238
Zar JH (1984). Biostatistical Analysis (2nd Ed.). Prentice Hall. New Jersey. p. 717

Medicinal Plants Scenario in Forest Vegetation of Baikunthpur (District-Koria) Chhattisgarh (India)

Mantosh Kumar Sinha[1] and Deepima Sinha[2]

[1]K.R. Technical College Ambikapur (Chhattisgarh) India.
[2]Vishwavidyalaya Engineering College Lakhanpur Ambikapur (Chhattisgarh) India.

The State of Chhattisgarh has about 44% of its geographical area covered with forests. The Koria district in Chhattisgarh lies between 22°58' to 23°49' North latitudes and 81° 33' to 82°45' East longitude. The average rainfall is 121.36 cm. The forest area is 81.23% of total district Area. The district Koria has a very rich flora exhibiting diversity especially of medicinal plants. There is no comprehensive description of the flora as well as vascular cryptogames of the district is available. Some plant species are on the verge of extinction. Keeping these points in view, the present investigation was planned. The present paper deals with diversity of the medicinal plants of the district and their ecological status. Vegetational analysis revealed some interesting observations on phytosociological characters. Shanon index of general diversity was calculated for tree shrub and herb layers. It was 4.21637 for trees, 4.6357 for shrubs and 4.8298 for herbs. A total of 108 angiospermic plant species of medicinal importance were found distributed in 46 dicot and 10 monocot families. Two medicinally important pteridophytes have also been reported.

Key words: Diversity, medicinal plants, phytosociology, Koria district.

INTRODUCTION

Koria district in Chhattisgarh is very rich in natural vegetation and biological wealth. The district lies between 22°58' to 23°49' North latitude and 81°33' to 82°45' East longitude. The average rainfall is 121.36 cm. The annual mean temperature is 24°C. The temperature varied between 16.2°C to 31°C. Geologically, the area is dominated by upper Gondwana rocks which are rich in coal deposit. The highest mountain ranges of the region occupy the northern part of the district and have a forest area of 81.23%.

The district has a sizeable tribal population using enormous range of plants for their basic needs, sustenance and livelihood. The district has very rich plant diversity, including medicinal plants. Many of them are on the verge of extinction due to over exploitation and destruction of their habitat. There has been no comprehensive study on the enumeration, distribution and the assessment of threat to the existing medicinal plants.

The vegetation particularly the forests have not been explored fully except a few reports from the forest department (Tiwari, 1992). There is no report on the rich forest flora of the district. Therefore, an attempt has been made to study the diversity of forest flora especially of medicinal plants.

MATERIALS AND METHODS

Extensive field survey was undertaken during 2005 to 2007. The floristic diversity of tropical dry deciduous forests was explored. The district comprises of five development block viz., Baikunthpur, Sonhat, Manendragarh, Khadgavan and Bharatpur. The present study was done in the Baikunthpur block of the district. The

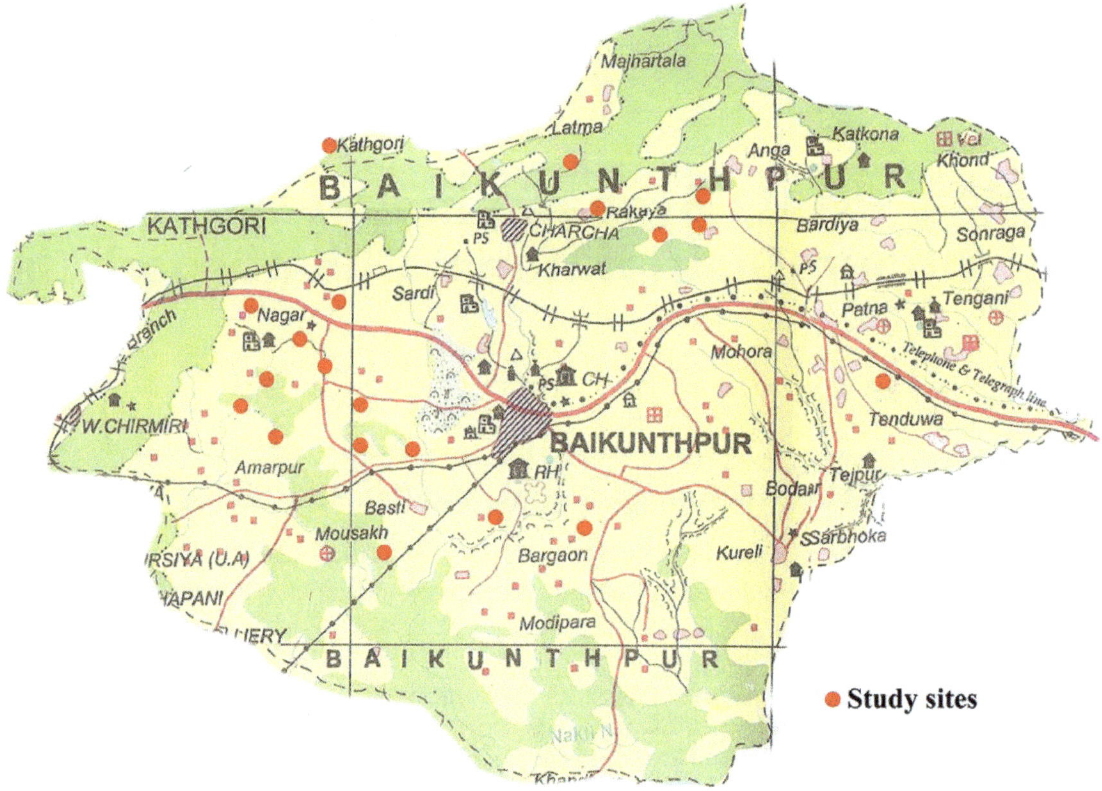

Figure 1. Baikunthpur block showing study sites.

phytosociological characters such as frequency, density and abundance were recorded as per the method described by Mishra (1968). Random quadrate sampling was done in Baikunthpur block taking 20 study site namely Shivpur, Bishunpur, Katghodi, Umghar, Nagar Pahadpara, Tilpandand, Itga, Dhudhania, Phulpur, Shankarpur, Ujiyarpur, Rakiya, Jamgahna, Jagdishpur, Kotaktall, Patrapali, Deori, Chilka and Ranai (Figure 1). The sites were widely separated from each other and cover an area of 20 to 40 km. The study sites were visited on frequent intervals and a thorough sampling was done to document the species diversity. Species diversity was calculated using the formula:

$\overline{H} = - \Sigma pi \log pi$ (Shanon and Weaver, 1963)

Where, pi is the proportion of individuals of its plant species and number of individuals of all the species.

The entire tree stands were grouped according to Girth (GBH) with an interval of 25 cm to understand the growth dynamics of the forest.

RESULTS AND DISCUSSION

Diversity

A total of 168 genera distributed in 224 species under 57 families and two pteridophytes with 3 species were also reported. Poaceae was found to be the most prominent family among monocots where as Rubiaceae and Euphorbiaceae were most prominent among the dicots (Table 1).

Phytosociological status

Quantitive vegetational analysis revealed a high degree of heterogeneity. The vegetation is Sal (*Shorea robusta*) dominated which showed cent percent occurrence in all stands. Maximum percentage frequency was also exhibited by *Randia dumetorum, Vernonia anthelminticum, Adhatoda vasica, Alangium lamarckii, Diospyros melanoxylon, Vicia sativa, Vanda roxburghai, Lawsonia inermis* and *Quisqualis indica*. These species were constantly present in the study sites.

Species showing high frequency and low abundance were attributed to a status of regular distribution and species showing low frequency and high abundance were attributed to a status of showing contagious distribution (Table 2).

Shanon index of general diversity

Shanon index was found to be 4.216 for trees, 4.635 for shrubs and 4.829 for herbs (Table 3). This clearly shows that the study area has great diversity.

The results show higher values than that of Supriya Devi and Yadav (2006) in a tropical forest of Manipur. In Indian forest H value ranged between 0.83 and 4.1 Parthasarathy et al. (1992); Singh et al. (1984); Visalakshi (1995) and Pascal (1988) reported that this ranged between 2.1

Table. 1 Family wise account of Diversity in the Baikunthpur forest area. (only medicinal important plants.

S/N	Family	Genera	Specie
1	*Asclepiadaceae*	5	5
2	*Apocynaceae*	3	4
3	*Acanthaceae*	6	9
4	*Anacardiaceae*	5	5
5	*Amaryllidaceae*	1	1
6	*Araceae*	1	1
7	*Aracaceae*	1	2
8	*Asteraceae*	6	7
9	*Apiaceae*	1	2
10	*Bixaceae*	1	1
11	*Boraginaceae*	2	2
12	*Bignoniaceae*	3	5
13	*Bombacaceae*	1	1
14	*Combretaceae*	5	9
15	*Celastraceae*	3	3
16	*Cornaceae*	1	1
17	*Convolvulaceae*	3	3
18	*Caesalpiniaceae*	5	7
19	*Cucurbitaceae*	2	2
20	*Cactaceae*	1	1
21	*Crassuliaceae*	1	1
22	*Cyperaceae*	1	4
23	*Dipterocarpaceae*	1	1
24	*Dioscoreaceae*	1	2
25	*Euphorbiaceae*	7	10
26	*Ebenaceae*	1	1
27	*Fumariaceae*	1	1
28	*Liliaceae*	5	7
29	*Papilionaceae*	6	10
30	*Piperaceae*	1	1
31	*Poaceae*	23	27
32	*Mimosaceae*	4	6
33	*Orchidaceae*	1	2
34	*Lythraceae*	3	3
35	*Lauraceae*	2	3
36	*Lamiaceae*	5	8
37	*Moraceae*	2	4
38	*Malvaceae*	4	4
39	*Myrsinaceae*	1	2
40	*Menispermaceae*	1	1
41	*Meliaceae*	4	4
42	*Myrpaceae*	2	4
43	*Nyctaginaceae*	2	2
44	*Oxalidaceae*	1	1
45	*Plumbiginaceae*	1	2
46	*Pedaliaceae*	1	1
47	*Polypodiaceae*	**2**	**3**
48	*Rubiaceae*	7	10
49	*Rutaceae*	4	5

Table 1. Contd.

50	*Sterculiaceae*	3	3
51	*Styreceae*	1	1
52	*Scitaminaceae*	1	1
53	*Tiliaceae*	1	3
54	*Verbenaceae*	5	8
55	*Vitaceae*	1	2
56	*Zingiberaceae*	3	3
57	*Zygophyllaceae*	2	2
	Total	**168**	**224**

Table 2. Phytosociological characters of different plant species of Baikunthpur block (Dist. Koria) C.G.

S/N	Name of plant Spp.	Phytocociological character			Distribution
		Frequency (%)	Density	Abundance	
1	*Andographis paniculata, Nees*	60	27.5	45.83	R*
2	*Pluchea lanceolata,oliver & Hiern*	60	15	25	R
3	*Randia dumetorum, Lank*	100	12.5	12.5	R
4	*Embelia ribes, Burm.F.*	5	0.5	10	C*
5	*Sphaeranthus indicus, L.*	90	80	88.88	R
6	*Adiantum lanulatum, Burm.*	20	5	20	
7	*Clerodendrum serratum, L.*	60	5	8.33	R
8	*Helicteres isora, L.*	60	12.5	20.83	R
9	*Nyctanthes arbortristis, L.*	50	12.5	25	R
10	*Gardenia lucida, Roxb.*	30	2.5	8.33	R
11	*Desmodium gangeticum, DC.*	60	20	33.33	R
12	*Vicoa auricalata Cass.*	95	62.5	65.78	R
13	*Hymenodiotyon excelsum, Wall.*	50	5	10	R
14	*Bauhinia Variegata L.*	85	10	11.76	R
15	*Grewia hirsuta, Vanb.*	25	2.5	10	R
16	*Asparagus recemosus, Willd.*	60	15	25	R
17	*Vernonia anthelminticum Willd.*	100	60	60	R
18	*Erythrina indica Lamk.*	30	5	16.66	R
19	*Abelmoschus moschatus, L.*	10	2.5	25	C
20	*Ocimum basilicum L.*	95	65	68.42	R
21	*Dryopteris crenata, Christ.*	75	15	20	R
22	*Acacia concinna, DC.*	40	2.5	6.25	R
23	*Symplocos racemosa Roxb.*	60	7.5	12.5	R
24	*Sterculia urens, Roxb.*	65	10	15.38	R
25	*Bixa orellana, L.*	20	1.5	7.5	R
26	*Leucas cephalotes Spreng.*	70	20	28.57	R
27	*Crotalaria juncea L.*	70	57.14	40	R
28	*Dioscorea daemona, Roxb.*	45	7.5	16.66	R
29	*Lannea grandis Roxb.*	75	5	6.66	R
30	*Croton tiglium L.*	10	1.25	12.5	C
31	*Adhatoda vasica, Nees*	100	15	15	R
32	*Grewia rotundifolia, Juss.*	10	0.75	7.5	R
33	*Thysanolaena agrostis Nees*	80	50	62.5	R
34	*Abutilon indicum, G.Don.*	60	25	41.66	R

Table 2. Contd

35	*Oxalis corniculata L.*	70	60	85.71	C
36	*Cassia glauca, Lam.*	60	5	8.33	R
37	*Celastrus paniculata, Willd.*	10	1	10	R
38	*Clitoria ternatea, L.*	60	17.5	29.16	R
39	*Ongeinia dalbergioides Benth*	75	5	6.66	R
40	*Bauhinia vahlii, W.& A.*	85	7.5	8.8	R
41	*Cassia sophera, L.*	90	150	166.6	C
42	*Acacia catechu Willd.*	50	10	20	R
43	*Terminalia tomentosa, W. & A.*	70	7.5	10.71	R
44	*Terminalia arjuna, W. & A.*	65	5	7.69	R
45	*Centella asiatica, L.*	30	12.5	62.5	C
46	*Alangium lamarckii , Thw.*	100	35	35	R
47	*Eclipta alba, Hassk.*	85	32.5	38.23	R
48	*Vernonia cinerea, Less.*	100	65	65	R
49	*Plumbago zeylanica, L.*	35	30	42.85	C
50	*Diospyros melanoxylon, Roxb.*	100	40	40	R
51	*Gymnema sylvestre, R. Br.*	20	15	30	C
52	*Hemidesmus indicus, Br.*	60	22.5	37.5	R
53	*Pergularia extensa N.E. Br.*	75	17.5	23.33	R
54	*Cordia myxa, L.*	60	7.5	12.5	R
55	*Bryophyllum Calycinum Salis.*	40	17.5	43.75	C
56	*Boerhaavia diffusa L.*	70	10	14.28	R
57	*Ficus infectoria, L.*	70	10	14.28	R
58	*Chlorophytum tuberosum, Baker*	25	2.5	12.5	R
59	*Smilex zeylanica, L.*	40	10	25	R
60	*Hygrophila augustifolia R.Br.*	75	120	150	C
61	*Luffa aegyptiaca Mill*	40	7.5	18.75	R
62	*Cissus quadrangularis L.*	55	50	62.50%	R
63	*Woodfordia fruticosa Kurz.*	90	40	50	R
64	*Vicia Sativa L.*	100	65	65	R
65	*Spilanthes acmella L.*	30	5	25	R
66	*Pterospermum acerifolium, bWilld.*	65	5	7.69	R
67	*Blumea lacera, DC.*	95	25	26.3	R
68	*Mucuna prurita, Hook.*	45	15	30	R
69	*Sida spinosa, L*	50	25	62.5	C
70	*Tribulus terrestris L.*	65	40	66.66	C
71	*Dioscorea bulbifera L.*	65	30	50	R
72	*Psoralea corylifolia, DC.*	75	60	85.71	C
73	*Ipomea paniculata, L.*	45	10	20	R
74	*Wedelia calandulacea Less.*	80	70	87.5	C
75	*Semecarpus anacardiun L.*	40	2.5	6.25	R
76	*Phyllanthus niruri L.*	95	80	84.21	R
77	*Cyperus rotundus L.*	70	10	35.71	R
78	*Tecoma undulata, G. Don.*	5	0.25	5	R
79	*Jatropha gossypifolia L.*	80	150	187.5	C
80	*Vanda roxburghai L.*	100	200	200	C
81	*Quisqualis indica, L.*	45	7.5	16.66	R
82	*Bryonia laciniosa Mong.*	60	20	33.33	R
83	*Convolvulus pluricaulis, Chois.*	75	22.5	30	R
84	*Salmalia malabaricum, DC.*	45	30	60	C
85	*Holorrhena antidysenterica, Wall.*	90	22.50%	22.5	R
86	*Abrus precatorius L.*	60	40	66.66	C

Table 2. Contd.

87	*Acorus Calamus L.*	25	30	100	C
88	*Carissa spinarum L*	80	100	125	C
89	*Barleria cristata, L.*	50	5	10	R
90	*Martynia diandra, Glox.*	30	10	33.33	C
91	*Shorea robusta Gaertn.*	100	65	65	R
92	*Curculigo orchioides Gaertn.*	5	0.25	5	R
93	*Vitex negundo L.*	95	35	36.83	R
94	*Dalbergia latifolia Roxb.*	70	70	100	C
95	*Cuscuta reflexa Roxb.*	90	80	88.88	R
96	*Hedychium coronarium, Koenig.*	20	3	15	R
97	*Rouwolfia serpentine, Benth.*	40	10	25	R
98	*Alostonia scholaris, Brown.*	60	50	83.33	C
99	*Curcuma angustifolia, Roxb.*	40	15	37.5	R
100	*Fumaria parviflora, Lamk.*	50	40	80	C
101	*Mimosa pudica, L.*	40	50	125	C
102	*Quisqualis indica, L.*	100	120	120	C
103	*Kaempferia rotunda, L.*	50	20	40	R
104	*Cinnamomum tamala, Fr. Nees.*	50	30	60	C
105	*Jatropha curcas, L.*	90	100	111.11	C
106	*Lawsonia inermis L.*	100	150	150	C
107	*Piper longum L.*	50	35	70	C
108	*Aloe barbadensis, Mill.*	100	10	10	R
*	R = Regular Distribution				
*	C = Contagious Distribution				

Table 3. Species diversity in different layers of plant communities in Baikunthpur (Dist. - Koria) C.G.

Strata in the forest	H
Trees	4.216
Shrubs	4.635
Herbs	4.829

and 4.3 in different forest ecosystems of Western Ghats. Thakur and Khare, 2008 reported Shannon Wiever diversity index (H) ranging from 2.22 to 3.66 in forest vegetation of Sagar (M.P.).

However, the results clearly indicate a high degree of diversity and the community is a tropical dry deciduous type of Sal forest.

Presently, the medicinal plant diversity was also recorded as very high. There are plant species showing very low population density which draws attention of researchers for conservation. The sampled area requires conservation because of its potential for natural regeneration and utility value as well as varied plant diversity.

REFERENCES

Mishra R. (1968). Ecology Work Book, Oxford and IBH Publication Co., New Delhi.

Parthasarathy NV, Kinhal, Kumar LP (1992). Plant species diversity and human impact in the tropical wet evergreen forest of Southern Western ghats, Indo-French Workshop as tropical forest ecosystem : Nov. 1992. Natural functioning and Anthropogenic Impact French Institute, Pondichery.

Pascal JP (1988). West evergreen forest of the western ghats of India; Ecology, structure, floristic compositions and succession. Inst. francisde Pondichery traw. De la Sci. et. tech. Tome XX bis. 345.

Shanon CE, Weaver W (1963). The mathematical theory of communication University of Illinois Press Urbana.

Singh JS Rawat YS, Chaturvedi OP (1984). Replacement of Oak forest with pine in the Himalaya affect the nitrogen cycle. Nat. 311: 54-56.

Supriya Devi L, Yadava PS (2006). Floristic diversity assessment and vegetation analysis of tropical semi-evergreen forest of Manipur, North East India.

Thakur AS, Khare PK (2008). Species Diversity and Composition of forest vegetation of Sagar district in Central India. The Ind. Forester 134 (6): 801-813.

Tiwari P (1992) Working plan for Koria forest division Surguja circle Vol.-I. Chhattisgarh Forest Department.

In vitro callus initiation of a 'threatened' Nigerian leafy vegetable, Gnetum africanum (WILW)

A. C. Iloh[1*], N. R. Isu[2] and D. D. Kuta[1]

[1]Biotechnology and Molecular Biology Advanced Laboratory, Sheda Science and Technology Complex, Abuja, Nigeria.
[2]Department of Biological Sciences, University of Abuja, Nigeria.

Gnetum africanum is a green leafy vegetable found in Nigeria, where it is a highly valued food source. Stocks of this plant in the wild are increasingly threatened by land transformation and harvesting methods are unsustainable. _In vitro_ callus initiation of _G. africanum_ has been developed. The surfaces of 3 to 4 day old leaf explants were sterilized before exposure to a range of different concentrations of plant growth regulators. _In vitro_ callus initiation of explants in single concentrations of auxins (2,4D: 2-4-dichloroxyphenoxyacetic and pichloram) did not initiate callus after 3 months of culture. However, combinations of cytokines (BAP: 6-benzylaminopurine, and kinetin) and auxins initiated callus. The highest percentage callus initiation response of 100% was observed in the combination of BAP (1.0 mg/l) + 2,4D (7 mg/l). There was a significant difference ($P<0.05$) in callus production of explants in response to different callus induction and initiation media. However, there was no significant difference ($P>0.05$) between the degree of callus response and callus size to these different media.

Key words: BAP (6-benzylaminopurine), callus initiation, *Gnetum africanum*, kinetin, 2,4D (2-4-dichloroxyphenoxyacetic), pichloram.

INTRODUCTION

Green leafy vegetables are plant bioresources that constitute an indispensable part of the human diet in Africa generally and West Africa in particular (Oguntona, 1988). Eneobong (1997) has defined bioresources as the total biological variation manifested as individual plants, animals or their genes which can be taken by man for use as drugs, food, live stock feed as well as environmental protection.

Gnetum is a leafy vegetable with lone genus in the family *Gnetaceae*; with about thirty species in the genus, occurring throughout the tropics in Asia, South America and in Central Africa up to Nigeria (Mialoundama and Paulet, 1986). The majority of the species of *Gnetum* are lianas or climbing vines, with two species native to Africa; *Gnetum africanum* and *G. buchholzianum*. These species are distributed in the humid tropical forests of Nigeria (10°N 8°E) to Cameroon (3° 52"N 11°31"E), Central African Republic (7°N 21°E), Gabon (1°N 11°45"E), DR of Congo (4°31"N 15°32"S) to Angola (12°30"S 18°30"E) (Lowe, 1984). Both species are understorey lianas, although in some cases individuals have been found to scramble into the crowns of emergent trees. Both species are very similar and can only be distinguished by the shape of the leaves and characteristics of the male reproductive parts (Mialoundama, 1993).

Both *Gnetum* species have significant value to many forest-based communities and have a number of vernacular and trade names such as Koko, Eru, Okok, Okazi and Afang (Bahuchet, 1990). *G. africanum* (Wilw) is a ceremonial delicacy in many parts of Eastern Nigeria up to the Southern part of Nigeria; its use has spread within Nigeria to towns and cities like Lagos, Abuja, Ibadan, and Makurdi. It is one of the vegetables in great demand by Nigerians in the Diaspora, although it is very expensive.

The plant's nutritional and ethno-medical effects are well documented by Okafor (2003), Okafor et al. (1996), Burkill (1994) and Watt and Breyer-Brandwijk (1962). It is for its food value that *Gnetum* is most prized. The leaves are either eaten raw or are finely shredded and added to soups and stews (Burkill, 1994). Okazi soup is popular to the people of Imo, Abia, Anambra and some parts of the Delta states and Afang soup to the people of Cross Rivers and Akwa Ibom; *Gnetum* is the only and major vegetable component in these soups.

*Corresponding author. E-mail: acfloh2002@yahoo.com.

The leaves of *G. africanum* are a very important article of trade in the central African region, particularly close to Cameroon. These leaves are harvested on a daily basis and sold in local and regional markets. As the leaves are evergreen, they are available all year round. The volume of export trade in these leafy vegetables has significantly increased in recent years (Shiembo, 1997). Irrespective of the high economic value of *Gnetum*, the plant is increasingly threatened as a result of industrialization and urbanization characterized by rapid deforestation, uncontrolled logging, burning and search for food.

In the wild, *Gnetum* grows and forms underground tubers or roots that store plant food reserves. These tubers can remain alive for many years when the vegetation and the *Gnetum* vines above ground are cleared and soil surface is laid bare. It has been reported that some local tribes in East Cameroon and Congo eat these tubers as wild yams, particularly during lean seasons (Bahuchet, 1990).

During vine collection, the trees on which *Gnetum* climbs are often felled, creating widespread damage. In the course of harvesting the vines, the buds on the tubers are damaged and it may take a long time for new buds to develop into vines. In some cases the forceful pulling of vines creates wounds on the tubers /roots, leaving them vulnerable to fungal attack that can cause rot disease. As a result, harvesting methods do not ensure the growth and supply of future *Gnetum* leaves, and are not sustainable (Shiembo, 1997). The unsustainable harvest increases the threat of genetic erosion of the species. For this reason, there is a great need to protect this plant.

Since man cannot do without exploiting plants and other available bioresources, there has to be a balance between use of these resources and their conservation. In this way an ecosystem, although altered, could retain its bioresources and provide food, as well as perform vital environmental functions on a long-term basis (Uyoh et al., 2003).

Biotechnology provides valuable gains in the research on conservation of plant bioresources (De-Smet, 1995). Although modem biotechnology is a newly introduced science, its impact has greatly excited the imagination and provoked the concern of almost every part of society worldwide (Eneobong, 2003). Many of our "orphan (Considered to be 'minor crops' that, although having high economic, livelihood and food value in developing countries, receive little attention from researchers or investors.)" crops are hard to propagate due to their production of seeds with little or defective endosperms, or the fact that they are frequently polyploids or aneupolids. Plant tissue culture (TC), a tool in biotechnology, is an option that provides a method for their mass clonal propagation, as well as serving as a tool for their germplasm collection and conservation (Uyoh et al., 2003; Opabode and Adeboye, 2005). The richness of plant and animal diversity in developing countries is a major asset in agricultural and sustainable development. Biotechnology can thus play an important role in the propagation and conservation of many plant bioresources.

In the light of these considerations, we report preliminary callus initiation studies of *G. africanum* as a stage in the development of a regeneration protocol for *G. africanum*.

MATERIALS AND METHODS

Collection of donor plants

G. africanum plants were collected from the wild in Issele-Azagba, Aniocha North Local Government Area in Delta state (6°12'00"N 6°43'12"E), by carefully uprooting them. Uprooted plants with forest soil were placed in buckets and an artificial forest condition was created for the plants to acclimatize: plants were shaded from the direct rays of the sun using oil palm fronds and were watered three times daily to maintain moist conditions of the forest understorey according to the method of Okafor (2003).

Those surviving acclimatized plants were then transferred to the Biotechnology Advanced Laboratory Sheda Science and Technology Complex, Abuja. These plants served as sources of leaf explants for the experiments (Plate 1).

Explants preparation and surface sterilization

Three to four day old young leaves were pruned off the mother plants using a sterile knife, for use as explants. In order to sterilize the surfaces of explants, they were washed in the liquid detergent Tween 20 (2 drops per 100 ml solution) for 3 min, rinsed in sterile distilled water for 20 min, soaked in 70% ethanol for 2 min and, lastly, soaked in 1% sodium hypochlorite for 20 min. Explants were rinsed twice with sterile water after ethanol treatments and three times after the sodium hypochlorite treatments, following the method of Gopi and Vatsala, (2006). All explants were then trimmed to small sizes of about 1 cm by 1 cm, after which they were cultured on culture medium.

Plant growth regulators (PGR)

Murashige and Skoog (1962) culture medium was supplemented with different concentrations of auxins (2,4D and Pichloram). In order to encourage callus initiation, low concentrations of cytokines (BAP and Kinetin) were combined with high concentrations of auxins according to the methods of Suahersan and Aboel- Nil (2002), Thomas and Puther (2004) and Balogun et al. (2007) as represented in Table 1.

In vitro callus initiation

Murashige and Skoog (1962) culture medium (MS medium) was supplemented with 30 mg/l sucrose, 0.1 mg/l inositol and 5 ml/l ascorbic acid. pH was adjusted to 5.8 using sodium hydroxide (1 mM) or dilute hydrochloric acid (1 mM). Agar (Phytagel) was added as gelling agent to the medium at a concentration of 3 g/l before the medium was sterilized at 121°C for 15 min. At the end of the sterilization, the medium was allowed to cool. It was aseptically poured into sterile Petri-dishes in the laminar flowhood and allowed to set for 24 h. Twenty sterile explants were cultured for each class of experiment by placing then about 2-4 cm apart on the culture medium. Petri dishes were sealed using parafilm "M" according to Sudhersan and Aboel-Nil (2002). All Petri dishes were wrapped in foil and stored in the dark in the growth room at 28°C. Cultures were moni-

Plate 1. *Gnetum africanum* plants in the screen house.

Table 1. Showing concentrations and combinations of PGR.

24D Mg/l							**Pichloram mM**			
	0	**1**	**3**	**5**	**7**	**0**	**0.01**	**0.05**	**0.1**	**0.5**
BAP	**CT_1**	**D_1**	**D_2**	**D3**	**D_4**	**CT_2**	**P_1**	**P_2**	**P_3**	**P_4**
0										
0.5			BD_1		BD_2				BP_1	BP_2
1.0			BD_3		BD_4				BP_3	BD_4
Kinetin	**CT_3**				**CT_4**					
0										
0.5				KD_1	KD_2				KP_1	P_2
1.0				KD_3	KD_4				KP_3	P_4

Key: CT; Control, **D**; 2,4D, **P**; Pichloram, **BD**; BAP + 2,4D, **BP**; BAP + Pichloram, **KD**; Kinetin + 2,4D, **KP**; Kinetin + Pichloram.

Cultures were monitored for callus initiation once every 2 weeks.

Measurement of callus response and size

Callus initiation response and size were evaluated visually. The criteria for scoring explants for callus according to Amoo and Ayisire (2005) was based on the number of explants producing callus per media. This is represented in Table 2. The percentage of explants that showed callus initiation was also calculated.

The degree of callus response and callus size was compared using the student's T-Test at 0.05 level of significance using the SPSS statistical software. The mean variance of degree of callus response, percentage callus response and callus size were also compared using SPSS software at 0.05 level of significance.

RESULTS AND DISCUSSION

The results of the *in vitro* callus initiation study are summarized in Table 3.

Although, callus production could be initiated using different auxins as reported by Tautorus et al. (1991), Amoo and Ayisire (2005), Eke et al. (2005), Suaherasn and Aboel-Nil (2002) and Puchooa (2004), it was not observed in this study on *G. africanum*. However, with the use of a combination of auxins and cytokines, callus production was initiated.

There was a significant difference ($P<0.05$) in the degree of response of explants and the size of callus pro-

Table 2. Showing callus criteria for scoring explants for callus response and size.

	Response	Size
	0	0
	1	I (1-3 mm)
Callus	2	III (3-6 mm)
	3	V (6-9 mm)
	4	

1-4: Order of callus response; I-V: Order of callus size.

Table 3. The results of callus initiation for various concentrations and combination of PGR.

PGR concentration	Number of explants	Degree of callus response[#$] Mean±SD	% Callus response*	Callus size[#Z] Mean±SD
2,4-D				
1mg/l	20	0.0±0.0	0	0.0±0.0
3 mg/l	20	0.0±0.0	0	0.0±0.0
5 mg/l	20	0.0±0.0	0	0.0±0.0
7 mg/l	20	0.0±0.0	0	0.0±0.0
Pichloram				
0.01mM	20	0.0±0.0	0	0.0±0.0
0.05mM	20	0.0±0.0	0	0.0±0.0
0.1mM	20	0.0±0.0	0	0.0±0.0
0.5mM	20	0.0±0.0	0	0.0±0.0
BAP + 2,4-D				
0.5mg/l + 5mg/l	20	0.0±0.0	0	0.0±0.0
0.5mg/l + 7mg/l	20	1.0±0.0	20	0.5±0.6
1.0mg/l + 5mg/l	20	2.75±0.5	70	2.5±1.0
1.0mg/l + 7mg/l	20	4.0±0.0	100	4.5±1.0
Kinetin + 2,4-D				
0.5mg/l + 5mg/l	20	0.0±0.0	0	0.0±0.0
0.5mg/l + 7mg/l	20	1.0±0.0	20	0.5±0.0
1.0mg/l + 5mg/l	20	2.0±0.0	65	2.5±1.9
1.0mg/l + 7mg/l	20	1.0±0.0	50	1.5±1.0
BAP + pichloram				
0.5mg/l + 0.1 mM	20	1.0±0.0	50	1.5±1.0
0.5mg/l + 0.5 mM	20	1.0 ±0.0	50	1.0±0.0
1.0mg/l + 0.1 mM	20	2.0±0.0	20	3.0±0.0
1.0mg/l + 0.5 mM	20	1.75±0.5	20	3.0±0.0
Kinetin + pichloram				
0.5mg/l + 0.1mM	20	2.0±0.0	70	3.0±0.0
0.5mg/l + 0.5mM	20	1.75±0.5	50	2.0±1.2
0.5mg/l + 0.1mM	20	1.25±0.5	50	1.0±0.0
1.0mg/l + 0.1mM	20	1.25± 0.5	20	2.0±1.2
Control No PGR	20	0.0±0.0	0	0.0±0.0

#: $P>0.05$: There is no significant difference between callus response and callus size; $: $P<0.05$: There was significant difference between the degree of callus response and other parameters tested; *: $P<0.05$: There was significant difference between the % callus response and other parameters tested.; Z: $P<0.05$: There was significant difference between callus size and other parameters tested.

duction for the different callus initiation media. However, there was no significant difference (p>0.05) between the degree of callus response and callus size, showing that the composition of the different media did not affect the callus size produced.

Explants of *G. africanum* cultured on MS medium formed callus at their cut end., Callus formation at the proximal end of the explants in this study conforms to reports on Peganum harmla (Saini and Jaiwal, 2000) as well as Holostemma ada-kodien (Martin, 2000).

Marks and Simpson (1994) suggested that callus formation may be due to the action of accumulated auxins at the basal cut ends which stimulates cell proliferation, especially in the presence of cytokines. This hypothesis seems to hold true for *G. africanum*, where callus was initiated with a combination of a cytokines and an auxin. Preece et al. (1991) reported that formation of callus at the basal cut explant on cytokines-enriched medium is frequent in species with strong apical dominance. This synergistic effect of a cytokine and an auxin has been demonstrated in many plants including Santolina canescens (Casado et al., 2002), Bupleurum fruticosum (Fraternale et al., 2002) and Acacia nilotica (Sane et al., 2001). Furthermore, a high degree of callusing was reported in single node explants of fluted pumpkin, cultured in medium containing both a cytokine (Kinetin) and an auxin (NAA) (Balogun et al., 2002). High callus formation from a combination of cytokine (BAP) and auxin (NAA) has also been reported by Balogun et al. (2007). What these observations imply is that the combination of auxin and cytokine can be used as an effective initiation me-dium for plant regeneration via somatic embryogenesis.

One of the serious constraints of this study was the browning of the callus and explants which led to the death of the cells. According to Mantel et al. (1998), browning can be attributed to the presence of secondary metabolities (Phenolic compounds). These compounds are oxidized when in contact with oxygen to form toxic compounds after wounding of leaf explants (Aliyu, 2005). Phenolic compounds are mostly secreted by woody plants which may explain why the explants and callus of *G. africanum* were able to produce these toxic metabolites. Death due to browning may have prevented formation of callus in the single PGR concentration experiment. However, explants in some media where callus was initiated showed necrosis after 5 days, Death of callus due to browning has been reported in cashew by Aliyu (2005), Sturt's Desert Pea (Suaherasn and Aboel-Nil, 2002) and Parkia biglobosa (Amoo and Ayisire, 2005).

Conclusion

There is no doubt that the exploration of plant bio-resources in the wild is having a negative impact on their natural populations and is threatening their survival. The provision of alternative sources of cultivation could go a long way in reducing the heavy dependence on these wild populations. Plant tissue culture, an indispensable part of biotechnology, has provided a sure pathway for the rescue of many "orphan" crops which are found in the wild and are extremely difficult to propagate. Their response to various plant growth regulators is a positive sign that a protocol for *in vitro* culture can be achieved, thereby helping to alleviate the economic exploitation of wild plants and enabling their conservation.

This study has successfully formulated a protocol for callus initiation in *G. africanum*. The reported results are expected to contribute to the scientific baseline data necessary for the somatic embryogenesis protocol of *G. africanum*; a sine qui non for its micropropagation and conservation.

DEDICATION

In memory of a great scientist, Dr. Danladi Dada KUTA.

REFERENCES

Amoo SO, Ayisire BE (2005). Induction of callus and somatic embryogenesis from cotyledon explants of Parkia biglobosa (Jocq). Afr. J. Biotechnol. 4(6):548-553.

Aliyu OM (2005). Application of tissue culture to Cashew (*Anacardium occidentale* L.) Breeding: An appraisal. Afr. J. Biotechnol. 4: 1485-1489.

Balogun MO, Akande SR, Ogunbodede BA (2007). Effects of plant growth regulators on callus shoot and root formation in fluted Pumpkin (*Telfairia occidentalis*). Afr. J. Biotechnol. 6 (4): 355-358.

Balogun MO, Ajibade SR, Ogunbodede BA (2002). Micropropagation of fluted pumpkin by enhanced auxiliary shoot formation. Niger. J. Hort. Sci. 6(1):85-88.

Bahuchet S (1990). The Akwa Pygmies: Hunting and Gathering in the labaye Forest. In Food and Nutrition in the African Rain Forest. Food Anthropol. Unit 263, UNESCO.

Burkill HM (1994). Useful plants of West Tropical Africa families E – T. Royal Botanical Gardens, Kew 2

Casado JP, Navarro MC, Utrilla MP, Martinez A , Jime NJ (2002). Micropropagation of Santolina Canescens Lagasca and *in vitro* Volatiles production by shoot explants. Plant cell issue, Organ Culture 69: 147 – 153

De-Smet K (1995). Germplasm conservation in banana and plantain in: Lectures on plant tissue culture and biotechnology. Afr. Biosci. network (ABN) Int. training course, Univ. of Niger. Nsukka.

Eke CR, Akomeah P, Asemota O (2005). Somatic embryogenesis of Date palm (*Phoenix dactylifere* L.) from apical meristem tissues from "Zebia" and "Loko" landraces. Afr. J. Biotechnol. 4(3): 244 – 246.

Eneobong EE (2003). Current issues in agricultural biotechnology. Quarterly Public lecture Nigerian Academy of Science 4: 23-29.

Eneobong EE (1997). Biotechnological Techniques for the conservation and use of plant genetic resources: Eneobong EE (Ed). Biological Conservation for sustainable Agricultural Production Federal University of Agriculture, Umudike, Nigeria pp: 72 – 73.

Fraternale D, Giamperi L, Ricci D , Rocchi MBL (2002). Micropropagation of Bupleurum fruticosum: the effect of triacontanol. Plant cell tissue organ culture 69: 135-140.

Gopi C, Vatsala TM (2006). *In vitro* studies on effects of plant growth regulators on callus and suspension culture biomass yield from Gymnema sylvestre. R. Br. Afr. J. Biotechnol. 5(12): 1215 – 1219.

Lowe J (1984). Gnetum in West African. Nigeria Field 49 (1-4): 99-104.

Mantel SH, Boggett B, Bessa AMS, Lemos EEP, Abdelhad AR, Mneney EE (1998). Micropropagation and Micrografting methods suitable for safe international transfer of Cashew. Proceed. of int. Cashew and Coconut conf., 1997, Dar-es-salaam pp: 95-107.

Marks TR, Simpson SE (1994). Factors affecting shoot development in

apically dominant Acer Cultivars. *in vitro* J. Hort. Sci. 69: 543:551.
Martin KS (2000). Rapid Propagation of Hocostemma ada-Kodien (schult) a rare medicinal plant through auxiliary bud multiplication and indirect Organogenesis. Plant cell Reprod. 21: 112-117.
Mialoundama F (1993). Nutritional and Socio-economic value of Gnetum leaves in central African forest. In Hladik. CM et al, Tropical Forests, people and Food: Biocultural interactions and applications to development. Carnforth UK, Parthenon Publishing group.
Mialoundama F, Paulet P (1986). Regulation of Vascular differentiation in Leaf primordial during the rhythmic growth of *Gnetum africanum.* Canadian J. Bot. 64(1): 208-213.
Murashige T, Skoog F (1962). A revised medium for rapid growth and bioassays with tobacco tissue culture. Physiologia Plantarum 15: 473-497.
Oguntona T (1988). Green Leafy Vegetable. Nutr. Qual. Plant Food 2:120-133
Okafor JC (2003). Conservation and use of traditional vegetables from woody forest species in South Eastern Nigeria. www.fao.org. /docrep/x2161e/x2161e13.htm.
Okafor JC, Okolo HC, Ejiofor MAN (1996). Strategies for enhancement of utilization potential of edible woody forest species of South – Eastern Nigeria. In: Vander Maesen (Ed) The Biodiversity of African Pants. Kluwer, The Netherlands pp. 684-695
Opabode JT, Adeboye OC (2005). Application of biotechnology for the improvement of Nigerian indigenous leaf Vegetables. Afr. J. Biotechnol. (3):138-142.
Preece JE, Huttreman CA, Ashby WC, Roth PL (1991). Micro and Cutting Propagation of Silver Maple I. Results with adult and Juvenile Progagules. J. Am. Soc. Hort. Sci. 116:142-148
Puchooa D (2004). *In vitro* regeneration of Lychee (Litchi chinensis Sonn.). Afr. J. Biotechnol. 3(11): 576-584.
Saini R, Jaiwal PK (2000). *In vitro* Multiplication of Peganum harmala an important medicinal plant. Indian J. Exp. Biol. 38: 499-503.
Sane D, Borgel A, Chevalier MH Gassama M, Dia YK (2001). Induction *in vitro* de l' enracinement de microbouture d' *Acacia tortilis.* Sub sp raddiana par traitement transitore auxine. Annu. Sci. 58: 43-47.
Shiembo PN (1997). The Sustainability of ERU (*Gnetum africanum* and *Gnetum buchholzianum*): Over-Exploited non-wood Forest product from the forests of central Africa. www.fao.org/docrep/x2161e06.htm.
Suaherasn C , Aboel – Nil, M. (2002). Somatic embryogenesis of sturt's Desert pea (*Swainsona formosa).* Current Sci. 83(9): 1074-1076.
Tautorus TE, Fowke LC, Dunstan DI (1991). Somatic Embryogenesis of Conifers. Canadian J. Bot. 69:1873-1890.
Thomas DT, Puther JT (2004). Thidazuron induced higher frequency shoot organogenesis in callus from *Kigelia pinnata* L. Bot. Bull. Sci. 45: 307-313.
Uyoh EA, Nkang AE, Eneobong EE (2003). Biotechnology, genetic conservation and sustainable use of bio-resources. Afr. J. Biotechnol. 2(12): 704-709.
Watt JMA, Breyer-Bradwijk MG (1962). The Medicinal and Poisonous Plants of Southern and Eastern Africa. Edinburgh: E and S Livingstone.

Permissions

All chapters in this book were first published in IJBC, by Academic Journals; hereby published with permission under the Creative Commons Attribution License or equivalent. Every chapter published in this book has been scrutinized by our experts. Their significance has been extensively debated. The topics covered herein carry significant findings which will fuel the growth of the discipline. They may even be implemented as practical applications or may be referred to as a beginning point for another development.

The contributors of this book come from diverse backgrounds, making this book a truly international effort. This book will bring forth new frontiers with its revolutionizing research information and detailed analysis of the nascent developments around the world.

We would like to thank all the contributing authors for lending their expertise to make the book truly unique. They have played a crucial role in the development of this book. Without their invaluable contributions this book wouldn't have been possible. They have made vital efforts to compile up to date information on the varied aspects of this subject to make this book a valuable addition to the collection of many professionals and students.

This book was conceptualized with the vision of imparting up-to-date information and advanced data in this field. To ensure the same, a matchless editorial board was set up. Every individual on the board went through rigorous rounds of assessment to prove their worth. After which they invested a large part of their time researching and compiling the most relevant data for our readers.

The editorial board has been involved in producing this book since its inception. They have spent rigorous hours researching and exploring the diverse topics which have resulted in the successful publishing of this book. They have passed on their knowledge of decades through this book. To expedite this challenging task, the publisher supported the team at every step. A small team of assistant editors was also appointed to further simplify the editing procedure and attain best results for the readers.

Apart from the editorial board, the designing team has also invested a significant amount of their time in understanding the subject and creating the most relevant covers. They scrutinized every image to scout for the most suitable representation of the subject and create an appropriate cover for the book.

The publishing team has been an ardent support to the editorial, designing and production team. Their endless efforts to recruit the best for this project, has resulted in the accomplishment of this book. They are a veteran in the field of academics and their pool of knowledge is as vast as their experience in printing. Their expertise and guidance has proved useful at every step. Their uncompromising quality standards have made this book an exceptional effort. Their encouragement from time to time has been an inspiration for everyone.

The publisher and the editorial board hope that this book will prove to be a valuable piece of knowledge for researchers, students, practitioners and scholars across the globe.

List of Contributors

Shahbaz Ahmad
Department of Agriculture-Entomology, University of Agriculture Faisalabad, Pakistan

Muhammad Ashfaq
Department of Agriculture-Entomology, University of Agriculture Faisalabad, Pakistan

Mansoor-ul-Hassan
Department of Agriculture-Entomology, University of Agriculture Faisalabad, Pakistan

Shahbaz Talib Sahi
Department of Agriculture-Entomology, University of Agriculture Faisalabad, Pakistan

Aminata Dolo Nantoumé
Institut d'Economie Rurale/CRRA-Sotuba, Programme Fruits andLégumes, BP 262 Bamako, Mali

Sidiki Traoré
Institut d'Economie Rurale/CRRA-Sotuba, Programme Fruits andLégumes, BP 262 Bamako, Mali

Jorgen Lindskrog Christiansen
Department of Agriculture and Ecology, Faculty of Life Sciences, University of Copenhagen, Thorvaldsensvej 40, 1871 Frederiksberg C, Denmark

Sven Bode Andersen
Department of Agriculture and Ecology, Faculty of Life Sciences, University of Copenhagen, Thorvaldsensvej 40, 1871 Frederiksberg C, Denmark

Brita Dahl Jensen
Department of Agriculture and Ecology, Faculty of Life Sciences, University of Copenhagen, Thorvaldsensvej 40, 1871 Frederiksberg C, Denmark

Zéphirin Tadu
Laboratory of Zoology, Faculty of Science, University of Yaounde I, Po. Box 812, Yaounde, Cameroon

Champlain Djiéto-Lordon
Laboratory of Zoology, Faculty of Science, University of Yaounde I, Po. Box 812, Yaounde, Cameroon

Regis Babin
CIRAD, UPR Bioagresseurs Analyse et Maîtrise du Risque, F-34398 Montpellier, France

Yede
Laboratory of Zoology, Faculty of Science, University of Yaounde I, Po. Box 812, Yaounde, Cameroon

Edith Blandine Messop-Youbi
Laboratory of Zoology, Faculty of Science, University of Yaounde I, Po. Box 812, Yaounde, Cameroon
Department of Zoology and Animal Physiology, Faculty of Science, University of Buea, PO Box 63 Buea, Cameroon

Abraham Fomena
Laboratory of Parasitology and Ecology, Faculty of Science, University of Yaounde I, P . Box 812, Yaounde, Cameroon

Kuldip S. Dogra
Department of Biosciences, Himachal Pradesh University Summer Hill Shimla (India) - 171005

Ravinder K. Kohli
Department of Botany, Panjab University Chandigarh (India) - 160014

Sarvesh K. Sood
Department of Biosciences, Himachal Pradesh University Summer Hill Shimla (India) - 171005

Praveen K. Dobhal
Department of Botany, Panjab University Chandigarh (India) - 160014

M. Muzila
University of Botswana Biological Sciences, Private Bag UB 00704 Gaborone Botswana

M. P. Setshogo
University of Botswana Biological Sciences, Private Bag UB 00704 Gaborone Botswana

S. W. Mpoloka
University of Botswana Biological Sciences, Private Bag UB 00704 Gaborone Botswana

Saurav Dwari
Plant Taxonomy, Biosystematics and Molecular Taxonomy Laboratory, Department of Botany and Forestry, Vidyasagar University, Midnapore-721 102, West Bengal, India

Amal Kumar Mondal
Plant Taxonomy, Biosystematics and Molecular Taxonomy Laboratory, Department of Botany and Forestry, Vidyasagar University, Midnapore-721 102, West Bengal, India

Muhammad Masood Tariq
Centre for Advanced studies in Vaccinology and Biotechnology University of Balochistan, Quetta, Pakistan

Masroor Ahmad Bajwa
Centre for Advanced studies in Vaccinology and Biotechnology University of Balochistan, Quetta, Pakistan

Ferhat Abbas
Centre for Advanced studies in Vaccinology and Biotechnology University of Balochistan, Quetta, Pakistan

Abdul Waheed
Department of Animal Breeding and Genetics, University of Agriculture, Faisalabad, Pakistan

Farhat Abbas Bokhari
Centre for Advanced studies in Vaccinology and Biotechnology University of Balochistan, Quetta, Pakistan

Majed Rafiq
Centre for Advanced studies in Vaccinology and Biotechnology University of Balochistan, Quetta, Pakistan

Kunle
Department of Medicinal Plant Research and Traditional Medicine, National Institute for harmaceutical Research and Development (NIPRD), Idu Industrial Layout Idu, PMB 21 Garki, Abuja, Nigeria

Oluyemisi Folashade
Department of Medicinal Plant Research and Traditional Medicine, National Institute for harmaceutical Research and Development (NIPRD), Idu Industrial Layout Idu, PMB 21 Garki, Abuja, Nigeria

Egharevba
Department of Medicinal Plant Research and Traditional Medicine, National Institute for harmaceutical Research and Development (NIPRD), Idu Industrial Layout Idu, PMB 21 Garki, Abuja, Nigeria

Henry Omoregie
Department of Medicinal Plant Research and Traditional Medicine, National Institute for harmaceutical Research and Development (NIPRD), Idu Industrial Layout Idu, PMB 21 Garki, Abuja, Nigeria

Ahmadu
Tianjin University of Traditional Chinese Medicine, Tianjin, China

Peter Ochogu
Tianjin University of Traditional Chinese Medicine, Tianjin, China

Jagadish Parajuli
WWF Nepal, Kathmandu, Nepal

Sanu Devi Joshi
Nepal Academy of Science and Technology (NAST), Nepal
Central Department of Botany, Tribhuvan University, Nepal

D. K. Patel
Department of Rural Technology, Guru Ghasidas Vishwavidyalaya, A Central University, Bilaspur, 495009

Rama Shankar
Ayurveda Regional Research Institute (CCRAS), Itanagar-Arunachal Pradesh, India-791111

Ramesh Babu Devalla
Central Council for Research in Ayurveda and Siddha, 61-65, Institutional Area, Opp. D- Block, Janakpuri, New Delhi-110058, India

Kiran Bargali
Department of Botany, Kumaun University, DSB Campus, Nainital -263002, Uttarakhand, India

Priyanka Bisht
Department of Botany, Kumaun University, DSB Campus, Nainital -263002, Uttarakhand, India

Abida Khan
Department of Botany, Kumaun University, DSB Campus, Nainital -263002, Uttarakhand, India

Y. S. Rawat
Department of Botany, Kumaun University, DSB Campus, Nainital -263002, Uttarakhand, India

Muhammad Hussnain Babar
Department of Agriculture Entomology, University of Agriculture, Faisalabad, Pakistan

Muhammad Ashfaq
Department of Agriculture Entomology, University of Agriculture, Faisalabad, Pakistan

Muhammad Afzal
Department of Agriculture Entomology, University College of Agriculture, University of Sargodha, Sargodha, Pakistan

Muhammad Hamid Bashir
Department of Agriculture Entomology, University of Agriculture, Faisalabad, Pakistan

Muhammad Asif Ali
Institute of Horticultural Sciences, University of Agriculture, Faisalabad, Pakistan

Simon A. Tamungang
Department of Animal Biology, University of Dschang, BP 146 Dschang, Cameroon

Robert A. Cheke
Natural Resources Institute, University of Greenwich at Medway, Central Avenue, Chatham Maritime, Kent ME4 4TB, UK

Ghislain P. Kougoum
Institut Supérieur du Sahel, Université de Maroua, BP 46 Maroua, Cameroon

Eric S. Ntiri
African Insect Science for Food and Health (ICIPE), P.O. Box 30772-00100, Nairobi, Kenya

Boudabous Raja
Laboratoire de Parasitologie- Mycologie code 99UR/08-05, Faculté de Pharmacie, Département de Biologie Clinique B, 1 Rue Avicenne, 5000 Monastir Tunisia, University of Monastir-Tunisia

K. Jaouadi
Laboratoire de Parasitologie- Mycologie code 99UR/08-05, Faculté de Pharmacie, Département de Biologie Clinique B, 1 Rue Avicenne, 5000 Monastir Tunisia, University of Monastir-Tunisia

N. Haouas
Laboratoire de Parasitologie- Mycologie code 99UR/08-05, Faculté de Pharmacie, Département de Biologie Clinique B, 1 Rue Avicenne, 5000 Monastir Tunisia, University of Monastir-Tunisia

H. Mezhoud
Laboratoire de Parasitologie- Mycologie code 99UR/08-05, Faculté de Pharmacie, Département de Biologie Clinique B, 1 Rue Avicenne, 5000 Monastir Tunisia, University of Monastir-Tunisia

S. Bdira
Service d'Hygiène de Monastir Tunisia

S. Amor
Service d'Hygiène de Monastir Tunisia

Babba Hamouda
Laboratoire de Parasitologie- Mycologie code 99UR/08-05, Faculté de Pharmacie, Département de Biologie Clinique B, 1 Rue Avicenne, 5000 Monastir Tunisia, University of Monastir-Tunisia

Subban Kamalraj
Centre for Advanced Studies in Botany, University of Madras, Maraimalai Campus, Guindy, Chennai-600 025, Tamil Nadu, India

Johnpaul Muthumary
Centre for Advanced Studies in Botany, University of Madras, Maraimalai Campus, Guindy, Chennai-600 025, Tamil Nadu, India

I. Kobeasy Mohamed
Department of Biochemistry, Faculty of Agriculture, Cairo University, Giza, Egypt

M. Abdel-Fatah Osama
Department of Biochemistry, Faculty of Agriculture, Cairo University, Giza, Egypt

Samiha M. Abd El-Salam
Department of Crops Technology Research, Food Technology Research Institute, Agriculture Research Center, Giza, Egypt

Zahrat El-Ola M. Mohamed
Department of Crops Technology Research, Food Technology Research Institute, Agriculture Research Center, Giza, Egypt

J. L. Seburanga
Faculty of Science, National University of Rwanda, University Avenue, Box 117, Huye, Rwanda

B. A. Kaplin
Department of Environmental Studies, Antioch University, New England, 40Avon St. Keene, NH 0343l, USA

E. Bizuru
Faculty of Science, National University of Rwanda, University Avenue, Box 117, Huye, Rwanda

E. N. Mwavu
Department of Forestry, Biodiversity and Tourism, Makerere University, P.O. Box 7062, Kampala, Republic of Uganda

Mohammad Nisar
Department of Botany1 and Biotechnology2, University of Malakand, Chakdara Dir (Lower), Khyber PakhtunKhwa, Pakistan

Naeem Khan
Department of Botany1 and Biotechnology2, University of Malakand, Chakdara Dir (Lower), Khyber PakhtunKhwa, Pakistan

Nausheen
Department of Botany1 and Biotechnology2, University of Malakand, Chakdara Dir (Lower), Khyber PakhtunKhwa, Pakistan
Plant Genetic Resources Program (PGRP), National Agricultural Research Centre (NARC), Islamabad, Pakistan

Zakia Ahmad
Department of Botany1 and Biotechnology2, University of Malakand, Chakdara Dir (Lower), Khyber PakhtunKhwa, Pakistan

Abdul Ghafoor
Plant Genetic Resources Program (PGRP), National Agricultural Research Centre (NARC), Islamabad, Pakistan

K. Srinivasa Murthy
Division of Insect Molecular Entomology, National Bureau of Agriculturally Important Insects, P.B.No. 2491, H. A. Farm Post, Bellary Road, Bangalore - 560 024, Karnataka, India

R. Rajeshwari
Division of Insect Molecular Entomology, National Bureau of Agriculturally Important Insects, P.B.No. 2491, H. A. Farm Post, Bellary Road, Bangalore - 560 024, Karnataka, India

S. K. Jalali
Division of Insect Molecular Entomology, National Bureau of Agriculturally Important Insects, P.B.No. 2491, H. A. Farm Post, Bellary Road, Bangalore - 560 024, Karnataka, India

T. Venkatesan
Division of Insect Molecular Entomology, National Bureau of Agriculturally Important Insects, P.B.No. 2491, H. A. Farm Post, Bellary Road, Bangalore - 560 024, Karnataka, India

J. Manjunathan
Centre for Advanced Studies in Botany, University of Madras, Guindy Campus, Chennai-600 025, Tamil Nadu, India

N. Subbulakshmi
Dhanalakshmi Srinivasan College of Arts and Science for Women, Perambalur-12, Tamil Nadu, India

R. Shanmugapriya
Dhanalakshmi Srinivasan College of Arts and Science for Women, Perambalur-12, Tamil Nadu, India

V. Kaviyarasan
Centre for Advanced Studies in Botany, University of Madras, Guindy Campus, Chennai-600 025, Tamil Nadu, India

Farzana Perveen
Department of Zoology, Shaheed Benazir Bhutto University (SBBU), Main Campus, Sheringal, Khyber Pakhtunkhwa (KP), Pakistan

Anzela Khan
Department of Zoology, Shaheed Benazir Bhutto University (SBBU), Main Campus, Sheringal, Khyber Pakhtunkhwa (KP), Pakistan

Gidey Yirga
Department of Biology, Mekelle University P. O. Box 3072, Mekelle, Ethiopia

Mekonen Teferi
Department of Biology, Mekelle University P. O. Box 3072, Mekelle, Ethiopia

Mezgebe Kasaye
Department of Biology, Mekelle University P. O. Box 3072, Mekelle, Ethiopia

D. Acquah - Lamptey
Department of Animal Biology and Conservation Science, P. O. Box LG67, University Of Ghana, Legon, Ghana

R. Kyerematen
Department of Animal Biology and Conservation Science, P. O. Box LG67, University Of Ghana, Legon, Ghana

E. O. Owusu
Department of Animal Biology and Conservation Science, P. O. Box LG67, University Of Ghana, Legon, Ghana

Suchitra Rout
Post Graduate Department of Botany, Utkal University, Vanivihar, Bhubaneswar-751004, Odisha, India

Sitaram Prasad Panda
Mangrove Forest Division (WL), Rajnagar, Kendrapara, Odisha, India

Hemant Kumar Patra
Post Graduate Department of Botany, Utkal University, Vanivihar, Bhubaneswar-751004, Odisha, India

Aseesh Pandey
Plant Tissue Culture and Molecular Biology Laboratory, Department of Biotechnology, Bhimtal Campus, Kumaun University, Nainital-263136, Uttarakhand, India

Susana Tamta
Plant Tissue Culture and Molecular Biology Laboratory, Department of Biotechnology, Bhimtal Campus, Kumaun University, Nainital-263136, Uttarakhand, India

Vivek Tripathi
Delhi Technological University, Shahbad, Daulatpur Bawana Road, Delhi-110042, India

Sandhya Goswami
Betel Vine Biotechnology Laboratory, National Botanical Research Institute,Lucknow-226 001, India

Majda Khalil Suleiman
Arid land Agriculture and Greenery Department, Kuwait Institute for Scientific Research, P. O. Box 24885, 13109 Safat, Kuwait

Narayana Ramachandra Bhat
Arid land Agriculture and Greenery Department, Kuwait Institute for Scientific Research, P. O. Box 24885, 13109 Safat, Kuwait

Mehdi Saleh Abdal
Arid land Agriculture and Greenery Department, Kuwait Institute for Scientific Research, P. O. Box 24885, 13109 Safat, Kuwait

Sameeha Zaman
Arid land Agriculture and Greenery Department, Kuwait Institute for Scientific Research, P. O. Box 24885, 13109 Safat, Kuwait

Sheena Jacob
Arid land Agriculture and Greenery Department, Kuwait Institute for Scientific Research, P. O. Box 24885, 13109 Safat, Kuwait

Rini Rachel Thomas
Arid land Agriculture and Greenery Department, Kuwait Institute for Scientific Research, P. O. Box 24885, 13109 Safat, Kuwait

Satish Balapure
Department of Environmental Sciences and Limnology, Barkatullah University, Bhopal, 462026, India

Sumana Dutta
Department of Environmental Sciences and Limnology, Barkatullah University, Bhopal, 462026, India

Vipin Vyas
Department of Environmental Sciences and Limnology, Barkatullah University, Bhopal, 462026, India

Clement A. Edokpayi
Department of Marine Sciences, University of Lagos, Akoka, Yaba, Lagos State, Nigeria

Aveez O. Olowoporoku
University of West of England, Bristol, United Kingdom

Roland E. Uwadiae
Department of Marine Sciences, University of Lagos, Akoka, Yaba, Lagos State, Nigeria

Mantosh Kumar Sinha
K.R. Technical College Ambikapur (Chhattisgarh) India

Deepima Sinha
Vishwavidyalaya Engineering College Lakhanpur Ambikapur (Chhattisgarh) India

A. C. Iloh
Biotechnology and Molecular Biology Advanced Laboratory, Sheda Science and Technology Complex, Abuja, Nigeria

N. R. Isu
Department of Biological Sciences, University of Abuja, Nigeria

D. D. Kuta
Biotechnology and Molecular Biology Advanced Laboratory, Sheda Science and Technology Complex, Abuja, Nigeria

$152.99

S